AMINO ACIDS TECHNOLOGY

AMINO ACIDS TECHNOLOGY

Recent Developments

J.C. Johnson

NOYES DATA CORPORATION
Park Ridge, New Jersey, U.S.A.
1978

Library of Congress Catalog Card Number: 78-54000
ISBN: 0-8155-0703-8
Printed in the United States

Published in the United States of America by
Noyes Data Corporation
Noyes Building, Park Ridge, New Jersey 07656

FOREWORD

The detailed, descriptive information in this book is based on 253 recent U.S. patents, all of which deal with amino acids, and which have been issued since January 1974.

This book serves a double purpose in that it supplies detailed technical information and can be used as a guide to the U.S. patent literature in this field. By indicating all the information that is significant, and eliminating legal jargon and juristic phraseology, this book presents an advanced, technically oriented review of amino acids.

The U.S. patent literature is the largest and most comprehensive collection of technical information in the world. There is more practical, commercial, timely process information assembled here than is available from any other source. The technical information obtained from a patent is extremely reliable and comprehensive; sufficient information must be included to avoid rejection for "insufficient disclosure." These patents include practically all of those issued on the subject in the United States during the period under review; there has been no bias in the selection of patents for inclusion.

The patent literature covers a substantial amount of information not available in the journal literature. The patent literature is a prime source of basic commercially useful information. This information is overlooked by those who rely primarily on the periodical journal literature. It is realized that there is a lag between a patent application on a new process development and the granting of a patent, but it is felt that this may roughly parallel or even anticipate the lag in putting that development into commercial practice.

Many of these patents are being utilized commercially. Whether used or not, they offer opportunities for technological transfer. Also, a major purpose of this book is to describe the number of technical possibilities available, which may open up profitable areas of research and development. The information contained in this book will allow you to establish a sound background before launching into research in this field.

Advanced composition and production methods developed by Noyes Data are employed to bring these durably bound books to you in a minimum of time. Special techniques are used to close the gap between "manuscript" and "completed book." Industrial technology is progressing so rapidly that time-honored, conventional typesetting, binding and shipping methods are no longer suitable. We have bypassed the delays in the conventional book publishing cycle and provide the user with an effective and convenient means of reviewing up-to-date information in depth.

The Table of Contents is organized in such a way as to serve as a subject index. Other indexes by company, inventor and patent number help in providing easy access to the information contained in this book.

15 Reasons Why the U.S. Patent Office Literature Is Important to You –

1. The U.S. patent literature is the largest and most comprehensive collection of technical information in the world. There is more practical commercial process information assembled here than is available from any other source.

2. The technical information obtained from the patent literature is extremely comprehensive; sufficient information must be included to avoid rejection for "insufficient disclosure."

3. The patent literature is a prime source of basic commercially utilizable information. This information is overlooked by those who rely primarily on the periodical journal literature.

4. An important feature of the patent literature is that it can serve to avoid duplication of research and development.

5. Patents, unlike periodical literature, are bound by definition to contain new information, data and ideas.

6. It can serve as a source of new ideas in a different but related field, and may be outside the patent protection offered the original invention.

7. Since claims are narrowly defined, much valuable information is included that may be outside the legal protection afforded by the claims.

8. Patents discuss the difficulties associated with previous research, development or production techniques, and offer a specific method of overcoming problems. This gives clues to current process information that has not been published in periodicals or books.

9. Can aid in process design by providing a selection of alternate techniques. A powerful research and engineering tool.

10. Obtain licenses — many U.S. chemical patents have not been developed commercially.

11. Patents provide an excellent starting point for the next investigator.

12. Frequently, innovations derived from research are first disclosed in the patent literature, prior to coverage in the periodical literature.

13. Patents offer a most valuable method of keeping abreast of latest technologies, serving an individual's own "current awareness" program.

14. Copies of U.S. patents are easily obtained from the U.S. Patent Office at 50¢ a copy.

15. It is a creative source of ideas for those with imagination.

CONTENTS AND SUBJECT INDEX

INTRODUCTION

Amino acid is the general term for any of more than 100 naturally occurring organic acids incorporating one or more amino groups ($-NH_2$). The 20 or so most important acids are the α-amino acids, so-called because of the position of the amino group relative to the carboxyl:

$$\begin{array}{c} \quad H \\ \quad | \\ R-C-COOH \\ \quad | \\ \quad NH_2 \end{array}$$

All proteins, regardless of their occurrence, are made up of α-amino acids and additional organic or inorganic prosthetic groups.

Of the more than 200 processes described in this review more than 30 are directly related to the use of amino acids or their peptides as foods or food additives, which is very important in the nitrogen metabolism of living organisms. There are 13 processes on the use of aspartyl peptides or amides as sweeteners and several on the production of protein hydrolyzates for oral or parenteral feeding.

In recent years α-amino acids closely related to those in proteins have been found to have valuable medicinal properties: 3-(3,4-dihydroxyphenyl)-L-alanine or L-dopa has been known for some time as an intermediate in the biological pathway from tyrosine to epinephrine. Recently L-dopa was proved to be a remarkably effective therapeutic agent in the treatment of Parkinson's disease.

Some 30 processes deal with miscellaneous medicinal uses of amino acids or their derivatives in the treatment of problems ranging from the imperfections of hair or fingernails to severe metabolic disorders. Other amino acids, particularly those having a phenyl moiety, are being investigated for use in treating disorders such as hypertension and depression. Several other processes deal with the preparation of aromatic amino acids useful as pharmaceuticals.

Microbiological processes play an important role in the synthesis of amino acids and their derivatives as well as in their resolution into optically active materials.

Over 60 processes described in this review make use of bacteria or enzymes. A list of culture depositories is given to serve as the source of various organisms cited.

Because some processes are concerned with the production and use of amino acid peptides, a table of the standard symbols for the amino acids used and their description is also included.

Culture Depositories

ATCC	American Type Culture Collection Washington, DC, USA
BRL	Beecham Research Laboratory Collection
CBS	Centraalbureau Voor Schimmelcultures Julianalaan 67 A, Delft (or Baarn), The Netherlands
FERM	Fermentation Research Institute Agency of Industrial Science and Technology Ministry of International Trade and Industry, Chiba, Japan
IAM	Institute of Applied Microbiology University of Tokyo, Tokyo, Japan
IFO	Institute for Fermentation, Osaka, Japan
INRA	Institute National de la Research Agronomique Paris, France
NCIB	National Collection Industrial Bacteria 135 Abbey Road, Aberdeen, UK
NRRL, NRRL-Y	United States Department of Agriculture Northern Utilization Research and Development Division Peoria, Illinois, USA
OUT	Osaka University, Faculty of Engineering, Osaka, Japan
R, A, B, C,	Collection de la Chaire de Genetique l'Ecole Nationale Superieure Agronomique Montpellier, France

Amino Acid Symbols

Amino Acid	Three-Letter Symbol	One-Letter Symbol
Alanine	Ala	A
Arginine	Arg	R
Asparagine	Asn	N
Aspartic acid	Asp	D
Asn + Asp	Asx	B
Cysteine	Cys	C
Glutamine	Gln	Q
Glutamic acid	Glu	E
Gln + Glu	Glx	Z
Glycine	Gly	G

(continued)

Amino Acid	Three-Letter Symbol	One-Letter Symbol
Histidine	His	H
Isoleucine	Ile	I
Leucine	Leu	L
Lysine	Lys	K
Methionine	Met	M
Phenylalanine	Phe	F
Proline	Pro	P
Serine	Ser	S
Threonine	Thr	T
Tryptophan	Trp	W
Tyrosine	Tyr	Y
Valine	Val	V

AMINO ACIDS—PREPARATIVE METHODS

AMINO ACIDS–SYNTHESIS

From Isocyanoacetate Esters or Salts

R.A. Damico and R.G. Laughlin; U.S. Patent 3,790,598; February 5, 1974; assigned to The Procter & Gamble Company have disclosed a process for preparing α-amino acids. This comprises reacting an isocyano acetate ester or salt with a strongly basic metalating agent to provide an α-metalated isocyano ester or salt carbanion, reacting the α-metalated isocyano ester or salt carbanion with an electrophilic reagent to provide an α-substituted isocyano acetate ester or salt; and thereafter hydrolyzing the α-substituted isocyano ester or salt to provide an α-amino acid.

The first step of the process may be summarized by the following equation

$$\begin{array}{c} \mathrm{H\ \ O} \\ |\ \ \ \| \\ \mathrm{H{-}C{-}C{-}OY} \\ | \\ \mathrm{NC} \end{array} + \mathrm{MZ} \longrightarrow \mathrm{M^{+}}\left(\begin{array}{c} \mathrm{H\ \ O} \\ |\ \ \ \| \\ \mathrm{C{-}C{-}OY} \\ | \\ \mathrm{NC} \end{array}\right)^{-} + \mathrm{HZ} \qquad (1)$$

where an isocyano acetate ester or salt is reacted with a basic metalating agent to form a metalated isocyano acetate ester or salt carbanion. The starting material is an ester or a salt as determined by Y. For example, if Y is an alkyl group, then the starting material is an isocyano acetate ester; on the other hand, if Y is an alkali metal ion, then the starting material is an isocyano acetate salt.

Isocyano acetate esters and salts are generally not readily available on the commercial market and therefore the starting material must itself be synthesized by a rather simple synthesis procedure which is described in detail in *Angewandte Chemie*, International Edition, vol. 4, pp. 472-485 (1965). Briefly, glycine is reacted with formic-acetic anhydride in the presence of small amounts of formic acid to yield N-formylglycine. The N-formylglycine is then oxygen alkylated by a combination of ingredients, comprising, for example, an alkyl sulfate and potassium hydroxide. The reaction is conducted in the presence of alcohol and the final reaction product is an ester of N-formylglycine. In the final step of the isocyano acetate ester synthesis, the ester of N-formylglycine is reacted with phosgene in the presence of triethylamine to yield an isocyano acetate ester.

Preferred strongly basic metalating agents of the formula MZ are those which enable production of the metalated isocyano acetate carbanion salts in a pure form by evolving a gas which may be collected. Alkali metal hydrides, and alkyls are examples of such materials, hydrogen and alkane gases being the side reaction products. The alkali metal hydrides are especially preferred, both from the standpoint of facility in undergoing reaction and ready commercial availability. Also, highly preferred as metalating agents are the alkali metal alkoxides. These are especially preferred metalating agents because the reaction proceeds easily at room temperature and the resulting carbanion is very stable in the alcoholic medium.

The α-metalated isocyano acetate ester or salt carbanions prepared in the step 1 reaction are prepared as a result of reacting an isocyano acetate ester or salt and a strongly basic metalating agent at -70° to 150°C, preferably at -40° to 80°C. Normally the reaction is effected in 0.5 to 24 hours, and preferably from 0.5 to 2 hours. If temperatures above 150°C are employed the α-metalated isocyano acetate carbanion is no longer stable and at temperatures below -70°C, the reaction time is so slow as to be impractical, and moreover, if a solvent is employed, the solvent may freeze.

The second step of the process can be summarized by the following equation.

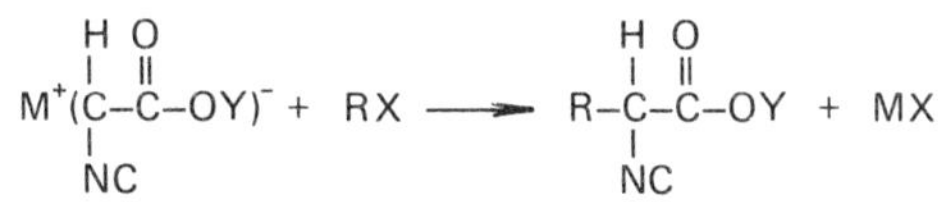

(2)

The α-metalated isocyano acetate ester or salt carbanion is reacted with an electrophilic agent, here an organic halide, represented by the formula RX to yield an α-substituted isocyano acetate ester or salt, plus as a by-product a salt represented by MX. Broadly, the phrase electrophilic agent here defines a compound which has a tendency to seek electrons, or in other words, a compound which will react with a carbanion. Where it is desired that the R moiety in the α-amino acid have a hydroxy group attached, i.e., have the formula

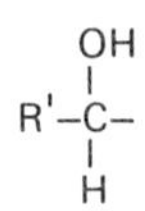

then the electrophilic reagent is an aldehyde and the step 2 equation becomes:

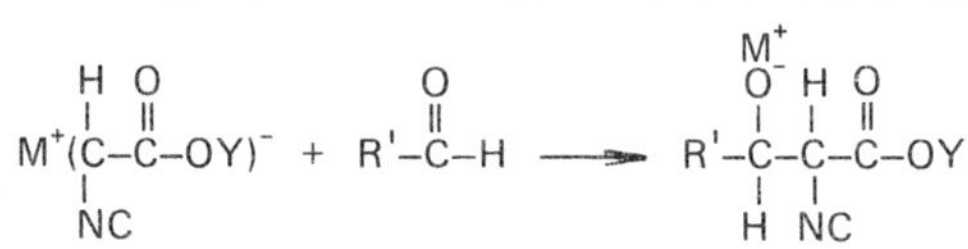

The final step comprises hydrolyzing the α-substituted isocyano acetate ester or salt to the α-amino acid.

Example: Preparation of Threonine from Acetaldehyde and Methyl Isocyano Acetate Using Sodium Methoxide as the Metalating Agent – The reaction vessel was a 3-neck flask fitted with a thermometer, a condenser, an addition funnel, and a continuous magnetic stirring device. The entire reaction vessel was flushed with N_2 prior to the commencement of the reaction. 100 ml of methyl alcohol was placed into the flask and 1.28 g of sodium (0.55 mol) was weighed into a beaker containing hexane. The sodium was removed from the hexane and dropped into the methyl alcohol which was continuously being stirred and which was maintained under a nitrogen blanket. Five grams (0.0505 mol) of methyl isocyano acetate was weighed into a beaker and diluted with monoglyme. Thereafter the monoglyme-methyl isocyano acetate mixture was syringed into the addition funnel and added slowly in a dropwise manner to the sodium-methyl alcohol mixture after the sodium and methyl alcohol had completely reacted at 40°C to yield sodium methoxide.

The methyl isocyano acetate and sodium methoxide strongly basic metalating agent mixture was allowed to react for 30 minutes at 40°C. This completed step 1. Thereafter 3.5 ml of acetaldehyde was syringed into the addition funnel and added dropwise over several minutes to the metalated methyl isocyano acetate ester carbanion. Because the reaction was exothermic, the temperature in the reaction vessel rose 13°C. The reaction was allowed to continue overnight (15 to 16 hours) at 40°C. This completed step 2.

A 10 ml aliquot portion of the reaction mixture was removed and heated for 2 hours at 90°C with 4 ml of concentrated hydrochloric acid and 12 ml of water. The acidified aliquot was evaporated to near dryness. At this point thin layer chromatography was run on a small portion of the residue and indicated the presence of threonine. Thereafter the remaining portion of the acidified dried residue was dissolved in isopropanol and aniline was added until the pH was approximately 7. The solution was stirred for 0.5 hour. Thereafter the solution was filtered and 0.4 g of a brown precipitate was obtained. Infrared analysis of the brown precipitate showed that it was threonine.

Thereafter the remaining portion of the acetaldehyde-carbanion salt mixture was refluxed with 100 ml of alcoholic hydrochloric acid (10 ml of concentrated hydrochloric acid and 10 ml of water and 80 ml of methyl alcohol). The refluxing was continued for 45 hours at 90°C. The solution was thereafter filtered and the filtrate was concentrated on a rotary evaporator to a dark brown acidic residue. This residue was taken up in isopropanol and the pH was adjusted to 7 utilizing aniline as the base.

At this point a large amount of white precipitate appeared. The precipitate was stored overnight at -5°C and thereafter collected and dried under vacuum. Nuclear magnetic resonance analysis of the dried precipitate weighing 3.5 g showed it consisted of only threonine. The total weight of threonine obtained, i.e., 3.5 g plus the 0.4 g from the previous aliquot portion, was 3.9 g. Based upon stoichiometric calculations, the theoretical amount of obtainable threonine was 6.0 g. The percentage of yield is 65%.

When in the above example the methyl isocyanate ester is replaced by the corresponding ethyl, propyl, butyl, octyl, and benzyl esters, similar formation of threonine occurs at the same yield level. Similar results are also obtained when the starting material is sodium isocyano acetate.

By Hydrolysis of Erythrocytes

A highly efficient process for preparing a nutritionally acceptable mixture of amino acids from erythrocytes has been developed by *M.A. Beigler, W.B. Benken and J.J. Nassar; U.S. Patent 4,003,992; January 18, 1977; assigned to Syntex (U.S.A.) Inc.* This process comprises the steps of denaturization, hydrolysis and neutralization, desludging, decolorization, deionization, concentration and sterilization as shown in detail by the steps of the following examples.

Example 1: Pretreatment – One hundred pounds of bovine erythrocytes are combined with 400 lb of water in a kettle and mixed with high shear agitation. Steam (20 psig) is injected directly into the erythrocyte slurry for 45 minutes to 1 hour. The heat denatured protein is pumped into a perforated basket centrifuge where the solubles are removed with the water. This is followed by cake washing with 100 lb of water. This results in a semidry cake containing 30 to 40% total solids which may be further dried, if desired, for use in the hydrolysis step.

Example 2: Acid Hydrolysis and Partial Neutralization – Twenty pounds of dry denatured erythrocyte solids and 60 to 80 lb of 8 N sulfuric acid are heated at 258°F or greater (16 psig) for 6 hours, not including a 20 minute preheating time. The mixture is allowed to cool to 150°C and a 20% w/w calcium hydroxide slurry slowly added to a final pH of 2.7 to 3.5. After cooling the mixture is desludged as in the following example.

Example 3: Desludging – The above mixture is placed in a kettle and the heavy crystalline solids are allowed to settle. The supernatant (containing 1 to 5% fine solids) is decanted and is filtered using a plate and frame press equipped with cotton filter cloths. The solids in the press and those left in the kettle are exhaustively water washed. Typical recoveries vary between 85 and 95% depending upon the thoroughness of the wash step.

In an alternate procedure, the mixture from Example 3 is pumped into a basket centrifuge, and the solution containing the amino acids and soluble salts is separated from the calcium sulfate cake. This cake is washed with hot (140°F) water, and the washings combined with the above solution.

Example 4: Decolorization – In a typical run 1.5 lb of powdered charcoal are combined with 100 lb hydrolysate in a kettle (approximately 5 to 6% amino acid solution w/v) and mixed for 30 minutes with a high shear agitator. The slurry is then filtered using a plate and frame press. The filtrate is examined by colorimetry indicating that 98% of the initial color has been removed. The filtrate is then used in the next step.

Example 5: Deionization – The resin used for the column is Duolite C-20 (40 to 80 mesh bead size). The height:diameter ratio for the column is between about 4:1 and 8:1. The C-20 resin is stored in the sodium form (pH $\geqslant$ 12). Prior to loading, the resin is converted to the hydrogen form by passing four bed volumes of 1.5 N sulfuric acid through the resin. After regeneration the resin is backwashed with water (4 to 5 bed volumes) and then is cocurrently rinsed with 2 bed volumes of water to compact the bed and prepare it for loading. The column is then loaded with erythrocyte hydrolysate (pH 3.5) by passing the hydrolysate through the column at a flow rate between about 1 and 5 gal/ft^3 min.

In such a manner, the resin binds between about 4 and 6 kg amino acids/ft^3 min. After loading, the column is washed cocurrently with water until sulfate-free. The column is then eluted at a flow rate of about 1 gal/ft^3 min with 0.1 N sodium hydroxide (0.4% w/w). Elution is monitored with a recording pH controller and amino acids are collected to pH 12.0 to maximize arginine recovery. Fifty-four equivalents of the sodium hydroxide solution are necessary to elute the amino acids from each cubic foot of C-20 resin. The final concentration of amino acids is between about 0.85 and 1.0% w/w with no cation other than sodium present. Recovery exceeds 95% of the amino acids loaded, with losses equally distributed across the amino acid profile.

Example 6: Concentration – The amino acid eluate from the previous step is reacidified to pH 3.0 with 1.5 N sulfuric acid (5 ml/l of solution). The amino acids are then loaded on a column identical to the one used for deionization. After loading the amino acids are washed sulfate-free with water, then eluted with 0.5 N sodium hydroxide (2% w/w). Elution again requires 54 equivalents of sodium hydroxide per cubic foot of resin. Amino acids are collected to pH 11.5 to afford a solution containing between about 2.9 and 4% w/w amino acids.

Example 7: Sterilization – A 3% solution of amino acids (pH 8.7) is steam injected at 280°F (holding time, 11 seconds) and then flashed into an evacuated chamber (124°F). This cycle is repeated 5 times. The residual ammonia after the 5 steam injection-flash evaporation sequences is approximately 50 ppm (0.005%).

Example 8: The final colorless amino acid solution prepared by the above process has an amino acid content between about 2.9 and 4% w/w.

Specifications

Theoretical Amino Acid Profile	Relative % of Total Protein
Aspartic acid	10.4
Threonine	4.1
Serine	3.3
Glutamic acid	7.7
Proline	3.4
Cysteine	0.3
Glycine	3.8
Alanine	8.4
Valine	9.6
Methionine	1.5
Isoleucine	0.3
Leucine	13.4
Tyrosine	2.4
Phenylalanine	8.0
Lysine	10.3
Histidine	7.2
Arginine	3.6

Chemical Specifications	Quantity*
Chloride, %	≤0.02
Ammonia, %	≤0.02
Sulfate, %	≤0.03

(continued)

Chemical Specifications	Quantity*
Heavy metals, ppm (Pb, Zn, Cu, Mn)	$\leqslant$10.0
Iron, ppm	$\leqslant$30.0
Na, %	$\leqslant$ 1.6

*Based on dry weight amino acids.

Fermentation with *Protaminobacter thiaminophagus* Mutant

A mixture of amino acids has been obtained by *K. Nakayama, M. Kohata, Y. Tanaka, T. Nomura and R. Katsumata; U.S. Patent 3,907,641; September 23, 1975; assigned to Kyowa Hakko Kogyo Co., Ltd, Japan* using a mutant strain of *Protaminobacter thiaminophagus.* The newly isolated mutant, *Protaminobacter thiaminophagus* K-217 (ATCC 21371) is capable of accumulating L-lysine, L-aspartic acid, L-alanine, L-valine, L-leucine and L-arginine in a considerable amount when cultured in a medium containing methanol as a source of carbon.

Generally, it is desirable that the concentration of methanol in the medium be maintained below 3% (v/v). Good results can be obtained when a medium initially having a low concentration, for example, 0.5 to 3% (v/v) of methanol is used and culturing is carried out while feeding methanol to the medium continuously, in an amount of 0.3 to 0.6% (v/v) based on the volume of the medium per hour, or intermittently in an amount of 0.5 to 2% (v/v) based on the volume of the medium at each feeding, as methanol is consumed by the microorganism.

As the nitrogen source, ammonium salts such as ammonium chloride, ammonium sulfate, ammonium phosphate and ammonium nitrate, ammonia and urea may be used. Casamino acid, peptone and yeast extract may also be used as the nitrogen source. Additionally, as inorganic materials, potassium phosphates, magnesium sulfate, iron and manganese salts may be used. As *Protaminobacter thiaminophagus* K-217, ATCC 21927 has a requirement for thiamine, thiamine or a natural product containing thiamine must be added to the medium.

Culturing is carried out under aerobic conditions at 20° to 40°C for 2 to 5 days. To obtain a high yield of the product, it is desirable that the pH of the medium be maintained at 4 to 9, preferably at around neutral during culturing. The pH may be adjusted with calcium carbonate, various buffer solutions or alkaline solutions. After the completion of culturing, the microbial cells are removed from the culture broth by filtration. The produced amino acids are recovered from the resulting filtrate by any of the well-known methods, such as ion exchange resin treatment, crystallization by concentration, etc.

Example: Protaminobacter thiaminophagus K-217 (FERM-P No. 2021) ATCC 21927 is inoculated in 10 ml of a seed medium having the following composition in a test tube and cultured at 30°C for 20 hours with shaking.

Methanol, ml	20
$(NH_4)_2SO_4$, g	6
KH_2PO_4, g	2
K_2HPO_4, g	7
$MgSO_4 \cdot 7H_2O$, g	0.5
$FeSO_4 \cdot 7H_2O$, mg	10

(continued)

$MnSO_4 \cdot 4H_2O$, mg	8
Thiamine hydrochloride, mg	1
Biotin, μg	10
Yeast extract, g	0.1
Phenol red (pH-indicator), mg	10
Water to make up the total volume (pH 7.2), liter	1

The resulting seed culture is inoculated into 10 ml portions of a fermentation medium having the same composition as the seed medium in test tubes in a ratio of 1 ml. Culturing is carried out at 30°C for 96 hours with shaking. Twenty-four hours, 48 hours and 72 hours after the start of the culturing, 2% of methanol is added to each culture broth (total amount of the added methanol: 6%). The pH of the culture broth is adjusted to about neutrality using 14% aqueous ammonia. After the completion of culturing, 0.2 mg/ml of L-aspartic acid, 0.8 mg/ml of L-alanine, 1.0 mg/ml of L-valine, 1.2 mg/ml of L-leucine, 0.7 mg/ml of L-arginine and 0.3 mg/ml of L-lysine (as L-lysine hydrochloride) are accumulated in the culture broth.

After the removal of the microbial cells by filtration, the L-aspartic acid, L-alanine, L-valine, L-leucine, L-arginine and L-lysine are recovered by a conventional ion exchange treatment. The yield of L-aspartic acid, L-alanine, L-valine, L-leucine, L-arginine and L-lysine from one liter of the culture is 0.14 g, 0.48 g, 0.6 g, 0.72 g, 0.49 g and 0.21 g, respectively.

Bacterial Hydrolysis of Nitriles

A. Commeyras, A. Arnaud, P. Galzy and J.-C. Jallageas; U.S. Patent 3,940,316; February 24, 1976; assigned to Agence Nationale de Valorisation de la Recherche (ANVAR), France have developed a technically simple process by which it is possible to hydrolyze almost all the nitriles under mild conditions.

In the process, an organic acid is produced from the corresponding nitrile by subjecting that nitrile, in aqueous solution, to the action of bacteria which show nitrilasic activity, and by subsequently separating the bacterial mass from the acid solution. The bacteria showing nitrilasic activity used for the process are preferably selected from the species Bacillus, Bacteridium as defined by Prevot, Micrococcus and Brevibacterium as defined by Bergey. More particularly, these bacteria are selected from the strains lodged in the Collection de la Chaire de Genetique de l'Ecole Nationale Superieure Agronomique de Montpellier (France), under the numbers R 332, R 340, R 341, A 111, B 222, A 112, A 13, A 141, A 142, B 211, B 212, B 221, C 211, R 21, R 22, R 311, R 312, R 331.

These strains are capable of hydrolyzing virtually every one of the following types of compounds: simple nitriles such as acetonitrile; α-aminonitriles such as α-aminopropionitrile, α-amino-γ-methyl thiobutyronitrile, α-aminobutyronitrile, aminoacetonitrile; α-hydroxy nitriles such as lactonitrile, hydroxy acetonitrile; β-aminonitriles such as amino-3-propionitrile; dinitriles such as malononitrile; α-unsaturated nitriles such as acrylonitrile; and α-benzonitriles, such as homoveratronitrile.

The nitrile solution is preferably adjusted to a slightly basic pH-value, for example, with potash or ammonia, before being exposed to the action of the bacteria.

In the process, the strains with nitrilasic activity, after culture in a nutrient medium, are suspended in an aqueous solution of the nitrile to be hydrolyzed for a period of a few hours. The pH-value of the solution is kept slightly basic (for example, pH 8) or neutral. In some cases, it is necessary in order to complete hydrolysis slightly to acidify the pH-value of the solution after about 1 hour. On completion of hydrolysis, i.e., generally after a few hours, the bacteria are eliminated by any method known in the science of biology, for example, by centrifuging. The acid is extracted from the solution by known methods, for example, by extraction or precipitation.

In this manner, racemic α-alanine was obtained from α-aminopropionitrile, β-alanine from amino-3-propionitrile and racemic methionine from α-amino-γ-methyl thiobutyronitrile sulfate.

Using Hydrogen-Oxidizing Bacteria

The cost of amino acids produced by fermentation depends largely on the cost of the carbon source used. In this respect, carbon dioxide is available at a low cost. Various autotrophic microorganisms are known to assimilate carbon dioxide, but none of them produces any amino acids from the assimilated carbon dioxide.

It has been found by *Y. Morinaga, A. Ishizaki and S.-I. Otsuka; U.S. Patent 3,943,038; March 9, 1976; assigned to Ajinomoto Co., Inc., Japan* that gram positive hydrogen-oxidizing bacteria, especially bacteria belonging to Arthrobacter, Brevibacterium and Mycobacterium, can produce a number of amino acids from carbon dioxide as the carbon source and hydrogen as the energy source. It is further found that amino acid production is remarkably improved by adding penicillin to the culture medium. Amino acids produced in accordance with this process include lysine, histidine, arginine, aspartic acid, threonine, serine, glutamic acid, proline, glycine, alanine, cysteine, valine, isoleucine, leucine, tyrosine, phenylalanine and methionine.

The bacteria useful in this process are: Brevibacterium sp. AJ 3588 (FERM P 2234); Mycobacterium sp. AJ 3589 (FERM P 2235); and Arthrobacter sp. AJ 3786 (FERM P 2638).

Example 1: An aqueous culture medium was prepared containing, per deciliter, 0.5 g $(NH_4)_2SO_4$, 0.03 g KH_2PO_4, 0.18 g $Na_2HPO_4 \cdot 12H_2O$, 0.02% $MgSO_4 \cdot 7H_2O$, 100 mg $FeSO_4 \cdot 7H_2O$, 100 mg $CaCl_2 \cdot 2H_2O$, and adjusted to a pH 7.2. Four hundred milliliter batches of the medium were placed in 1 liter fermentation vessels and sterilized with steam. Each batch of medium was inoculated with one of gram positive hydrogen-oxidizing bacteria, i.e., Brevibacterium sp. AJ 3588, Mycobacterium sp. AJ 3589 and Arthrobacter sp. AJ 3786, or one of gram negative hydrogen-oxidizing bacteria, i.e., *Alcaligenes eutrophus* ATCC 17697, *Pseudomonas facilis* ATCC 11228, and *Pseudomonas ruhlandii* ATCC 15749, and held at 30°C with agitating at 1,200 rpm.

During the cultivation, hydrogen, carbon dioxide and air were introduced at the rate of 18 ml/min, 5 ml/min, and 22 ml/min, respectively. The pH of the medium was maintained at 6.8 with gaseous ammonia. After 96 hours cultivation or culturing, the amino acids produced in the culture broth were determined by an amino acid autoanalyzer. The results are shown in Table 1.

TABLE 1

	. . . . . . .Gram Positive Bacteria			. . Gram Negative Bacteria. . .		
	Brevibac-terium sp. Ferm P-2234	Mycobac-terium sp. Ferm P-2235	Arthro-bacter sp. Ferm P-2638	*Alcali-genes eutrophus* ATCC 17697	*Pseudo-monas facilis* ATCC 11228	*Pseudo-monas ruhlandii* ATCC 15749
Amino Acid	. . . .		mg/dl.			
Lysine	0.86	0.54	0.72	0.03	0.02	0.02
Histidine	0.45	0.31	0.40	0.02	0.02	0.03
Arginine	1.45	0.98	1.33	0.08	0.04	0.03
Aspartic acid	9.48	8.25	6.24	0.13	0.09	0.18
Threonine	5.10	2.60	1.52	0.06	0.03	0.02
Serine	2.23	1.97	1.04	0.02	0.02	0.04
Glutamic acid	25.46	18.41	14.37	0.45	0.42	0.69
Proline	3.87	2.33	1.04	0.03	0.04	0.03
Glycine	5.06	2.48	1.80	0.05	0.05	0.08
Alanine	12.77	8.50	3.20	0.04	0.08	0.09
Cysteine	0.48	0.37	0.22	0.01	0.01	0.01
Valine	4.39	13.29	3.01	0.05	0.06	0.04
Isoleucine	1.89	1.47	0.89	0.02	0.02	0.02
Leucine	3.96	6.61	1.20	0.03	0.04	0.04
Tyrosine	1.97	0.32	0.33	0.02	0.01	0.02
Phenylalanine	2.50	1.57	0.85	0.02	0.02	0.02
Methionine	1.15	0.80	0.40	0.01	0.01	0.01

Example 2: The hydrogen-oxidizing bacteria of Example 1 were cultured in the same manner as described in Example 1. Penicillin G (50 IU/ml) was added to the media of Brevibacterium sp. AJ 3588 and Mycobacterium sp. AJ 3589 after 68 hours cultivation and to the media of Arthrobacter sp. AJ 3786, *Alcaligenes eutrophus* ATCC 17697, *Pseudomonas facilis* ATCC 11228, and *Pseudomonas ruhlandii* ATCC 15749 after 48 hours. Aspartic acid and glutamic acid accumulated in the culture broth are shown in Table 2.

TABLE 2

	Brevibac-terium sp. Ferm P-2234	Mycobac-terium sp. Ferm P-2235	Arthro-bacter sp. Ferm P-2638	*Alcali-genes eutrophus* ATCC 17697	*Pseudo-monas facilis* ATCC 11228	*Pseudo-monas ruhlandii* ATCC 15749
	. . . .		mg/dl.			
Aspartic acid	27.60	24.37	19.70	0.14	0.12	0.29
Glutamic acid	205.8	179.5	159.4	0.50	0.58	0.88

Preparation of L-Carbamyl Amino Acids

Dihydropyrimidinase, prepared from calf liver according to Donald P. Wallach and Santiago Grisolia (*J. of Biol. Chem.* 226, 277 (1957), hydrolyzes 4,5 dihydrouracil, dihydrothymine and hydantoin, i.e., compounds having no asymmetrical carbon atom. Surprisingly, it has been found that if this enzyme acts on racemic compounds (5-hydantoins), it hydrolyzes only the L form according to a strictly selective way. Similar behavior is shown by other classes of enzymes which are known to hydrolyze only one stereoisomeric form. That is, these enzymes, too, when used in presence of the racemic compound, generally hydrolyze only one of the two forms present in the racemic compound in a strictly selective way. A process disclosed by *D. Dinelli, F. Morisi and D. Zaccardelli; U.S. Patent 3,964,970; June 22, 1976; assigned to Snam Progetti, SpA, Italy* provides for the preparation of only one of the two stereoisomeric carbamyl amino acids wherein certain heterocyclic compounds having an asymmetrical carbon, such as 5-substituted hydantoins, are hydrolyzed in the presence of the above enzymes. The corresponding amino acids are also provided by hydrolyzing the carbamyl amino acids in the presence of these same enzymes.

For the 5-hydantoins, the hydrolysis is carried out by keeping the pH at 6 to 11; in fact at pH lower than 6 or higher than 11 the reaction is very slow. The hydrolysis occurs according to the scheme below.

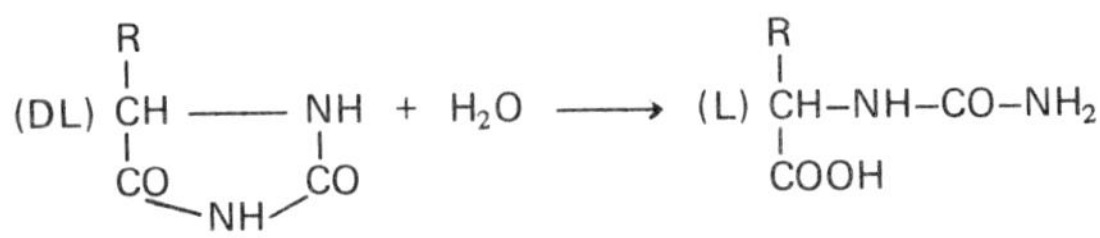

Because the L-carbamyl amino acid formed in the hydrolysis has an acid nature, the pH is maintained at the starting value by adding alkali while the reaction is occurring. The separation of L-carbamyl amino acid from D-5-hydantoin, which is not hydrolyzed, may be carried out according to various methods. Very often, by cautiously and under mild stirring adding an acid amount equal to the employed alkali and cooling at 0°C, it is possible to obtain the separation of substantially all the L-carbamyl amino acid whereas D-5-hydantoin remains in solution.

It has also been found that, if the waters containing D-5-hydantoin and some L-carbamyl amino acid are brought to a pH higher than 7, the rotatory power decreases to a zero value or even to a little opposite value, because D-5-hydantoin racemizes while L-carbamyl amino acid does not change. The racemization rate of D-5-hydantoin is a function of pH and the temperature. The racemization rate increases as the temperature increases and also increases as the pH rises to values higher than 7. Optically inactive 5-hydantoin, formed in this way, can be utilized in following operations.

Finally it has been found, that L-carbamyl amino acid, dissolved in water and heated up to boiling, decomposes according to the reaction below.

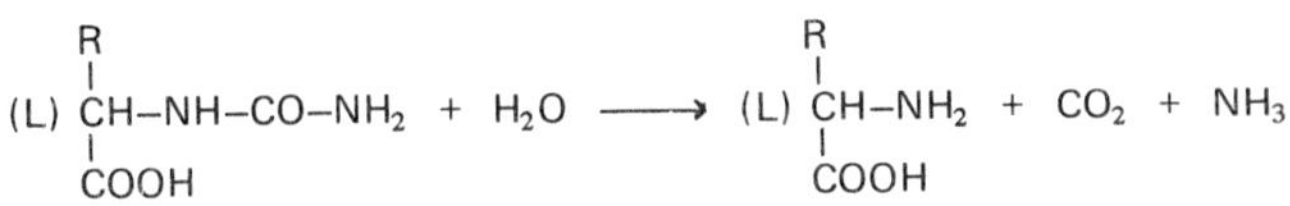

Example: One liter of a solution containing 142 g of 5-isopropylhydantoin brought to pH of 8.0 by adding sodium hydroxide was poured into a 2 liter vessel provided with a pH regulator and a stirrer. After the stirrer has been started, 10 cm^3 of a solution of dihydropyrimidinase obtained from calf liver were added. At the same time the pH-regulator was activated to keep the pH constant at about 8.0 by controlling the addition of a 5 N sodium hydroxide solution. In the beginning the reaction is very fast but decreases as L-5-isopropylhydantoin is consumed.

The apparatus was allowed to operate overnight and, as a consequence, the reaction continued until 100 cm^3 of the sodium hydroxide solution were consumed. In the morning the sodium hydroxide addition was stopped and an amount of 5 N hydrochloric acid corresponding to the total amount of previously used sodium was slowly added. Some white crystals began to form, the separation of which was completed by cooling the vessel at 0°C over some hours. By filtration and washing with 100 cm^3 of ice water, 60 to 70 g of solid product (A) were obtained and also the washing waters (B) and the mother liquors (C) which were treated separately.

The solid product was moist L-carbamyl valine containing traces of sodium chloride and protein substances. It was suspended in 200 cm^3 of water and steam heated up to boiling while an equimolecular amount of 10% HCl was being added. The product dissolved and carbon dioxide was developed. Using vacuum evaporation, a residue was obtained consisting of L-valine and ammonium chloride, wherefrom L-valine was isolated according to a conventional method. The product had a rotatory power $[\alpha]_D^{20}$ = -28.3 (c = 8, 6 N HCl) and a melting point of 310° to 315°C (in a closed capillary). By treating the mother liquors it was possible to obtain additional small amounts of the product. The washing waters could be employed directly in a following operation.

The mother liquors had a pH of about 2.5 and a remarkable rotatory power since they contain D-isopropylhydantoin. When brought to pH of 8.5 to 9 and kept at 35° to 40°C for 24 to 36 hours they lost the rotatory power since 5-isopropylhydantoin racemized. Then they were brought to pH of 7 by adding hydrochloric acid and evaporated under vacuum until dryness. The solid residue was treated again with water at 60°C and filtered. The solution contained, besides sodium chloride, about 70 g of 5-isopropylhydantoin and 10 to 20 g of L-carbamyl valine as the sodium salt thereof.

This solution, after the addition of 72 g of 5-isopropylhydantoin, could be used for further operation. This process is also more economic if use is made of the enzyme insolubilized by means of any technique employed for the enzyme insolubilization such as the techniques of chemically binding the enzymes to carriers or the technique comprising englobing the enzymes in fibrous structures.

BLOCKED AMINO ACIDS AND PEPTIDES–PREPARATIVE PROCESSES

9-Fluorenylmethyloxy Carbonyl Derivatives of Amino Acids

Blocking agents for peptide synthesis which can be removed under mildly alkaline conditions are disclosed by *L.A. Carpino and G.Y. Han; U.S. Patent 3,835,175; September 10, 1974; assigned to Research Corporation*. These agents are 9-fluorenylmethyloxy carbonyl compounds which may be unsubstituted or substituted with one or more reaction inert substituents. A substituent, if present, may be on the alpha carbon atom, that is the carbon atom attached to the 9-position of the fluorene nucleus, or it may be on one or both of the fused phenyl rings of the nucleus. The term "reaction inert" means that the substituent remains attached to the alpha carbon atom or the fluorene nucleus during the reaction being used.

Typical compounds within the scope of this process include compounds of the formula:

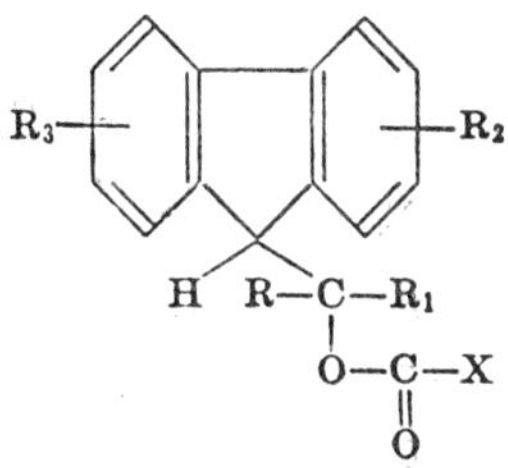

where each of R and R_1 are hydrogen, alkyl or aryl containing up to 8 carbon atoms; each of R_2 and R_3 are hydrogen, fluorine, chlorine, bromine, iodine or nitro; and X is chlorine, fluorine, bromine, cyano, thioalkyl or thioaryl where the alkyl or aryl group contains up to 10 carbon atoms, substituted and unsubstituted aryloxy containing up to 10 carbon atoms where the aryl group if substituted is preferably substituted on at least one position with an electron attracting substituent, hydrazino, $-N_3$,

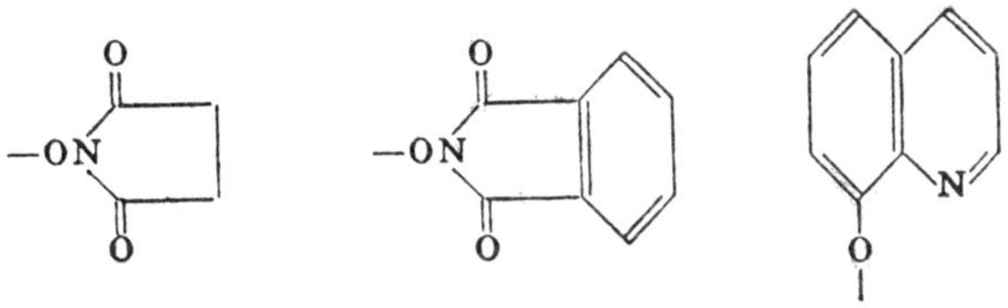

and $(AlkO)_2PO-$

where Alk is a lower alkyl group containing up to 6 carbon atoms. These compounds can be used as blocking agents for amino and carboxylic groups of all types of amino acids and peptides including neutral, acidic and basic. In the following examples the 9-fluorenylmethoxy carbonyl group is referred to as FMOC and the amino acids are sometimes indicated by their conventional abbreviations.

Example 1: Preparation of FMOC-Amino Acids – A solution containing 1 mmol of the amino acid and 2.5 mmol of Na_2CO_3 in 2.7 ml of water is cooled in an ice bath and treated dropwise with a solution containing 1 mmol of FMOC chloride in 2 ml of dioxane. The solution is stirred at room temperature for 2 to 3 hours, 50 ml of water added and the solution extracted with ether. The aqueous layer is acidified with concentrated HCl to Congo Red paper and the precipitate recrystallized from the appropriate solvent (see the following table for recrystallization solvent, MP, etc.).

Preparation of FMOC-Amino Acids

Compound	Time (hr)	MP (°C)	Recrystallization Solvent	Yield (%)	. . .Optical Rotation Data. . . α_D	°C	c	Solvent
FMOC-glycine	2	174-175	CH_3NO_2	89	–	–	–	–
FMOC-DL-alanine	3	176-178	CH_3NO_2	88	–	–	–	–
FMOC-L-alanine	2	144-145	$EtOAc-Et_2O$	94	-3.48	28.6	2.5	EtOAc
FMOC-β-alanine	3	145-147	EtOAc	91	–	–	–	–
FMOC-L-leucine	2	155-156	EtOAc	90	-4.44	28.3	2.5	EtOAc
FMOC-L-phenylalanine	2	183-185	EtOAc	92	+11.6	28.2	1.2	EtOAc

FMOC-glycine ethyl ester, MP 178° to 179° (CH_3CN) in 92% yield and FMOC-glycylglycine, MP 110° to 112° (EtOAc-Et_2O) in 94% yield are similarly prepared. Similar results are obtained with substituted FMOC compounds including 1-Br-FMOC-chloride; 2-Cl-FMOC-chloride; 2,7-diiodo-FMOC-chloride; and 2,7-dinitro-FMOC-chloride.

Example 2: This example illustrates one general procedure by which FMOC groups are removed using ammonia. A mixture of 1.55 g (4 mmol) of FMOC-L-phenylalanine in approximately 250 ml of liquid ammonia is stirred for 10 hours and then allowed to evaporate to dryness. The mixture is treated with ether to remove dibenzofulvene and the insoluble material dissolved in the minimum amount of water (a trace amount of dibenzofulvene polymer is removed by filtration). Evaporation of the water solution gives 0.66 g (100%) of phenylalanine $[\alpha]_D^{27}$ -33.20 (1.992, H_2O), identified by its infrared spectrum. The phenylalanine from which the above FMOC-derivative is prepared has $[\alpha]_D^{26.7}$ -33.24 (1.962, H_2O).

Trifluoroacetylation of Amino Acids

The trifluoroacetylating reagents most widely used are trifluoroacetic acid and trifluoroacetic anhydride. However, as these reagents are strongly acidic materials, they have been known to cause peptide bond cleavage during attempts to trifluoroacetylate peptides, and can cause undesired racemization in amino acids.

C.A. Panetta and T.G. Casanova; U.S. Patent 3,985,723; October 12, 1976; assigned to The University of Mississippi have now found that N-trifluoroacetylation of various amino compounds can be effected without the above disadvantages by reacting the amino compound with sym-trichlorotrifluoroacetone in a dimethyl sulfoxide solvent. The reaction can be carried out at room temperature and under mild and substantially neutral conditions. Although the exact mechanism of the reaction has not been determined, the sym-trichlorotrifluoroacetone apparently is cleaved between the carbonyl carbon atom and the trichloromethyl moiety so as to form the N-trifluoroacetylated derivative of the amino compound, in accordance with the following reaction equation.

$$Cl_3C\overset{O}{\overset{\|}{C}}CF_3 + R\underset{NH_2}{\underset{|}{C}}HCOOH \xrightarrow{DMSO} R\underset{NH\underset{O}{\underset{\|}{C}}CF_3}{\underset{|}{C}}HCOOH + CHCl_3$$

The reaction between the amino compound and the sym-trichlorotrifluoroacetone is carried out in a dimethyl sulfoxide solvent under substantially neutral conditions, preferably using a molar excess of the sym-trichlorotrifluoroacetone. In the preferred practice of the process, the molar ratio of the sym-trichlorotrifluoroacetone to the amino compound in the reaction mixture is approximately 3:1. The reaction may be carried out at room temperature, and at such temperature will be substantially complete in about 24 hours.

Example: A mixture of 7.4 mmol of L-valine, 10 ml of dimethyl sulfoxide and 23 mmol of sym-trichlorotrifluoroacetone was stirred at room temperature for 24 hours in a flask protected from atmospheric moisture. A complete solution was obtained within the first hour.

The progress of the reaction was followed by taking samples periodically and submitting them to gas-liquid chromatographic analysis. The reaction mixture was poured into 50 ml of ice water and the resultant mixture was extracted thrice with n-butanol and chromatographed on a column packed with silica gel. N-trifluoroacetylvaline, melting at 88° to 89°C, was obtained in a yield of 94%.

Vinyloxycarbonyl Derivatives

R.A. Olofson and Y.S. Yamamoto; U.S. Patents 3,835,109; September 10, 1974; and 3,711,458; January 16, 1973; both assigned to Research Corporation have disclosed the use of the vinyloxycarbonyl group((VOC):

$$\mathrm{CH_2{=}CHO\overset{\overset{O}{\|}}{C}-}$$

as a temporary blocking material in peptide syntheses. VOC amino acids or peptides can be readily prepared by treatment of an amino acid or peptide with vinyl chloroformate and base.

$$\mathrm{CH_2{=}CHO\overset{\overset{O}{\|}}{C}Cl} + \mathrm{H_3{}^{+}N\overset{\overset{R}{|}}{C}HCO_2{}^{-}} \xrightarrow{\text{Base}} \mathrm{CH_2{=}CHO\overset{\overset{O}{\|}}{C}NH\overset{\overset{R}{|}}{C}HCO_2H}$$

Vinyl chloroformate is readily synthesized in 40 to 45% yield by passing the bischloroformate of ethylene glycol through a hot tube. Details of the production of vinyl chloroformate can be found in U.S. Patent 2,377,085 dated May 29, 1945 and in a paper by L.H. Lee, *J. Org. Chem*, 30, 3943 (1965).

Although vinyl chloroformate is the preferred material, other sources of the VOC group may be used. For example, other vinyl haloformates and in general species such as

$$\mathrm{CH_2{=}CHOCOX}$$

where X is a reactive group may be used in place of vinyl chloroformate. In addition substituted vinyloxycarbonyl groups should also be useful as temporary protecting groups in peptide syntheses according to this process. These can be introduced by treatment of an amino acid or peptide with a substituted vinyl haloformate or X-formate and removed in the same way as the vinyloxycarbonyl group has been removed.

Production of the VOC amino acids or peptides is readily accomplished in yields averaging 40 to 95% by treatment of a naturally occurring amino acid or peptide with a source of the VOC residue, such as vinyl chloroformate in base using known procedures. The VOC group is stable under standard peptide coupling conditions using, for example, dicyclohexylcarbodiimide (hereinafter DCC) or N-ethyl-5-phenylisoxazolium-3'-sulfonate (hereinafter NEPIS) as the amide bond forming reagent. VOC-L-phenylalanyl-L-leucine methyl ester is obtained, for example, in 80% yield using DCC as the coupling reagent, while VOC-L-asparaginylglycylglycine ethyl ester is isolated in 85% yield using the Woodward-Olofson reagent, NEPIS, as the coupling agent.

In general, peptide yields from VOC amino acids are the same as those obtained for similar non-VOC peptides using the same peptide forming reagent.

Several techniques are available for removal of the VOC group from a product after peptide condensation. For example, the VOC group may be split off by titration of the resultant VOC peptide with bromine in an inert solvent followed by addition of methanol and removal of the solvent and the $BrCH_2CH(OMe)_2$ at reduced pressure to produce the peptide hydrobromide. Alternatively the VOC group may be split off by treatment of the resultant VOC peptide with a mild acid. Moreover, the VOC group may be split off from the VOC by mercuric ion induced hydrolysis.

Example: Vinyloxycarbonylglycine – A 50 ml three-necked round bottom flask, charged with glycine (1.5 g, 0.02 mol) in 15 ml water, is equipped with a magnetic stirring apparatus and a dropping funnel containing vinyl chloroformate (3.2 g, 0.03 mol) in 8 ml dioxane. The flask is immersed in an ice bath and the solution stirred rapidly. Magnesium oxide powder (1.2 g, 0.03 mol) is added in 3 portions while the chloroformate is dripped in (30 minutes). The ice bath is then removed and the solution stirred for an additional 4 hours at room temperature.

The reaction mixture is next cooled to 0°C, acidified to Congo Red with 6 N hydrochloric acid, and finally extracted with three 50 ml portions of ethyl acetate. The combined extracts are washed with two 25 ml portions of water, dried over anhydrous sodium sulfate and the solvent stripped off under vacuum. The solid residue, vinyloxycarbonylglycine, is recrystallized from chloroform. Yield: 2.35 g, 81%; MP 94.5° to 95.5°. By a similar procedure the following vinyloxycarbonyl derivatives (abbreviated VOC derivatives) can be synthesized. They are listed in the following table.

Compound	Crystallized From	MP, °C
VOC-L-asparagine	Ethyl acetate-pentane	157-159
VOC-L-proline	2-butanone-petroleum ether	93-94.5
VOC-nitro-L-arginine	Dissolved in ether precipitated with pentane	~66*
VOC-L-alanine		Obtained as oil
VOC-L-asparatic acid-β-methyl ester		Obtained as oil
VOC-S-benzyl-L-cysteine		Obtained as oil
VOC-L-glutamine		Obtained as oil
VOC-L-glutamic acid-γ-methyl ester		Obtained as oil
VOC-L-leucine		Obtained as oil
N-α-VOC-N-ε-tBOC-L-lysine**		Obtained as oil
VOC-L-methionine		Obtained as oil
VOC-L-phenylalanine		Obtained as oil
VOC-L-serine		Obtained as oil
VOC-L-tyrosine		Obtained as oil
VOC-L-valine		Obtained as oil
VOC-glycine ethyl ester		Liquid, BP 72° at 0.2 mm

*Amorphous **tBOC = tert-butoxycarbonyl

Peptide Synthesis in Presence of Isonitriles

Active esters were previously prepared from amino acids and oligopeptides by condensation in the presence of an N,N'-disubstituted carbodiimide. The latter forms substantial amounts of urea derivatives which are not readily removed from the desired product and are reconverted to the carbodiimide derivative only with much difficulty when they can be isolated from the reaction mixture. *N. Yasuda, Y. Ariyoshi and K. Toi; U.S. Patent 3,872,099; March 18, 1975; assigned to Ajinomoto Co., Inc., Japan* have found that isonitriles generally are as effective as the known carbodiimide derivatives in promoting the condensation of amino acids and oligopeptides to active esters, that the yields of the desired products are better, and that any by-products which may be formed are readily separated from the desired active ester.

In this process an amino acid having a masked amino group and an available carboxyl group, (also referred to as an N-protected amino acid) is contacted with at least one compound having either an active hydroxyl group or an amino acid having a masked carboxyl group and an available amino group, (also referred to as a carboxyl-protected amino acid) in a liquid reaction system in the presence of an isonitrile until a peptide or an ester of an amino acid derivative active for peptide synthesis is formed.

The mol ratio of the reactants in the method of this process is not critical. They usually react in equimolar amounts, and any excess present may remain unchanged. The NC group is the active moiety in the isonitriles employed as condensing agents, and any isonitrile free from obviously interfering functional groups is suitable. Alkyl, alkenyl, aryl, and aralkyl isonitriles are preferred.

The reaction is carried out in a liquid medium, and the mixture of reactants and of the isonitrile may constitute such a medium, any excess of the isonitrile providing improved liquidity. The reaction also takes place in aqueous solution, but it is preferably carried out in a suitable amount of inert organic solvent such as methylene chloride, chloroform, ethylene dichloride, ethyl ether, tetrahydrofuran, dioxane, ethyl acetate, acetonitrile, or dimethylformamide. The sequence in which the ingredients of the reaction mixture are combined is not critical and the reaction occurs over the wide range of temperatures in which the reaction system is liquid, and the time required to achieve a desired conversion rate is normally inversely proportional to the temperature.

Example: A solution of 6.0 g (20 mmol) N-carbobenzoxy-L-phenylalanine and 2.1 g (30 mmol) isopropylisonitrile (isopropyl isocyanide, 2-isocyanopropane, or isopropylcarbylamine) in 40 ml methylene chloride was refluxed at 45° to 50°C. A solution of 2.06 g (20 mmol) glycine ethyl ester in 40 ml methylene chloride was divided into three equal portions which were added to the refluxing material after 3, 16, and 21 hours respectively. Refluxing was continued for 6 hours after the last addition.

The reaction mixture was stripped of solvent, N-isopropylformamide, and unreacted isopropyl isonitrile by vacuum distillation, and the residue was dissolved in 150 ml ethyl acetate. The solution was washed sequentially with 1-N hydrochloric acid, 4% sodium bicarbonate solution, and water, dried with desiccated sodium sulfate, and evaporated to dryness. The residue was crystallized from ethyl acetate and petroleum ether, and 4.7 g N-carbobenzoxy-L-phenylalanylglycine ethyl ester was obtained (61% yield). When recrystallized from ethyl acetate and petroleum ether, it melted at 105° to 107°C and had a specific rotation of $[\alpha]_D^{20}$ = −17.1° (c = 2.1 in ethanol).

Using different solvents and conditions, the yields of N-carbobenzoxy-L-phenylalanyl-glycine ethyl ester obtained are indicated as follows: n-propyl isonitrile in methylene chloride for 70 hours at 20° to 25°C produced a yield of 67%; phenyl isonitrile in chloroform at 40°C for 28 hours gave a yield of 64%; benzyl isonitrile in dioxane at 50° to 60°C gave a 51% yield in 20 hours; and ethoxycarbonylmethyl isonitrile in acetonitrile produced a yield of 63% in 40 hours at 20° to 25°C. Peptides of proline, lysine, glycine, aspartic acid, alanine, valine, cysteine in various combinations were also prepared by this process.

Lactone Intermediates in Peptide Synthesis

M. Bodanszky; U.S. Patent 3,957,760; May 18, 1976; assigned to E. R. Squibb & Sons, Inc. has found that α-amino acids can be both protected and activated by reacting the amino acid with a single compound and then forming a lactone. These lactones readily react with α-amino acid esters to form condensation products from which the protecting-activating moiety is readily removed to leave the desired peptide. Briefly, the method comprises reacting an α-amino acid with an active carbonyl compound which forms a Schiff's base and then the Schiff's base is treated with a condensing agent to induce cyclization and yield a lactone of the formula

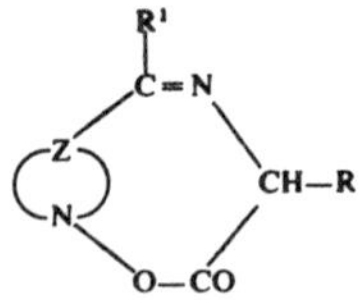

where R is the residue of an α-amino acid, R^1 is hydrogen or lower alkyl, and Z taken with N is a heterocyclic ring or a condensed ring system wherein a heterocyclic ring is condensed with a benzene ring.

Examples of suitable aliphatic carbonyl compounds are α-formyl-N-hydroxysuccinimide; N-hydroxy-β-formylpyrrole; N-hydroxy-β-formyl-2,5-dimethylpyrrole; 4-acetyl-N-hydroxypiperidine; and N-hydroxy-3-formylphthalimide. The amino acid is reacted with the aldehyde or ketone in an inert solvent, preferably an inert organic solvent such as dimethylformamide, dioxane, pyridine, etc. under conventional conditions for forming a Schiff's base. Then, after the Schiff's base has formed, without isolating the base unless desired, a condensing agent is added to form the lactone. An alcohol may also be used as the reaction medium in which the Schiff's base is formed, but in this case the solvent must be removed before adding the condensing agent.

Any condensing agent which promotes the formation of acyl esters may be used. These include, for example, dicycloalkylcarbodiimides, e.g., dicyclohexylcarbodiimide; dialkylcarbodiimides, e.g., dipropylcarbodiimide; diarylcarbodiimides, e.g., diphenylcarbodiimide; alkoxyacetylenes, e.g., ethoxyacetylene, diphenylketene, etc. Dicyclohexylcarbodiimide and ethoxyacetylene are preferred. About 1 mol or more of condensing agent per mol of Schiff base may be used. A mol ratio of about 1:1 may be used when the condensing agent is a carbodiimide. An excess of the condensing agent is preferable when other condensing agents are used. Sufficient time must elapse for the Schiff base to form before the condensing agent is introduced. A period of 1 to 24 hours is generally adequate. Stirring and/or heating will accelerate the reaction. Temperatures of 15° to 60°C may be used.

The cyclization reaction which occurs upon addition of the condensing agent must be effected in an anhydrous medium and the solvent must be one which has no reactive hydroxy or carboxy group. Media such as those named above, except alcohols, may be used for the cyclization as well. Room temperature up to 60°C may be used, with a temperature in the lower end of the range preferred for carbodiimides and a temperature at the higher end of the range for alkoxyacetylenes. The active lactones will then react with another amino acid, preferably a simple ester, such as a lower alkyl ester whereupon the ring opens and the amino acid is acylated. Treatment with aqueous acid, e.g., with hydrochloric acid, dilute sulfuric acid or the like, removes the protecting groups and the desired peptide is obtained.

Example: 2-Hydroxy-5-Nitrobenzylidene-L-Leucine and Its Lactone — L-leucine (2.62 g) and 5-nitrosalicylaldehyde (5.0 g) are added to a mixture of absolute ethanol (750 ml) and methanol (50 ml). The mixture is stirred at room temperature for several hours. When all the leucine is dissolved, the solvents are removed in vacuo, the residue extracted with ether, the ether extracts concentrated and diluted with hexane. A crystalline solid separates. It is filtered and washed with hexane. The crude Schiff base (decomposing at ca 175° to 189°, completely melting at ca 200°) is used in the cyclization step without additional purification.

The benzylidene derivative (1.40 g) is dissolved in tetrahydrofuran (40 ml) and dicyclohexylcarbodiimide (1.03 g) is added to the solution. After about 2 hours at room temperature, the completion of the reaction is checked with the IR spectrum of a sample (disappearance of the strong band at 4.8 μ, corresponding to the CN bond of the diimide, and appearance of the carbonyl band of the active ester at 5.65 μ). The by-product, mainly N,N'-dicyclohexylurea, is removed by filtration and washed with tetrahydrofuran (20 ml). The filtrate and washings are concentrated in vacuo to dryness, the residue is dissolved in ether, and the solution is diluted with hexane. The active ester is obtained in good yield in the form of crystals with strong double refraction. The product has a poorly defined melting point.

The lactone obtained above and glycine ethyl ester are reacted in chloroform. Dilute hydrochloric acid is added to the reaction mixture and L-leucyl glycine ethyl ester is obtained as the product.

Production of Peptides Using Pepsin

The process of *Y. Isowa, T. Nagasawa, K. Kuroiwa and K. Narita; U.S. Patent 3,972,773; August 3, 1976; assigned to Sagami Chemical Research Center and Nitto Boseki Co., Ltd., both of Japan* provides a synthesis of polypeptides, especially higher than tripeptides, by using the enzyme pepsin which is free from the defects of the known processes.

Briefly, the desired peptide has the formula

X–A–B–C–Y

where A represents alanine (Ala), glutamine (Gln), asparagine (Asn), ω-BOC-lysine (Lys), leucine (Leu), glycine (Gly), glutamic acid (Glu), MeO–Glu or proline (Pro); or A can be A_1–A_2, where A_1 represents a hydrophilic amino acid residue which can be Ala, Gln, Asn, ω-BOC-Lys, Leu, Gly, Glu, MeO–Glu or Pro, and A_2 represents valine (Val), methionine (Met), Leu, or Gln; B represents phenylalanine (Phe), tyrosine (Tyr), Leu, Met, Glu, aspartic acid (Asp), Gln, Asn, or tryptophan (Trp); C represents Phe, Leu, isoleucin (Ile), Tyr, cysteine (Cys)-SBzl, serine (Ser)-OBzl, Trp, or Met; X represents an α-amino acid protective group, an amino acid residue having an N-terminal protective group or a peptide residue having an N-terminal protective group; and Y represents a carboxyl protective group, an amino acid residue having a C-terminal protective group or a peptide residue having a C-terminal protective group.

In this process the peptide having the formula X–A–B–OH (hereinafter referred to as the C-component) is used as a starting material. Generally, it is preferred to use as a starting material peptides (including polypeptides and oligopeptides) having at least slight solubility in the medium, since insoluble peptides are inactive. The C-component should have a partial structure of condensation such that 2 or 3 specific amino acids in a specific order of bonding, i.e., the bond –A–B– in the formula above, are in the reactive position. As long as this specific partial structure is included, various peptides can be used.

The terminal group of the C-component, i.e., the α-position of the amino group, should be protected by an amino-protecting group such as a carbobenzoxy group (Z); a substituted carbobenzoxy group, e.g., p-methoxy benzyloxycarbonyl group (pMZ); a tert-butyloxycarbonyl group (BOC) or a tosyl group (TOS), etc.

The amino acid or peptide having the formula HCY (hereinafter referred to as the N-component) which is the other starting material, should have at the reactive position specific amino acids such as phenylalanine, leucine, isoleucine, tyrosine, S-benzylcysteine, O-benzyl serine, tryptophan, or methionine. The terminal carboxyl group of the N-component should be protected by a carboxyl protecting group such as a methyl ester, an ethyl ester, a benzyl ester, a tert-butyl ester or p-nitrobenzyl ester group; an amide group, a substituted amide group, e.g., a 2,4,6-trimethyl benzylamide group (TMB), a hydrazide or a derivative thereof. The N-component can be free or in the form of a salt, such as a hydrochloride, hydrobromide, trifluoroacetate, p-toluenesulfonate, or some other inorganic or organic salt.

In this process the C-component and the N-component can be used in equivalent amounts. However, it is also possible to react them using an excess of one of them. The molar ratio of the N-component to the C-component is usually 1:5 to 5:1, preferably 1:2 to 2:1. In the reaction, it is preferable to use the N-component by dissolving it in an alkaline medium containing an alkali such as an alkali metal hydroxide or an alkaline earth metal hydroxide. It is necessary to carry out the reaction in a buffer solution having a pH of 2 to 6, at a temperature of 20° to 50°C. Suitable buffer solutions include citric acid buffer solution, Michaelis buffer solution, McIlvaine buffer solution, and other buffer solutions having a pH of 2 to 6. Buffer solutions having a pH of lower than 2 or higher than 6 should not be used because the yield is low. The preferred pH is 3 to 5, and the optimum pH is 4.

The pepsin used is a digestive endopeptitase enzyme. Pepsin having a pepsin titer of 1:5,000; 1:10,000; and 1:60,000 can be easily obtained. A crude pepsin can be used as long as it has high activity. A catalytic amount of pepsin is employed, preferably 0.4 to 400 mg of pepsin per 1 mmol of the starting materials.

Example: A solution of 0.753 g (1.5 mmol) of HCl·H-Phe-Gly-Leu-Met-NH_2 (MW 502.1) in 40 ml of citric acid buffer solution (pH = 4.0) was added to a solution of 1.26 g (2.5 mmol) of α,ω-BOC-Lys-Phe-OH (MW 505.6) in 5 ml of 1 N NaOH and 30 ml of water was further added with stirring. Then, 0.2 g of pepsin (1:5,000 manufactured by Mikuni Kagaku Sangyo K. K.) was added to the mixture with stirring at 40°C for 24 hours to react them. The resulting white precipitate was filtered by a glass filter (G-3) and was sequentially washed with 5% ammonia water, 5% citric acid aqueous solution and water. The product was dried above P_2O_5 under reduced pressure at 50°C to obtain 1.25 g of α,ω-BOC-Lys-Phe-Phe-Gly-Leu-Met-NH_2 having a melting point of 207° to 216°C (decomposition) and $[\alpha]_D^{20}$ = -38.8 (c = 0.5 DMF). The yield was 88.2%.

RESOLUTION AND RACEMIZATION OF AMINO ACIDS

Using a Fixed Mold Acylase

Molds such as *Aspergillus oryzae* produce acylase with the ability to hydrolyze the L-isomer of acylamino acid into the free amino acid. Further it is also known that the alpha-N-acyl-DL-amino acid can be asymmetrically hydrolyzed by conventional acylase whereby the racemate is readily resolved into its optically active enantiomers. After the hydrolysis has been completed, the reaction solution is boiled and/or acidified to precipitate the enzyme, and the enzyme precipitate is filtered off. Thus, mold acylase can be used only once and must be discarded.

I. Chibata, T. Tosa, T. Sato and T. Mori; U.S. Patent 3,816,254; June 11, 1974; assigned to Tanabe Seiyake Co., Ltd., Japan have developed the process whereby the mold acylase is treated with a substantially water-insoluble haloacetyl-polysaccharide. The resultant substantially water-insoluble mold acylase preparation is incubated with a solution of an alpha-N-acyl derivative of a racemic acid to form the free L-amino acid.

Bromoacetyl-cellulose, bromoacetyl-dextran, bromoacetyl-agarose, chloroacetyl-cellulose, chloroacetyl-dextran, chloroacetyl-agarose, iodoacetyl-cellulose, iodoacetyl-dextran and iodoacetyl-agarose are suitably used as the water-insoluble haloacetyl-polysaccharide of this process. The haloacetyl-polysaccharide contains 4 to 40 w/w % of halogen atom, preferably 8 w/w % of chlorine, 18 w/2 % bromine or 31 w/w % of iodine.

The water-insoluble acylase preparation can be prepared by contacting one of the haloacetyl-polysaccharides with mold acylase in the presence of a salting out agent of protein. Inorganic sulfate such as ammonium sulfate and sodium sulfate are suitably used as the salting out agent. The amount of the salting out agent which is added to the reaction mixture should be 0.6 to 3.6 mmol, preferably 1.0 to 3.2 mmol, per ml of the reaction mixture. When a haloacetyl-polysaccharide containing the above-mentioned concentration of halogen is used, the preferable amount of mold acylase to be introduced into the polysaccharide is 0.1 to 0.25 g per g of haloacetyl-polysaccharide.

The resulting water-insoluble acylase preparation is collected by filtration or centrifugation, and, if required, the preparation may be further purified by washing with physiological saline solution. As the source of acylase, enzyme preparations containing acylase obtained from microorganisms such as *Aspergillus oryzae,*

Aspergillus meleus, Aspergillus midulans, Penicillium vinaceum or *Penicillium corymbiferum* are preferred. Many N-acyl derivatives of racemic amino acids may be used as the substrate or starting material. Examples of the N-acyl amino acids are: alpha-N-acetyl-amino acids such as N-acetyl-alanine, N-acetyl-valine, N-acetyl-leucine, N-acetyl-isoleucine, N-acetyl-serine, N-acetyl-threonine, N-acetyl-cysteine, N-acetyl-methionine, N-acetyl-phenylalanine, N-acetyl-tyrosine, N-acetyl-aspartic acid, N-acetyl-glutamic acid, N-acetyl-histidine, alpha-N-epsilon-N-diacetyllysine, alpha-N-delta-N-diacetyl-ornithine and alpha-N-acetyl-epsilon-N-benzoyl-lysine. Other acyl derivatives, such as N-formyl-, N-chloroacetyl-, N-bromoacetyl-, N-propionyl-, N-butyryl- or N-benzoyl derivatives of alpha-amino acids may also be used.

Example: A solid culture prepared by the cultivation of *Aspergillus oryzae* on wheat bran-hull medium is extracted with 10 times its volume of water under stirring for 2 hours. Ammonium sulfate is gradually added to the extract until the ammonium sulfate concentration reaches a level of 0.3 saturation in the extract, and the resultant precipitates are removed by centrifugation. Then, ammonium sulfate is added to the supernatant liquid until its concentration in the solution reaches 0.5 of saturation. The precipitates thus obtained are collected by centrifugation and dissolved in water. The solution is dialyzed against running water and then lyophilized. The acylase activity of the lyophilized enzyme is 4 u/mg.

One hundred milligrams of the enzyme are dissolved in 50 ml of 0.2 M-phosphate buffer solution (pH 8.5). Ten grams of ammonium sulfate and 1 g of bromo-acetyl-cellulose containing 18.1% bromine by weight are added to the solution. The mixture is stirred for 24 hours at 7°C and the insoluble materials are collected by filtration; 2.5 ml of a water-insoluble enzyme preparation are obtained. Acylase activity: 48 u/ml.

A mixture of 2.5 ml of the enzyme preparation and 2.5 ml of cellulose powder is charged into a 1 x 6.3 cm column. An aqueous solution (pH 7.0) containing 0.2 M concentration of N-acetyl-DL-methionine and 5 x 10^{-4} M concentration of cobalt ion is continuously passed through the column at 37°C at flow rates of 40 ml/hr and of 20 ml/hr. The concentration of L-methionine in the effluent is colorimetrically assayed by the ninhydrin method from samples obtained at intervals, and the conversion rate of N-acetyl-DL-methionine to L-methionine is calculated therefrom. The results obtained are shown in the table below.

Conversion Rates of N-Acetyl-DL-Methionine to L-Methionine (%)

Operation Time (hr)	Flow Rate 40 ml/hr	20 ml/hr
2	68.3	100
21	68.6	100
48	68.3	100
75	68.5	100
96	68.6	100
120	68.5	100

Continuous Resolution Using Acylase

S. Asai, S. Ishii, T. Tabira, E. Tsuchiya and H. Tazuke; U.S. Patent 3,841,966; October 15, 1974; assigned to Ajinomoto Co., Inc., Japan have developed a process for obtaining an enantiomorph of an amino acid in an increased yield based on the amount of acylase used. It has been found that the by-products of the method, that is, the carboxylic acid of the original acyl moiety and the N-acyl derivative of the undesired amino acid enantiomorph, repress acylase activity, and that this accounts for many of the difficulties encountered heretofore in preparing an optically active amino acid from an acyl derivative of its racemic form.

A high resolution ratio, that is, a high mol ratio of formed desired amino acid enantiomorph to initially employed acyl derivative of the racemic amino acid, can be obtained only by either using a very large amount of costly acylase or by operating in a very dilute aqueous medium which requires the handling and ultimate removal of a large amount of water.

However, acylase can be used practically to exhaustion of its enzyme activity for preparing an enantiomorph of an amino acid from an N-acyl derivative of the racemic form of the acid, more specifically a salt of the derivative, by contacting the salt in an aqueous medium containing an effective amount of acylase until the desired enantiomorph is formed by the acylase simultaneously with the forming of a corresponding amount of the acyl derivative of the other enantiomorph of the amino acid salt. The aqueous medium thereafter is contacted with crystals of the acyl derivatives of the racemic amino acid salt, while it contains the aforementioned acyl derivative of the other enantiomorph.

If the solubility of the acyl-racemate is greater than that of the corresponding derivatives of the enantiomorphs, the medium preferentially dissolves from the crystals the acyl derivative salt of the one enantiomorph so that the residual crystals are enriched with the acyl salt of the other enantiomorph. It is relatively simple to recover the one enantiomorph of the amino acid separately in high optical purity from the medium, and the acyl derivative of the other amino acid enantiomorph salt is separated from the medium without difficulty and is not severely contaminated with the salt of the one enantiomorph. After removal of the two products, the mother liquor is again ready for further selective deacylation by the acylase still present.

The number of cycles possible without replenishing the acylase or otherwise rejuvenating the liquid medium depends largely on the nature and amount of the acylase initially added to the hydrolysis mixture, on the composition of the acylamino acid salt, and the like. Generally, at least 3 and often as many as 10 cycles can be performed with a single charge of acylase. Because the liquid medium is recycled, it is not necessary that the resolution ratio in each cycle be very high, and the amount of acylase originally supplied may be quite small, further contributing to the high yields of L-amino acid per weight unit of acylase which is achieved by the method of this process.

Example: Concentrated ammonium hydroxide solution was added to a slurry of 650 ml water and 140 g N-acetyl-DL-phenylalanine until a solution of pH 6.5 was formed. 2.8 g acylase (15,000 u/g) was added to the solution which was stirred at 37°C for 30 hours. It was then cooled to 15°C with continued stirring, and a crystalline precipitate of L-phenylalanine was filtered off. The recovered L-phenylalanine weighed 27 g and had an optical purity of 100%. The mother liquor was mixed with 145 g N-acetyl-DL-phenylalanine ammonium salt. The mixture was heated briefly to 40°C, then cooled to 15°C with stirring, and undissolved crystals were filtered off. They consisted of N-acetyl-D-phenylalanine ammonium salt having an optical purity of 96.3% and weighed 68 g. The filtrate was again stirred at 37°C for 30 hours to start a new cycle. The procedure outlined above was repeated through 9 cycles and yielded 324 g L-phenylalanine per gram of the acylase initially introduced into the reaction system.

Phase Separation of Enantiomorphs

A process of selectively deacylating N-acyl-DL-amino acids in an aqueous medium in such a manner that the reaction products and the enzyme are recovered separately with minimal losses has been disclosed by *T. Uzuki, M. Takahashi, M. Noda, Y. Komachiya and H. Wakamatsu; U.S. Patent 3,907,638; September 23, 1975; assigned to Ajinomoto Co., Inc., Japan.*

This is achieved by contacting the N-acyl-DL-amino acid with acylase in an aqueous medium while the medium is in contact with an organic solvent not miscible with water, the solvent being chosen so that the concentration in the solvent of the N-acyl-DL-amino acid and of the carboxylic acid from which the initial acyl group is derived is at least equal to the concentration in the aqueous medium when solution equilibrium between the solvent and the aqueous medium is reached. It is furthermore important that the acylase be readily soluble in water. If the deacylated amino acid is only sparingly soluble in the solvent, continuous operation in a very simple manner is possible.

In this process, water, N-acyl-DL-amino acid, and acylase are stirred or otherwise held in contact with the organic solvent at a temperature and pH below 7, suitable for acylase activity, until the desired, selective deacylation has taken place. It is then found that the acylase is entirely in the aqueous liquid, the optically active N-acylamino acid and the carboxylic acid formed by deacylation are in the organic solvent phase, and the desired, deacylated amino acid is in the aqueous phase as a solute or precipitate in the solid state.

The solvent may be chosen conveniently to permit the optically active N-acylamino acid to be racemized by heating while dissolved in the solvent so that the resulting solution or mixture of solvent and racemic starting material may be fed to the reaction zone without isolating the N-acetyl-DL-amino acid. The acid formed by the enzyme must be stripped at least partly from the recycled solvent since it must not accumulate in the reaction system.

Solvents meeting the more restrictive conditions include the triesters of phosphoric acid having at least 5, and preferably at least 7 carbon atoms in the combined alcohol moieties, such as triethyl, tripropyl, tributyl, triisobutyl, trichloroethyl, methyldibutyl, and methylpropylbutyl phosphate. Also suitable are the lower alkyl esters of the lower alkanoic acids, the term lower alkyl and its analogs relating to carbon chains of up to 4 members. Examples of such fatty acid esters are methyl acetate, ethyl acetate, butyl acetate, and ethyl propionate, these ester solvents being chosen mainly for their ready commercial availability and low cost.

Other suitable solvents include the dialkyl ketones having at least 4 carbon atoms, such as methyl ethyl ketone, diethyl ketone, methyl isobutyl ketone, and alcohols having at least 4 carbon atoms, such as n-butanol and amyl alcohol (1-pentanol).

The phosphoric acid esters are preferred because they can be handled safely through the entire processing cycle including thermal racemization of the optically active N-acylamino acid. The concentration of the N-acylamino acid in the racemization mixture is preferably higher than in the deacylation zone so that a portion of the solvent is preferably distilled off prior to heat treatment. In order to facilitate such distillation, the boiling point of the organic solvent employed should not be too high. Trialkyl phosphates having fewer than 15 carbon atoms in their combined alcohol moieties, such as tributyl and triisobutyl phosphate, and the dialkyl ketones having 4 to 6 carbon atoms have been found most advantageous among cheaply available commercial solvents in combining desirable properties.

The method is employed to particular advantage in the continuous processing of N-acyl derivatives of amino acids which are only sparingly soluble in water at pH values beneficial to acylase activity. Such amino acids include valine, leucine, isoleucine, methionine, aspartic acid, glutamic acid, phenylalanine, phenylglycine, 3,4-methylenedihydroxyphenylalanine, 3,4-dimethoxyphenylalanine, and tryptophan.

Example: Ninety grams of N-acetyl-DL-phenylalanine was dissolved in 250 ml water and enough ammonium hydroxide solution to make the solution neutral. Its ultimate volume was 300 ml, and it was transferred to a 2 liter reactor together with 900 ml tributyl phosphate containing 8.73 g L-phenylalanine, and 30.6 g N-acetyl-DL-phenylalanine. When equilibrium was established by stirring, the aqueous phase has a pH of 6.03. At this stage, 71.4 mg $CoCl_2 \cdot H_2O$ and 6.03 g acylase (19,000 u/g) were added and the contents of the reactor were stirred while a constant temperature of 37°C was maintained.

As the enzymatic deacylation of the N-acetyl-L-phenylalanine proceeded, crystals of L-phenylalanine precipitated. After 48 hours, the organic solvent phase was drawn off, and the L-phenylalanine crystals were filtered from the aqueous phase which was saturated with L-phenylalanine. The crystals were washed with a small amount of ice-cold water and dried. 40.43 g L-phenylalanine of 100% optical purity was recovered, corresponding to an 84.1% yield based on the entire N-acetyl-DL-phenylalanine initially present.

The organic solvent phase was found to contain 5.21 g acetic acid and 25.0 g N-acetylphenylalanine which was D-enantiomorph of 72.6% optical purity. The acetic acid and a portion of the solvent were distilled off in a vacuum, and the residue was held at about 180°C until the N-acetyl-D-phenylalanine was racemized. The resulting suspension of N-acetyl-DL-phenylalanine was recycled.

The aqueous mother liquor obtained from the crystallization of the L-enantiomorph was returned to the reactor together with 900 ml tributyl phosphate and 29.1 g N-acetyl-DL-phenylalanine. The aqueous layer had a pH of 5.97. The two-phase mixture was stirred at 37°C for 12 hours, whereupon the two liquids were separated, and 11.6 g crystalline L-phenylalanine was recovered from the aqueous phase. The procedure outlined above was repeated 4 times, and a total of 44.5 g crystalline L-phenylalanine was harvested.

Use of Transition Metal Complexes

An improved chromatographic method of separation of racemates into constituent optical isomers has been developed by *S.V. Rogozhin and V.A. Davankov; U.S. Patent 3,853,906; December 10, 1974.* This method is particularly useful for the separation of optically active isomers of amino acids, hydroxy acids, amino alcohols and their derivatives.

This is accomplished by chromatographic separation of the optical isomers of compounds capable of forming complexes with transition metal ions, which comprises passing a solution of the mixture of the isomers through the bed of an asymmetric sorbent, e.g., an ion exchange resin, which is capable of forming complexes with transition metal ions, wherein the sorbent is pretreated with a transition metal salt solution.

The preferred asymmetric sorbents consist of products obtained by reacting chloromethylated styrene copolymers crosslinked by divinylbenzene or other crosslinking agents with optically active α- or β-aminocarboxylic or aminophosphonic acids, and amides or esters thereof. Use can also be made of sorbents consisting of products prepared by polycondensation of optically active complexing compounds, e.g., the products obtained by reacting formaldehyde with tyrosine or derivatives thereof. Also useful are asymmetric sorbents based on natural polymeric products, e.g., sorbents derived from chitosan, alginic acid, etc.

The salts of transition metals for treating the sorbents are selected from the group consisting of mineral and organic salts of copper, zinc, nickel, cobalt, cadmium, chromium, manganese, iron, lead, silver, platinum, palladium, rhodium, gold, mercury, etc. The asymmetric sorbent is treated with a solution of a transition metal salt in water or organic solvent, charged into a chromatographic column and washed with water or organic solvent. Next into the column is fed the solution of a racemic compound in water or organic solvent, followed by washing the column with the solvent used until one of the optical isomers being separated is completely displaced from the column.

The second isomer, which forms with the sorbent a more stable complex, is eluted from the column by subsequent passage of the solvent, or desorbed as a result of varying the pH of the solvent or feeding into the column displacing agents, or else by changing the solvent or raising the column temperature. Evaporation of respective eluate fractions yields optically pure isomers.

Example: The asymmetric sorbent (ion exchange resin) is prepared from a mixture of 98.5% styrene and 1.5% divinylbenzene subjected to suspension polymerization, and the copolymer fraction having a particle size of 0.05 to 0.10 mm is then chloromethylated. The resultant product containing 20.5% Cl is treated with L-proline in an amount of 2.25 mols per mol of chloromethyl group and then with sodium iodide in an amount of 0.30 mol per mol of chloromethyl group; the reaction being carried out in an excess of a dioxane-methanol mixture (6:1 by volume) at a temperature of 60°C for a period of 15 hours.

The analytical exchange capacity of the sorbent equals 2.2 milliequivalents/gram. The sorbent is noted for its chemical stability and mechanical strength. Then the sorbent is treated with an excess of 0.1 N solution of copper sulfate in 0.5 N ammonia (30 ml of the solution per 1 g of the sorbent). 13 g of the dark-blue sorbent is charged into the column (d = 9 mm; l = 500 mm) and washed with 0.5 N ammonia solution and water until the washings are neutral. Placed after the principal column is a second small column (d = 9 mm; l = 100 mm) packed with 2 g of the aforesaid sorbent not treated with the copper salt solution. A solution of 0.5 g of D,L-proline in 5 ml of water is introduced into the column and then the system is washed with distilled water at a water feed rate of 10 ml/hr. Collected at the outlet side of the system (after the second column) is the eluate which gives a positive reaction with ninhydrin. Eluate evaporation yields 0.25 g of optically pure L-proline, $[\alpha]^{20}_{D}$ = 80.5° (c = 1, water).

D-proline is eluted from the columns by passing therethrough 100 ml of a 1 N solution of ammonia, the separation of D-proline from the eluate being also effected by the evaporation technique. The yield of optically pure D-proline equals 0.25 g, $[\alpha]^{20}_{D}$ = 80.5° (c = 1, water). The sorbents in the principal and auxiliary columns are regenerated by washing with distilled water until the washing shows neutral reaction. After 3 chromatographic cycles, the sorbent contained in the auxiliary column is saturated with copper ions, as evidenced by sorbent coloration, and sorbent regeneration is accomplished by 0.5 N hydrochloric acid which readily decomposes the copper complex. The copper ions thus eluted are again absorbed in the principal column after adjusting the copper solution to pH $>$ 6 with ammonia.

The same procedure is also used to completely separate the optical isomers from the following racemic compounds: valine, valinamide, α-alanine, isovaline, leucine, isoleucine, methionine, allothreonine, β-phenyl-β-alanine, and 1-amino-2-propanol.

Using Salts of Optically Active Acids with Amino Acid Esters

Optically active amino acids and their esters are frequently of use in the synthesis of physiologically active compounds in which optical isomerism is significant in relation to activity. In particular, important antibiotics of the penicillin and cephalosporin series have acylamido side chains derived from such acids. Thus D-2-amino-2-arylacetic acids may be used in the synthesis of 6β-(D-2-amino-2-phenylacetamido) penicillanic acid (ampicillin), 6β-[D-2-amino-2-(p-hydroxyphenyl)acetamido] penicillanic acid (amoxycillin), 7β-(D-2-amino-2-phenylacetamido)-3-methylceph-3-em-4-carboxylic acid (cephalexin), 3-acetoxymethyl-7β-(D-2-amino-2-phenylacetamido)-ceph-3-em-4-carboxylic acid (cephaloglycin), 7β-[D-2-amino-2-(1,4-cyclohexadien-1-yl)-acetamido] -3-methyl-ceph-3-em-4-carboxylic acid (cephradine), and others.

Furthermore, naturally occurring α-amino acids are incorporated into nutritional media for intravenous administration in medical conditions where normal oral nutrition is not possible. In order to minimize the solids content of such solutions, it is essential that the α-amino acids be in the optically active form (usually the L-form) which can be utilized by the body in protein synthesis.

J.C. Clark and J. Elks; U.S. Patent 3,976,680; August 24, 1976; assigned to Glaxo Laboratories Limited, England have developed a process whereby the desired enantiomer is obtained as a salt while the unwanted isomer is racemized in situ. This process comprises reacting an ester of the opposite enantiomer of the α-amino acid (which may be in admixture with an ester of the desired enantiomer) with an optically active acid and an aldehyde or ketone, whereby an ester of the desired enantiomer separates out in the form of the salt. The free α-amino acid and/or its salts may be obtained by hydrolysis of the ester thus formed. In this way, one enantiomer can be converted into the opposite enantiomer and a racemic mixture may thus be converted into a single optically active form. In practice, this racemization/resolution process may be rather slow and in some cases it may be preferable simply to enrich the product with respect to one optical isomer. It was found, however, that α-amino acids carrying an aryl substituent, particularly in the α-position, are converted especially rapidly and in this case virtually all of the DL-starting material can be converted into the required isomer in an economical time. The starting material may be the L-isomer, when the D-isomer is the required product or vice versa or a mixture of the two isomers may be used.

This process is based on the finding that Schiff bases of the above esters are more readily racemized than the parent esters and usually exist in equilibrium therewith. The Schiff bases tend not to form insoluble salts with the resolving acids such as tartaric acid and in consequence, if a DL-ester is partially converted in solution into a Schiff base in the presence of a resolving acid, such as tartaric acid, an equilibrium is eventually reached out of which the required optically active amino acid ester is separated in the form of its salt. In the equilibrium process the Schiff base of the other isomer however will continuously be racemized.

The α-amino acids which are conveniently used in this process include compounds of the formula:

$$R^1\text{–}\underset{\displaystyle NH_2}{\underset{|}{CH}}\text{–COOH}$$

where R^1 represents an optionally substituted alkyl, aralkyl, aryl or cycloalkyl group. The process is generally applicable to the esters of α-amino acids including for example the esters of naturally occurring α-amino acids such as alanine, valine, leucine, phenylalanine, tyrosine, serine, cysteine, methionine, tryptophan, aspartic acid, glutamic acid, lysine and histidine.

In view of the excellent results which may be obtained with aromatic amino acids, however, R^1 preferably represents phenyl and phenyl substituted by, for example, hydroxy; alkoxy, e.g., methoxy; acyloxy, e.g., acetoxy or benzyloxycarbonyloxy; halogen, e.g., chlorine; or lower alkyl, e.g., methyl. Thus R^1 may be, for example, m- or p-hydroxyphenyl, m- or p-methoxyphenyl, m- or p-chlorophenyl, m-tolyl, etc., or a di- or polycyclic group such as α- or β-naphthyl. Particularly preferred R^1 groups are unsubstituted phenyl and p-hydroxyphenyl. Before the racemization/resolution process is effected it may be necessary to protect any reactive substituents in the R^1 group, and subsequently remove such protecting groups.

The aldehyde or ketone may in general be represented by the formula R^2–CO–R^3 where R^2 and R^3, which may be the same or different, each represents hydrogen or an aliphatic, araliphatic, aromatic or heterocyclic group, e.g., an alkyl group or a monocyclic aralkyl or aryl group, which may carry substituents such as nitro or alkoxy groups. Examples include acetaldehyde, benzaldehyde, anisaldehyde, p-nitrobenzaldehyde, formaldehyde, 2-formylthiophene, furfural, pyridoxal, acetophenone, methyl ethyl ketone, methyl isobutyl ketone and acetone. R^2 and R^3 may also be joined together as in a cycloalkanone such as cyclohexanone or cyclopentanone.

Simple derivatives of aldehydes and ketones such as acetals, hemiacetals, ketals and hemiketals may be used in place of the aldehyde or ketone. The optically active resolving acid is preferably tartaric acid, more especially (+)-tartaric acid but diesters such as dibenzoyltartaric acid may also be used. Other suitable resolving acids include (+)-mandelic acid and (+)-5-acetoxy-4-methylpentanoic acid.

By selecting the right conditions a D-isomer may be prepared from the corresponding L-isomer, the racemic mixture, or any other mixture of D- and L-isomers. Furthermore by selecting the right conditions an L-isomer may be prepared from either the corresponding D-isomer or a mixture of D- and L-isomers. The starting material may be the wrong isomer or a mixture of isomers in the form of a salt with the resolving acid.

Example 1: Resolution with Racemization of Methyl DL-Phenylglycinate Using Benzaldehyde – A solution of methyl DL-phenylglycinate (4.86 g, 29.5 mmol) and (+)-tartaric acid (4.9 g, 32.6 mmol, 1.1 equiv.) in ethanol containing benzaldehyde (51 ml, containing 3.16 ml PhCHO, 1.05 equiv.) at about 70°C was added. A solid crystallized almost immediately. The mixture cooled to about 25°C in 30 minutes, and was stirred at about 25°C for 5 days.

The solid was filtered off, then washed with solvent (10 ml), displacement washed with solvent (10 ml), and washed with ethanol (10 ml). It was dried at 28°C/3 min for 4 hours to give methyl D-phenylglycinate (+)-hemitartrate (9.46 g, 89%), as a monoethanol solvate, MP 142° to 144°, $[\alpha]_D^{25}$ –63° (c = 2.5, H_2O).

Example 2: Following the general procedure of Example 1, but using different aldehydes to form the Schiff base, D-phenylglycinate (+)-hemitartrate esters were prepared as set out in the table below.

Resolution with Racemization of Esters of DL-Phenylglycine in Ethanol with Various Aldehydes

Aldehyde	Equivs. Aldehyde	Ester	Conc. Ester % w/v	Time Hrs.	Temp. °C	$[\alpha]_D$°	Product Hemitartrate % Optical Purity	Yield %
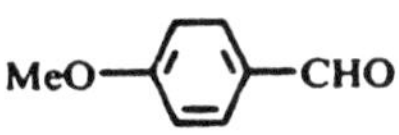	1	Et	10	19	20	–46	95	76
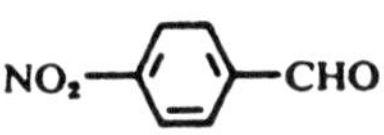	1	Et	10	19	20	–46	95	51
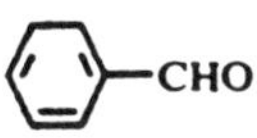	1	Et	10	18	20	–46	95	73
CH_3CHO	1	Et	10	19	20	–44.5	93	68.5
CH_2O	1	Et	10	42	20–26	–46	95	38.5
PhCHO	1	Et + Me = 57:43	20	70	20–25	–53.5	96	82.5
PhCHO	1	Et + Me = 55:45	10	24	20	–52	94	59

In these examples, optical purity is defined as a percentage taking a racemic mixture as having an optical purity of 0%. The yields given are percentages independent of the amount of the desired optical isomer in the starting material. This definition, for example, can give a yield of 100% in a complete resolution where all the unwanted isomer in a racemic starting material has been converted to the desired isomer. The rotations and percentage yields given are values calculated to the unsolvated material.

Racemizing in Inert Organic Solvents

When racemic amino acids are produced synthetically, and only one of the optically active enantiomorphs is wanted, it is common practice to resolve the synthetic product to recover the desired enantiomorph, to racemize the unwanted antipode, and to recycle the racemization product to the resolution step. Racemization methods widely used rely on melting of the optically active material either in a gaseous medium or in an inert liquid in which the material is insoluble.

T. Uzuki, M. Takahashi, M. Noda, Y. Komachiya and H. Wakamatsu; U.S. Patent 3,991,077; November 9, 1976; assigned to Ajinomoto Co., Inc., Japan have found that optically active N-acylamino acids can be racemized quickly, completely, and without significant decomposition at elevated temperatures in the presence of organic solvents in which they are soluble at the racemization temperature. Such solvents include triesters of phosphoric acid with alkanols or haloalkanols having up to 8 carbon atoms, phenol, halophenols, or lower-alkyl phenols. The dialkyl ketones are also suitable solvents.

The following table shows the percentage of N-acetyl derivatives of various L-amino acids racemized when held at 180°C for 30 minutes in mixtures with tri-n-butyl phosphate at respective weight ratios of 1:2 and 1:20.

L-Amino Acid	Racemization Rate, % 1:2	1:20
Phenylalanine	100	22.0
Methionine	93.9	33.5
Leucine	99.1	20.4
Valine	94.1	5.1
Aspartic acid	68.2	32.3
Tryptophan	94.6	21.7
Alanine	63.5	3.7

The racemic N-acylamino acid may be recovered from the racemization mixture by methods conventional in themselves. When the solvent employed is insoluble or only sparingly soluble in water, the product may be extracted from the mixture with water, and crystallized from the aqueous extract. When the solvent is readily soluble in water, the diluted mixture may be adjusted to the isoelectric point of the N-acylamino-DL-acid to induce crystallization of the latter. The solvent, if volatile at sufficiently low temperature, may be removed from the mixture, and other methods of recovery will be exemplified below and will be obvious to those skilled in the art.

Example 1: Mixtures of 2 g N-acetyl-L-phenylalanine and 1 g tri-n-butyl phosphate were held at 160°, 170° and 180°C respectively, and the optical rotation of the mixtures, which rapidly became homogeneous at the temperatures chosen, was monitored. Complete racemization occurred at 160°C within 25 minutes, at 170°C within 15 minutes, and at 180°C within 10 minutes. When N-acetyl-L-phenylalanine was held at the same temperatures in the absence of the solvent, only 81.4% of the compound was racemized after 150 minutes at 150°C, 91.1% after 20 minutes at 170°C, and racemization was complete only after 10 minutes at 180°C, the MP of the compound being 172°C.

When N-acetyl-L-phenylalanine was dispersed in twice its weight of refined paraffin oil (Nujol) in which it is insoluble, and the mixture was held at 180°C for 60 minutes, the amino acid was completely racemized, but decomposed enough to make the mixture dark and opaque. Recovery of pure N-acetyl-DL-phenylalanine was difficult.

Example 2: A mixture of 5 g N-acetyl-L-leucine and 10 g tri-n-butyl phosphate was heated with stirring for 30 minutes in an open 50 ml vessel immersed in an oil bath at 180°C, quickly cooled, diluted to 100 ml with tri-n-butyl triphosphate, and tested for its optical rotation. A specific rotatory power of $[\alpha]_D^{25} = -0.15°$ was found, indicating 99.1% racemization. N-acetyl-DL-leucine was extracted from the racemization mixture with 3% sodium hydroxide and precipitated in crystalline form when the extract was adjusted to pH 1.0. The crystals weighed 4.66 g, melted at 155°C, and did not show measurable rotatory power.

AMINO ACIDS AND PEPTIDES—USES

AMINO ACIDS FOR ORAL FEEDING

A drawback associated with the oral administration of amino acids is that they exhibit a particularly intense obnoxious taste and can, in fact, be orally ingested only with great psychological stress. Prior attempts at solving this problem have involved adding to these compositions various flavoring materials in large quantities whereby they exhibit an intense flavor note of their own sufficient to overwhelm that of the amino acids. In the main, only a few flavors have produced an intensely flavored composition which is less objectionable than the unflavored amino acid compositions and, in these few cases, the intense substituted flavor note has, without exception, been found to be organoleptically oppressive. This problem is particularly acute when a patient must subsist for an extended length of time on such dietary compositions.

Bland Tasting Amino Acids Containing Oil of Ginger

A dietary composition is provided by *K. Mohammed; U.S. Patent 3,821,432; June 28, 1974; assigned to Johnson & Johnson* comprising at least one amino acid and enough oil of ginger to reduce the obnoxious taste of the amino acids when dispersed in water. The oil of ginger interacts organoleptically to reduce the amino acid intensity without itself producing an intense ginger flavor. The composition is suitable for being dispersed in water to obtain a beverage or may be combined with thickening agents and dispersed in water to obtain a dietary pudding. Other nutrients such as carbohydrates, fats, vitamins and edible mineral sources may be added to produce a nutritionally complete dietary composition and moderate quantities of nonginger flavoring material may be added to produce an organoleptically pleasing food product.

Oil of ginger, a yellowish, viscid, volatile oil, sparingly soluble in alcohol and almost completely insoluble in water, is derived from the rhizome of the root stock of *Zingiber Officinale Roscoe,* known commercially as ginger root. The plant is commercially cultivated, the chief source being Jamaica and India. The oil, usually obtained by the water and steam distillation of the dried, freshly ground rhizomes, comprises a mixture of sesquiterpenoids, monoterpenoids and associated compounds, the major constituent being the sesquiterpenoid, zingiberene.

While the interaction between the amino acids and ginger occurs over a substantial range of relative proportions of oil of ginger to amino acids, it has been determined that the most effective suppression of unwanted flavor notes occurs at ratios within the specific range of 0.04 to 1.2 milligrams of oil of ginger per gram of amino acids. Within this range the combined intensity of the unwanted flavor notes of both components is minimized. By reducing the intensity of these flavor notes to a minimum, the composition need be flavored with only moderate quantities of various flavorings to produce an aesthetically pleasing dietary composition.

Example: Four sample solutions are prepared using tap water and containing the components and proportions tabulated below:

Sample	Oil of Ginger (mg/1,000 ml water)*	Component Amino Acids (g/1,000 ml H_2O)	Oil of Ginger/Amino Acid Ratio (mg/g)
1	0.0	0.0	-
2	4.24	0.0	-
3	0.0	25.5	0.0
4	4.24	25.5	0.166

*On a carrier-free basis.

The ginger component is a spray-dried form of oil of ginger on an inert carrier wherein 20% by weight is the oil. The amino acid mixture used when provided in sufficient quantities to a human, will supply complete nitrogen nutritional requirements. The mixture, having a ratio of dispensable to indispensable amino acids of 0.667, has the following composition, in the following proportions:

Component	Parts by Weight
L-isoleucine	3.69
L-leucine	4.79
L-lysine monohydrochloride	4.56
L-phenylalanine	2.69
DL-methionine	2.73
L-threonine	2.36
L-tryptophan	0.64
L-valine	3.46
L-tyrosine ethyl ester monohydrochloride	2.55
L-arginine monohydrochloride	4.43
L-histidine monohydrochloride	1.13
L-alanine	0.58
L-aspartic acid	1.96
L-glutamine	5.03
L-proline	0.99
L-serine	1.24

A panel of qualified judges, in a blind test, appraised these samples organoleptically with respect to both the intensity of the ginger flavor and the intensity of the amino acid flavor. 90% of the judges correctly responded to Sample 1 which contained only water by reporting no ginger or amino acid flavor at all. 60% of the judges reported that Sample 2, containing only ginger and water, exhibited only a slight ginger flavor. 30% reported this sample as having a moderate ginger flavor intensity. Only 10% reported a distinct ginger flavor. The uncertaintly in detecting a ginger flavor in this sample is further indicated by the fact that 30% of the judges reported that the sample exhibited a slight amino acid flavor, whereas in fact no amino acid was present.

In contrast with the judges' difficulty in identifying a distinct ginger flavor in Sample 2, Sample 3, containing only amino acids, received a response of at least a strong amino acid flavor intensity from 90% of the judges. 60% characterized the amino acid intensity as extreme. When presented with Sample 4, the sample containing both the amino acids and the ginger, 90% of the judges reported a distinct decrease in the intensity of the amino acid flavor relative to that of Sample 3.

Palatable Predigested Protein Compositions

A.M. Gans, A.J. Goren and E.M. Gorenberg; U.S. Patents 4,025,650; May 24, 1977; 4,042,688; August 16, 1977 and 4,053,589; October 11, 1977; all assigned to Control Drug Inc. disclose a method of providing high-protein nutrition which comprises ingestion of predigested protein composition containing all of the essential amino acids and having a palatable taste and odor. These compositions are particularly useful in preventing body wasting during the oncologic treatment of cancer patients, for use before and immediately after surgery, or in the treatment of nutritional deficiencies due to other illnesses.

The orally-ingestible, predigested protein composition comprises a sweet-tasting and readily ingestible gelatin hydrolysate formulation containing all of the essential amino acids and being free of undesirable triglycerides and fats.

In its liquid form, the composition comprises an aqueous composition preferably containing a gelatin hydrolysate, sorbitol, a palatable acid, such as citric acid, fumaric acid or adipic acid, for maintaining the necessary acid pH, tryptophan, a synthetic sweetener, and flavoring and coloring agents when desired. Preferably, one or more preservatives is included for stabilization purposes. The flavoring agent is one which is highly palatable and sweet, preferably cherry, orange, green apple, or the like, while the preservative agent may be any fit for human consumption, such as potassium sorbate, sodium benzoate, methyl parabem and propyl parabem, either alone or in mixture.

The gelatin hydrolysate used is made by hydrolyzing animal collagen, preferably a collagen derived from the skin of pork bellies, by enzymatic hydrolysis, and then spray drying the solution. It is necessary to use enzymatic hydrolysis rather than acid or base hydrolysis because the enzyme converts the gelatin to the more palatable small peptides (i.e., mono-, di-, or tripeptides) rather than to the more unpalatable amino acids. In addition, it produces fewer distasteful impurities. The preferred enzymes used are those safe for human consumption. The most preferable are bromelin and papain, although pepsin and trypsin may be used.

As enzymatically hydrolyzed gelatins do not contain the essential amino acid tryptophan, tryptophan is added to constitute 0.02 to 0.75 part by weight of the composition. Other adjustments in amino acid content may also be made for special purposes. Although the above gelatin hydrolysate has a far better taste and odor than other gelatin hydrolysates, it does retain a certain amount of acridity. In the case of other gelatin products, it usually is necessary to add large amounts of sugar to at least partially mask the underlying acrid taste. In such instances, artificial sweeteners, such as sodium saccharin or the like, cannot readily be used because the aftertaste of these artificial sweeteners, when combined with the acrid taste of the gelatin product, would make the product relatively unpalatable.

In this product, however, even though a certain small amount of residual acridity might remain, it is capable of being readily masked by artificial sweeteners, with no serious aftertaste problem except such as is normally present in the sweeteners themselves. Even this, however, is overcome by the inclusion of the sorbitol, which not only tastes sweet itself, but also acts to coat the palate and gullet surfaces with a lubricating effect so that the composition is permitted to slide easily into the stomach. In addition, the sorbitol coats the taste buds, further masking any possible residual acrid taste.

In this manner, a product is produced which is a sugarless, lipid-free composition, that is free of those carbohydrates that effect rapid rise of blood glucose levels. In this respect, sorbitol, unlike sugar, only slowly affects the blood glucose. In general, the liquid composition comprises the following range of proportions:

Components	Parts by Weight
Hydrolyzed gelatin	5-75
Tryptophan	0.02-0.75
Acid	0.3-10
Sorbitol (70%)	7-25
Artificial sweetener (e.g., sodium saccharin)	0.1-2
Lubricant (glycerin or the like)	0.4-1
Preservatives	0.4-1
Dye and flavor	0-0.4
Water	15-80

Example: 400 pounds of water was charged into a stirrer-equipped electrically-heated kettle and heated to 70°C, and then 9 pounds of L-tryptophan was added and dissolved. The solution was then transferred to a (stirrer-equipped) 500 gallon holding tank. Next the kettle was charged with 800 pounds of water, 2,028 pounds of gelatin hydrolysate (powder), 160 pounds of citric acid and the mixture heated to 70°C with mixing continued until full solution resulted. The solution was then added to the material already in the holding tank.

To the stirred mixture in the holding tank was added: 4½ pounds sodium benzoate dissolved in 24 pounds of water; 4½ pounds potassium sorbate dissolved in 24 pounds of water; 4½ pounds of methyl paraben and 2 pounds of propyl paraben dissolved in 50 pounds of glycerin; 102 grams of a red dye dissolved in 10 pounds of water; 2,000 ml cherry flavor; 5 pounds of sodium saccharin dissolved in 20 pounds of water; 800 pounds of sorbitol (70% solution); and sufficient water to bring the volume to 500 gallons. The above product comprises a liquid form of the composition.

Two tablespoonfuls (one ounce) of a typical gelatin hydrolysate made in the above manner has the following approximate composition (providing 15 grams of soluble protein hydrolysate per 30 cubic centimeters):

Amino Acids	Mg
L-Alanine	1,300
L-Arginine	1,200
L-Aspartic acid	900
L-Cystine	10
L-Glutamic acid	1,500
Glycine	3,500
L-Histidine	110
L-Hydroxyproline	1,000
L-Leucine	450
L-Isoleucine	200
L-Lysine	650
Hydroxylysine	150
L-Methionine	110
L-Phenylalanine	350
L-Proline	2,300
L-Serine	1,000
L-Threonine	300
L-Tryptophan	65
L-Tyrosine	100
L-Valine	350

Powders, capsules, granules and tablets, suitable for ingestion, can be prepared by conventional methods using mixtures as represented by the following table:

Components	Parts by Weight
Hydrolyzed gelatin	5-75
Tryptophan	0.2-0.5
Dispersant	0.1-10.0
Artificial sweetener	0.1-2
Dye and flavor	0-2.0
Preservatives	0.1-2.0
Lubricant (glycerin or the like)	0.1-5.0
Soluble edible polymer	0-5.0
Solid carrier	5-100.0

The dispersant may be any of a number of materials such as, for example, cornstarch, silica gel, etc.; the edible polymer may also be any of a number of materials such as, for example, PVP, methylcellulose, etc.; the solid carrier may be any of a number of materials in pulverulent (granular) form, such as, for example, dicalcium phosphate, mannitol, etc.; and the preservative may be any of a number of materials such as, for example, sodium benzoate, methyl paraben, etc.

Preparation of Low-Phenylalanine Plastein

Phenylketonuria is a genetic, abnormal metabolic disease which causes retardation. Newly born babies suffering from this disease are congenitally lacking an enzyme, phenylalaninehydroxydase, which converts phenylalanine to tyrosine. Therefore, the disease is characterized by enormous accumulation of phenylalanine in the blood and excretion of the abnormal metabolic products, such as phenylpyruvic acid and its derivatives, phenyllactic acid, phenylacetic acid and phenylacetylglutamine or the like.

The main object of the process developed by *M. Fujimaki, S. Arai and M. Watanabe; U.S. Patent 4,016,147; April 5, 1977; assigned to Fuji Oil Company, Ltd., Japan* is to furnish a synthetic protein-like substance, like natural protein in its physical and nutritional properties and which contains no or very little phenylalanine.

It was found that albumin or globulin proteins such as meat, soybean, egg and microorganism proteins are decomposed to oligopeptides by degradation with an endopeptidase having an affinity for the peptide bonds near the aromatic amino acid residues, such as pepsin or chymotrypsin; and this hydrolysate is further hydrolyzed to the aromatic amino-acid-free peptides by the second hydrolysis with the exopeptidase. The second hydrolysate can be purified through gel-filtration and then polymerized to give an aromatic-amino-acid-free, protein-like substance (so-called plastein) in good yield according to a usual plastein synthesis.

It was also possible to prepare phenylalanine-free, tyrosine and/or tryptophan-enriched plastein where the deficient, aromatic, essential amino acids excepting phenylalanine are incorporated into the peptide linkages in the plastein, by adding tyrosine and/or tryptophan in the activated form to the substrate in the plastein synthesis.

Since these plasteins are protein-like polymers having a mean molecular weight of about 2,000, and the constituent amino acids are linked with each other with peptide bonding, they have physicochemical and nutritional properties similar to those of natural proteins, which makes the product valuable as an adequate diet for the treatment of phenylketonuria. This process can be summarized as follows:

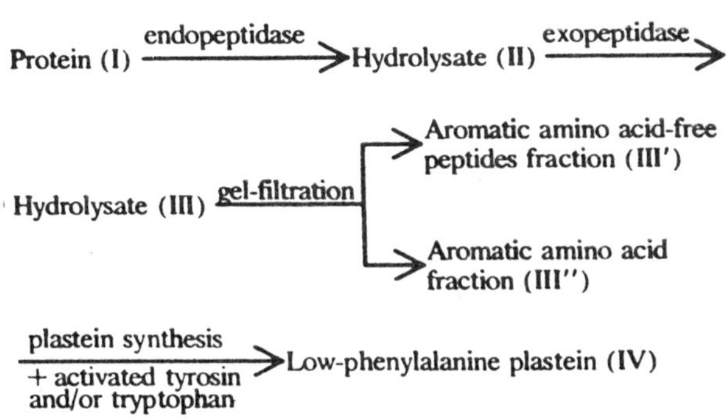

The exopeptidase used in this process is classified into two genera: carboxypeptidase and aminopeptidase according to their attacking position, i.e., either the terminal carbons or the terminal nitrogen atoms in the peptide linkages. However, in this process, mixture of both of the above genera is preferable in view of the fact that the hydrolysate (II) will probably contain both free acid and free amino groups. Carboxypeptidase A is the pure carboxypeptidase and leucine-aminopeptidase is the pure aminopeptidase, respectively. However, these pure enzyme preparations are expensive and are not suitable for practical use, and therefore, the use of a crude enzyme containing both enzymes is more practical. Pronase (Kaken Chemical Co., Ltd., Tokyo) is a commercially available enzyme preparation which can break the peptides of hydrolysate (II) down to smaller peptides having a mean molecular weight of about 800 which can be favorably used for the subsequent plastein synthesis.

Example: To 10 grams of a fish meat protein known as Protane (Sopharma SA, Co., Switzerland) (I) was added with an amount of water to bring the total volume up to 1 liter and then adjusted to pH 1.5 with concentrated hydrochloric acid to form a homogeneous solution. To the solution 10 grams of pepsin (Difco Co.) was added and the mixture was stirred at 37°C for 48 hours. The reaction mixture [hydrolysate (II)] showed 67.8% of the degree of the hydrolyzation according to a measurement with 10% trichloroacetic acid. The mixture was then readjusted to pH 6.5 and 10 mg Pronase was added and the mixture further incubated for 5 hours. During this incubation, the pH was kept at 6.5 with concentrated sodium hydroxide solution. The hydrolysate (III) thus obtained was then freeze dried.

To 5 grams of this freeze dried powder [hydrolysate (III)] was added 10 ml of 0.5 M acetic acid and the mixture centrifuged to exclude any insoluble materials. The supernatant including 4.64 grams of the insoluble solids was then passed into a column (5 cm i.d. x 75 cm length) filled with Sephadex G-15. Thereafter, the column was eluted with 0.5 M acetic acid at a velocity of 6 ml per minute at 25°C and divided into several fractions of 20 ml each with a fraction collector. Every fraction thus obtained was measured for its optical density at 280 nm with a spectrophotometer. The patterns of the gel-filtration are as shown in Figure 2.1.

FIGURE 2.1: PREPARATION OF LOW-PHENYLALANINE PLASTEIN

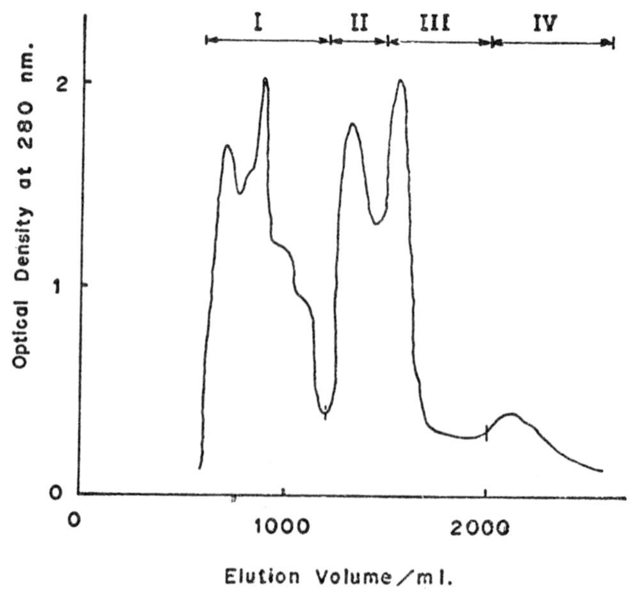

Source: U.S. Patent 4,016,147

The results are shown in the table below where the amino acid composition of each fraction is indicated in percent by weight.

Amino acids	Fractions I	II	III	IV
Arg	8.18 %	2.64 %	0.90 %	0.00 %
His	2.98	0.64	0.46	0.00
Ile	5.69	7.57	1.83	0.00
Leu	3.39	18.49	0.47	0.00
Lys	13.48	2.42	0.44	0.00
Thr	4.94	3.78	2.18	0.00
Trp	0.00	0.00	0.00	98.25
Val	7.34	3.13	2.13	0.00
S-containing amino acids (total)	3.35	—	—	—
[Met	1.92	—	—	—]
[½Cys	1.43	—	—	—]
Aromatic amino acids (total)	0.00	31.89	66.79	1.38
[Phe	0.00	28.25	8.21	0.00]
[Tyr	0.00	3.64	58.58	1.38]
Ala	6.51	2.28	3.31	0.00
Asp	12.06	4.07	5.50	0.00
Glu	20.89	5.92	9.46	0.00
Gly	4.36	2.14	3.02	0.00
Pro	2.37	1.57	1.04	0.00
Ser	3.80	2.57	1.79	0.00

As indicated in the table, Fraction I includes neither tryptophan, phenylalanine nor tyrosine. These amino acids are concentrated into the Fractions II, III and IV. In particular, it should be noted that tryptophan is concentrated into Fraction IV and that tyrosine occupies the major portion of Fraction III. Thus, Fractions III and IV can be used as a source of tryptophan and/or tyrosine per se or for the next plastein synthesis after the reesterification thereof. The yield of each fraction is shown in the table below, in which the yield of each fraction is calculated by the nitrogen recovery rate on the basis of the weight of the hydrolysate (III).

Hydrolysate (III)	Soluble portion (total)		92.8 %
		Fraction I	78.9 %
		Fraction II	8.7 %
		Fraction III	4.3 %
		Fraction IV	0.9 %
	Insoluble portion		7.2 %

These fractions were then freeze dried for the next procedures.

2 grams of Fraction I [hydrolysate (III')], 200 mg of tyrosine ethyl ester and 39 mg of tryptophan ethyl ester were dissolved in 2 ml of aqueous acetone. The solution was then adjusted to pH 6.5 with aqueous sodium hydroxide solution (but, this pH adjustment is not essential) and subjected to the plastein synthesis with the addition of 20 mg of papain (Difco Co.) at 37°C for 48 hours. The reaction mixture was then diluted with 100 ml of 10% aqueous ethanol, homogenized and agitated for 5 hours at room temperature in order to split the ester linkages of the incorporated amino acid residues.

Thereafter, the reaction mixture was readjusted to pH 7.0 and centrifuged (the amount of the precipitates was negligible). To this supernatant was added 100 ml of 10% aqueous ethanol and the mixture was then filtered through an Amicon Type 402 ultrafilter so as to obtain a fraction having a molecular weight of more than 500. This fraction was distilled in vacuo to remove the solvent (ethanol) and freeze dried to give the final product (78.9% on the basis of the raw protein material). This product is completely tasteless and odorless when subjected to a panel test.

AMINO ACIDS FOR PARENTERAL FEEDING

In medical practice, parenteral intravenous feeding is effected in all cases in which the patient is prevented from food intake by mouth because of a grave disease or after an operation. The decision whether parenteral feeding has to be administered depends on the patient's general condition. In case the general condition is bad and an oral food intake cannot be expected to be possible for a longer period, it is necessary to supply all essential nutritive substances intravenously. Water, vitamins, electrolytes, trace elements, carbohydrates, amino acids and fats are essential for a complete parenteral feeding and, in some cases, alcohol may be required as a source of calories.

Stabilized Amino Acid-Fat Emulsions

As basis for a complete parenteral feeding three types of infusion solutions are known, viz carbohydrate solutions, amino acid solutions and fat emulsions. To these solutions there are added vitamins, electrolytes and trace elements which constitute, as far as their amount is concerned, only a small part of the nutrition as compared to the three main food carriers.

It would be an important simplification if it would be possible to combine fats and amino acids in one single emulsion. Another advantage of such a combined infusion solution would be that the supply of water to the organism would be reduced.

However, fat emulsions or, generally, oil-in-water emulsions, physically and chemically, constitute a metastable system which is converted by external disturbances, for example, by the effects of temperature or the addition of electrolytes or colloids, into an energetically stable state. This process is manifested in the separation of the phases, designated as creaming. Since amino acids also may be regarded as electrolytes because of their dipolar-ionic character, it could be expected that the stability of fat emulsions containing amino acids would be considerably reduced.

J. Schnell; U.S. Patent 3,793,450; February 19, 1974; assigned to B. Braun Melsungen AG, Germany has disclosed a process for the preparation of stable infusion compositions in which essential amino acids are present in the form of free bases in a fat emulsion, such as soybean oil. The emulsified compositions also contain soy phosphatide or egg yolk phosphatide as stabilizers. These stable fat emulsions which have no pyrogenic effect and can be administered intravenously contain 1 to 10% of amino acids and 1 to 15% of fat. The particle size is always below 1 micron, which excludes the danger of fat embolism. These emulsions which had been prepared, for example, by using fractionated soy phosphatides as emulsifiers in a high-pressure homogenizer, could be sterilized by heat and behaved as usual commercial pure fat emulsions when stored at 60°C.

Furthermore, it has been found that the addition of glycerin or carbohydrates, for example, saccharin, alcohol, sorbite and xylite, in quantities from 1 to 15%, preferably from 5 to 10%, has a favorable effect on the preparation and the stability of the fat emulsions containing amino acids. The addition of these carbohydrates is advantageous because it has a favorable effect on the fat metabolism. Moreover, it increases the caloric value of the emulsion.

Example: The following substances were dissolved in 5 liters of aqua bidestillata, while using a stirrer:

	Grams
DL-isoleucine	16.9
L-leucine	26.9
DL-threonine	16.9
DL-valine	22.9
DL-tryptophan	10.0
DL-phenylalanine	22.9
DL-methionine	27.5
L-lysine	6.0
L-histidine	11.0
L-diaminovaleric acid	8.5
Glycine	320.0
Glycerin	500.0

To this solution there were added 75 grams of fractionated soy phosphatides, consisting essentially of soybean lecithin, obtained by extraction from soy meat with ethyl alcohol, and the solution was then mixed. Then, 500 grams of purified soybean oil were added and aqua bidestillata was added until a volume of 10 liters was attained. The mixture was then preemulsified by means of a high speed stirrer (e.g., an Ultraturrax) and subsequently homogenized in a high-pressure homogenizer at a pressure of 400 kg/cm^2. The finished emulsion was filled, under nitrogen, into glass bottles of the first hydrolytic grade, and the closed bottles were sterilized by heat. The particles of an emulsion thus prepared were smaller than 1 micron. The emulsion was sterile and pyrogen-free.

Tailored Injectable Amino Acid Solutions

In principle the amino acid compositions developed by *H. Ghadimi; U.S. Patent 3,832,465; August 27, 1974* are based on the fact that intravenously administered solutions bypass the liver. Consequently, such solutions, unlike orally ingested protein, avoid the major catabolic pathways of this organ. Therefore, the amino acid compositions of this process are designed so that their administration does not change the pattern of the free amino acids of the peripheral blood. In addition, unlike existing products, these amino acid compositions take cognizance of the radical difference in the requirements of newborn versus older patients and can be tailored to meet these requirements.

The following simplification provides guide lines for formulating the amino acid compositions of this process. Accordingly the amino acids can be divided into three categories, as far as the ingestion and the role of the liver is concerned.

Group 1: Glutamine, glutamic acid and aspartic acid, which are kept to a minimum when the intravenous route is used.

Group 2: Branched-chain amino acids (leucine, isoleucine, and valine). The liver seems to play a minor role in the economy of these three substances. The major metabolic pathways for these amino acids thus are beyond the liver. Therefore, they can be administered in abundance when the intravenous route is used.

Group 3: The remaining amino acids, namely methionine, cysteine, lysine, threonine, glycine, serine, histidine, phenylalanine, tyrosine, arginine, alanine, proline and tryptophan have their major metabolic pathways in the liver. In general, the amino acid compositions for newborns, prematures, and young infants, i.e., compositions of the GF-1 type are composed and administered as shown in Example 1.

Example 1: Amino Acid Composition for Neonatal Period and Early Infancy – The composition contains the amounts of pure crystalline amino acids in their L-configuration as given in the table below.

Amino Acid	mg/100 ml	mg/kg of Body Weight on Administration of 25 ml/kg
Methionine	60	15
Cysteine	340	85
Lysine	480	120
Threonine	240	60
Glycine	400	100
Serine	400	100
Leucine	1,600	400
Isoleucine	800	200
Valine	800	200
Histidine	200	50
Phenylalanine	400	100
Tyrosine	50	12.5
Glutamic acid	50	12.5
Aspartic acid	50	12.5
Arginine	1,000	250
Alanine	400	100
Proline	200	50
Tryptophan	120	30
Ornithine	0	0
Total	7,590	1,897.5

The third column of this table indicates the approximate amount of amino acid in mg/kg of body weight when an average dose of 25 ml/kg of the amino acid solution is administered. Each 100 ml of this amino acid composition which is designated as GF-1, is equivalent to 7.6 grams of protein and will supply 30 calories. 20 to 60 ml/kg of body weight administered daily are equal to 1.6 to 4.8 grams of protein per kilograms of body weight daily. Administration of 25 ml/kg of body weight will give sufficient amino acids to maintain anabolic pathway of growing newborns. For prematures under 1,000 grams birth weight the recommended dose is 10 ml/kg of body weight for the first day. This dose can be built up to 28 or 30 ml/kg of body weight within a week.

In general the dose may be adjusted depending on the degree of malnutrition and clinical state of the patient. When using higher doses, it is advisable to monitor the plasma amino acids by quantitatively analyzing the free amino acids. Suitable analytical techniques involve the use of the amino acid analyzer built according to the modifications described by Piez et al in *Analyt. Biochem.* volume 1, page 187 (1960) of the technique originally described by Spackman et al in *Analyt. Chem.* volume 30, page 1190 (1958).

The amino acids are dissolved in distilled water to which 60 mg/100 ml of potassium metabisulfite may be added. The osmolarity of the solution is about 650 milliosmols per liter.

Example 2: Amino Acid Composition for Children and Adults – The composition contains the amounts of pure crystalline amino acids in their L-configuration as given in the table below.

Amino Acid	mg/1,000 ml	mg/day for Adult (70 kg of Body Weight)
Methionine	1,600	800
Cysteine	800	400
Lysine	2,400	1,200
Threonine	3,000	1,500
Glycine	3,000	1,500
Serine	3,000	1,500
Leucine	12,000	6,000
Isoleucine	10,000	5,000
Valine	10,000	5,000
Histidine	3,000	1,500
Phenylalanine	5,000	2,500
Tyrosine	500	250
Glutamic acid	400	200
Aspartic acid	0	0
Arginine	8,000	4,000
Alanine	3,000	1,500
Proline	2,000	1,000
Tryptophan	1,200	6,000
Ornithine	0	0
Total	68,900	34,450

The third column of the table indicates the approximate amount of amino acid in mg/day for an adult of 70 kg of body weight when an average dose of 500 milliliters per day of the amino acid solution is administered. Each 1,000 ml of this amino acid composition which is designated as GF-2, is equivalent to 69 grams of protein and 10 grams of nitrogen and will supply 275 calories. The dose for adults is 500 to 1,000 ml/day and for children, for every 10 pounds of body weight, 200 ml up to 40 kg of body weight. The dose can, of course, be increased according to clinical condition.

The amino acids are dissolved in distilled water to which 600 mg/1,000 ml of sodium bisulfite may be added. The osmolarity of the solution is about 500 milliosmoles per liter.

Amino Acid Therapy Without Carbohydrates

J.-P. Flatt and G.L. Blackburn; U.S. Patent 3,920,838; November 18, 1975 have disclosed a method of treating patients during periods of severe negative caloric balance due to dysfunction or disuse of the gastro-intestinal tract for reasons such as trauma, burn, sepsis, other illness or physician's prescription. The method is based on the parenteral application of amino acids while substantially withholding the patient's intake of carbohydrates. This method allows for the development of starvation ketosis in the patient, facilitating the utilization of fat stores and substantially reducing, even totally eliminating, net nitrogen losses. This method is a break from the common practice used in treatment of such patients which was designed to prevent starvation ketosis by supplying at least 100 grams of glucose per day to the patient. The validity of this approach to protein sparing therapy is established by the following test procedures.

Nineteen surgical patients requiring intravenous fluid replacement were treated with the following parenteral solutions: (1) 5% dextrose solution (D5W) to deliver 100 grams of dextrose per 24 hours; (2) 3% crystalline L-amino acid solutions containing essential and nonessential amino acids delivering 90 grams of amino acids per 24 hours; and (3) solutions containing 3% L-amino acids and 5% dextrose to deliver 70 grams of amino acids plus 100 grams of dextrose per 24 hours. The compositions of the L-amino acid solution is as follows:

Amino Acid	g/100 ml	g in 2.5 l of Isotonic (3%) Solution
Isoleucine	0.21	5.25
Leucine	0.27	6.75
Lysine	0.27	6.75
Methonine	0.16	4.00
Phenylalanine	0.17	4.25
Threonine	0.12	3.00
Tryptophan	0.05	1.25
Valine	0.20	5.00
All essential	**1.45**	**36.3**
Alanine	0.21	5.25
Argininine	0.11	2.75
Histidine	0.08	2.00
Proline	0.33	8.25
Serine	0.18	4.50
Glycine	0.63	15.75
Cysteine	0.01	0.25
All amino acids	**3.0**	**75.0**
Amino acid N, g	0.45	11.3
Protein equivalent	2.8	70.0
Calories	120	300

Urine was collected over a successive 24 hour period beginning immediately after the 7:00 A.M. blood specimen was obtained. The volumes were measured and aliquots were frozen at -20°C until analysis of total nitrogen, urea nitrogen, and creatinine were completed.

Glucose, lactate, pyruvate, β-hydroxybutyrate and acetoacetate were determined enzymatically after deproteinization with perchloric acid. Free fatty acids were titrated acidimetrically. Insulin was measured by radioimmunoassay and urinary nitrogen by a micro-Kjeldahl method.

Daily urinary nitrogen excretion was determined on 24 hour urine collections. The ΔNu corresponds to the urinary nitrogen loss when on total fast or on dextrose infusions, or to the difference between the grams of nitrogen administered in the form of infused amino acids and the urinary nitrogen excretion. The overall nitrogen balance was calculated by adding 1½ grams of nitrogen to the urinary nitrogen loss to account for fecal and cutaneous losses. The following table represents a clinical description of 10 patients receiving 2 to 3 liters of parenteral infusion per day for 10 to 14 days.

Weight, kg	Age	Sex	
58.6	56	M	Obstructing cancer of colon
56.3	52	F	Pelvic abscess
61.2	69	F	Small bowel obstruction
54.5	52	M	Reactive hepatitis, subhepatic abscess
76.2	28	M	Stab wound of liver
89.5	64	M	Gunshot to abdomen with peritonitis
64.3	37	F	Subhepatic abscess with perforated duodenal ulcer
54.5	47	M	Emphysema, chronic lung disease, ileus
60.0	42	M	Pancreatitis and pseudocyst
32.0	56	F	Subtotal gastrectomy and dumping syndrome

The following table represents a clinical description of the remaining nine patients receiving a similar therapy but who were generally more severely ill and demonstrated a higher protein catabolism.

Weight, kg	Age	Sex	
-	84	F	40% burn
54.1	54	M	Pancreatitis
106.4	46	M	Multiple trauma
63.6	26	M	Pancreatitis
55.9	36	F	Cholongitis
55.9	73	M	CA of sigmoid colon
61.4	40	M	Perforated duodenal ulcer, subphrenic abcess, pelvic abcess
109.1	25	F	Subphrenic abcess
61.8	74	F	Entero cutaneous fistula

No adverse effects could be attributed to the intravenous therapy in any patient; no infections, cellulitis, thrombosis or pyrogenic responses occurred. Except in the fourth patient with reactive hepatitis, serum enzymes (SGOT and LDH), bilirubin and ammonia levels remained within normal limits.

The first ten patients represented moderate protein catabolism, averaging a daily nitrogen loss of 7.1±3.5 grams on administration of dextrose in water. This is equivalent to loss of 44 grams of protein per day (6.25 grams of protein/g N), which, if this protein was mobilized from muscle tissue, represents a loss of a about 0.5 pound per day of muscle tissue.

Infusion of an amino acid solution, instead of the routinely used 5% glucose infusate, resulted in a reduction in the nitrogen loss from -8.5 to -1.0 g N/day. The infusion of a mixture of glucose and amino acids resulted in some improvement in the nitrogen balance but not to the degree observed with amino acid infusion alone. In the subsequent study of the nine additional patients, the infusion of amino acids instead of glucose resulted in a reduction in nitrogen loss from -11.8 to -1.0 g/day.

Amino Acid Formulations for Patients with Liver Disease

By restricting protein intake, the tendency of liver diseased patients to develop hepatic encephalopathy may be reduced. However, such limitation of protein intake further contributes to the malnutrition of the patient. Studies on intravenous feeding of amino acids have shown that the content of amino acids in the plasma of liver diseased patients is seriously distorted, the branched-chain amino acids (isoleucine, leucine, and valine) being lower than normal, while methionine and the aromatic amino acids phenylalanine and tryptophan are higher than normal.

It has been suggested that an excess of phenylalanine may inhibit the transport of tyrosine to the brain. Further, it is known that tyrosine is important for the synthesis of some of the normal catecholamine neurotransmitters in the brain. It has also been suggested that the level of serotonin in the brain may be related to the association between plasma tryptophan, as opposed to plasma phenylalanine, tyrosine, and the branched-chain amino acids.

The amino acid formulations developed by *J.E. Fischer, N.N. Yoshimura, T.L. Westman and F.H. Deindoerfer; U.S. Patent 3,950,529; April 13, 1976; assigned to Massachusetts General Hospital and American Hospital Supply Corporation* can be used to provide nutritional support for liver diseased patients while overcoming the distorted amino acid content of the plasma.

In the process, the amount of phenylalanine alone, or phenylalanine and tyrosine in combination, or tyrosine alone is controlled in relation to the total of the essential branched-chain amino acids, specifically isoleucine, leucine, and valine to achieve metabolically acceptable levels of plasma phenylalanine and tyrosine. It appears that these formulations permit adequate transfer of tyrosine to the brain, and that there is no serious inhibition of such transport due to excessive plasma phenylalanine. Further, the amount of phenylalanine administered may be reduced, or in some embodiments eliminated entirely, if the phenylalanine in the amino acid mixture is partially replaced by tyrosine. Such replacement is limited by the low water solubility of tyrosine. The tyrosine can be supplemented by more water-soluble tyrosine derivatives, providing the derivative is convertible by the body to tyrosine.

The experimental work leading to these formulations has also indicated that the relative proportions of the essential branched-chain amino acids should be controlled in relation to the proportion of tryptophan administered. By proper balancing of the proportions of isoleucine, leucine, and valine to tryptophan, it is believed that the transfer of excessive tryptophan to the brain can be avoided. This is desirable since tryptophan is converted by the brain to serotonin, which would be expected to complicate hepatic encephalopathy.

In this formulation, the ratio of the combined molar proportions of isoleucine, leucine, and valine to (a) the molar proportion of tryptophan should be from 40 to 300, and to (b) the molar proportion of phenylalanine, or the combined molar proportion of phenylalanine and tyrosine, should be from 15 to 135. Although the complete optimization of the formulation within the specified ranges has not yet been fully defined, nutritionally and/or therapeutically applicable molar proportions for the total of the branched-chain amino acids (isoleucine, leucine, and valine) to the molar proportion of tryptophan is from 50 to 90, and to the molar proportion phenylalanine, or phenylalanine and tyrosine, from 30 to 50.

The formulations are desirably free of ammonia. When prepared from crystalline amino acids, the resultant formulation will be low in free ammonia. In general, the formulations preferably contain less than 2 mmols of ammonia per each 800 mmols of amino acids, including all of the amino acids present (essential and nonessential). The formulations may be advantageously prepared in the form of sterile aqueous solutions adapted for intravenous administration.

Example: A sterile, nonpyrogenic, stable solution suitable for intravenously infusing into liver diseased patients is prepared from pure crystalline amino acids, which are dissolved in distilled water in the following concentrations:

Amino Acid	Grams/Liter	Mols/Liter
L-isoleucine	9.0	0.0686
L-leucine	11.0	0.0838
L-valine	8.4	0.0717
L-tryptophan	0.75	0.00367
L-phenylalanine	1.0	0.00605
L-lysine acetate	8.6*	0.0417
L-methionine	1.0	0.0067
L-threonine	4.5	0.0378
L-alanine	7.65	0.0858
L-arginine	6.0	0.0344
L-histidine	2.4	0.0155
L-proline	8.0	0.0695
L-serine	5.0	0.0476
Glycine	9.0	0.120
L-cysteine·HCl·H_2O	0.2**	0.00114

*Free base, 6.09 g/l
**Free base, 0.14 g/l

In the foregoing formula, the ratio of the essential branched-chain amino acids to phenylalanine is about 37 and to tryptophan is about 61. To this solution is added 1.152 g/l of 85% phosphoric acid to adjust the pH to a more physiological pH (approximately 6.8), and to serve as a source of phosphorus, an important element for good nutrition. The volume is then brought to the desired volume with distilled water. Sodium bisulfite U.S.P. grade, 1.0 g/l, is added and stirred until dissolution is complete. Two grams of activated charcoal is then added and stirred for an additional 10 minutes. The solution is then filtered and filled into appropriate containers for intravenous fluids and steam sterilized at 250°F for 10 minutes.

Solubilizing Agents for Total Infusion Solutions

An infusion solution for parenteral feeding which comprises water-soluble amino acids, polyols, minerals and vitamins and water-insoluble components comprising phosphatides or phosphatides and vitamins A, E, D and K, has been developed by *H. Mader, W. Fekl and K. Schultis; U.S. Patent 4,005,190; January 25, 1977; assigned to J. Pfrimmer & Co., Germany.*

This parenteral infusion solution also contains a minor amount, sufficient to solubilize the water-insoluble lipids and fatty components, of ethanol and at least one member selected from propanediol, butanediol and derivatives of hydroxy-substituted cholanic acid. Use of these materials as solvent intermediaries for the normally water-insoluble lipids and fatty components results in a clear, pyrogen-free aqueous solution. The quantity of ethanol generally should not exceed 50 grams per liter, while the divalent alcohols, that is, propanediol and butanediol, can be present individually or together in amounts of up to 100 grams per liter of infusion solution.

The infusion solution preferably contains ethanol and a hydroxy-substituted cholanic acid derivative since the latter also will solubilize free fatty acids, especially essential fatty acids, or esters (such as the monoester) of salts thereof as well as fat soluble vitamins, such as vitamins A, E, D and K. It has been found that these free fatty acids and fat soluble vitamins form corresponding water-soluble molecule compounds with the hydroxy-substituted cholanic acid derivatives in the presence of the alcohol and thus can also become a component of the total parenteral nutritive solution. The preferred hydroxy-substituted cholanic acid derivatives are 3α,12α-dihydroxy-5β-cholanic acid and 3α,7α,12α-trihydroxy-5β-cholanic acid or their salts, and mixtures thereof.

The hydroxy-substituted cholanic acid derivatives are present in the infusion solution in an amount sufficient to yield a weight ratio of normally water-insoluble phosphatides to hydroxy-substituted cholanic acid derivatives of from 1:1 to 3:1. If the solution contains free fatty acids as water-insoluble fatty components, e.g., linoleic acid, a higher amount of hydroxy-substituted cholanic acid derivatives may be necessary to keep the solution clear.

Example: An infusion solution was formed from the following components using conventional mixing techniques. The phospholipid used was a purified soybean fraction containing 95% phosphatidylcholine and 5% phosphatidylethanolamine.

	Grams
Sodium glycerophosphate·$5H_2O$	1.000
Sodium chloride	1.227
Sodium acetate·$3H_2O$	1.905
Potassium chloride	1.490
Potassium acetate	0.489
Sodium hydroxide	0.740
Potassium hydroxide	0.561

(continued)

	Grams
Magnesium acetate·$4H_2O$	0.643
L-isoleucine	0.780
L-leucine	1.090
L-lysine HCl	1.250
L-methionine	1.050
L-phenylalanine	1.090
L-threonine	0.490
L-tryptophan	0.220
L-valine	0.750
L-histidine	0.490
L-arginine	1.990
L-alanine	3.000
L-proline	3.490
L-glutamic acid	4.500
Glycocoll	4.990
Sorbitol	70.000
Xylitol	40.000
Inositol	0.200
Nicotinic acid amide	0.020
Pyridoxine hydrochloride	0.015
Riboflavin-5'-phosphoric acid ester, sodium salt	0.002
Ethanol	21.300
Phospholipid	5.0
$3\alpha,7\alpha,12\alpha$-trihydroxy-5β-cholanic acid	4.8

The substances were combined with water for injection at 1,000 ml, filled into infusion bottles at 250 ml and were sterilized. The solutions were used for continuous dripping over several weeks and showed good compatibility.

OTHER FOOD PRODUCTS USING AMINO ACIDS

Food Binding Compositions

Compounds like alginic acid, casein, polyphosphoric acid, polyacrylic acid and carboxymethylcellulose have been used as coating agents in processing foods. However, these agents do not always form a coating which is resistant to oxidation, moisture, putrefaction, etc. Similar compounds of higher molecular weight have also been used in binding food components. However, in some cases, the binding is weak and these compounds are not suitable for binding components containing a high percentage of water.

In the process developed by *A. Yamamoto, T. Ikemoto and R. Shimizu; U.S. Patent 3,840,676; October 8, 1974; assigned to Kyowa Hakko Kogyo Co., Ltd., Japan* the coating and binding of foods is accomplished by the combined use of prolamins and amino acids. Prolamins are proteins found in grains. For example, gliadin, zein, rice-gliadin and kaffirin are obtained respectively from wheat, corn, rice and kaffir by extraction with aqueous alcohol or slightly acidic or alkaline aqueous solutions. The prolamins are insoluble in water and show remarkably high viscosity when dissolved in alcohol or alcoholic solutions. When prolamin is contacted with an amino acid under aqueous conditions, a highly complex gelatinous mixture is formed.

In this process, a prolamin is dissolved or converted into a homogeneously emulsified suspension in a 20 to 80% aqueous alcohol solution or dissolved in an organic acid solution having a pH of 2 to 6. The material to be coated is first dipped into the prolamin solution or suspension and then into an aqueous or aqueous alcoholic solution of an amino acid. Alternatively, the dipping into the amino acid solution may be carried out in advance of the dipping into the prolamin solution. In other words, the order of the dipping is optional. The coated material is then dried at room temperature or elevated temperatures to remove the solvent and settle the coating. Generally, temperatures of 70° to 150°C may be used for drying. As a result of this treatment, a glossy thin film is formed on the surface of the material.

Alcohols such as ethanol, methanol and propanol may be used as the solvent. Where the amino acid or prolamin is only slightly soluble in the solvent, a suspension of the reactant may be used. When an organic acid solution is used as the solvent carrier, organic acids such as citric acid, malic acid, ascorbic acid, acetic acid, etc., may be used.

The coating film is effective in preserving foodstuffs against oxidation and putrefaction as well as in maintaining freshness by preventing exposure of the foodstuff to air and moisture.

Example 1: Gliadin (5 grams) and fat (2 grams) are mixed with 18 ml of 50% ethanol and the mixture is emulsified. The emulsion is uniformly applied to the surface of a piece of molded raw kamaboko (fish-paste) which is preliminarily coated to give a smooth glossy surface. The resulting raw kamaboko is then dried in the air for four hours. After drying, a solution of 2 grams of L-alanine in 10 ml of water is applied to the foodstuff. An emulsified aqueous solution of 5 grams of sodium alginate and 2 grams of fat is applied to the surface of a second piece of molded raw kamaboko which also is preliminarily coated to give a smooth glossy surface. The two pieces of raw kamaboko are baked at 150°C for 60 minutes. The baked kamaboko pieces are cooled and allowed to stand exposed to air at room temperature for a week. The alginic acid-coated kamaboko, which is used as a control, is discolored and putrefied on its surface whereas the gliadin-coated kamaboko shows almost no discoloration.

Example 2: The procedures of Example 1 are repeated except that various prolamins and amino acids or salts thereof shown in the following table are used. The products are allowed to stand exposed to air at room temperature for a week and show neither discoloration nor putrefaction.

Prolamin	Amino Acids or Acid Salts	Prolamin	Amino Acids or Acid Salts
Rice-gliadin	L-lysine hydrochloride	Gliadin	L-threonine
Kaffirin	L-lysine hydrochloride	Gliadin	L-cysteine
Gliadin	L-arginine	Gliadin	L-methionine
Gliadin	L-histidine hydrochloride	Gliadin	L-phenylalanine
Gliadin	L-histidine	Gliadin	L-tyrosine
Gliadin	Glycine	Gliadin	L-tryptophan
Gliadin	L-valine	Gliadin	L-proline
Gliadin	L-leucine	Gliadin	L-hydroxyproline
Gliadin	L-isoleucine	Gliadin	Monosodium L-aspartate
Gliadin	L-serine	Gliadin	Monosodium L-glutamate

Color Preservatives for Meat

The attractive color of meat and other animal tissue colored by hemoglobin or myoglobin turns brownish during storage because of chemical changes in the coloring material. It has been common practice to apply nitrates and/or nitrites to animal tissue to preserve the color, but the nitrates are effective mainly by their conversion to nitrite in the meat, and the nitrites have been found to be toxic in amounts not much greater than those used.

It has been found by *H. Hara, M. Nakano and S. Kamiura; U.S. Patent 3,911,146; October 7, 1975; assigned to Seinankaihatu Co. L.T.D., Japan* that the water-soluble ingredients of edible vegetal matter, when contacted with animal tissue preserve the color of the tissue in subsequent cooking and that the effect achieved can be as good as that of nitrite. Being derived from edible vegetal matter, the color-preserving agents of the process are nontoxic in the necessary amounts.

Approximately 15 to 100 ml of the natural plant juice separated from the plant solids or an aqueous extract obtained from plant material which would yield 15 to 100 ml juice under pressure when comminuted, is required for preserving the color of 1 kilogram of meat or other animal tissue. In most instances, 30 to 60 ml juice or the corresponding amount of extract is effective. If the water-soluble ingredients are recovered as a solid material free from insoluble ingredients, the application rate is approximately 1 to 10 grams per kilogram of animal tissue, and 1.5 to 6 grams are usually effective.

The action of the vegetal ingredients on the color of stored animal tissue can be enhanced by certain adjuvants which may not have any color preserving effect of their own. They include vitamins, such as L-ascorbic acid, its nontoxic salts, and nicotinamide, but also erythorbic acid and its nontoxic salts, particularly the sodium salt. Amino acids generally enhance the effects produced by the vegetal ingredients, and aspartic acid, glutamic acid, alanine, valine, leucine, proline, γ-aminobutyric acid are merely examples of the amino acids found useful in this process. Pentoses such as d-xylose, l-arabinose and d-ribose have been used successfully in combination with the vegetal ingredients.

The amounts of the adjuvants needed to enhance the color preserving effect varies. Approximately 0.1 to 0.5% by weight based on the plant juice or its equivalent is preferably used of ascorbic acid, erythorbic acid, their salts or nicotinamide. The amino acids and pentoses are used to best advantage in a ratio of 0.5 to 1.0% of the dissolved solids in the plant juice or its equivalent.

Example 1: Fresh samples of spinach (leaves and roots), Chinese cabbage, lettuce, onions, potatoes, oranges (peel), turnips (leaves and roots), carrots and cabbage were comminuted in a Waring Blendor. Each comminuted sample was divided into two portions. One portion was subjected directly to centrifuging at high speed to separate a juice practically free from solids from a residue which was discarded. The second portion was heated on a steam bath at about 100°C for 30 minutes, and solids removed by centrifuging. The samples were sufficient to yield at least 9 ml juice from each portion.

300-gram specimens of fresh pork were sprayed with 9 ml of respective juices and further sprinkled with 6 grams table salt. Each meat specimen was then stored in a refrigerator at 5°C for 4 to 8 days, and then soaked in water at 70°C for 30 minutes to develop and fix the color of the cured, cooked meat. The color difference between the original meat and the cured and cooked specimens was then assigned a value by an experienced meat inspector on an arbitrary, but readily reproducible scale on which (—) indicates deterioration characteristic of a control which was merely treated with salt, (+) indicates slight improvement, (++) good color retention, and (+++) a superior result.

Each meat specimen was also tested for initial nitrate and nitrite content immediately after being treated with juice and salt and prior to cold storage. The testing method used was that standardized for animal feed by the Association of Official Agricultural Chemists. The effect of each juice was tested after 4 days and after 8 days of cold storage, and the nitrates and nitrites present after 4 and 8 days were determined. The nitrate and nitrite values listed in the table below are calculated as ppm potassium nitrate and sodium nitrite respectively. Trace amounts are indicated by x, and 0 indicates that none was found.

	. Initial, ppm. .		. . . . After 4 Days			. . . After 8 Days		
Vegetal Matter	KNO_3	$NaNO_2$	Color	KNO_3	$NaNO_2$	Color	KNO_3	$NaNO_2$
Spinach (root)	38	2.0	+++	15	3.2	+++	5	1.2
Spinach (leaves)	33	2.0	++	20	2.0	+++	5	1.5
Chinese cabbage	17	x	++	5	x	+++	0	x
Turnip (root)	25	0.8	+++	10	0.4	+++	5	x
Turnip (leaves)	25	1.2	+++	10	1.2	+++	5	1.6
Carrot	13	x	++	5	x	+++	0	x
Cabbage	9	x	++	0	x	+++	0	x

Example 2: Cabbage was washed in water until clean of foreign matter, drained, and shredded, and a juice was recovered from the shredded cabbage by applied pressure. The juice was cooked in a jacketed kettle for 30 minutes, and separated from suspended solids in a filter press.

The clear liquid so obtained was used as a base liquor which was applied to 300-gram-specimens of pork together with 6 grams table salt in the manner described in Example 1. The specimens were stored in the refrigerator for 6 days and then checked for color retention after cooking as described above. The results obtained were comparable with those listed in the above table.

In comparison tests, batches of the base liquor were mixed respectively with L-ascorbic acid, L-sodium ascorbate, erythorbic acid, sodium erythorbate, and nicotinamide in amounts corresponding to 0.3% of the base liquor weight. When applied to pork as described above, the results were uniformly superior to those from the base liquor alone.

Even better results were achieved when mixtures of aspartic acid with d-xylose 1:3, of alanine with valine and d-xylose 1:1:6, of leucine with glutamic acid and l-arabinose 1:1:6, of γ-aminobutyric acid with proline and d-ribose 1:1:6, and aspartic acid with proline and d-xylose 1:1:6 were added to the base liquor in amounts corresponding to 0.8% of the weight of solids content in the liquor.

As a control, a meat specimen was treated with 3 grams of a 1:3 mixture of aspartic acid and d-xylose and 6 grams salt. The meat color deteriorated as much as with salt alone, showing the lack of color-preserving-effect of the amino acids and pentoses in the absence of the vegetal matter.

Production of Crumb for Milk Chocolate

In the manufacture of milk chocolate, an intermediate crumb prepared from the milk components, sugar and cocoa constituents has been used. This mixture is heated which causes a brown coloring and flavor change which is important to the final product formed by addition of cocoa fat. The Maillard reaction which causes desired changes of the color and the flavor in the mixture, is a so-called carbonyl/amino reaction and comprises reaction of reducing sugars present with amino groups in amino acids and proteins present in the milk and other substances.

S. Turos; U.S. Patent 3,900,578; August 19, 1975; assigned to Semper AB, Sweden has developed an improved continuous process for producing such crumb in a much shorter period of time. Amino acids are used with the milk to react with the reducing sugars in the Maillard reaction. The cocoa constituents are added at a later stage of milk chocolate manufacture. The amino acids which may be added in the process are such which are active in the Maillard reaction and include glycine, alanine, valine and lysine. Lysine is preferred since this amino acid is the most reactive one in the Maillard reaction. The apparatus used in the process is shown in the following Figure 2.1.

FIGURE 2.1: PRODUCTION OF CRUMB FOR MILK CHOCOLATE

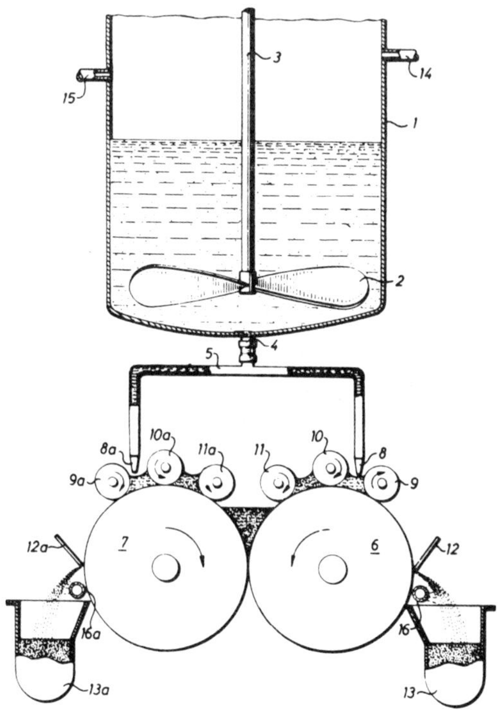

Source: U.S. Patent 3,900,578

The apparatus of Figure 2.1 consists of a mixing vessel **1** with inlet openings **14** and **15** and propeller stirrer **2** driven via shaft **3**. Centrally located on the bottom of the mixing vessel is outlet **4** connected to line **5** which is forked over drying rollers **6** and **7**. The outlet openings **8** and **8a** provided with means (not shown) for controlling the liquid amount fed therethrough terminate somewhat above the surfaces of the drying rollers, viz, between the outer rollers **9** and **9a** of the series of rollers, **9, 10, 11, 9a, 10a** and **11a** which are arranged in contact with the drying rollers **6** and **7** and somewhat separated from each other.

The series of rollers form intermediary sump pairs in which the dry solids content of the mixture is successively increased. A third concentration step is performed in the third sump formed between the drying rollers **6** and **7** above their contact line. In the above apparatus, the mixture fed from the mixing vessel is concentrated from an initial dry solids content of 45 to 55% in the first sump, whereas a dry solids content of 60 to 65% has been achieved in the second sump. The third sump concentrates to a dry solids content of 70 to 85% by weight. From the third sump the concentrated mixture is fed onto the surfaces of the drying rollers by controlling the distance between the rollers in the form of a film having a thickness of usually 0.2 to 0.3 mm for heat treatment and simultaneous drying.

With conventional scraping knives **12** and **12a** and suitably a so-called film guide, i.e., tubes **16** and **16a** arranged over the whole width of the roller through which warm or cold air is blown against the underside of the scraped-off film so as to facilitate the correct removal of the same, the product is scraped off and stored in containers **13** and **13a** for further treatment such as grinding and sieving.

Examples 1 through 6: In the apparatus of Figure 2.1, first flows consisting of skimmed milk concentrated to the stated dry solids content and second flows containing the further components in the amounts stated in the table were fed according to Examples 1 through 6 below. In the first experiment no second flow was used. Examples 1 through 3 are comparative experiments. In the first sump the mixture was concentrated to a dry solids content of 50 to 55% and in the second sump to a dry solids content of 60 to 65%, whereas the mixture in the third sump was concentrated to 70 to 85%. The table below gives the conditions used and the properties of products obtained.

Ex. No.	Composition . Dry Solids	kg	Dry Solids Content (%)	Roller Temp. (°C)	Film Temp. (°C)	Reaction Time (sec)	Product Properties
1	DSM*	180	43	125	105	4	Somewhat yellowish color; taste similar to that of roller dried skim-milk powder
2	DSM Dextrose	180 3	43	125	104	4	Same as Ex. 1
3	DSM Dextrose	180 6	43	123	100	4	Color somewhat more brown than Ex. 1; taste more round but product not acceptable
4	DSM Lysine	180 3	43	122	98	4	Brownish color with good taste
5	DSM Lysine	180 6	42	125	102	4	Browner color than Ex. 4; taste somewhat too heavy
6	DSM Dextrose Lysine	180 3 3	42	125	102	4	Same color as in Ex. 5

*Dry skimmed milk.

Improved Cheese Flavors

In prior investigations, it was known that mixtures of amino acids can contribute to a ripe cheese flavor, in particular the cheese taste. However, it is not clear which amino acids, and what proportion were necessary for obtaining the desired effect.

A process has been disclosed by *K.H. Ney, I.P.G. Wirotana and W.G. Freytag; U.S. Patent 3,922,365; Nov. 25, 1975; assigned to Lever Brothers Co.* for preparing a foodstuff having the desired ripe cheese flavor. This process includes introducing into a food composition a mixture of volatile cheese flavor ingredients and a mixture of at least three amino acids selected from glutamic acid, glycine, lysine and methionine (whereby glycine can be replaced partially or completely by alanine and/or proline). In addition, up to 10% of other free protein amino acids may be mixed with the food composition.

During the ripening of cheese many low-molecular weight compounds are formed, such as alkanoic acids, 2-alkanones, 2-alkanols, aldehydes, amino acids, 2-oxoalkanoic acids and many other compounds which contribute to the typical flavor of the different varieties of cheese.

In this process, volatile cheese flavor ingredients which are added to food compositions can be flavor concentrates obtained in a known manner by steam distillation and subsequent concentration of the distillate from ripe or old cheese, for example Cheddar and Emmental cheese. However, volatile compositions which are mixtures of synthetic compounds corresponding to the flavor ingredients occurring in natural ripe or old cheese can also be used. Such volatile mixtures are described in U.S. Patent 3,520,699 and J. Merory, *Food Flavorings,* 2nd Ed., 191-192.

Example 1: From 100 grams young cheese, 64 grams water and 3 grams melting salt (a mixture of commercially used polyphosphates) a processed cheese was prepared in a casserole under stirring and careful heating up to 80°C. Stirring was continued for 10 minutes and then 0.1 gram of a flavor concentrate, obtained in the known manner from old Cheddar cheese by steam distillation and subsequent concentration of the distillate, was added to the melted mass which was cooling. In this way a processed cheese with a Cheddar taste was obtained which, however, was not full and which, when the taste was judged by a panel of experts, was only given 4 out of the possible maximum of 8 points.

Example 2: The following mixture of amino acids were prepared by mixing:

	A	B	C	D	E
Monosodium L-glutamate	2	5	1	40	435
L-lysine·HCl	2	4	1	10	377
DL-methionine	1	2	–	–	188
L-alanine	–	–	2	–	–
L-proline	–	–	–	0.6	–
Glycine	–	–	–	49.4	–

The figures give the parts by weight of amino acid mixed with each other. The monosodium L-glutamate content of composition C is 25%. From 100 grams young cheese a processed cheese was prepared and flavored as in Example 1.

However, in addition, during the melting 0.8 gram (i.e., about 5,300 mg/kg) of a mixture of amino acids A was added. The processed cheese obtained had a full taste and was given 7 points for taste. In a similar manner, mixture B of the above table is used to produce old Cheddar flavor, C and D for Emmental flavor and E for blue cheese flavor.

Amino Acid Phosphates as Nutritional and Cariostatic Agents

A food composition is disclosed by *R.H. Anderson and A.L. Saari; U.S. Patent 3,932,608; January 13, 1976; assigned to General Mills, Inc.* which contains an amino acid phosphate, namely, an alkali metal amino acid phosphate or an alkaline earth metal amino acid phosphate. The amino acid phosphate serves as a cariostatic agent and as an agent for nutritional fortification.

This cariostatic composition may be prepared by reacting an amino acid in the free base form with phosphoric acid to form an amino acid phosphate. The amino acid phosphate may be then reacted with a base or basic salt of an alkali metal or an alkaline earth metal to form an alkaline amino acid phosphate. In some instances the cariostatic agent may be prepared by reacting the free base amino acid directly with an alkali metal phosphate or an alkaline earth metal phosphate. The term alkaline amino acid phosphate as used here includes alkali metal amino acid phosphates, alkaline earth metal amino acid phosphates and their diphosphates.

Any amino acid may be used in preparing this cariostatic composition providing the amino acid will react to form the alkaline amino acid phosphate. However, the amino acid is preferably an essential amino acid for purposes of nutrition. The amino acids used typically include lysine, ornithine, arginine, tryptophan, phenyl alanine, leucine, isoleucine, threonine, methionine, valine, hydroxyproline and glycine. The preferred amino acid is L-lysine. The amino acid is preferably in the free base form when used to prepare the cariostatic composition. Certain of the free base amino acids, typically L-lysine, are unstable and therefore may be produced from the hydrochloride immediately prior to use in producing the cariostatic composition. The free base amino acid may be prepared by passing a water solution of the amino acid hydrochloride through a resin column in the OH^- ionic form.

The food composition containing this cariostatic agent may be any cereal or sugar-containing food product such as bread, breakfast cereal (either presweetened or unsweetened), various baked goods (i.e., cakes, pies, cookies and the like), casseroles (i.e., noodles and the like), candy, chewing gum and soft drinks. For cariostatic purposes the alkaline amino acid phosphate will generally be present in an amount of at least 0.05% phosphorus based on the total weight of the food product. For economic reasons the alkaline amino acid phosphate will rarely be present in an amount greater than 0.2% phosphorus based on the total weight of the food product and the preferred level is 0.1%. For nutritional purposes, i.e., amino acid fortification of protein-containing food products, the alkaline amino acid phosphate is prepared from the desired essential amino acid and is present in a range of 0.5 to 3.0%, preferably 1 to 2% based on the total weight of the protein present.

The alkaline amino acid phosphate may be combined with the food product in any of various ways such as spraying a solution onto the surface of the food product, mixing a dry form with the dry finished food product, or mixing with the other ingredients prior to baking.

Example 1: Sodium lysine diphosphate was prepared by passing 300 mmols of L-lysine hydrochloride through a resin column to produce free base lysine. The resin used was Rexyn AGI in the OH^- form. To the free base L-lysine was added 300 mmols of sodium dihydrogen phosphate. Then 300 mmols of orthophosphoric acid were added with thorough stirring. The resulting solution was evaporated in a vacuum until a milky appearance was acquired and then cooled to 5°C. The material was filtered and a tacky crystalline material was obtained and air dried. The material was identified as Na-lysine-$H_5(PO_4)_2$.

Example 2: An unsweetened breakfast cereal was prepared by adding a 40% aqueous solution of the sodium lysine diphosphate of Example 1 to the cornflakes in an enrober by spraying. Sufficient solution was applied to provide the flakes with an added phosphorus content of 0.08% by weight. The flavor of the corn flakes was not impaired.

Amino Acid Esters as Feed Supplements for Ruminants

Ruminants obtain their requirements of essential amino acids largely from the digestion of protein in the abomasum and small intestine. This protein comprises dietary protein which has escaped fermentation in the rumen and microbial protein synthesized in the rumen.

K.A. Ferguson and N.T. Hinks; U.S. Patent 4,000,318; December 28, 1976; assigned to the Commonwealth Scientific and Industrial Research Organization, Australia have disclosed means of promoting the absorption from the rumen of animals such as cattle, sheep and goats of biologically useful amino acids (other than methionine) or hydroxy analogues of such biologically useful amino acids. This method comprises orally administering to the animal a feed supplement which comprises a lower alkyl ester of a biologically useful amino acid other than methionine or a salt thereof, or a lower alkyl ester of a hydroxy analogue of a biologically useful amino acid other than methionine or a salt thereof.

These lower alkyl esters are rapidly absorbed from the rumen of the animal before significant breakdown by the bacteria and protozoa in the rumen has occurred. The esters are rapidly hydrolyzed in the blood so that the effectiveness of the method with respect to any particular amino acid can be monitored simply by determining the blood concentration of the amino acid, at different times after oral administration to the ruminant animal.

Example: Figure 2.2 illustrates the plasma amino acid concentration responses to the intraruminal administration of the methyl esters of L-leucine (●), L-threonine (□), L-valine (x), L-phenylalanine (△) and L-lysine (○) in a dose of 5 grams amino acid equivalent.

Figure 2.2a shows that the plasma concentrations of L-leucine, L-threonine, L-valine and L-lysine are increased severalfold within four hours of administration of the methyl esters into the rumen.

Figure 2.2b shows that administration of equivalent amounts of the corresponding free acids into the rumen is without significant effect. The administration

FIGURE 2.2: AMINO ACID ESTERS AS FEED SUPPLEMENTS FOR RUMINANTS

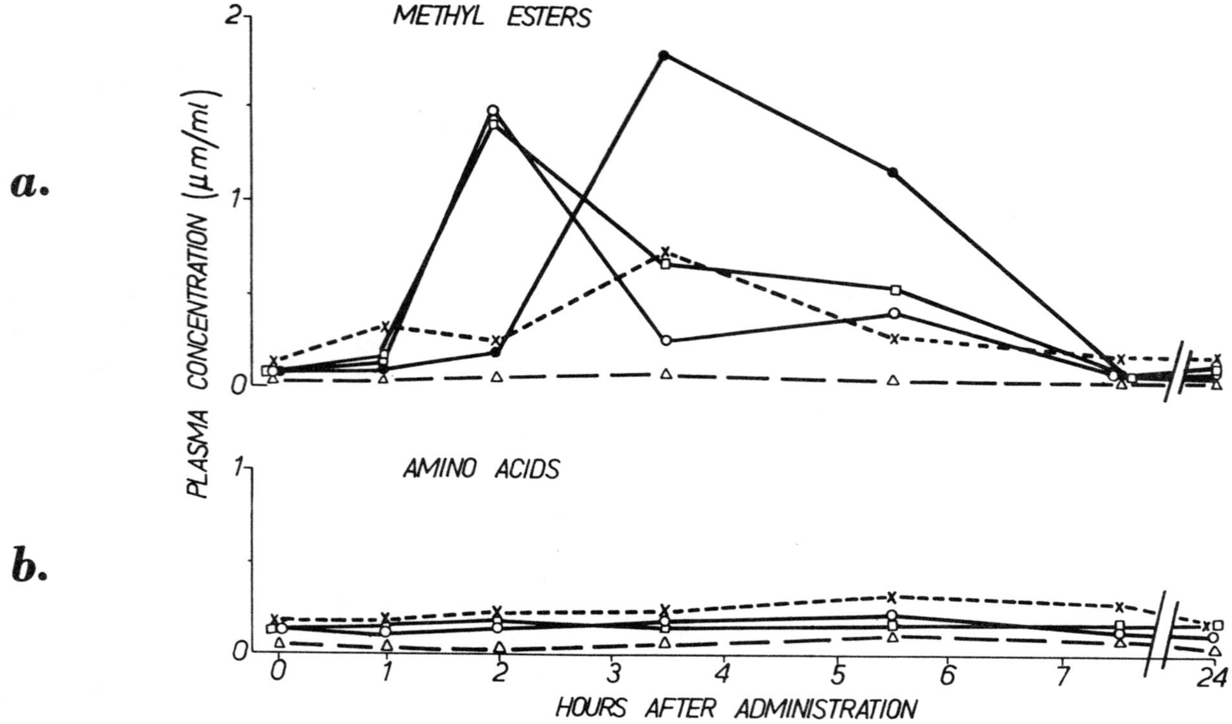

Source: U.S. Patent 4,000,318

of the methyl ester of L-phenylalanine into the rumen did not raise the plasma concentration indicating that the hydrophobic side chain of this amino acid may prevent absorption through the rumen. However, even in cases where the absorption of an amino acid from the rumen is not facilitated by ester formation, the derivative may still be relatively resistant to microbial degradation in the rumen and be capable of absorption from the small intestine. The amino acid esters were prepared by the method of M. Brenner and W. Huber, (*Helvetica Chimica Acta* 36, 1109, 1953).

The alcohol is chilled by the addition of solid carbon dioxide (dry ice), the appropriate amount of thionyl chloride is added slowly so as to keep the temperature below -5°C and finally the amino acid is stirred into the mixture. The temperature is allowed to rise to ambient temperature and the mixture is distilled slowly to remove most of the alcohol. The remainder is removed at reduced pressure by rotary evaporation. By redissolving the residue in a minimum amount of the alcohol and adding 4 to 5 volumes of ether followed by stirring and chilling to 0°C the solid hydrochloride is obtained.

Fish and Crustacea Food

One of the primary problems encountered in culturing fish and crustacea is the problem of poor feed efficiency; expressed as the ratio of feed expended to the feed consumed. This is generally attributable to the failure of the aqua life to actually find and/or consume the food supplied to them.

M.A. Beigler and R.J. Amen; U.S. Patents 3,993,746; November 23, 1976 and 3,903,260; September 2, 1975; both assigned to Syntex (U.S.A.) Inc. have found that the addition of one or more selected amino acids to the food composition induces a feeding response in fish and crustacea. Briefly, the process comprises methods of inducing a feeding response in fish and/or crustacea by applying a combination of lysine and glycine; or water-soluble salts thereof, or a combination of monosodium glutamate and an alkali metal aspartate, in the vicinity of the fish and/or crustacea. Alternatively, the respective amino acids or salts can be used separately but are not nearly as effective.

Example 1: This example illustrates methods of preparing an attractant feed suited for fish. A dry mixture containing 7 pounds of fish meal, 3.2 pounds high lysine corn, 1.36 pounds of mixed minerals (including calcium, phosphorus, magnesium, sodium, potassium, chlorine, iron, zinc and others naturally present in the ingredients), 0.8 pound of rolled oats, 0.64 pound of mixed vitamins (i.e., vitamins A, C, D, E, K and B complex), 0.05 pound of monosodium glutamate, 0.05 pound monosodium aspartate, 0.02 pound of sodium benzoate and 0.02 pound potassium sorbate is dry blended in a commercial mixer at room temperature.

To the blend is then added 1.2 pounds of wheat germ oil, 0.8 pound of whey and 0.75 pound of corn oil. The mixture is then blended and mixed until a homogeneous mixture is obtained. 84.11 pounds of water is then added to the mix and the composition mixed until a homogeneous slurry is obtained. The mixture is then passed through a commercial roller dryer resulting in the formation of a thin flaked product which readily releases the amino acid attractants (i.e., monosodium glutamate and monosodium aspartate) when introduced into an aqueous medium.

Example 2: The feed of Example 1 is tested with various species of tropical fish by feeding samples of the feed to aquaria containing two members of the species, of each of the species of fish identified in the table below. In each case the fish became unusually excited upon the application of the feed to the aquarium water. The excitation is substantially greater than observed for conventional commercial tropical fish feed and led to aggressive behavior and an active feeding pattern.

Species	Common Name	Family	Common Family Name
Hyphessobrycon cardinalis (Cheirodon axelrodi)	Cardinal tetra	Characidae	Characins
Cholisa lalia	Dwarf gourami	Anabantidae	Labryinth fish anabantids
Peterophyllum scalare (Pterophyllum eimekei)	Black or black lace angel fish	Cichlidae	Cichlidae
Mollienisia sphenops	Marbled or orange tail mollie	Poeciliidae	Gambusinos or live bearing tooth carps
Barbus nigrofasciatus	Black ruby barb	Cyprinidae	Carps or minnows
Xiphophorus hellerii	Swordtail	Foeciliidae	Gambusinos

AMINO ACIDS IN PHARMACEUTICAL PREPARATIONS

Intermediates for β-Adrenergic Agents

The amino acid derivatives as disclosed by *W. Winter, M. Thiel, K. Stach, K. Dietmann and W. Bartsch; U.S. Patent 3,793,365; February 19, 1974; assigned to Boehringer Mannheim GmbH, Germany* are compounds useful for the preparation of β-adrenergic pharmaceutical compositions. These amino acid derivatives have the formula

(1)

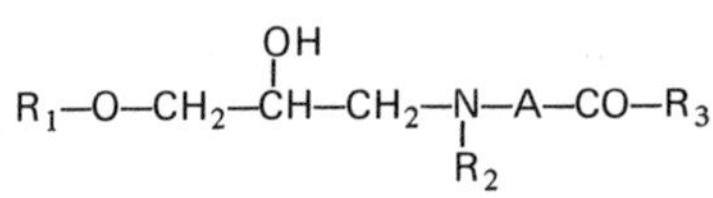

where R_1 may be aryl, such as naphthyl or phenyl, which may be substituted by halogen or hydroxyl, alkoxy, alkenyl, alkenyloxy, alkyl, alkoxyalkyl, trifluoroalkyl, alkylmercapto, alkenylmercapto, amino, alkylsulfonylamino or acylamino; R_1 may be indenyl or hydrogenated indenyl or hydrogenated naphthyl; R_2 is hydrogen or lower alkyl or aralkyl; R_3 is hydroxyl or amino or alkoxy, alkylamino or dialkylamino, or an N-piperazyl radical, which may be N'-substituted; and A is an alkylated, or arylated lower alkylene chain, which can carry a further carboxyl or carbalkoxy radical.

These compounds can be prepared, for example, in one of the following ways (in the following, R_1, R_2, R_3 and A have the same meaning as above unless otherwise defined):

(a) reaction of epoxides of the formula

(2)

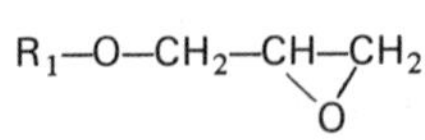

with amino acid derivatives of the formula

(3) $R_2\text{—HN—A—CO—}R_3$

and, when R_3 is a hydroxyl group, this is first converted into a salt by reaction with an organic or inorganic base; or

(b) reaction of compounds of the formula

(4) $R_2\text{—O—CH}_2\text{—CH(OH)—CH}_2\text{—Z}$

in which Z is a reactive ester group, with amino acid derivatives of formula (3); or

(c) reaction of amines of the formula

(5) $R_1\text{—O—CH}_2\text{—CH(OH)—CH}_2\text{—NH—}R_2$

with compounds of the formula

(6) $\text{Z—A—CO—}R_3$

in which Z is as defined as in formula (4); or

(d) reaction of tetrahydro-oxazinones of the formula

(7)

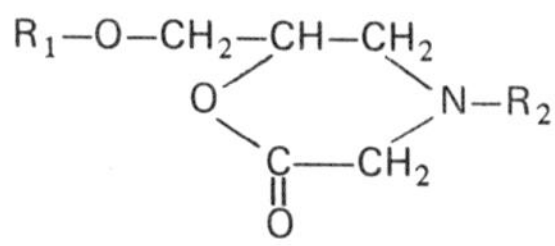

with compounds of the formula $\text{H—}R_3$,

in which when R_2 is hydrogen, it may be N-alkylated in known manner, or when R_2 is a benzyl radical, it may be debenzylated in known manner and the grouping $\text{—CO—}R_3$ in the product obtained, is, if desired, changed by saponification, esterification, transesterification or amidation and free bases or acids of formula (1) are, if desired, converted into pharmacologically compatible salts by reaction with acids and bases.

All the above preparations are preferably carried out in a polar solvent. Solvents which can be used for this purpose include, in particular, dimethylformamide and tetrahydrofuran, as well as isopropanol, n-propanol and other lower alcohols; however, since the intermediates as well as the end products, are not very sensitive, most of the reactions can be carried out in the melt, without the use of a solvent.

Example: 13.35 grams (0.15 mol) L-α-alanine and 6.0 grams (0.15 mol) sodium hydroxide were heated together in 125 ml dimethylformamide until complete dissolution was obtained. 32.8 grams 2,4-dichlorophenoxy-glycidyl ether were then added dropwise, whereafter the reaction mixture was heated for 4 hours to 120°C. The reaction mixture was then evaporated, the precipitated sodium salt was dissolved in water and the solution neutralized with 1 N hydrochloric acid. The precipitate which formed was filtered off with suction, dried, boiled with 75 ml methanol and filtered off. There were obtained 35 grams (75.8%

of theory) N-[1-(2,4-dichlorophenoxy)-2-hydroxy-propyl]-L-α-alanine, (compound 1) which has a melting point of 224°C (decomposition). The compounds set out in the table below were prepared in an analogous manner.

Compound Number	Compound Chemical Name	MP (°C)	Yield (%)
2	N-[1-(3-methylphenoxy)-2-hydroxypropyl]-D,L-α-alanine	208*	60
3	N-[1-(3-methylphenoxy)-2-hydroxypropyl]-L-α-alanine	218-220	55
4	N-[1-(3-methylphenoxy)-2-hydroxypropyl]-D,L-α-leucine	216-217	58
5	N-[1-(3-methylphenoxy)-2-hydroxypropyl]-D,L-α-isoleucine	208-210	55
6	N-[1-(3-methylphenoxy)-2-hydroxypropyl]-D,L-α-valine	218-219	69
7	N-[1-(3-methylphenoxy)-2-hydroxypropyl]-L-α-phenylalanine	208-209	53
8	N-[1-(2-allyloxyphenoxy)-2-hydroxypropyl]-D,L-α-alanine	198	48
9	N-[1-(2-allylphenoxy)-2-hydroxypropyl]-D,L-α-alanine	190	52
10	N-[1-(2-bromophenoxy)-2-hydroxypropyl]-D,L-α-alanine	203-204	71
11	N-[1-(2,5-dichlorophenoxy)-2-hydroxypropyl]-D,L-α-alanine	201-202	91
12	N-[1-(2-bromophenoxy)-2-hydroxypropyl]-L-α-phenylalanine	206-208	68
13	N-[1-(2,3-dichlorophenoxy)-2-hydroxypropyl]-D,L-α-alanine	202	58
14	N-[1-(α-naphthyloxy)-2-hydroxypropyl]-D,L-α-alanine	189-190	74
15	N-[1-(α-naphthyloxy)-2-hydroxypropyl]-L-α-alanine	202-203	77
16	N-[1-(α-naphthyloxy)-2-hydroxypropyl]-D-α-alanine	177-178	48
17	N-[1-(α-naphthyloxy)-2-hydroxypropyl]-D,L-α-phenylalanine	201-203	52
18	N-[1-(α-naphthyloxy)-2-hydroxypropyl]-D,L-α-valine	215-217	50
19	N-[1-(α-naphthyloxy)-2-hydroxypropyl]-D,L-α-leucine	202-204	58
20	N-[1-(α-naphthyloxy)-2-hydroxypropyl]-D,L-α-isoleucine	213-215	63
21	N-[1-(α-naphthyloxy)-2-hydroxypropyl] glycine	110**	43
22	N-[1-(α-naphthyloxy)-2-hydroxypropyl]-D,L-α-aminobutyric acid	211-212	65
23	N-[1-(α-naphthyloxy)-2-hydroxypropyl]-sarcosine	180	45
24	N-[1-(α-naphthyloxy)-2-hydroxypropyl]-N-isopropylglycine	**	84
25	N-[1-(α-naphthyloxy)-2-hydroxypropyl]-D,L-β-aminobutyric acid	169-171***	52
26	N-methyl-N-[1-(α-naphthyloxy)-2-hydroxypropyl]-D,L-α-alanine	†	61
27	N-isopropyl-N-[1-(α-naphthyloxy)-2-hydroxypropyl]-D,L-α-alanine	†	53
27a	N-[1-(α-naphthyloxy)-2-hydroxypropyl]-α-aminoisobutyric acid	243-244	66.7
27b	N-[1-(2-allylphenoxy)-2-hydroxypropyl]-α-aminoisobutyric acid	202-203	67.0

*Decomposes **Amorphous ***Oxalate †Amorphous, no salt

The compounds of this process were tested for cardiac β-receptor blocking activity. These compounds possessed either a greater effectiveness, a comparative protective effect at a lesser dosage, or a lower toxicity, than the comparison compound. The comparison compound, 1-p-nitrophenyl-2-isopropylamino-ethanol, (Inpea), does not protect at a dosage of 10 mg/kg IP. Only at a dose of 50 mg/kg, IP, does it show an incomplete degree of protection against the impoverishment of glycogen occurring under hypoxia stress.

N-Substituted Amino Acids to Combat Influenza

S. Toyoshima, S. Kanao, T. Toyoda and T. Suyama; U.S. Patents 3,801,633; April 2, 1974 and 3,845,097; October 29, 1974; both assigned to Ajinomoto Company, Inc., Japan have disclosed numerous N-substituted amino acids which were found to combat infection with influenza virus in vivo.

In U.S. Patent 3,801,633, it was found that the compounds listed below were effective in vivo against influenza virus in amounts in which their toxicity is low.

N-α-naphthylacetylleucine
N-4-nitrophenoxyacetyl-4-nitrophenylalanine
N-carboxy methylphenylalanine
N-benzylphenylalanine

(continued)

N-furfurylphenylalanine
N-β-naphthylaminomethylalanine
N-β-naphthylaminomethylmethionine
6-(β-naphthylaminomethyl)-amino-caproic acid
N-lauroylphenylalanine
N-4-toluenesulfonyltryptophan
N-2-fluorenesulfonyl-β-alanine
N-2-fluorenesulfonylvaline
N-2-fluorenesulfonyl-4-nitrophenylalanine

These compounds may be used in the free acid form or in the form of their pharmaceutically acceptable salts, particularly the alkali metal salts, the alkaline earth metal salts, organic amine salts and the like, such as the sodium, potassium, magnesium, calcium, ammonium, monoethanolamine and diethanolamine salts which are generally more soluble in water than the free acids, and are not more toxic than the free acids in equimolecular amounts. Those compounds which exist either in optically active (D or L) forms or in the optically inactive racemic form are equally effective in all forms.

Example: A mixture of 9.3 grams α-naphthaleneacetic acid and 4.6 grams phosphorus trichloride was heated for about one hour on a water bath maintained at 90°C. The mixture was then cooled, the supernatant liquid was decanted from the precipitated α-naphthylacetyl chloride, and the latter was dissolved in 30 ml anhydrous ether.

The ether solution was added dropwise over a period of 20 minutes with stirring to a solution of 6.3 grams L-leucine in 20 ml 10% sodium hydroxide solution mixed with a solution of 15.9 grams sodium carbonate in 100 ml water at room temperature. Stirring was continued for three hours, and the reaction mixture was then made acid to Congo red with 6 N hydrochloric acid. The precipitated crystals of N-α-naphthylacetyl-L-leucine were filtered out, washed with water, and recrystallized from aqueous ethanol. 11.8 grams columnar crystals (78% yield) melting at 152°C were obtained $[\alpha]_D^{25} = -22.07°$ (c = 1.675, absolute ethanol).

In U.S. Patent 3,845,097, the following compounds were also cited as antiinflammatory agents as well as exhibiting antitumor effects in mice.

N-β-naphthylaminomethylleucine
N-β-naphthylaminomethylurethane
N-furfurylaminomethylphenylalanine
N-furfuryl-4-nitrophenylalanine
N-benzylvaline
N-2-fluorenesulfonylmethionine
N-2-fluorenesulfonylphenylalanine
N-lauroylleucine
N-ethylcarbaminomethylisoleucine
N-β-naphthylaminomethylthreonine
N-9-fluoroenylacetylphenylalanine
N-myristoylisoleucine
N-β-naphthalenesulfonyltryptophan
N-propionylvaline

Example: A solution of 14.3 grams β-naphthylamine in 100 ml 60% aqueous ethanol was added to a mixture of 13.1 grams L-leucine, 60 ml 37% formaldehyde solution and 250 ml 60% aqueous ethanol. The mixture so obtained was refluxed at 80°C for one hour on a water bath. A small amount of oily matetrial precipitated and was filtered off. The filtrate was concentrated in a vacuum to precipitate crystals of N-β-naphthylaminomethyl-L-leucine which were filtered off, washed with acetone, and recrystallized from absolute ethanol. The purified, very small crystals weighed 18.9 grams (66% yield) and melted at 199°C.

Peptides for Combating Allergic Encephalomyelitis

Experimental allergic encephalomyelitis (EAE) is an autoimmune disease affecting the brain. EAE serves as a model disease for multiple sclerosis, post-infectious encephalitis and subacute panencephalitis. Curative agents suitable for the treatment of EAE are of interest for the possible treatment of these further diseases.

D. Teitelbaum, A. Meshorer, T. Hirshfeld, R. Arnon and M. Sela; U.S. Patent 3,849,550; November 19, 1974; assigned to Yeda Research and Development Co. Ltd., Israel have disclosed a composition adapted to cure or at least reduce the severity of EAE, bringing about either a complete cure if applied at an early stage, or at least decreasing the overall mortality of the disease. Alternatively, the agent can also be used to immunize against EAE.

The compositions used for the treatment of EAE comprise a synthetic water-soluble copolymer consisting of a number of amino acids, the molecular weight of the copolymer being in excess of 10,000 and preferably above 18,000, and the copolymer being characterized by a net positive electrical charge and by a content of a lesser quantity of a negative electrical charge. The copolymer may contain as a further component a neutral amino acid, and if desired also an amino acid adapted to provide immunogenic properties.

These compositions comprise a basic copolymer with a net positive electrical charge, comprising a suitable quantity of an amino acid of positive electrical charge, such as lysine or arginine, in combination with a lesser quantity of an amino acid with a negative electrical charge, such as glutamic acid or aspartic acid, possibly in combination with an electrically neutral amino acid such as alanine or glycine, serving as a filler, and possibly with an amino acid adapted to confer on the copolymer immunogenic properties, such as an aromatic amino acid.

A preferred copolymer comprises in combination alanine, glutamic acid, lysine and tryosine, of net overall positive electrical charge and of a molecular weight of 20,000 to 25,000. The composition of these was about as follows: (molar ratio of amino acids in copolymer); alanine/glutamic acid/lysine/tryosine = 6/2/4.5/1. Similar results were obtained with a soluble copolymer comprising tyrosine, aspartic acid, alanine and lysine in a molar ratio of 1/1.5/5/3.5; and with another such copolymer comprising glutamic acid, alanine and lysine in a molar ratio of about 1.5/5/3.5.

Copolymers of the above type were effective in suppressing EAE in guinea pigs when administered in saline as late as 5 days after challenge with basic protein (BE), and reduced the incidence of EAE from 64% in the control group to 22%; the histological lesions were also substantially decreased both in prevalence and in severity.

A similar protective effect was obtained with rabbits, EAE, induced by 500 μg of BE per rabbit was suppressed by three injections of 8 mg each of the first copolymer solution in 1 ml saline, injected 1, 6 and 11 days after the challenge with BE. The incidence of the disease was decreased from 70% in the control group to 19% in the treated group. Groups of 20 animals each were used. As these copolymers are devoid of encephalitogenic activity and as they are not immunorepressive, they are considered to be of advantage as compared with previous curative agents. These copolymers are easily prepared by conventional procedures. The first of the above copolymers was prepared from the N-carboxyanhydrides of tyrosine, alanine, γ-benzyl glutamate and ϵ-N-trifluoroacetyllysine.

The polymerization was carried out at ambient temperature in anhydrous dioxane with diethylamine as initiator. The deblocking of the γ-carboxyl group of the glutamic acid was effected with hydrogen bromide in glacial acetic acid and was followed by the removal of the trifluoroacetyl groups from the lysine residues by 1 M piperidine.

Disubstituted Phenethylamides of Amino Acids as Hypotensive Agents

Numerous amino acid amides of disubstituted phenethylamines have been disclosed by *J.H. Biel and I.L. Klundt; U.S. Patent 3,907,864; September 23, 1975; assigned to Abbott Laboratories* as renal vasodilators and hypotensive agents. These compounds are represented by the formula

(1)

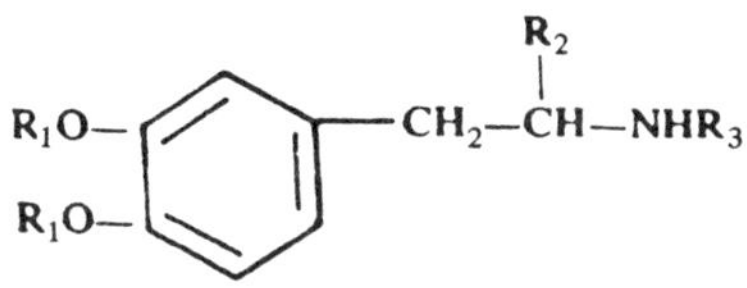

where each R_1 is hydrogen or benzyl; R_2 is hydrogen or lower alkyl and R_3 is an amino acid selected from β-alanine, α-methylalanine, phenylalanine, 3,4-dihydroxyphenylalanine, 4-chlorophenylalanine, glycine, tyrosine, valine, leucine, isoleucine, serine, threonine, cysteine, methionine, tryptophan, aspartic acid, glutamine, arginine, lysine, histidine, γ-aminobutyric acid, pyroglutamic acid, alanine, proline, sarcosine, N-methylpiperazinobutyric acid or

(2)

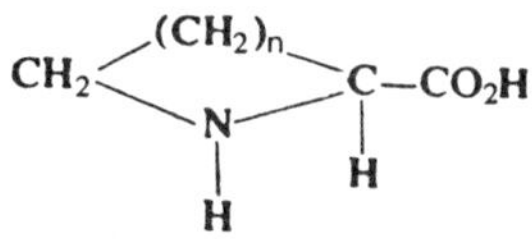

where n is from 1 to 4, or R_3 is one of the above amino acids protected by either carbobenzyloxy or t-butyloxycarbonyl; and the pharmaceutically acceptable acid addition salts thereof, with the limitation that when R_3 is an amino acid other than that represented by formula (2) above, R_1 cannot be hydrogen. The term lower alkyl refers to both straight and branched-chain C_{1-6} alkyl groups.

When R_1 is hydrogen, representative compounds include the following:

N-3',4'-dihydroxy-β-phenethyl-L-3,4-dihydroxyphenylalanine amide
N-3',4'-dihydroxy-α-methyl-β-phenethyl-L-leucine amide tartrate
N-3',4'-dihydroxy-β-phenethyl-L-threonine amide
N-3',4'-dihydroxy-α-methyl-β-phenethyl-D-lysine amide, etc.

When R_1 is benzyl and the amino acid is protected by a carbobenzyloxy group, representative compounds include:

[N'-carbobenzyloxy-L-alanyl] -N-(3',4'-dibenzyloxy-α-methyl-β-phenethyl)amine
[N'-carbobenzyloxy-L-alanyl] -N-(3',4'-dibenzyloxy-β-phenethyl)amine
[N'-carbobenzyloxy-L-glycyl] -N-(3',4'-dibenzyloxy-β-phenethyl)amine, etc.

These compounds are renal vasodilators and have been found to significantly increase renal blood flow in test animals. The compounds are particularly useful in the treatment of renal hypertension.

Example: N-3′,4′-Dihydroxy-β-Phenethyl-L-Alanine Amide Hydrochloride – A solution of 11.1 grams (0.03 mol) of 3,4-dibenzyloxy phenethylamine hydrochloride in 100 ml of methanol was treated with 1.62 grams (0.03 mol) of dry sodium methoxide, the solvent was removed in vacuo and the residue dissolved in 100 ml of acetonitrile, filtered and evaporated to dryness. This was treated with 20 ml of acetonitrile and 40 ml of dry tetrahydrofuran (THF) and warmed until solution was complete. To the warm mixture was added 4.46 grams (0.02 mol) of carbobenzyloxy-L-alanine. The solution was stirred until it was clear and then 4.95 grams (0.024 mol) of dicyclohexylcarbodiimide was added.

An immediate exothermic reaction took place and a white precipitate started to form. The reaction was allowed to stand at room temperature for 3 hours and then filtered to remove 5.2 grams (96.6% of theory) of dicyclohexylurea. The filtrate was diluted with 10 ml of THF, warmed to 40°C and 2 ml of acetic acid was added to the reaction. The solvent was removed in vacuo and the solid residue dissolved in chloroform, washed twice with 100 ml of 2 N hydrochloric acid, washed with 100 ml of saturated sodium chloride, washed twice with 100 ml of saturated sodium bicarbonate, washed with 100 ml of saturated sodium chloride and dried over anhydrous magnesium sulfate. Removal of the solvent gave 10.1 grams (94%) of a white solid. This was recrystallized from 125 ml of ethanol to give 4.8 grams of [N'-carbobenzyloxy-L-alanyl]-N-[3',4'-dibenzyloxy-β-phenethyl]amine, MP 151.5° to 152.5°C, $[\alpha]_D^{23}$ = -8.6° (1.6% $CHCl_3$).

To 250 ml of methanol containing 0.5 ml (6.2 mmols) of concentrated hydrochloric acid was added 2.7 grams (5.0 mmols) of the above-prepared amine and 300 mg of 10% palladium on carbon. The solution was hydrogenated at atmospheric pressure. Theoretical uptake was 380 ml, actual uptake 365 ml (96%). The starting material was not soluble but as the reaction progressed the material dissolved. The catalyst was removed by filtration and the solvent was evaporated in vacuo. The solid was azeotroped dry with ethanol (2 x 50 ml), triturated with 50 ml of hot acetonitrile and filtered to yield 1.15 grams of N-3',4'-dihydroxy-β-phenethyl-L-alanine amide hydrochloride, MP 214° to 218°C (decomposed) $[\alpha]_D^{24}$ = -36.4° (2.08% H_2O).

N,N′-Dichloro Amino Acids as Bactericides and Fungicides

N-chloro nitrogen-containing carboxylic acids have been found by *S.J. Barer, R.F. Stockel and P.C. Valenti; U.S. Patent 3,950,536; April 13, 1976* to be useful as microbiocides in topical treatment of the skin and mucous membranes. These compounds have the formulas

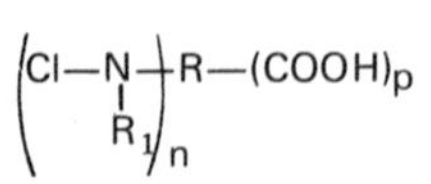

or

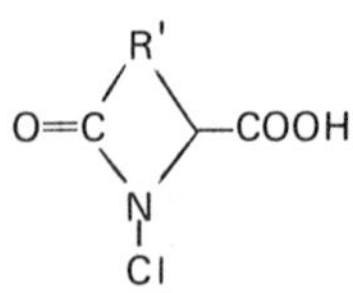

where R is a divalent saturated aliphatic hydrocarbon; R_1 is a monovalent saturated aliphatic hydrocarbon, acyl, or chlorine; n is 1 or 2; p is 1 or 2; R' is a divalent saturated aliphatic hydrocarbon having 2 to 4 carbons to complete the heterocyclic ring. In preferred embodiments, R is a divalent saturated aliphatic hydrocarbon (e.g., alkylene or alkylidene) of 1 to 17 carbons; R_1 is alkyl of 1 to 4 carbons; alkanoyl of 2 to 6 carbons; or chlorine; and R' is ethylene

$-\!\!(CH_2CH_2)\!\!-$, trimethylene $-\!\!(CH_2CH_2CH_2)\!\!-$, or tetramethylene $-\!\!(CH_2CH_2CH_2CH_2)\!\!-$. Preferred subclasses of N-chlorocarboxylic compounds include, for example, N-chloroaminoalkanoic acid having 2 to 11 carbons and C_{2-5} alkane bis(N-chloroamino) C_{2-5} alkanedioic acid. Preferred compounds include, N,N-dichloro-β-alanine, N,N-dichloro-11-aminoundecanoic acid, N-chloro-2-pyrrolidone-5-carboxylic acid, and N-chloro-N-acetyl-β-alanine. The compounds can be applied as such, in mixture with water, in mixture with triacetin, in mixture with water and triacetin, in creams, ointments or together with other conventional pharmaceutical adjuvants.

The N-chloro carboxylic compounds can be formed in conventional fashion by reacting a positive halogen source such as a hypochlorite, e.g., t-butyl hypochlorite, isopropyl hypochlorite, ethyl hypochlorite, calcium hypochlorite, lithium hypochlorite, sodium hypochlorite, hypochlorous acid, etc.; an N-chloromide, e.g., N-chlorosuccinimide, trichloroisocyanuric acid, etc.; or chlorine gas; with an aminocarboxylic acid or derivative such as glycine, sarcosine, α-aminoisobutyric acid, 4-aminobutyric acid, alanine, β-alanine, etc.

Example: 19 grams of β-alanine is added to 150 ml methanol. The suspension is stirred and cooled to -25°C. 26 ml of t-BuOCl is added dropwise while maintaining the temperature at -20° to 30°C. The gradual disappearance of most of the insoluble starting material indicates the progression of the reaction. The reaction is permitted to warm to room temperature with constant agitation and the insoluble β-alanine filtered. The methanol solution of N,N-dichloro-β-alanine is stripped of solvent to yield a green viscous liquid product in 94% yield and 91% purity.

Briefly, the efficacy of such compounds is shown by the number of viable organisms after exposure to varying concentrations of N,N-dichloro-β-alanine in a 14:1 mixture of water and triacetin, at the end of each time period (1, 5 and 10 minutes) as set forth in the following table.

N,N-Dichloro-β-Alanine (wt %)	Activity After Test (min) 1	5	10
	Staphylococcus aureus 10 x 10^8		
0.1	<100	<100	<100
0.01	17 x 10^4	57 x 10^4	39 x 10^3
	Saccharomyces cerevisiae 14 x 10^7		
1.0	<100	<100	<100
0.5	<100	<100	<100
0.1	<100	<100	<100

Amino Acids in Prevention of Gastric Lesions

K. Takagi and S. Okabe; U.S. Patent 3,988,466; October 26, 1976; assigned to Kyowa Hakko Kogyo Co., Ltd., Japan have found that gastric lesions brought about by the administration of nonsteroidal antiinflammatory agents are prevented by administering at least one amino acid. Amino acids which may be used, include for example, L-lysine, L-arginine, L-histidine, L-serine, L-valine, L-threonine, glycine, L-alanine, L-glutamine, D-glutamine, L-leucine, L-isoleucine, L-glutamic acid, L-asparagine, L-proline, L-hydroxyproline, L-methionine, L-phenylalanine, DL-tryptophan and their derivatives. Further, pharmaceutically acceptable salts of these amino acids and their derivatives may also be used. The amino acid is administered in at least 1 gram, preferably at least 2 gram, amounts in mixture with the antiinflammatory agent or separately therefrom. The foregoing dosage of the amino acid is prescribed regardless of the amino acids and the kinds or doses of the antiinflammatory agent. It is considered that when an increased dose of the antiinflammatory agent is administered, an increased dose of the amino acid may provide good results.

However, it has been found that a dose of more than 10 grams of the amino acid is meaningless. For example, when the antiinflammatory agent is aspirin and the amino acid is L-glutamine, for the single dose of aspirin, which is usually in a range of 0.3 to 1.0 grams, L-glutamine effectively prevents gastric lesions at a dose of 2 to 5 grams. Further, in the case of indomethacin and L-glutamine, for the usual single dose of 25 to 50 mg of indomethacin, an administration of 2 grams of L-glutamine considerably prevents gastric lesions.

The amino acids may be prepared in any convenient dosage form together with the nonsteroidal antiinflammatory agent, for example, in a compressed tablet, capsule, granule or powder according to any of the well-known methods.

Example: Effect of L-Glutamine on Aspirin-Induced Gastric Lesions in Pylorus-Ligated Rats — In this example, six groups of male Donryu strain rats are used. The animals are fasted for 24 hours but water is allowed ad libitum. Under ether anesthesia, the pylorus is ligated according to the standard method of Shay et al.

Five groups are treated with graded doses of L-glutamine suspended in 1% carboxymethylcellulose solution by oral administration immediately after the pylorus ligation. A sixth group (control) is treated with 5 ml/kg of 1% carboxymethylcellulose solution alone. Following a control period of 10 minutes, the animals are treated with 100 mg/kg of aspirin suspended in 1% carboxymethylcellulose solution by oral administration.

After a 7-hour interval during which the animals are not allowed feed or water, the animals are killed by an overdose of ether. Ten minutes prior to the killing, the animals are treated, under ether anesthesia, with 1 ml of 5% solution of Pontamine Sky Blue 6BX in saline (pH 7.2 adjusted with 0.5 N HCl) by intravenous injection through the femoral vein.

The stomach is removed and slightly inflated by injection of 1% formalin solution through the esophageal junction. Then the stomach is maintained in 1% formalin solution for 10 minutes for fixation of the inner and outer layers of the gastric wall. The stomach is opened along the greater curvature and the length of lesions (dark blue areas against pale blue background) in the glandular portion is measured under a dissecting microscope (10X) provided with a square grid.

The results are shown in the table below. The ulcer index expressed by mm indicates the mean value of the sum of the length of the lesions per animal of each group. The level of significance is calculated according to Student t-test. The stomachs of three out of twelve animals treated with 1,000 mg/kg of L-glutamine are quite clear and no mucosal defect is detected at all. Although the gastric lesions preventive effect of L-glutamine may be apparent from the table below, it is noted that no such effect is observed at a dose of 62.5 mg/kg of L-glutamine.

Effect of L-Glutamine on Aspirin-Induced Gastric Lesions in Pylorus-Ligated Rats

	Dose (mg/kg)	No. of rats	Average body weight (g)	Ulcer index (mm)	Inhibition* (%)	Level of significance
Control		10	188	54.7±3.4		
L-glutamine	1,000	12	189	3.5±1.5	93.6	<0.001
	500	12	186	18.0±3.1	67.1	<0.001
	250	12	182	15.0±3.4	72.6	<0.001
	125	10	179	27.9±4.7	49.0	<0.001
	62.5	10	189	57.6±5.6	−5.0	—

$$\text{*Inhibition (\%)} = \frac{\text{Control ulcer index} - \text{Test ulcer index}}{\text{Control ulcer index}} \times 100$$

Ureidoacetohydroxamic Acid for Preventing Hypertension

This process disclosed by *D.C. Fessler, G.A. Heavner and T.H. Massey; U.S. Patent 4,028,401; June 7, 1977; assigned to Morton-Norwich Products, Inc.* is concerned with (substituted)ureidoacetohydroxamic acids of the formula

where R is isobutyl, hydrogen, methyl, benzyl, 4-aminobutyl or hexyl and R_1 is hydrogen or benzyl.

The members of this series are potent inhibitors of the enzyme responsible for the conversion of angiotensin I to the powerful pressor agent angiotensin II associated with hypertension. They are thus capable of interrupting the biological pathway leading to the formation of that active pressor substance. For example, they inhibit the pure converting enzyme isolated from rabbit lung tissue at levels of from 0.4 to 3.2 x 10^{-5} mol per liter. The methods preferred for the preparation of these compounds are shown by the following example.

Example: t-Butyloxycarbonyl-O-Benzyl Glycine Hydroxamic Acid — O-benzylhydroxylamine hydrochloride (15.0 grams, 93.8 mmols) was dissolved in methanol (150 ml) and added to a solution of potassium hydroxide (5.27 grams, 93.8 mmols) in methanol (50 ml). The resulting mixture was cooled, filtered, evaporated, dissolved in tetrahydrofuran (25 ml) and filtered as solution A. t-Butyloxycarbonyl glycine (16.4 grams, 93.8 mmols) was dissolved in tetrahydrofuran (75 ml), triethylamine (13.2 ml, 93.8 mmols) added and the solution cooled to -10°C with dry ice/ethanol. Isobutyl chloroformate (12.2 ml, 93.8 mmols) was added and the solution stirred cold 2 minutes, before adding solution A dropwise. The mixture was stirred cold 1 hour and at room temperature 2 hours, filtered, evaporated, dissolved in ethyl acetate (100 ml), washed with water (3 x 50 ml), dried over magnesium sulfate and evaporated to an oil.

O-Benzyl Glycine Hydroxamic Acid Trifluoroacetate — t-Butyloxycarbonyl-O-benzyl glycine hydroxamic acid (17.5 grams, 62.5 mmols; see step above) was dissolved in trifluoroacetic acid (25 ml), stirred at room temperature 30 minutes and ether (200 ml) added. The product was filtered and dried to yield: 9.0 grams, MP 122° to 124°C.

Benzyl-N-[N¹-(N²-Benzyloxycarbamoylmethyl)Carbamoyl] DL-Leucinate – Methyl-L-leucinate hydrochloride (3.63 grams, 0.02 mol) was suspended in toluene (75 ml) and heated to 130°C while bubbling nitrogen through the solution. Phosgene was bubbled into the 130°C solution for 30 minutes. The solution was allowed to cool to room temperature while flushing with nitrogen, and evaporated to an oil. A solution of O-benzyl glycine hydroxamic acid trifluoroacetate (5.9 grams, 0.02 mol) in pyridine (50 ml) was added to the oil and allowed to set at room temperature overnight. The solution was evaporated to an oil, dissolved in ethyl acetate (50 ml), washed with water (3 x 25 ml), dried over magnesium sulfate, evaporated to an oil and crystallized from ethyl acetate/petroleum ether. The product was recrystallized from ethyl acetate/1:1 ether:petroleum ether to yield: 4.0 grams, MP 100° to 102°C.

N-[N¹-(N²-Hydroxycarbamoylmethyl)Carbamoyl]-DL-Leucine – Benzyl-N-[N¹-(N²-benzyloxycarbamoylmethyl)carbamoyl]-DL-leucinate (3.45 grams, 9.8 mmols) was dissolved in methanol (25 ml), 10% Pd/C catalyst (0.5 gram) added, and the mixture stirred in a hydrogen atmosphere about 6 hours. The mixture was filtered, an equal volume of 1 N sodium hydroxide (40 ml) added, and the solution stirred at room temperature 1 hour. The methanol was evaporated and the resulting solution was chromatographed on a 20 gram CG-120 (H^+) column. Fractions containing product were lyophilized to yield: 1.0 gram, MP 80° to 87°C (decomposed).

N,N'-carbonyl-glycine-glycine hydroxamic acid monohydrate, N,N'-carbonyl-DL-alanine-glycine hydroxamic acid hemihydrate, N,N'-carbonyl-DL-phenylalanine-glycine hydroxamic acid, N,N'-carbonyl-L-alanine glycine hydroxamic acid, N,N'-carbonyl-L-lysine glycine hydroxamic acid, etc., are other inhibitors produced by these methods.

N-(Mercaptoacyl)Amino Acids for Metabolic Disorders

Compounds useful for prophylaxis and therapy in treating a metabolic disorder, such as nosotoxicosis due to a heavy metal radiation disorder, diabetes or hepatitis are disclosed by *I. Mita, S. Okumura, S. Yamabe, Y. Funae and J. Matumoto; U.S. Patents 3,971,828; July 27, 1976 and 3,897,480; July 29, 1975; both assigned to Santen Pharmaceutical Co. Ltd., Japan.* These compounds comprise

$$R{-}\underset{\displaystyle SH}{\underset{|}{CH}}{-}CONH{-}\underset{\displaystyle R'}{\underset{|}{CH}}{-}(CH_2)_n{-}COOH$$

and intermediates where R is selected from alkyl, aryl and aralkyl radicals; R' is selected from hydrogen atom and alkyl, aryl and aralkyl radicals and n is an integer of 0 to 4; when n is an integer of 1 to 4, R and R' are CH_3 and H respectively; when n is zero, R is CH_3 and R' is selected from CH_3, C_6H_5, $C_6H_5CH_2$ and $(CH_3)_2CHCH_2$, or R is a member selected from the group consisting of $(CH_3)_2CH$, C_6H_5, p-Cl—C_6H_4 and $C_6H_5CH_2$ and R' is H.

The known compound, 2-mercaptopropionylglycine was used in this test as control. The mice were divided into groups of thirty mice each. The members of groups were administered with test compounds in the dose indicated, respectively. After the administration, the mice were irradiated with fluorescence x-ray to give the total dose of 800 R as γ-rays and 30 days after the irradiation, the number of surviving mice was counted. The test results are shown in the table below.

Test Compounds	Percent of Mice Surviving 30 Days	Dose (mg/kg)	Partition Coefficient*
None	17.1	–	–
N-(2-mercaptopropionyl)-glycine	43.0	20	0.17
	31.0	10	–
N-(2-mercaptoisovaleryl) glycine	20.0	10	0.92
N-(2-mercaptophenylacetyl)-glycine	54.5	10	1.03
	31.5	5	–
N-(2-mercaptophenylpropionyl) glycine	51.5	10	2.54
	25.7	5	–
N-(2-mercapto-p-chlorophenyl-acetyl) glycine	11.4	10	3.98
N-(2-mercaptopropionyl-β-alanine)	48.0	10	0.21
N-(2-mercaptopropionyl)-leucine	24.0	10	1.88
N-(2-mercaptopropionyl-5-amino-caproic acid)	60.0	10	0.63
N-(2-mercaptopropionyl)phenyl-glycine	46.0	10	1.73
N-(2-mercaptopropionyl)phenyl-alanine, MP 131° to 132°C	52.0	10	3.68
2-Mercaptopropionylphenyl-alanine, MP 105° to 108°C	48.0	10	3.55

*Butanol/buffer, pH 7.4

These compounds may be prepared by various known processes, but the process shown in the example is the most convenient because it gives the highest yield. This process comprises reacting a compound

$$R-\underset{X}{\underset{|}{CH}}-COX' \quad \text{with} \quad NH_2-\underset{R'}{\underset{|}{CH}}-(CH_2)_n-COOH$$

in a solvent in the presence of an alkaline agent to form

$$R-\underset{X}{\underset{|}{CH}}-CONH-\underset{R'}{\underset{|}{CH}}-(CH_2)_n-COOH$$

This compound is reacted with a salt of thiobenzoic acid or xanthogenic acid to form a compound represented by the formula

$$R-\underset{M}{\underset{|}{CH}}-CONH-\underset{R'}{\underset{|}{CH}}-(CH_2)_n-COOH$$

which is subjected to hydrolysis. In the above formulas, R, R' and n are as defined above; X and X' are independently a halogen atom; and M is a thiobenzoic acid or ethylxanthogenic acid residue.

Example: 200 ml of thionyl chloride was added to 62.9 grams (0.293 mol) of α-bromophenylacetic acid and the mixture refluxed for 2 hours. After the completion of the reaction, excess thionyl chloride was removed from the mixture under reduced pressure and then the mixture distilled to yield 62.8 grams of 2-bromophenylacetic chloride (BP 105°C/5 mm Hg). Then, 7.5 grams (0.10 mol) of glycine was neutralized with 50 ml of 2 N NaOH and to the resulting solution were added dropwise over 1 hour 20.0 grams (0.085 mol) of 2-bromophenylacetyl chloride and 43 ml of 2 N NaOH while stirring. During the addition, the reaction mixture was maintained below 5°C and adjusted to a pH of 8. After the completion of the addition, the mixture was stirred at room temperature for 4 hours to complete the reaction. To the resulting solution, without isolating a bromine-containing compound, was added the filtrate which had been prepared by neutralizing 13.8 grams (0.10 mol) of thiobenzoic acid with 50 ml of 2 N KOH and the thus obtained solution was filtered to remove undissolved solids, and then the mixture was allowed to stand overnight at room temperature while adjusting the pH to 8.

On the next morning, the solution was acidified with hydrochloric acid while ice-cooling to precipitate crystals. The crystals were recovered by filtration, washed with water, dried and recrystallized from benzene to give 23.9 grams of 2-benzoylmercaptophenylacetylglycine as white crystals (MP 133° to 138°C).

9.9 grams (0.03 mol) of the 2-benzoylmercaptophenylacetylglycine was added to 50 ml of concentrated aqueous ammonia and the mixture was stirred at room temperature for one hour. During stirring and about 30 minutes after the addition, the formation of benzamide was observed. After the completion of the reaction, the benzamide was removed by extraction with ether. After removing the ammonia under reduced pressure, the aqueous layer was acidified with hydrochloric acid and extracted with ethyl acetate. The extract was washed with water and dried over anhydrous sodium sulfate. After the ethyl acetate had been distilled off, a small amount of benzene was added to the extract to give N-(2-mercaptophenylacetyl)glycine as white crystals. Yield: 5.6 grams (MP 91° to 93°C).

Treating Impaired Consciousness with L-Pyroglutamyl-L-Histidyl-L-Prolinamide

Y. Takahashi and K. Takeuchi; U.S. Patent 4,059,692; November 22, 1977; assigned to Takeda Chemical Industries, Ltd., Japan have disclosed a therapeutic means for improving and restoring impaired consciousness. It was found that the administration of L-pyroglutamyl-L-histidyl-L-prolinamide (thyrotropin-releasing hormone, TRH) and its physiologically acceptable salts was unexpectedly found to significantly improve impaired consciousness.

According to this process, TRH may be employed as the free base or in the form of a physiologically acceptable salt such as an acid addition salt, e.g., a mineral acid salt (hydrochloride, sulfate, etc.) or an organic acid salt (acetate, tartrate, etc.). TRH tartrate is especially convenient for the administration. TRH or its salts are generally administered by the parenteral routes such as intravenous or intramuscular routes to the patients. The therapeutic dosage depends upon the route and upon the severity of impaired consciousness in each patient, and it may generally be within the range of from 500 to 5,000 μg in terms of free TRH per adult human patient per day, when parenterally, e.g., intravenously or intramuscularly administered.

In case of those patients having swallowing capacity TRH or its salts may be orally administered. In this case, the dosage may be, of course, chosen from within the above range, but it may be increased over that range, for example, up to about 150 mg in terms of free TRH per adult human patient per day.

Pharmaceutical compositions containing TRH or its salts can be prepared by conventional techniques for the preparation of injections, powders, capsules, tablets, pills, and the like. For intravenous or intramuscular injections, an injectable solvent such as physiological saline is used. When intravenous drip infusion is applied, an optional carrier such as an aqueous solution of glucose may be employed. If desired, the medicament of this process may be administered in combination with other drugs for impaired consciousness, such as cytidine-5'-diphosphate choline and the like.

The following clinical data is illustrative of a sample case: A nine year old boy was in a postoperative state of haematoma of right frontal lobe caused by cranial trauma. Before administration, the patient was able to respond to only painful stimuli with a verbal expression of pain, otherwise being unconcious with a slight palpebral opening. Such conscious state lasted without modification for a further 10 days. He was given a daily intramuscular injection of 292 μg of TRH tartrate (dissolved in 0.8 ml of a sterile physiological saline; 200 μg in terms of free TRH) for 10 successive days. On the 5th day of the treatment, there was produced a prompt verbal reply with a smile to the interview by the doctor. On the 7th day, he was able to ride and sit by himself on a wheelchair. The next day, he was able to go to a stool with support and he was able to sit by himself on the bed and to have a conversation with others on one day after the termination of the daily administration.

Example 1: Tablets containing 2.928 mg of TRH tartrate (2.000 mg in terms of free TRH) and having the following composition are prepared by known techniques:

	Mg per Tablet
TRH tartrate	2.928
Lactose	65.072
Cornstarch	39.800
Hydroxypropylcellulose	2.000
Magnesium stearate	0.200
	110.000

Example 2: An injection for intramuscular use is prepared by dissolving 730 μg of TRH tartrate (500 μg in terms of free TRH) in 1 ml of a sterile physiological saline containing 5% of sorbitol.

Example 3: An injection for intravenous drip infusion is prepared by dissolving 3,650 μg of TRH tartrate (2,500 μg in terms of free TRH) in 500 ml of an electrolyte solution.

ALIPHATIC AMINO ACIDS

GLYCINE–PREPARATION AND ISOLATION

Aminoacetic acid or glycine is the simplest of the amino acids and has no asymmetric carbon atom. There are several known methods for preparing amino carboxylic acids such as glycine. A favorable synthesis route to glycine is a cyanomethylation process. According to this process, glycinonitrile (H_2NCH_2CN) is formed by the reaction of ammonia and glycolonitrile ($HOCH_2CN$) or a mixture of formaldehyde and hydrogen cyanide. Alkaline saponification of this reaction mixture yields the sodium glycinate salt which is converted to glycine upon the removal of the sodium ion, e.g., by acidification.

Via this route, competitive side reactions occur producing iminodiacetonitrile during the cyanomethylation of ammonia, which is subsequently saponified to disodium iminodiacetate. In the subsequent acidification steps, the disodium iminodiacetate is converted to iminodiacetic acid or monosodium iminodiacetate, depending on the pH. In other known methods used for preparing glycine, e.g., the ammoniation of chloroacetic acid, glycine products are formed which also have various amounts of iminodiacetic acid impurities. Iminodiacetic acid (IDA) is an undesirable contaminant of the glycine and in certain uses (e.g., in foods, drugs, animal feeds, etc.), the presence of the IDA impurity is unacceptable.

Preparation of Glycine from Glycolonitrile

N. Mihara, O. Furuya and K. Wada; U.S. Patent 3,875,221; April 1, 1975; assigned to Showa Denko KK, Japan have developed an improved commercial process for preparing either glycine or alanine from either glycolonitrile or lactonitrile. This is accomplished by carrying out in successively disposed vessels the following steps:

(1) Feeding continuously to a reactor an aqueous cyanohydrin solution of glycolonitrile or lactonitrile, and at least 3 mols of gaseous ammonia per each mol of the cyanohydrin and carrying out the reaction below 70°C under autogenous pressure to form a reaction solution containing the corresponding glycino-

nitrile or alpha-aminopropionitrile, while in the meantime transferring the reaction solution successively to a separate vessel, where the pressure is released and reverted to atmospheric pressure, and recycling to the foregoing reaction the excess ammonia gas that is flashed from the solution;

(2) Contacting the above aqueous solution of either glycinonitrile or alpha-aminopropionitrile at a temperature lower than the boiling point of the aqueous solution under atmospheric pressure or slightly less, with an aqueous alkaline solution containing a stoichiometric amount or slight excess of a caustic alkali to obtain by hydrolysis an aqueous solution of an alkali metal salt of the corresponding lower amino acid;

(3) Treating the aqueous solution of an alkali metal salt of a lower amino acid obtained in (2) with a stoichiometric amount of a reagent selected from sulfuric acid, hydrochloric acid and the ammonium salts thereof to obtain an aqueous solution containing the corresponding alkali metal salt of a mineral acid and free lower amino acid; and

(4) Submitting the aqueous solution obtained in (3) to fractional crystallization in the presence of ammonia to isolate and recover the lower amino acid.

One of the features of step (1) of this process is that, differing from the previously used ammonia water or liquid ammonia, gaseous ammonia is used in forming the glycinonitrile. The gaseous ammonia that is fed continuously to the reactor along with the glycolonitrile, is fed in an amount of at least 3 mols per each mol of the glycolonitrile. The use of the ammonia in an amount of less than 3 mols does not give a satisfactory rate of selectivity in the conversion of glycolonitrile to glycinonitrile.

Generally speaking, a solution of high concentration of the glycolonitrile and hence a small content of water is to be preferred as the starting material. However, another feature of this process is that even a relatively dilute solution of about 30 to 60% concentration of the glycolonitrile can be advantageously used.

Example: A continuous reaction apparatus of 0.1 cubic meter capacity was continuously fed gaseous ammonia and an aqueous glycolonitrile solution (48.5 weight percent) at a molar ratio of ammonia/glycolonitrile of 4.0. The reaction was carried out for an average dwell time of one hour at 60°C and pressure of 4.3 kg/cm^2 gauge. The starting materials were fed at the following rates: ammonia 1.94 kmol/hr, glycolonitrile 0.485 kmol/hr, water 1.63 kmol/hr.

The concentration of glycinonitrile at the outlet of the reaction apparatus was 28.6 weight percent, and the per-pass yield, based on the glycolonitrile, was 94.6%. When one cubic meter of the solution leaving the reaction apparatus outlet was flashed at atmospheric pressure at 60°C, 185 kg (10.9 kmol) of ammonia was recovered. As a remainder, there was obtained 695 kg of an aqueous glycinonitrile solution (purity 37.0 weight percent, containing saturated ammonia).

The aqueous glycinonitrile solution was mixed with 643 kg of a 30 weight percent aqueous caustic soda solution containing 193 kg of caustic soda, after which the mixture was reacted for one hour at 90°C and atmospheric pressure followed by neutralization with 242 kg of 98% sulfuric acid.

The composition of the resulting solution was as follows: 345 kg glycine, 342 kg sodium sulfate, and 862 kg water, etc. Next, the fractional crystallization operation was carried out. The foregoing neutralized reaction solution was concentrated by heating it at the boil to evaporate 380 kg of the water. Then, ammonia gas was introduced until the solution became saturated. This was followed by hot-filtering the solution at 80°C while continuing the introduction of ammonia gas. 164 kg of crude sodium sulfate was obtained.

The filtrate was cooled to 30°C, and the glycine crystallizing out was recovered by centrifuging. 168 kg of light brown crude glycine was obtained. The purity was 98.2%. The crude glycine was dissolved in 60°C hot water and then passed through a column packed with an anion-exchange resin and a cation-exchange resin to remove the sulfate radicals and sodium ions. The solution was then decolorized with active carbon and cooled to 30°C to crystallize out glycine, which was separated by filtration and dried. Thus, there was obtained 75 kg of purified glycine of a purity of 99.6%.

Separation of Glycine from Iminodiacetic Acid

Pure glycine, free of contaminants, may be prepared economically using the ion-exchange process disclosed by *N.S. Marans; U.S. Patent 3,813,434; May 28, 1974; assigned to W.R. Grace & Co.* In addition, the iminodiacetic acid impurity may be isolated and recovered. Broadly, the process comprises treating an aqueous solution containing glycine, its alkali metal salts or mixtures thereof, and iminodiacetic acid, its alkali metal salts or mixtures thereof, as impurities, with an anion-exchange resin which retains the iminodiacetic acid component, thereby leaving a purified glycine solution. After separating the purified glycine containing solution from the anion-exchange resin, the solvent is removed and a pure glycine product is recovered.

The term glycine component or iminodiacetic acid (IDA) component refers to glycine in the acid form or alkali metal salt form; or iminodiacetic acid in the acid form or alkali metal salt form, respectively. The term glycine refers to glycine in the acid form, either the H_2NCH_2COOH structure or the zwitterion structure, $^+NH_3CH_2COO^-$.

The anion-exchange (i.e., basic ion exchange) resin used in the purification process is preferably a strongly basic resin in the chloride (Cl^-) form. This Cl^- form is the more stable form, and permits long term reuse without decomposition of the resin.

Operable anion-exchange resins are those having pK_b value of <5. Generally, it is desirable that these anion-exchange resins have a maximum through-put efficiency, i.e., possess a relatively large particle size, have a high degree of porosity and a rigid structure. Anion-exchange resins having particle sizes of 16 to 50 mesh (U.S. standard screen size) and wet volume capacities of 1.0 to 6.0 meq/ml, preferably 1.35 to 5 meq/ml, are well known and commercially available. Particularly suitable are the strongly basic ion-exchange resins having a quaternary ammonium functionality and generally based on a styrene-divinylbenzene copolymer matrix.

Examples of commercially available quaternary ammonium-type ion-exchange resins include macroreticular-type, such as Amberlite IRA-900, Amberlite IRA-910 or gel-type, such as Amberlite IRA-400, Amberlite IRA-402, Amberlite IRA-410 or Amberlite IRA-458 (an acrylic-divinyl benzene based resin). Other highly suitable strongly basic resins incorporating quaternary ammonium functionality include Dowex 1, Dowex 2, Dowex 11, Dowex 21K and Dowex SAR.

Particularly suitable are resins having a relatively high degree of crosslinkage, e.g., from 2 to 8% crosslinking agent such as divinylbenzene. Highly satisfactory resins are Dowex 1X8 or 2X8, i.e., containing 8% divinylbenzene crosslinking agent.

In order to have efficient separation and purification, no soluble inorganic salts, e.g., NaCl or Na_2SO_4, should be present in the mixtures containing glycine and IDA. The presence of sodium chloride is particularly undesirable since, when the solution containing the salt comes in contact with the basic anion-exchange resin, simultaneous regeneration of the anion-exchange resin occurs, thus preventing effective separation of the glycine component. Thus, when glycine is prepared according to the cyanomethylation route, it is necessary to remove the metal salt by-product formed upon acidification of the glycinate salt. This removal can be done by conventional methods such as fractional crystallization or use of other bases, e.g., barium hydroxide in place of sodium hydroxide which, upon acidification of the glycinate with a suitable acid, e.g., H_2SO_4, will form an insoluble salt which can be easily separated from the glycine and IDA mixture.

Example: A glass column 18 inches long and 1.5 inches in diameter, equipped with a stopcock attachment, was packed first with Pyrex wool, then 100 g of a strongly basic anion-exchange resin, Dowex 2X8, and finally with sea sand. This strongly basic anion-exchange resin, a diethanolamine-substituted polystyrene-divinyl benzene copolymer resin in the chloride form, had a wet volume exchange capacity of 1.4 meq/ml, a dry capacity of 3.3 meq/g and a particle size of 20 to 40 mesh (wet).

The resin column was prewashed with 100 ml of deionized water, and eluate was discarded. Thereafter, 100 ml of a 1% aqueous solution of a 90:10 mixture of glycine acid:iminodiacetic acid, respectively, was passed slowly over the resin column. The resin column was then washed with two successive portions, 100 ml each, of distilled water. The first 200 ml eluted fraction was collected and, after being evaporated to dryness in an oven at 60°C and 5 inches Hg pressure, gave 0.85 g of a solid product. An NMR analysis of the solid using deuterated water (D_2O) as the solvent showed the product to be pure glycine free of iminodiacetic acid impurity.

The iminodiacetic acid impurity retained on the ion-exchange resin column could be removed by washing the column with 100 ml of a 5% aqueous sodium chloride solution which also simultaneously regenerated the ion-exchange resin column. Thereafter, the eluate, which contained the sodium salt derivative of iminodiacetic acid, was acidified with hydrochloric acid, and the iminodiacetic acid was recovered using fractional crystallization. The purity of isolated iminodiacetic acid, as determined by an NMR analysis in D_2O solution, was about 90+ percent.

Recovery of Glycine from Sodium Salt Solutions

Commercially acceptable methods have not yet been developed for the separation of sodium sulfate or sodium chloride from amino acids such as glycine and alanine which have high water solubility. Workers at W.R. Grace have developed three related processes for recovery of substantially pure glycine from such solutions. In the first process disclosed by *J.C. Thunberg, R.W. Bragdon and W.P. Moore; U.S. Patent 3,904,585; September 9, 1975; assigned to W.R. Grace & Co.*, a solution containing at least 5% of the amino acid is used as the starting material.

The starting aqueous solution has a temperature above 33°C, a pH of 4.5 to 8.5, and a mol ratio of amino acid to sodium sulfate of 1–5:1. The process comprises:

(a) Forming a first slurry having a temperature effective for preventing precipitation of the amino acid, the first slurry being a mixture of precipitated sodium sulfate and a first mother liquor, the first mother liquor being a second aqueous solution consisting essentially of sodium sulfate and the amino acid, by evaporating water from the starting solution while maintaining its temperature within a range (e.g., from 60° or 70°C up to the normal boiling point) effective for preventing the precipitation of the amino acid;
(b) Separating the first mother liquor from the precipitated sodium sulfate;
(c) Cooling the separated first mother liquor to a temperature within a range (e.g., 33° to 40°C) effective for precipitating the amino acid to form a second slurry, the second slurry being a mixture of precipitated glycine or β-alanine and a second mother liquor, the second mother liquor being a third aqueous solution of sodium sulfate and the amino acid; and
(d) Separating and recovering the precipitated amino acid.

Example 1: An ammoniacal solution of glycinonitrile containing 2,804 g (50 mols) of glycinonitrile was fed into an aqueous sodium hydroxide solution (20% sodium hydroxide) containing 2,040 g (51 mols) of sodium hydroxide at 50° to 60°C. The resulting mixture was then boiled until free of ammonia. The resulting ammonia-free sodium glycinate solution was cooled to 80°C, bleached with hydrogen peroxide (5 ml of 36% H_2O_2), treated with charcoal and filtered to remove color bodies, cooled to room temperature (ca 25°C), and diluted to 12.13 kg with water.

The resulting aqueous solution consisted essentially of 40% sodium glycinate and 0.3% sodium hydroxide. The solution was substantially free of disodium iminodiacetate and it contained only trace quantities of color bodies and other undesired side products of the hydrolysis. A starting aqueous solution having a pH of 6 and consisting essentially of water, sodium sulfate and glycine was formed by adding 93% sulfuric acid (2,662 g) to the above 12.13 kg of sodium glycinate solution. The starting aqueous solution was designated Starting Solution No. 1.

Example 2: A quantity (1,183 g) of Starting Solution No. 1 containing 4 mols (300 g) of glycine was boiled at atmospheric pressure to evaporate water therefrom to form a first slurry weighing 1,129 g and consisting essentially of crystalline anhydrous sodium sulfate and a solution of glycine sodium sulfate and water. This slurry was centrifuged at 80°C. The separated sodium sulfate was dried, weighed (67 g), and then discarded. The hot separated mother liquor (first mother liquor) was cooled to 35°C to form a second slurry consisting essentially of precipitated (crystalline) glycine and a solution of glycine, sodium sulfate, and water. This second slurry was stirred at 35°C for 2 hours and centrifuged. The separated glycine was dried and weighed (132 g).

The mother liquor from the precipitated glycine can be recycled in the above process with additional starting solution.

J.C. Thunberg, R.W. Bragdon and W.P. Moore; U.S. Patent 3,947,496; March 30, 1976; assigned to W.R. Grace & Co. have also modified their process with the first step comprising cooling to precipitate the glycine. The multiple steps of this modified process comprise:

(a) Forming a first slurry consisting of a first lot of precipitated glycine and a first mother liquor consisting of water, dissolved glycine, and dissolved sodium sulfate by cooling the aqueous starting solution to a temperature above 33°C (e.g., 34°C or 35°C) effective for precipitation of glycine;

(b) Separating the first mother liquor from the first lot of precipitated glycine (e.g., by centrifugation, decantation, or filtration) at 34° to 35°C and recovering the separated glycine;

(c) Forming a second slurry consisting of a first lot of precipitated sodium sulfate and a second mother liquor consisting of water, dissolved glycine, and dissolved sodium sulfate by evaporating from the first mother liquor an amount of water effective for causing sodium sulfate to precipitate while maintaining the resulting second slurry at a temperature effective for preventing the precipitation of solid glycine [e.g., between 60°C and the normal boiling point of the second slurry (e.g., 70° to 100°C or 80° to 95°C)] ;

(d) Separating the second mother liquor from the first lot of precipitated sodium sulfate while maintaining the second slurry at a temperature effective for preventing the precipitation of solid glycine [e.g., within 60°C and the normal boiling point of the second slurry (e.g., 70° to 100°C or 80° to 95°C)] ;

(e) Forming a first solution consisting essentially of water, dissolved glycine, and dissolved sodium sulfate by mixing the separated second mother liquor (which is preferably hot, i.e., between 60°C and the normal boiling point of the second mother liquor) with water and a second lot of aqueous starting solution, the water being in an amount effective for preventing the precipitation of sodium sulfate where the first solution is cooled in a later step;

(f) Forming a third slurry consisting of a second lot of precipitated glycine and a third mother liquor consisting of water, dissolved glycine, and dissolved sodium sulfate by cooling the first solution to above 33°C effective for causing the second lot of glycine to precipitate (e.g., within the 34° to 55°C, or 35° to 40°C, or 36° to 44°C, or to any temperature above about 33°C and below about 60°C);

(g) Separating the third mother liquor from the second lot of precipitated glycine (e.g., by centrifugation, decantation, or filtration) at 34° to 55°C, or 35° to 40°C, or 36° to 44°C, or any temperature above 33°C and below 60°C, and recovering the separated glycine;

(h) Forming a fourth slurry consisting of a second lot of precipitated sodium sulfate and a fourth mother liquor consisting of water, dissolved glycine, and dissolved sodium sulfate by evaporating from the third mother liquor an amount of water effective for causing sodium sulfate to precipitate while maintaining the resulting fourth slurry at a temperature effective for preventing the precipitation of solid glycine [e.g., within 60°C and the normal boiling point of the second slurry (e.g., 70° to 100°C or 80° to 95°C)], and

(i) Separating the fourth mother liquor from the second lot of precipitated sodium sulfate while maintaining the fourth slurry at a temperature effective for preventing the precipitation of glycine [e.g., within 60°C and the normal boiling point of the second slurry (e.g., 70° to 100°C or 80° to 95°C)].

Steps e through i above can be repeated indefinitely by mixing the separated mother liquor obtained in step d with water and starting solution as in step e and then proceeding as in steps f through i, the water being provided in an amount effective for preventing the precipitation of sodium sulfate when the resulting mixture of mother liquor and water is cooled to precipitate glycine.

Where carrying on a long series of such runs (where steps e through i are repeated many times), it is generally preferred to remove a small portion of the separated mother liquor (e.g., about 1 to 10% of such mother liquor) separated in step i to prevent the build up of color bodies and other undesired side-products which are present in small amounts in the starting aqueous solution. This removed portion is not mixed with starting aqueous solution in a repetition of step e. It (the removed portion) can be discarded or processed separately to produce crude solid glycine which can be used as such or purified by conventional techniques such as recrystallization.

J.C. Thunberg, R.W. Bragdon and W.P. Moore; U.S. Patent 3,985,801; Oct. 12, 1976; assigned to W.R. Grace & Co. have also disclosed a multistep crystallization process for separating glycine from sodium chloride in aqueous solution. Such a system results where glycine is formed from glycinonitrile by hydrolyzing the nitrile with sodium hydroxide and treating the resulting sodium glycinate with hydrochloric acid to convert the sodium salt (sodium glycinate) to the free amino acid (glycine).

The starting solution should have a temperature above 0°C (e.g., between 0.5°C and about its normal boiling point), a pH of 4.5 to 8.5, a mol ratio of glycine to sodium chloride of 0.7 to 5:1, the process comprising:

(a) Forming a first slurry consisting of a first lot of precipitated glycine and a first mother liquor consisting of water, glycine, and sodium chloride by cooling the aqueous starting solution to a temperature effective for precipitating glycine (e.g., 0.5° to 55°C) if the aqueous starting solution is above such a temperature;

(b) Separating the first mother liquor from the first lot of precipitated glycine, and recovering the separated glycine;

(c) Forming a second slurry consisting of a first lot of precipitated sodium chloride and a second mother liquor consisting of water, glycine, and sodium chloride by evaporating water from the first mother liquor while maintaining the resulting second slurry at a temperature effective for preventing the precipitation of glycine (e.g., within 60°C and the normal boiling point of the second slurry);

(d) Separating the second mother liquor from the first lot of precipitated sodium chloride while maintaining the second slurry at a temperature effective for preventing the precipitation of glycine (e.g., within 60°C and the normal boiling point of the second slurry);

(e) Forming a third slurry consisting of a second lot of precipitated glycine and a third mother liquor consisting of water, glycine, and sodium chloride by mixing the separated second mother liquor with a second lot of the aqueous starting solution and cooling the resulting mixture to a temperature effective for causing the second lot of glycine to precipitate from the resulting mixture (e.g., 0.5° to 55°C) if the resulting mixture is above such temperature;

(f) Separating the third mother liquor from the second lot of precipitated glycine and recovering the separated glycine;

(g) Forming a fourth slurry consisting essentially of a second lot of precipitated sodium chloride and a fourth mother liquor consisting essentially of water, glycine, and sodium chloride by evaporating water from the third mother liquor while maintaining the temperature of the resulting fourth slurry at a temperature effective for preventing the precipitation of glycine (e.g., within 60°C and the normal boiling point of the second slurry); and

(h) Separating the fourth mother liquor from the second lot of precipitated sodium chloride while maintaining the temperature of the fourth slurry at a temperature effective for preventing the precipitation of glycine (e.g., within 60°C and the normal boiling point of the second slurry).

If the aqueous starting solution used in step a of this summary is so dilute that glycine does not precipitate on cooling, an evaporated system can be formed by evaporating from the aqueous starting solution an amount of water effective for causing glycine to precipitate when the evaporated system is cooled. The evaporated system can then be cooled to cause glycine to precipitate therefrom and to form the first slurry of this summary.

Steps e through h, above, can be repeated indefinitely by mixing the separated fourth mother liquor obtained in step h with a third (or subsequent) lot of starting aqueous solution (as recited in step e) and proceeding as recited in steps e through h.

Where carrying on a long series of such runs (where steps e through h are repeated many times), it is generally preferred to remove a small portion of the fourth mother liquor (e.g., 1 to 10% or 3 to 6% of the fourth mother liquor) to prevent the build-up of color bodies and other undesired side-products which are present in small amounts in the starting aqueous solution. This removed portion is not mixed with starting aqueous solution in a repetition of step e. It (the removed portion) can be discarded or processed separately to produce crude solid glycine which can be used as such or purified by conventional techniques such as recrystallization.

GLYCINE AND DERIVATIVES–USES

Glycine-KCl Solutions

It is the object of the disclosure by *A.E. Corker; U.S. Patents 3,822,344; July 2, 1974; and 3,708,574; January 2, 1973; both assigned to Arthur H. Cox & Co., Ltd., England* to provide pharmaceutical compositions which contain potassium chloride in an orally administrable form, for the treatment of hypokalemia and, at the same time, not causing nausea or ulceration or stenosis of the wall of the alimentary tract.

These solutions, containing potassium chloride, are obtained by dissolving in water or an aqueous medium compositions comprising a mixture of a solid, water-soluble hydrochloride of a physiologically innocuous nitrogen-containing organic compound, which dissociates on addition to water to yield an acid solution containing hydrochloric acid and the organic base, e.g., glycine hydrochloride, and at least one of potassium bicarbonate and potassium carbonate.

Preferred hydrochlorides incorporated in these compositions include betaine hydrochloride, the hydrochlorides of glycine, N-methyl-, N-ethyl-, N,N-dimethyl- and N,N-diethyl-glycine and the methyl and ethyl esters of these five amino acids, alanine or β-phenylalanine hydrochloride, glutamic acid hydrochloride, glutamic acid mono methyl or ethyl ester hydrochloride, glutamic acid dimethyl or diethyl ester hydrochloride, N-methyl-glutamic acid hydrochloride, alanine amide hydrochloride and glycine amide hydrochloride.

Example: Tablets are formed in the usual way, each having the composition:

Glycine hydrochloride, g	0.750
Potassium bicarbonate, g	0.675
Calcium cyclamate, g	0.05
Peppermint oil, g	0.004
Polyvinyl pyrrolidone (MW 25,000), g	0.035
Lubricants, percent by weight	
Stearic acid	1
Magnesium stearate	1

The stipulated amount of glycine hydrochloride may be replaced by 0.855 g of glycine methyl ester hydrochloride, 1.325 g of methyl glutamate hydrochloride, or 1.420 g of dimethyl glutamate hydrochloride.

N-[1-(3-Trifluoromethylphenyl)Prop-2-yl] Glycine and Derivatives

L. Beregi, P. Hugon and J.-C. Le Douarec; U.S. Patents 3,856,857; December 24, 1974 and 3,759,979; September 18, 1973; both assigned to Science Union et Cie Societe Francaise De Recherche Medicale, France have disclosed N-substituted glycines which possess anorexigenic, analgesic and lipid metabolism regulating properties. These compounds have the formula:

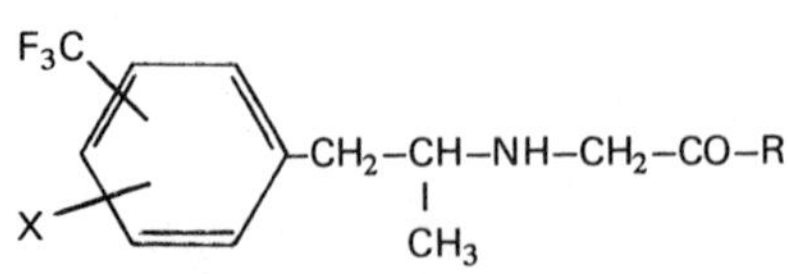

where –X is hydrogen or halogen such as chlorine or fluorine; –R is a hydroxyl radical, a C_{1-4} alkyloxy radical, a $-NR_1R_2$ radical where $-R_1$ and $-R_2$ are the same or different and each is selected from hydrogen, a C_{1-4} alkyl, a cyclic radical selected from phenyl and pyridyl radicals, optionally substituted by one or more halogen atoms, C_{1-4} alkyl or alkyloxy radicals, or a trifluoromethyl radical, $-R_2$ is also a $-CONH_2$ radical provided that $-R_1$ represents a hydrogen atom, and $-R_1$ and $-R_2$ may be joined together to form, with the nitrogen atom to which they are attached, a heterocyclic radical selected from a morpholinyl, pyrrolidinyl, piperidyl, indolinyl and isoindolinyl radical, or a $-HN-NR_3R_4$ radical where R_3 and R_4 are the same or different and each is selected from hydrogen and a C_{1-4} alkyl.

Preparation of the above four types of compounds are shown in the following examples.

Example 1: Ethyl [1-(3-Trifluoromethylphenyl)Prop-2-yl] Aminoacetate – Twenty-five parts of ethyl bromoacetate were added to a solution of 61 parts of 1-(3-trifluoromethylphenyl)-2-amino propane in 120 parts of anhydrous benzene. The mixture was refluxed for 4 hours. After being allowed to cool at room temperature, the salt was filtered off. The distillation of the residual liquid yielded 31 parts of ethyl [1-(3-trifluoromethylphenyl)prop-2-yl] aminoacetate having BP/0.3 mm Hg: 102° to 104°C. The corresponding hydrochloride, recrystallized in ethyl acetate, melted at 203°C.

Example 2: N-[1-(3-Trifluoromethylphenyl)Prop-2-yl] Glycine – A mixture of 20 parts of ethyl [1-(3-trifluoromethylphenyl)prop-2-yl] aminoacetate hydrochloride and 300 parts of concentrated hydrochloric acid (d=1.19) was refluxed for 3 hours. The crude product was collected by filtration, washed with 100 parts of water and dried. There was obtained 16 parts of N-[1-(3-trifluoromethylphenyl)prop-2-yl] glycine hydrochloride, MP 200°C with decomposition.

Example 3: N-Methyl [1-(3-Trifluoromethylphenyl)Prop-2-yl] Aminoacetamide – To a solution of 40.6 parts of 1-(3-trifluoromethylphenyl)-2-aminopropane in 150 parts of anhydrous benzene there were added 10.8 parts of N-methyl chloroacetamide. The addition required ten minutes and the reaction mixture was then refluxed for 10 hours.

The hydrochloride of the starting amine which separated was collected by filtration, the solvent was evaporated and the residue taken up by a mixture of 100 parts of isopropanol and 8.75 parts of methanesulfonic acid. The solid which separated was collected by filtration. It was crystallized from 300 parts of isopropanol to give 21 parts of N-methyl [1-(3-trifluoromethylphenyl)prop-2-yl] aminoacetamide methanesulfonate.

Example 4: [1-(3-Trifluoromethylphenyl)Prop-2-yl] Aminoacetylhydrazine – To 25 parts of hydrazine hydrate (98%) in 250 parts of ethanol there were added, during ten minutes, 29 parts of ethyl [1-(3-trifluoromethylphenyl)prop-2-yl] aminoacetate. The mixture was refluxed for 1 hour.

After evaporation of the solvent, the residue was taken up by 200 parts of anhydrous ethanol and an excess of hydrochloric ether. The salt separated in a well crystallized form was dried; 30 parts of [1-(3-trifluoromethylphenyl)prop-2-yl]-aminoacetylhydrazine dihydrochloride, MP 172° to 173°C (Micro Kofler) were obtained.

The compounds of the process and their physiologically tolerable addition salts possess valuable pharmacological and therapeutic properties. They may be used in particular as an anorexigenic, analgesic and lipid metabolism regulator. Their toxicity is weak and the LD_{50} in mice varies from 125 to 300 mg/kg by the intraperitoneal route and from 375 to 750 mg/kg by the oral route.

Synergistic Antimicrobial Materials Containing Dodecyldi(Aminoethyl)Glycine

Dodecyldi(aminoethyl)glycine (DAG) is an amphoteric electrolyte, normally known as ampholyte soap. These substances, which are thus both anionic and cationic, combine the bactericidal properties of the cationic substances with the surface active and soil dissolving properties of the anionic substances. Contrary, for example, to the invert soaps, the ampholyte soaps together with protein result only in negligible precipitations or none at all. It is known that, in weak concentrations and over short periods of exposure, DAG is effective against gram negative bacteria as well as against fungus. There is, however, no record of the effect of the compound on oral microflora and the possibility of using it in the practice of odontology.

It is also known that chlorohexidine is effective against gram positive and gram negative bacteria as well as against fungus. Chlorohexidine, which is a base, can be used in the form of salts, such as digluconate (easily soluble in water), diacetate (less soluble in water than digluconate), dichloride (soluble in water to only 0.06%) or monofluorophosphate.

S.-G. Hesselgren; U.S. Patent 3,932,607; January 13, 1976 has found that DAG, in combination with chlorohexidine or salts thereof, produces an antimicrobial composition or inhibitor which has a synergistic effect on the organisms and materials mentioned above.

The following are tests performed by using DAG and chlorohexidine diglyconate, 1,1'-hexamethylene bis[5-(p-chlorophenyl)biguanide] digluconate, separately and with compositions containing DAG and chlorohexidine digluconate. Equal parts (by weight) of the two components were used in the combined compositions unless otherwise stated.

The tests were carried out in the following manner: 1 ml of the homogenized bacteria suspensions or the homogenized test material was mixed with 1 ml inhibitor solution in 8 ml sterile phosphate buffer or 8 ml sterilized saliva. After expiration of the exposure time, a suitable amount of the mixture was membrane filtered and washed with sterile water to remove the inhibitor from the filter.

In order to calculate the number of reproducible cells, incubation was carried out at 37°C for 24 to 48 hours with the membrane filter placed on blood agar plates (5% horse blood). The number of washes required to eliminate the inhibitor was determined by separate experiments with the strongest concentration of the inhibitors used and with a bacteria suspension of known density.

The results given in the tables show the number of surviving organisms expressed in logarithmic numbers after 4 minutes inhibitor exposure. A indicates DAG; B, chlorohexidine digluconate and C, the combination of these two substances. Control means no inhibitor in the solutions.

Table 1 shows the results using washed bacteria suspensions of *Escherichia coli* in phosphate buffer having a pH value of 7.0

TABLE 1

Inhibitor	. . . . . Amount of Active Substance, mg/ml				Control
	0.001	0.01	0.1	1.0	
A	2×10^6	2×10^6	1×10^2	0	2×10^6
B	2×10^4	2×10^2	0	0	2×10^6
C	6×10^5	2×10^0	0	0	2×10^6

Tests were carried out on homogenized dental plaque in phosphate buffer with a pH value of 7.0. The plaque material was subjected to aerobic cultivation and anaerobic cultivation. The results obtained are shown in Table 2 (aerobic cultivation) and Table 3 (anaerobic cultivation).

TABLE 2

Inhibitor	. . . . . Amount of Active Substance, mg/ml				Control
	0.001	0.01	0.1	1.0	
A	1×10^7	7×10^6	1×10^6	3×10^2	1×10^7
B	1×10^7	1×10^6	5×10^4	7×10^2	1×10^7
C	1×10^7	1×10^5	1×10^3	1×10^1	1×10^7

TABLE 3

Inhibitor	. . . . . Amount of Active Substance, mg/ml				Control
	0.001	0.01	0.1	1.0	
A	7×10^7	9×10^6	2×10^6	2×10^2	7×10^7
B	7×10^7	7×10^6	8×10^4	3×10^3	7×10^7
C	7×10^7	7×10^5	6×10^4	1×10^0	7×10^7

N-(2-Mercaptopropionyl)Glycine for Treating Respiratory Diseases

J.-P. Buret; U.S. Patent 3,857,951; December 31, 1974; assigned to Laboratories Cassenne, France has found that N-(2-mercaptopropionyl)glycine and its alkaline salts possess the pharmacological properties of fluidifying mucus coming from the rhinopharyngeal and tracheobronchial membranes. This activity has not been reported for these compounds in the prior literature.

By their ability to fluidify mucus coming particularly from the rhinopharyngeal and tracheobronchial mucous membranes, N-(2-mercaptopropionyl)glycine and its sodium salt can be used in the treatment of bronchitis, rhinitis, rhinopharyngitis and sinusitis, to dissolve the mucus (plugs) that obstruct the bronchia, in the treatment of respiratory insufficiencies, and in all the syndromes of hypersecretion and blockage of the various air passages.

N-(2-mercaptopropionyl)glycine and its alkaline salts can be administered through the digestive tract (orally) by parenteral means or by direct contact with the rhinopharyngeal and tracheobronchial mucous membranes. The daily dosage can be, for example, between 100 to 500 mg taken orally by a human. The pharmaceutical compositions containing N-(2-mercaptopropionyl)glycine or its alkaline salts can be either solid or liquid and can be incorporated with the excipients usually used in pharmaceutical compositions. There has thus been prepared, for example, tablets dosed with 100 mg of N-(2-mercaptopropionyl)glycine, an aqueous injectable solution dosed with 100 mg of the sodium salt of N-(2-mercaptopropionyl)glycine and suppositories dosed with 300 mg of the sodium salt of N-(2-mercaptopropionyl)glycine.

The following clinical observations illustrate the process. The sodium salt of N-(2-mercaptopropionyl)glycine was administered to adults intravenously in the form of ampoules containing 100 mg of the active principle in 2 cm^3 of solvent. The resulting observations with respect to the adults are shown in the table below.

Subject	Sex	Age	Diagnosis	Posology	Results	Tolerance	Conclusion
A	Male	73	Showed acute respiratory insufficiency in a chronic bronchitis, marked emphysema	i.v. 3 amp/day dosed with 100 mg for 8 days	good	good	very good effect on expectoration which has increased in volume; it is fluidified.
B	Male	55	Asthma complicated with chronic bronchitis. Dusty lungs. (The man was a miner and then a founder.)	i.v. 3 amp/day dosed with 100 mg for 7 days	fairly good on expectoration	good	distinct increase of bronchial secretion
C	Male	68	Large chronic respiratory insufficiency	i.v. 3 amp/day dosed with 100 mg for 9 days	very good expectoration—fairly good clinically	excellent	important increase in expectoration

Similar good to excellent results were also obtained treating adults with pills containing 100 mg of the active principle and in the treatment of children with syrups (125 mg active material per 125 cm^3 syrup) and the aerosol form containing 4% active principle.

Glycine Derivatives as Potentiators for Tetracyclines

The effectiveness of the tetracycline antibiotics in most of their therapeutic and/or prophylactic applications is largely dependent upon the degree to which they are absorbed into the blood system. However, it has long been known that the tetracycline antibiotics have an affinity for metals present in the digestive system, the most prevalent being calcium. This affinity is exhibited even at low concentrations of metal, such as those normally found in the digestive tracts of warm-blooded animals. Accordingly, a substantial portion of the antibiotic is bound up in the digestive system, and its uptake is severely inhibited.

Extensive attempts have been made to overcome the difficulty associated with oral administration of the tetracycline antibiotics. Notably, the calcium content of feeds has been diminished as low as is consistent with feed manufacturing practices, and certain well-known chelating agents, most notably ethylenediaminetetraacetic acid or its sodium salt, have been evaluated in feeds containing the antibiotics. None of these prior investigations has proved satisfactory.

C. Johnston; U.S. Patent 3,863,009; January 28, 1975; assigned to The Dow Chemical Company has discovered a greatly improved method and composition for the oral administration of a tetracycline antibiotic. This method comprises administering orally to a warm-blooded animal an effective amount of a tetracycline antibiotic and a potentiating agent in an amount sufficient to potentiate (i.e., enhance) the tetracycline antibiotic. The potentiating agent to be used is a compound of the formula:

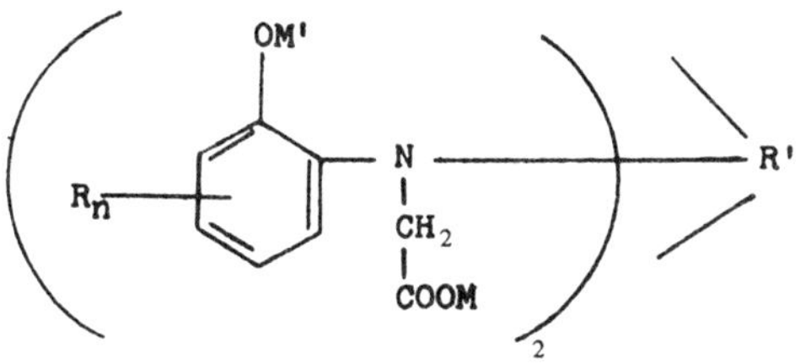

where M is hydrogen, alkali metal, or ammonium, and M being hydrogen, M' is hydrogen, or, M being alkali metal or ammonium, M' is hydrogen or the same moiety represented by M; R in each of its n occurrences independently represents halo, hydroxy, loweralkyl, or loweralkoxy; n is an integer of 0 to 2, and R' is ethylene, propylene, or trimethylene.

The terms loweralkyl and loweralkoxy designate radicals of 1 to 4 carbon atoms; the term alkali metal designates sodium and potassium, only; and the terms halide and halo designate bromine, chlorine, and iodine, only.

The products to be used as the potentiating agent are prepared in known procedures. In these procedures, an o-aminophenol is reacted with an alkylene dibromide to obtain the 2,2'-(alkylenediimino)diphenol which is then reacted with chloroacetic acid to obtain the potentiating agent wherein each of M and M' represents hydrogen. Under even slightly acidic conditions, however, this substance forms a lactone. Conveniently and preferably, therefore, the present potentiating agent is used as the tetrasodium, tetrapotassium, or tetraammonium salt; however, it is also possible to prepare and use the potentiating agent as the disodium, dipotassium, or diammonium salt. Also, it is preferred to use N,N'-ethylenebis[N-(o-hydroxyphenyl)glycine] or its tetrasodium, tetrapotassium or tetraammonium salt as the potentiating agent.

The amount of the agent which is effective to potentiate the tetracycline antibiotic varies. Generally, good results are obtained when the potentiating agent is present in the animal feed in a concentration of 0.010% (90 grams per ton) to 0.40% (3,600 grams per ton). A preferred amount of the potentiating agent is that representing a concentration of 0.10 to 0.20% of the animal feed, or, more preferred, 0.15% of the animal feed.

However, the appropriate amount varies, depending upon, among other factors, the animal concerned. With fowl, the preferred practice has been found to comprise the administration of an animal feed containing the antibiotic in a concentration of 200 grams per ton and the potentiating agent in a concentration of about 0.15%.

Example: In numerous evaluations, a modified poultry feed designated the modified basic ration was used. This modified basic ration was obtained by deleting dicalcium phosphate and ground limestone from the following standard formula for poultry feed:

	Percent of Ration
Yellow corn meal	58.65
Corn gluten meal	2.50
Soybean meal	24.75
Fish meal	6.25
Stabilized animal fat	3.75
Alfalfa meal	1.25
Limestone, ground	0.75
Dicalcium phosphate	1.25
Salt (NaCl)	0.30
Vitamin supplement	0.50
Mineral supplement	0.05
	100.00

Upon analysis, it was found that the modified basic ration contained approximately 0.39% calcium and 0.50% phosphorus. No antibiotic of any type was present in the modified basic ration.

In a first operation, two groups of young chicks, of an age of about three weeks, were used. Each group comprised approximately equal numbers of each sex. All of the chicks were in good health. One group, the control group, was fed a diet consisting of the modified basic ration to which there had been added tetracycline hydrochloride, only, in an amount to provide a concentration of 400 grams per ton of resulting feed.

The second group of chicks, the treated group, was fed a diet which consisted of the modified basic ration to which there had been added tetracycline hydrochloride as well as N,N'-ethylenebis[N-(o-hydroxyphenyl)glycine] as the tetrasodium salt. In this latter diet, the treated diet, the additives were present in a concentration of 400 grams of tetracycline hydrochloride per ton of ultimate treated diet and 0.15% of the N,N'-ethylenebis[N-(o-hydroxyphenyl)glycine], the percentage expressed as weight of potentiating agent by weight of the ultimate treated diet.

Feeding of the two groups was begun simultaneously, both groups being maintained under the same conditions, with unrestricted access to water, their conditions differing only in the composition of the diet. Feeding was continued for 5 days. At the end of this period, all of the chicks were bled, and the blood from each chick analyzed to determine the level of tetracycline antibiotic present in the blood. It was found that the average blood level of tetracycline antibiotic for the treated group of chicks was 0.90 microgram per milliliter, whereas in the control group, the tetracycline antibiotic level in the blood was found to be only 0.33 microgram per milliliter.

Glycine Formulations for Treatment of Scours in Animals

One of the most destructive and debilitating diseases affecting calves and other farm animals is scours. This is a severe form of diarrhea or dysentery which causes dehydration resulting in weight loss and frequently results in the death of the animal.

M.A. Beigler, S. Saperstein and S.S. Shastri; U.S. Patent 3,898,328; August 5, 1975; assigned to Syntex (U.S.A.) Inc. have developed a method for treating calf scours and piglet scours using an aqueous solution comprising, based on the dry weight, a major portion of glycine and glucose in combination with a minor portion of mineral salts and having an osmolality of 280 to 600 milliosmols per kilogram of water.

The rehydration transfer mixture preferably contains, based on weight percent, 18 to 22% glycine, 51 to 56% glucose, and 13 to 18% pharmaceutically acceptable electrolytes. The composition and quantity of rehydration mixture added is adjusted to provide 0.7 to 2 grams of sodium ion per liter of aqueous solution. The upper limit of sodium ion should be carefully controlled since an excess of sodium ion may cause the animal (e.g., calf) to develop hypernatremia. In addition, the transport mixture can also contain various other pharmaceutically compatible agents such as, for example, vitamins, mineral salts, buffers (e.g., sodium bicarbonate), neutral amino acids, antibiotics, or other pharmaceutically compatible agents useful for the particular animal or animals being treated. Suitable electrolyte ions which can be used include, for example, sodium, potassium, calcium, magnesium, manganese, chloride, and the like.

The particular dosage range can vary over relatively wide ranges but typically with respect to calves will be in the range of 2 to 4 liters per day per calf, and in the case of piglets will typically be in the range of one-fourth to one-half liter per day per piglet. The aqueous composition is administered orally and thus represents a further advantage over many of the other scours medicaments which are required to be administered parenterally. Typically, the composition is administered, in the case of calves, two to four times daily in equal regimens, and in the case of piglets, two times daily or ad libitum. Also, the composition can be administered in conjunction with the administration of antibiotics which are active against one or more scours-producing organisms, for example, penicillin, penicillin derivatives, tetracycline, tetracycline derivatives, nitrofurazone, nitrofurazone derivatives and the like.

Example: In this example, 50 male Holstein calves, about seven days old and weighing 105 to 130 pounds, at the onset of scours, were divided into four groups and treated for at least five days until complete cessation of scours as follows:

Group 1, containing 12 calves was fed a test rehydration composition of this process [about 180 grams (dry) daily] plus antibiotic [200 mg ampicillin trihydrate IM given b.i.d. (twice daily) with 1 oz nitrofurazone powder per os given b.i.d.] and commercial milk replacer given ad lib.

Group 2, containing 13 calves, was treated in the same manner as Group 1 but was not given milk replacer.

Group 3, containing 12 calves, was given only the test rehydration composition as in Group 1 but was not given antibiotics or milk replacer.

Group 4, containing 13 calves, was given antibiotics and milk replacer as in Group 1 but was not given the test rehydration diet.

Also, each group of calves was allowed to feed ad lib on calf feed (known as Startena). At the time scours began in each calf, fecal samples were obtained and examined for Coliform bacteria. The presence of Coliform bacteria was found in eight of the calves in Group 1; nine calves in Group 2; nine calves in Group 3; and eight calves in Group 4. The remaining calves all scoured but no positive diagnosis of a Coliform infection could be made. One calf in Group 4 died on the third day from infectious scours (Coliform). The calves were periodically examined over the test period for general appearance, dehydration, body weight and weight loss, and clinical condition at the end of five days.

The dehydration parameter was determined by an investigator by observation. The results were:

Days After Start of Scours

	1st				3rd				5th			
Group	**Sv***	**M***	**Sl***	**N***	**Sv**	**M**	**Sl**	**N**	**Sv**	**M**	**Sl**	**N**
1	2	9	1	0	0	7	5	0	0	4	5	3
2	5	7	1	0	2	9	2	0	0	2	6	5
3	2	9	1	0	1	9	2	0	0	2	6	4
4	0	11	1	0	5	7	0	0	2	10	0	0

*Sv means severe, M means moderate, Sl means slight and N means none.

ALANINE AND DERIVATIVES

L-alanine or L-2-aminopropionic acid is classified as a nonessential amino acid.

L-Alanine from Aspartic Acid Using Immobilized Microorganisms

It is known that L-aspartic acid β-decarboxylase (Enzyme Classification No. 4-1-1-12) has the ability to convert L-aspartic acid or its salt into L-alanine. To prepare L-alanine having high purity, after the enzymatic reaction is completed, the solution is boiled and/or acidified to precipitate the enzyme or an L-aspartic acid β-decarboxylase-producing microorganism and the precipitate is filtered off. Thus, the L-aspartic acid β-decarboxylase or the L-aspartic acid β-decarboxylase-producing microorganism can be used only once and must be discarded.

I. Chibata, T. Tosa, T. Sato and K. Yamamoto; U.S. Patent 3,898,128; August 5, 1975; assigned to Tanabe Seiyaku Co. Ltd., Japan have now found that an L-aspartic acid β-decarboxylase-producing microorganism, immobilized with a semipermeable membrane, can be advantageously used in the production of L-alanine.

According to the process, L-alanine can be prepared by polymerizing at least one monomer selected from acrylamide, N,N'-lower alkylenebis(acrylamide) and bis-(acrylamidomethyl) ether in aqueous suspension of an L-aspartic acid β-decarboxylase-producing microorganism to produce an immobilized L-aspartic acid β-decarboxylase-producing microorganism, and subjecting the immobilized L-aspartic acid β-decarboxylase-producing microorganism to enzymatic reaction with L-aspartic acid, DL-aspartic acid or a salt thereof.

The polymerization is preferably carried out in the presence of a polymerization initiator and a polymerization accelerator. Potassium persulfate, ammonium persulfate, vitamin B_2 and methylene blue are suitable as the polymerization initiator. On the other hand, β-(dimethylamino)propionitrile and N,N,N',N'-tetramethylethylenediamine are used as the polymerization accelerator. It is preferred to carry out the reaction at 10° to 40°C. The reaction may be completed within 10 to 60 minutes. Microorganisms which produce L-aspartic acid β-decarboxylase include *Acetobacter rancens* OUT No. 8300, *Achromobacter pestifer* IAM No. 1446, *Achromobacter pestifer* ATCC No. 23584, *Alcaligenes faecalis* ATCC No. 25094, *Pseudomonas dacunhae* IAM No. 1152.

L-alanine or a mixture of L-alanine and D-aspartic acid can be prepared by contacting the resultant immobilized microorganism with L-aspartic acid, DL-aspartic acid or a salt thereof. Suitable examples of the salts of L- and DL-aspartic acid include the ammonium, potassium, sodium salt and magnesium salts. When DL-aspartic acid or its salt is used in the enzymatic reaction, D-aspartic acid is also obtained because D-aspartic acid in DL-aspartic acid is not affected by the immobilized L-aspartic acid β-decarboxylase-producing microorganism.

Example: Pseudomonas dacunhae IAM No. 1152 is inoculated into 200 ml of a nutrient medium (pH 7.0) containing 0.5 w/v % ammonium fumarate, 1.0 w/v % sodium fumarate, 0.55 w/v % corn steep liquor, 1.8 w/v % peptones, 0.05 w/v % potassium dihydrophosphate and 0.01 w/v % magnesium sulfate·7 hydrate. The medium is cultivated at 30°C for 24 hours under shaking. After the cultivation, the microbial cells of *P. dacunhae* IAM No. 1152 are collected by centrifugation. The microbial cells are suspended in 20 ml of physiological saline solution. 3.75 g of acrylamide, 0.2 g of N,N'-methylenebis(acrylamide), 2.5 ml of 5% β-(dimethylamino)propionitrile and 2.5 ml of 2.5% potassium persulfate are added to the suspension.

Then, the suspension is allowed to stand at 25°C for 10 minutes. The immobilized microbial cells are ground into pieces and then washed with physiological saline solution. 40 g of an immobilized preparation of *P. dacunhae* IAM No. 1152 are obtained.

40 g of the immobilized preparation of *P. dacunhae* No. 1152 are charged into a 1.6 cm x 19 cm column. 500 ml of a 1 M ammonium L-aspartate aqueous solution (pH 5.5) containing 10^{-4} M concentration of pyridoxal phosphate are continuously passed through the column at 37°C at a flow rate of 6 ml/hr. 500 ml of the effluent are concentrated to 150 ml. 150 ml of methanol are added to the concentrated effluent. The crystalline precipitate thus formed is collected by filtration and then washed with 30 ml of cold methanol. 40.1 g of L-alanine are obtained, MP 270°C (dec.), $[\alpha]_D^{23}$ = +14.4° (c = 6.46, 1 N HCl).

Preparation of D-Alanine Using *Corynebacterium fascians*

D-alanine is useful as a biochemical or a starting material in the synthesis of peptides and antibacterial agents. It is also useful to promote the growth of lentils and wheat. Various methods for the fermentative production of L- or DL-alanine are known. However, no selective fermentative production of the D-alanine isomer has been reported.

In the process disclosed by *I. Chibata, S. Yamada, M. Wada, H. Maeshima and N. Izuo; U.S. Patent 3,871,959; March 18, 1975; assigned to Tanabe Seiyaku Co., Ltd., Japan,* D-alanine can be produced by cultivating a strain of *Corynebacterium fascians* in an aqueous nutrient medium under aerobic conditions, and recovering the accumulated D-alanine from the fermentation broth. The fermentation of a D-alanine-producing strain of *C.fascians*, such as *C.fascians* ATCC No. 21950, may be conducted by either shaking or submerged fermentation under aeration. It is preferred to carry out the fermentation at a pH of 5.0 to 9.0, especially at pH 7.0. When the pH of the nutrient medium is adjusted within the above range prior to the fermentation, readjustment of the pH of the medium may be unnecessary because it scarcely varies during the fermentation. The preferred temperature for the fermentation is 25° to 30°C. The aqueous nutrient medium should contain a carbon source and a nitrogen source.

Suitable sources of carbon include polyalcohols (e.g., glycerol), sugar alcohols (e.g., sorbitol) and monosaccharides (e.g., glucose, sorbose, and fructose). Inorganic ammonium salts, such as ammonium phosphate and ammonium sulfate, are suitable as the sources of nitrogen. In some cases, sodium phosphate, potassium phosphate, calcium carbonate and the like may be further added to the nutrient medium. The preferred amount of the carbon source, which is added to the medium, is within 3 to 10%. On the other hand, the preferred amount of the nitrogen source is 1 to 8%.

In carrying out the fermentation, D-alanine-productivity of the above microorganism may be enhanced by addition of corn steep liquor. For this purpose, it is preferred to add 0.05 to 0.5% of corn steep liquor to the nutrient medium. Alternatively, a mixture of an organic nitrogen source, vitamins and minerals may be used instead of corn steep liquor in order to enhance the D-alanine-productivity of the microorganism. The fermentation can be accomplished in 24 to 120 hours. D-alanine is accumulated in the fermentation broth. After the fermentation is completed, cells and other solid compositions are removed from the fermentation broth by conventional procedures such as filtration or centrifugation.

Example: A loopful of *Corynebacterium fascians* ATCC No. 21950 is inoculated into 100 ml of an aqueous nutrient medium (pH 7.0) containing 1% glycerol, 1% ammonium phosphate and 0.15% corn steep liquor. The medium is cultivated at 30°C for 2 days under shaking. A seed culture is obtained.

100 ml of an aqueous nutrient medium (pH 7.0) containing 5% glycerol, 4% ammonium phosphate and 0.4% corn steep liquor are charged into a 500 ml shaking flask, and the medium is sterilized at 120°C for 10 minutes by autoclaving. 5 ml of the seed culture are added to the medium. The medium is then cultivated at 30°C for 5 days under shaking. The fermentation medium thus obtained contains 7 mg/ml of D-alanine.

One liter of the fermentation medium is filtered by centrifugation. The supernatant solution is treated with an active carbon and then passed through a column of cation exchange resin (H-form, Amberlite IR-120). After washing with water, the column is eluted with an aqueous 2 N ammonia solution. The eluate is evaporated to dryness under reduced pressure. Ethanol is added to the residue obtained, and the crystalline precipitate is collected by filtration. The precipitate is recrystallized from an aqueous ethanol. 5.6 g of D-alanine are obtained, $[\alpha]_D^{25} = -14°$ (c = 2, 5 N HCl).

3-Fluoro-D-Alanine and Deutero Analogs

3-Fluoro-D-alanine and its deutero analogs are potent antibacterial agents valuable in inhibiting the growth of pathogenic bacteria of both the gram-positive and gram-negative type.

In the process developed by *D.F. Reinhold; U.S. Patents 3,929,576; Dec. 30, 1975; and 3,976,689; August 24, 1976; both assigned to Merck & Co., Inc.,* 3-fluoro-D-alanine and its deutero analogs are prepared from fluoropyruvic acid by asymmetric synthesis using an optically active amine.

In accordance with the process, a D-optically active amine compound, such as D-α-methylbenzylamine, R(D)-phenylglycine, 1-amino-(S)-2-[(R)-1-hydroxyethyl]-indoline and the like, are reacted with fluoropyruvic acid, preferably in solution in a lower alkanol such as ethanol. The reaction is initially conducted in the cold and is allowed to warm to room temperature under which conditions the reaction is substantially complete in 1 hour.

The resulting ketimine, such as 2-(D-α-methylbenzylimino)-3-fluoropropionic acid, 2-(D-α-carboxybenzylimino)-3-fluoropropionic acid and the like, is then reacted with hydrogen under pressure, preferably at 40 psi in the presence of a hydrogenation catalyst such as palladium on carbon; the hydrogenation is continued until uptake ceases, the catalyst is removed by filtration, and the hydrogenation product is recovered from the filtrate if desired to give N-(D-α-methylbenzyl)-3-fluoro-D-alanine, N-(D-α-carboxybenzyl)-3-fluoro-D-alanine and the like.

A solution of this N-substituted derivative in aqueous ethanol is then reacted with hydrogen at elevated pressure, e.g., 40 psi, using a hydrogenation catalyst, such as palladium hydroxide-on-charcoal, thereby hydrogenolyzing the N-substituent. After hydrogen uptake ceases, the catalyst is filtered, the filtrate is evaporated to dryness in vacuo, and the residual material is recrystallized from aqueous isopropanol to give 3-fluoro-D-alanine.

Example 1: A cold solution of 12.1 g of D-α-methylbenzylamine in 100 ml of ethanol is slowly added to a solution of 5.3 g of fluoropyruvic acid in 250 ml of ethanol, while maintaining the resulting solution at 0°C. The solution is allowed to warm to 25°C and kept at 25°C for 1 hour. The solution containing 2-(D-α-methylbenzylimino)-3-fluoropropionic acid is then reacted with hydrogen at 40 psi using 5.0 g of 10% palladium-on-carbon catalyst, the hydrogenation being continued until uptake ceases; the catalyst is removed by filtration, and the filtrate is evaporated to half-volume.

About 100 ml of water is added to form an aqueous ethanolic solution containing the N-(D-α-methylbenzyl)-3-fluoro-D-alanine, and the methylbenzyl group is hydrogenolyzed at 40 psi using 5.0 g of 10% palladium hydroxide-on-charcoal catalyst. After hydrogen uptake ceases, the catalyst is removed by filtration, the filtrate is evaporated to dryness in vacuo, and the residual material is recrystallized from 50% isopropanol-water, and then from water, to yield substantially pure 3-fluoro-D-alanine.

Example 2: A cold solution of 12.1 g of D-α-methylbenzylamine in 100 ml of ethanol is slowly added to a solution of 5.3 g of fluoropyruvic acid in 250 ml of ethanol, while maintaining the resulting solution at 0°C. The solution is allowed to warm to 25°C and kept at 25°C for 1 hour. The solution containing 2-(D-α-methylbenzylimino)-3-fluoropropionic acid is then reacted with deuterium at 40 psi using 5.0 g of 10% palladium-on-carbon catalyst, the deuteration being continued until uptake ceases; the catalyst is removed by filtration, and the filtrate is evaporated to half-volume. About 100 ml of water is added to form an aqueous ethanolic solution containing the N-(D-α-methylbenzyl)-2-deutero-3-fluoro-D-alanine, and the methylbenzyl group is hydrogenolyzed utilizing palladium hydroxide-on-charcoal catalyst. After hydrogen uptake ceases, the catalyst is removed by filtration, the filtrate is evaporated to dryness in vacuo, and the residual material is recrystallized from 50% isopropanol-water, and then from water, to yield substantially pure 2-deutero-3-fluoro-D-alanine.

Instead of using D-α-methylbenzylamine in the reaction with fluoropyruvic acid, other optically active amino compounds as mentioned above may also be used, or alternatively, an amino group may be transferred enzymatically, using a D-amino acid oxidase or a D-amino acid specific transaminase, to the fluoropyruvic acid from a D-amino acid such as D-alanine, D-2-aminobutyrate, D-proline, D-phenylalanine, D-methionine, either in their optically pure form, or in admixture with their L-counterpart. For example, 150 μg/ml of crystalline hog kidney D-amino acid oxidase are added to a solution containing 0.1 M sodium pyrophosphate buffer, pH 8.5, 40 mmol ammonium sulfate, 40 mmol sodium 3-fluoropyruvate and 8 mmol D-proline, and the system is flushed with nitrogen to exclude all oxygen.

The mixture is incubated at 25°C for 4 hours, and the enzyme is inactivated by heating at 95°C for 2 minutes. The reaction solution is diluted with isopropanol and adjusted to 4.8 by addition of aqueous ammonia; the precipitated material is recovered by filtration and dried to give 3-fluoro-D-alanine.

In the process developed by *J. Kollonitsch; U.S. Patents 3,956,367; May 11, 1976; and 3,839,170; October 1, 1974; both assigned to Merck & Co., Inc.*, 3-fluoro-D-alanine, its pharmacologically acceptable esters and its salts are prepared by fluorination of D-alanine with fluoroxyperfluoroalkanes or fluoroxypentasulfur.

This process comprises treating the substrate with a fluoroxyperfluoroalkane or fluoroxypentafluorosulfur under the influence of a free radical initiator such as light which includes ultraviolet light, ionizing radiations such as γ-rays or microwaves, or chemical chain initiators such as azo compounds, for example, azobisisobutyronitrile or combinations of such free radical initiators. The preferred mode of operation is to dissolve the substrate in a suitable solvent which is inert to the fluorination reaction such as fluorotrichloromethane or other similar halogenated alkane, or a strong acid such as liquid hydrogen fluoride, fluorosulfonic acid, trifluoroacetic acid or sulfuric acid; expose the solution to the free radical initiator; with vigorous stirring and maintenance of temperature, admit the required amount of the fluoroxy reagent slowly to the reaction mixture; and continue agitation and irradiation until reaction is complete.

Use of one of the strong acids as solvent is particularly advantageous in, but not limited to, those instances wherein the substrate carries one or more basic functional groups such as amino, or alcohol or the like. In addition, if desired, a strong acid can be used in combination with one of the nonacidic solvents. Because of the low boiling point of the reagents, it is convenient to conduct the reaction at temperatures as low as –80°C, in which case, the reaction proceeds at atmospheric pressure.

A suitable reaction vessel for atmospheric pressure reactions is one machined from a Kel-F rod equipped with an ultraviolet-transparent window. Alternatively, the reaction can be conducted in a pressure vessel, such as a Hastelloy bomb or a steel bomb with a platinum lining, in which case, higher temperatures up to 100°C can be used. In such case, the use of γ-rays or x-rays as free radical initiator is convenient, as these high energy rays penetrate the wall of the reactor. The above process can be performed by conventional batch techniques, or can be run in a continuous manner in a tubular reactor either with or without packing such as Raschig rings, saddles or the like through which the substrate or its solution and the fluorinating agent are pumped, preferably in a countercurrent fashion while being exposed to radical generating radiation.

This method of operation is particularly advantageous in cases where the particular substrate is subject to reaction with the fluoroxy compound even in the absence of radical generating conditions. In this case, by using the above continuous technique, it becomes possible to accelerate the radical type reaction, while leaving unchanged the rate of the nonradical reaction, thus raising the yield of the product formed by radical reaction.

A convenient source of radiation for radical generation was found to be a Hanovia mercury-xenon arc lamp No 9778-1, run by a 1000 W power supply. The lamp was mounted in a Schoeffel IH 15 1-N Projector equipped with a quartz condenser lens and a heat filter (water).

Example: 3-Fluoro-D-Alanine – Into a solution of 1.822 g of D(–)-alanine in 45 ml of liquid HF, 0.6 g of fluoroxytrifluoromethane gas was passed over a period of about 1 hour while being magnetically stirred, cooled in a dry ice-acetone bath and irradiated by ultraviolet light. After 80 minutes, further ultraviolet irradiation, 2 g more of fluoroxytrifluoromethane gas was passed in while being ultraviolet irradiated over 1½ hours, followed by another 1 hour ultraviolet irradiation.

The solvent was removed by blowing through it a stream of nitrogen gas. The residue was dissolved in ice-water and a sample of it was analyzed in the Spinco-Beckman amino acid analyzer, indicating a 41% yield of 3-fluoro-D-alanine, and 32% of unreacted starting material. For isolation, the mixture was chromatographed on Dowex 50X8 cation exchange resin (H^+ form, a polystyrene nuclear sulfonic acid resin). For elution, 2 N HCl was used. From the appropriate fractions, by evaporation in vacuo, pure 3-fluoro-D-alanine hydrochloride was obtained. 3-Fluoro-D-alanine was liberated from the hydrochloride in water-pyridine-isopropanol mixture, MP 166° to 168°C (dec), $[\alpha]_D = -9.3°$ (1 N HCl).

This process disclosed by *U.-H. Dolling, E.J.J. Grabowski, E.F. Schoenewaldt and M. Sletzinger; U.S. Patent 3,972,921; August 3, 1976; assigned to Merck & Co., Inc.* is concerned with the production of racemates of 2-deutero-3-fluoro-alanine and its salts. A salt of 3-fluoro-pyruvic acid, or hydrate thereof, is subjected to reductive amination with an alkali metal borodeuteride, thereby forming directly a salt of racemic 2-deutero-3-fluoro-alanine.

This reductive amination is conveniently conducted by first equilibrating a salt of fluoro-pyruvic acid hydrate (e.g., lithium 2,2-dihydroxy-3-fluoro-propionate) in aqueous ammoniacal solution with formation of the corresponding salt of 2-imino-3-fluoro-propionic acid hydrate (e.g., lithium or sodium 2-hydroxy-2-amino-3-fluoro-propionate); the equilibrium ratio of the hydrated imine to hydrated pyruvate (i.e., 2,2-dihydroxy-3-fluoro-propionate) is a function of ammonia concentration and, in concentrated aqueous ammonia (13 N), the ratio is approximately 95 to 5. Although not ordinarily preferred, lower aqueous ammonia concentrations may be used; for example, with 6.5 N aqueous ammonium hydroxide, the equilibrium ratio is approximately 90 to 10. Upon reduction, this mixture would necessarily result in a reduced yield of the equilibrium 2-deutero-3-fluoro-alanine and increased amount of 2-deutero-3-fluoro-lactate by-product. The equilibration at 37°C is a pseudo-first-order reaction with half-life of 15 minutes; equilibration at 37°C for 90 minutes provides 6 half-lives, and an effective ratio of hydrated imine to hydrated pyruvate of 95 to 5. Although the hydrated 2-imino-3-fluoro-propionate salt is not isolated from the ammonia solution, its presence, in a purity of 95%, is readily demonstrated by NMR (nuclear magnetic resonance) measurement.

The hydrated imine (the 2-hydroxy-2-amino-3-fluoro-propionate) is itself in equilibrium, by loss of water, with a minor proportion of the 2-imine (the 2-imino-3-fluoro-propionate); and the hydrated pyruvate (the 2,2-dihydroxy-3-fluoro-propionate) is likewise in equilibrium by loss of water, with a minor proportion of the 2-ketone or carbonyl (i.e., the fluoro-pyruvate or 2-keto-3-fluoro-propionate). In the alkali metal borodeuteride reduction operation, it is the 2-imine and carbonyl (not the hydrated forms) which undergo reduction. As imine and carbonyl are reduced, the hydrated imine and hydrated carbonyl are rapidly converted to the imine and carbonyl, respectively.

Example: To about 150 ml of concentrated aqueous ammonium hydroxide is added, with good agitation and at room temperature, 18.35 g of lithium fluoropyruvate hydrate. The resulting suspension is heated to 35° to 37°C (substantially all of the solid material dissolves), and the solution is maintained at that temperature for about 1.5 hours. The resulting solution which may be dark in color is cooled to 10°C, and to this is added 1.785 g of sodium borodeuteride. The resulting solution is placed under vacuum with stirring and vigorous subsurface nitrogen flow to remove dissolved ammonia.

Temperature of the solution is maintained at 10° to 13°C for about 1 hour, then gradually raised to 25°C over a 1 hour period, and held at 28° to 33°C for about 1.5 hours. The reaction solution is evaporated in vacuo at 35°C until water distills and solution is essentially free of ammonia, and the resulting solution is then acidified with 80 ml of 2.5 N aqueous hydrochloric acid solution. The acidified reaction solution is stirred with 2.5 g of activated charcoal (Darco KB) for 15 minutes and filtered.

The filtered solution is slowly passed through a column containing 850 ml of acid prewashed, strongly acidic, cation exchange resin (Dowex 50W-X4). The column is washed with deionized water until the eluate is no longer acidic (about 4 liters water required), and the column is then eluted with 0.5 N aqueous ammonium hydroxide solution. The ninhydrin-positive fractions are combined, and evaporated in vacuo at a temperature not exceeding 30°C to give about 400 ml of an ammonia-free solution.

This solution is stirred at room temperature with 2.5 g of activated charcoal; the charcoal is removed by filtration, the filtered solution is again stirred with an additional 1.5 g of activated charcoal, and the slurry is again filtered. The filtered solution is evaporated to dryness in vacuo at a temperature not exceeding 30°C to give 7.3 g of crude material. This material is dissolved in 33 ml of water at 60°C; about 27 ml of isopropanol (preheated to 60°C) is added; the aqueous isopropanol solution is seeded with crystals of 2-deutero-3-fluoro-DL-alanine; and the resulting mixture is cooled slowly first to room temperature and then to 0°C. The crystalline slurry is allowed to stand at 0°C for 1 to 2 hours, the slurry is filtered, and the crystalline material on the filter is washed with two 5-ml portions of 90% aqueous isopropanol, then with 5 ml portions of isopropanol, and finally with hexane. The washed material is dried in vacuo at a temperature of 50° to 60°C to give 5.6 g of 2-deutero-3-fluoro-DL-alanine.

Resolution of Alanine and Fluoro Derivatives

In the process disclosed by *J.M. Chemerda and G. Gal; U.S. Patent 4,048,224; September 13, 1977; assigned to Merck & Co., Inc.*, DL-alanine, 3-fluoro-DL-alanine and 2-deutero-3-fluoro-DL-alanine are resolved by reaction with a resolving base and a β-dicarbonyl compound. The resulting azomethine diastereomeric salts are easily separated. The masking group is readily removed under mildly acidic conditions.

The process uses a resolving base such as quinine, quinidine, brucine, strychnine, d- or l-α,N,N-trimethylbenzylamine or the like and a β-dicarbonyl compound of the formula

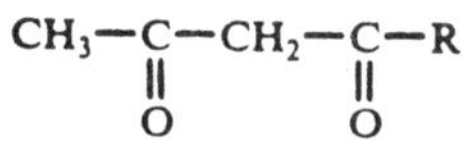

where R is methyl, phenyl, or alkoxy of 1 to 5 carbon atoms, especially ethoxy. This resolution process is of general applicability to all other α-amino acids. This reaction is conducted using approximately equimolar amounts of the three reactants. A lower alkanol of 1 to 5 carbon atoms is used as a solvent, preferably methanol. The reaction takes place at reflux of the solvent, and the reaction time is from ½ to 3 hours. The reaction mixture is then cooled to about room temperature and allowed to age for a brief time, about 15 minutes to 1 hour. The two diastereoisomeric salts are then easily separable.

Example 1: To a solution of 17.2 g quinine and 5.5 g 2,4-pentanedione in 100 ml of methanol is added 5.35 g D,L-3-fluoroalanine. The mixture is refluxed in an inert atmosphere for 60 minutes. The resulting solution is cooled to 20° to 22°C, aged for 1 hour, then the crystalline quinine salt of L-N-[1-methyl-2-acetylvinyl]-α-amino-β-fluoropropionic acid is filtered, washed with cold methanol and dried in vacuo; weight, 9.9 g.

The filtrate is concentrated in vacuo and treated with 30 ml of ethyl acetate. After keeping it at around 0°C for several hours the solid was filtered, washed with EtOAc and dried in vacuo; weight, 3.0 g. The solids are combined and dissolved in 20 ml water. To separate the quinine, 26.5 ml of 1 N sodium hydroxide solution is added slowly with stirring. The mixture is extracted four times with chloroform (at 25 ml). To the aqueous phase is added 15 ml 2 N acetic acid and 19 ml of 1 N hydrochloric acid. After 10 minutes, the cleaved 2,4-pentanedione is separated by extracting the aqueous solution with chloroform.

The acidic solution is passed through a Dowex 50WX4 (H^+) column. The column is washed with distilled water until the eluate is no longer acidic. The L-3-fluoroalanine is eluted with 0.5 N ammonium hydroxide. The total eluate is concentrated under good vacuum such that the solution temperature does not exceed 25°C. Concentration is continued until a slurry is obtained with volume of 10 ml or less. L-3-fluoroalanine is filtered, washed with a small amount of cold water and dried in vacuo; yield = 1.4 g (52.4%); MP = 165° to 167°C (dec); and $[\alpha]_D$ = +10.28° (c = 6 in 1 N HCl). The combined ethyl acetate filtrate and washes is concentrated in vacuo. The glassy residue is dissolved in 20 ml of water and the D-3-fluoroalanine is isolated by the same method as described above; yield = 1.2 g (45%); MP = 165° to 167°C (dec); and $[\alpha]_D$ = 10.21° (c = 6% in 1 N HCl).

Example 2: To a solution of 3.24 g quinine and 1.10 g 2,4-pentanedione in 30 ml of methanol is added 0.891 g D,L-alanine. The mixture is heated under reflux until a solution is formed (90 minutes). The methanol is removed in vacuo and the glass residue is recrystallized from 100 ml ethyl acetate. The yield of the quinine salt of N-[1-methyl-2-acetylvinyl]-L-α-aminopropionic acid is 2.3 g (93%). The L-alanine is liberated from this salt by the method described in Example 1. Yield of L-alanine = 382 mg (86%); MP = 297°C (dec); and $[\alpha]_D^{22}$ = 8.6 (conc = 6% in 1 N HCl).

Example 3: 100 g of 2-deutero-3-fluoro-DL-alanine are treated as described in Example 1. The following results were obtained: 2-deutero-3-fluoro-D-alanine–first crop, yield 57% $[\alpha]_D$= –10.68°; second crop, yield 7.6% $[\alpha]_D$= –10.30°; and 2-deutero-3-fluoro-L-alanine–first crop, yield 68% $[\alpha]_D$= +10.1°; second crop, yield 1.56% $[\alpha]_D$= +4.87°.

Resolution of DL-Alanine with Maleic Acid

DL-alanine is one of the amino acids which are capable of being synthesized at low cost, but the optically inactive synthetic product has little practical use whereas the optically active L-enantiomorph finds an expanding field of application in the production of polymers. Known methods for optical resolution of synthetic alanine which are based on seeding of supersaturated solutions, on differences in the solubilities of diastereoisomeric salts with optically active acids and bases, and on biological processes are not well suited for operation on an industrial scale.

The object of the disclosure by *S. Asai, H. Tazuke and H. Kageyama; U.S. Patent 3,897,484; July 29, 1975; assigned to Ajinomoto Co., Inc., Japan* is providing a method of optically resolving DL-alanine which is simple, low in cost, and without the production of waste which would be difficult to dispose of.

It has been found that mono-DL-alanine maleate, the acid addition salt of equimolecular amounts of DL-alanine and maleic acid, is more soluble in water than the optically active enantiomorphs of monoalanine maleate, and that the DL-form is a racemic mechanical mixture of the optically active forms. It is possible, therefore, to crystallize one or even both optically active forms from a supersaturated solution of the racemate by contact with seed crystals. The monoalanine maleate in all its forms is readily decomposed for separate recovery of the alanine moiety and of the maleic acid moiety, and the latter may be returned to the resolution process to form additional monoalanine maleate with DL-alanine.

Monoalanine maleate is the only salt of alanine and maleic acid that can be precipitated from a supersaturated solution of the two acids. Even in the presence of a large excess of alanine, a dialanine maleate could not be recovered in solid form from an aqueous solution. The crystals of monoalanine maleate are prismatic and grow very quickly. Because of their favorable volume-to-surface ratio, they absorb little of the mother liquor and can readily be obtained in a state of high purity. The infrared absorption spectra and X-ray powder spectra of the optically active forms and of the racemate are identical.

The optically active enantiomorphs of monoalanine maleate may be decomposed for separate recovery of the alanine and maleic acid moieties by means of ion exchange resins or by differential extraction with organic solvents. These known methods, however, are not well suited for industrial operation.

It has now been found that the monosodium, monocalcium, hemicalcium, and monobarium salts of maleic acid are only sparingly soluble in aqueous solutions of suitable pH, and may thus be precipitated from solutions of the optically active monoalanine maleates by the addition of the hydroxides of sodium, calcium, and barium or other sources of sodium, calcium, or barium ions at a pH at which these maleates in the solid state are stable in contact with the solution as will presently become apparent. The mother liquor yields optically active alanine most economically by crystallization at pH of approximately 7 to 8 after partial evaporation of the water, but other recovery methods may be resorted to in an obvious manner since the liquor is practically free of other compounds except small amounts of maleic acid or its salts.

Example 1: A mixture of 62.4 g (0.7 mol) DL-alanine, 81.1 g (0.7 mol) maleic acid, and 70 ml water was heated to form a homogeneous solution which was then adjusted to 49°C and stirred while 10 g mono-L-alanine maleate in the form of crystals larger than 24 mesh was added. The mixture was cooled at a rate of 1°C per 10 minutes. When the temperature reached 47°C, 2 g mono-D-alanine maleate was added with continuing stirring in the form of a fine powder. Cooling was continued at the same rate to 30°C, and this temperature was maintained for about 1 hour until no further crystallization took place, the solution having become saturated.

The precipitated crystals were separated from the mother liquor, washed with methanol and ether, and then dried. They weighed 53 g and were fractionated by means of a 48 mesh screen. The retained crystals weighed 32.3 g and had a specific rotatory power $[\alpha]_D = +6.46°$ (c = 5, 1 N HCl) corresponding to mono-L-alanine maleate 81.6% optically pure. The crystals passing through the screen weighed 20.6 g and were found to be mono-D-alanine maleate having an optical purity of 76.7%.

Example 2: A solution of 250 g mono-L-alanine maleate in 660 g water was stirred with 90 g calcium hydroxide at 25°C to precipitate 190 g monocalcium maleate monohydrate of the formula $Ca(C_4H_2O_4)\cdot H_2O$. The mixture was filtered; 10 ml 35% hydrochloric acid was added to the filtrate which was thereafter partly evaporated to produce two crops of L-alanine crystals jointly weighing 99 g (91% yield).

VALINE

L-2-amino-3-methylbutanoic acid or L-valine is one of the essential amino acids not synthesized by the human body.

Production Using Mutants Resistant to 2-TA

T. Tsuchida and F. Yoshinaga; U.S. Patent 3,893,888; July 8, 1975; assigned to Ajinomoto Co., Inc., Japan have found that some bacteria having a resistance to 2-thiazolalanine (2-TA) produce a large amount of L-valine when cultured in a nutrient medium. Also, a mutant strain having a resistance to 2-TA together with a certain nutrient requirement has a better ability to produce L-valine than a strain having only resistance to 2-TA or having only a certain nutrient requirement. The required nutrient may be leucine, isoleucine, threonine, or isoleucine and methionine, and the other reagents α-amino-β-hydroxy valeric acid and ethionine. Representative strains useful for the process include:

(1) 2-TA resistant mutants such as *Corynebacterium glutamicum* FERM P-1768 whose parent strain is *C. glutamicum* ATCC 13032, *C. acetoacidophilum* FERM P-1314 whose parent strain is *C. acetoacidophilum* ATCC 13870 and *B. lactofermentum* FERM P-1945 whose parent strain is *B. lactofermentum* ATCC 13869.

(2) Mutants having nutrient requirement in addition to 2-TA resistance such as *C. glutamicum* FERM P-1967 (leucine requiring), FERM P-1968 (threonine requiring) and FERM P-1966 (isoleucine requiring) which were obtained by a conventional artificial mutant inducing method from *C. glutamicum* FERM P-1768 (2-TA resistant mutant), *B. lactofermentum* FERM P-1964 (leucine requiring), FERM P-1963 (isoleucine requiring) and FERM P-1965 (threonine requiring) which were obtained by a conventional artificial mutant inducing method from *B. lactofermentum* FERM P-1945 (2-TA resistant mutant), and *B. lactofermentum* FERM P-1845 (isoleucine plus methionine requiring mutant having resistance to 2-TA) which were induced from *B. lactofermentum* FERM P-1858 (isoleucine plus methionine requiring mutant).

The method of obtaining mutant strains used in the process is a conventional artificial mutant inducing method such as by use of nitrosoguanidine.

Example: A culture medium whose composition is shown hereunder was prepared, and 20 ml batches of the medium were each placed in a 500 ml shaking flask.

Composition of the medium:	
Glucose, %	10
$(NH_4)_2SO_4$, %	4
KH_2PO_4, %	0.1
$MgSO_4 \cdot 7H_2O$, %	0.04
Fe^{++}, ppm	2
Mn^{++}, ppm	2
Biotin, γ/l	50
Thiamine HCl, γ/l	300
Soybean protein hydrolyzate, ml/dl	1
$CaCO_3$ (sterilized separately), %	5
(pH 7.0 adjusted by KOH)	

The eight strains shown in the following table, which had previously been cultured on respective bouillon slants at 30°C for 24 hours, were introduced respectively into each medium supplemented with a required amino acid in an amount shown in the following table, and cultivation was carried out at 31.5°C for 72 hours. L-valine produced in each cultured broth was determined by bioassay and shown in the following table.

Strains Employed*	Amino Acids Required: Amino Acid	Amino Acids Required: Amount (mg/dl)	L-Valine Produced (g/dl)
Brevibacterium lactofermentum			
FERM P-1945	–	–	1.50
FERM P-1963	Isoleucine	15	2.51
FERM P-1964	Leucine	30	2.05
FERM P-1965	Threonine	30	1.96
Corynebacterium glutamicum			
FERM P-1768	–	–	1.20
FERM P-1966	Isoleucine	15	2.30
FERM P-1967	Leucine	30	1.70
FERM P-1968	Threonine	30	1.51

*All strains have a resistance to 2-TA.

One liter each of the broth of the strain FERM P-1964 and strain FERM P-1966 was centrifuged to remove bacterial cells, each supernatant was passed through a column packed with a cation exchange resin, and after washing with water, L-valine was eluted. From the broths, 9.3 g and 11.2 g of crystalline L-valine were recovered, respectively.

LEUCINE

L-Leucine (2-amino-4-methylvaleric acid or 2-amino-4-methylpentanoic acid) is one of the essential amino acids, and has been used in food additives. It has been customary heretofore to hydrolyze natural proteins and to isolate L-leucine from the resulting mixture of amino acids by fractionation.

Production with Brevibacterium and Corynebacterium Mutants

Some bacteria belonging to the genera Brevibacterium and Corynebacterium, and resistant to a leucine antagonist have been found by *S. Okumura, F. Yoshinaga, K. Kubota and H. Kamijo; U.S. Patent 3,865,690; February 11, 1975; assigned to Ajinomoto Co., Inc., Japan* to produce a large amount of leucine. The L-leucine producing mutants may be isolated from natural sources, or can be derived by conventional mutant inducing procedures, such as X-ray radiation, ultraviolet light irradiation, or treatment with nitrosoguanidines, diethylsulfate or nitrite.

L-leucine producing bacteria include *Brevibacterium flavum* AJ-3226 (FERM P-420) which has been derived from *B. flavum* ATCC 14067, *B. lactofermentum* AJ-3427 (FERM P-1769) which has been derived from *B. lactofermentum* ATCC 13869, *Corynebacterium acetoacidophilum* AJ-3228 (FERM P-421) which has been derived from *C. acetoacidophilum* ATCC 13870, and *C. glutamicum* AJ-3426 (FERM P-1768) which has been derived from *Micrococcus glutamicus* ATCC 13032 (*Micrococcus glutamicus* is now referred to as *C. glutamicum*).

Leucine antagonists include 2-thiazolealanine, 4-azaleucine, 5',5',5'-trifluoroleucine, D-leucine, α-amino-isoamylsulfonic acid, norvaline, norleucine, methallylglycine, α-amino-β-chlorobutyric acid, δ-chloroleucine, β-hydroxynorleucine, β-hydroxyleucine, cyclopentanealanine, 3-cyclopentene-1-alanine, 2-amino-4-methylhexanoic acid.

The following table shows the growth of bacteria when cultured on media containing varying amounts of 2-thiazolealanine or 4-azaleucine.

		ATCC 14067		AJ-3226		ATCC 13870		AJ-3228	
		OD	Specific turbidity	OD	Specific turbidity	OD	Specific turbidity	OD	Specific turbidity
Initial suspension*		0.075		0.070		0.115		0.108	
Additive									
2-TA**	None	0.640	100	0.425	100	0.560	100	0.465	100
	50	0.440	69	0.434	102	0.285	51	0.450	97.
	100	0.325	51	0.420	99	0.170	30	0.400	86
	250	0.228	36	0.465	109	0.120	21	0.345	74
	500	0.185	29	0.490	115	0.099	18	0.300	65
	1000	0.088	14	0.450	106	0.082	15	0.264	57
	2000	0.043	7	0.365	86	0.048	9	0.265	57
4-AL***	50	0.415	65	0.430	101	0.320	57	0.445	96
	100	0.310	48	0.420	99	0.205	37	0.440	95
	250	0.211	33	0.465	109	0.145	26	0.485	104
	500	0.160	25	0.455	107	0.100	18	0.485	104
	1000	0.103	16	0.470	111	0.050	9	0.495	106
	2000	0.077	12	0.470	111	0.037	7	0.515	111

*Optical density of initial suspension was determined for suspension diluted to 26 times initial volume

**2-Thiazolealanine ***4-Azaleucine

The above table shows that *B. flavum* AJ-3226 and *C. acetoacidophilum* AJ-3228, strains resistant to 2-thiazolealanine and 4-azaleucine, can grow on media containing 2-thiazolealanine or 4-azaleucine, by which growth of respective parent strains *B. flavum* ATCC 14067 and *C. acetoacidophilum* ATCC 13870 is inhibited.

The culture medium used to produce L-leucine may be conventional. It includes a carbon and nitrogen source, both assimilable, and the usual minor nutrients. The fermentation of the process is performed at a pH between 5 and 9, at 24° to 37°C under aerobic conditions for 2 to 7 days. The pH of the culture medium can be adjusted by adding sterile calcium carbonate, aqueous or gaseous ammonia, mineral acid or organic acid during the fermentation.

Example 1: A culture medium containing 10 g/dl glucose, 0.1 g/dl KH_2PO_4, 0.04 g/dl $MgSO_4 \cdot 7H_2O$, 4 g/dl $(NH_4)_2SO_4$, 100 γ/l biotin, 200 γ/l vitamin $B_1 \cdot HCl$, 2 ppm Fe and Mn ions, 1 ml/dl Aji-Eki (soybean protein hydrolyzate) and 5 g/dl $CaCO_3$ of pH 7.0 was prepared; 300 ml batches of the medium were each placed in a small glass jar-fermentor, and sterilized. The medium was inoculated with *Brevibacterium flavum* AJ-3226 (FERM P-420) which had previously been cultured on a bouillon slant at 30°C for 24 hours, and cultured at 31°C for 48 hours with stirring and aerating.

The culture broth was found to contain 1.11 g/dl of L-leucine. One liter of the broth was centrifuged to remove bacterial cells, the supernatant was passed through a column packed with a cation exchange resin (Doulite C x 20), and after washing with water, L-leucine was eluted with 1 N NH_4OH solution. The eluate was concentrated to remove ammonia, treated with active charcoal, and pure crystalline L-leucine was obtained in an amount of 5.2 g.

Example 2: Corynebacterium acetoacidophilum AJ-3228 (FERM P-421) was inoculated on a culture medium containing 10 g/dl sucrose, 0.1 g/dl KH_2PO_4, 0.04 g/dl $MgSO_4 \cdot 7H_2O$, 4 g/dl $(NH_4)_2SO_4$, 200 γ/l biotin, 200 γ/l vitamin $B_1 \cdot HCl$, 2 ppm Fe and Mn ions and 5 g/dl $CaCO_3$ of pH 7.5 and cultured at 30°C for 72 hours with shaking. The culture broth was found to contain 0.66 g/dl of L-leucine.

Example 3: *Brevibacterium lactofermentum* AJ-3427 (FERM P-1769) and *Corynebacterium acetoacidophilum* AJ-3426 (FERM P-1768) also produced L-leucine in a similar manner.

T. Tsuchida, H. Momose and Y. Hirose; U.S. Patent 3,970,519; July 20, 1976; assigned to Ajinomoto Co., Inc., Japan have also found that organisms which are characterized by requiring at least one of isoleucine, threonine or methionine as a growth nutriment, and that are resistant to feedback inhibition by leucine and analogs of leucine are useful for producing L-leucine.

Particularly suitable mutant strains for use in the process are *Brevibacterium lactofermentum* AJ-3718 (FERM P-2516) which is resistant to 2-thiazolealanine and β-hydroxyleucine and requires isoleucine and methionine, *B. lactofermentum* AJ-3452 (FERM P-1965) which is resistant to 2-thiazolealanine and β-hydroxyleucine and requires threonine, and *B. lactofermentum* AJ-3719 (FERM P-2517) which is resistant to 2-thiazolealanine and β-hydroxyleucine and requires isoleucine and methionine. These three mutants were all derived from the parent strain *B. lactofermentum* ATCC 13869.

Corynebacterium glutamicum AJ-3453 (FERM P-1966) which is resistant to 2-thiazolealanine and β-hydroxyleucine and requires isoleucine, *C. glutamicum* AJ-3455 (FERM P-1968) which is resistant to 2-thiazolealamine and β-hydroxyleucine and requires threonine, and *C. glutamicum* AJ-3720 (FERM P-2518) which is resistant to 2-thiazolealeanine and β-hydroxyleucine and requires isoleucine (all of which were derived from *C. glutamicum* ATCC 13032) are also particularly useful in this process.

Example: A culture medium containing 10% glucose, 4% $(NH_4)_2SO_4$, 0.1% KH_2PO_4, 0.04% $MgSO_4 \cdot 7H_2O$, 2 ppm Fe^{++}, 2 ppm Mn^{++}, 100 γ/l biotin, 350 γ/l thiamine hydrochloride, 2 ml/dl soybean protein hydrolysate, and 5% $CaCO_3$ (sterilized separately) at pH 7.0 was prepared; 20 ml aliquots of the medium were placed in 500 ml shaking flasks, and autoclaved at 100°C for 5 minutes. The media were supplemented with the required amino acids and thereafter inoculated with the strains shown in the table below.

Strains employed	Resistance	Amino Acids Required: Amino Acids	Amino Acids Required: Amount (mg/dl)	L-leucine Produced (g/dl)
Brevibacterium lactofermentum FERM P-1841	—	Threonine	20	0
Brevibacterium lactofermentum FERM P-1858	—	Isoleucine & Methionine	10 10	0
Brevibacterium lactofermentum FERM P-2516	2-TA, β-HL	Isoleucine & Methionine	10 10	2.21
Brevibacterium lactofermentum FERM P-1965	2-TA, β-HL	Threonine	20	2.10
Brevibacterium lactofermentum FERM P-2517	2-TA, β-HL	Isoleucine & Methionine	20 30	2.20

(continued)

Strains employed	Resis-tance	Amino Acids Required: Amino Acids	Amino Acids Required: Amount (mg/dl)	L-leucine Produced (g/dl)
Coryne. glutamicum ATCC 13032	—	Isoleucine	—	0
Coryne. glutamicum ATCC 14296	—	Threonine	—	0
Coryne. glutamicum FERM P-1966	2-TA, β-HL	Isoleucine	15	1.80
Coryne. glutamicum FERM P-1968	2-TA, β-HL	Threonine	15	1.70
Coryne. glutamicum FERM P-2518	2-TA, β-HL	Isoleucine	15	1.95

Note: 2-TA is 2-thiazolealanine; β-HL is β-hydroxyleucine

After 72 hours' cultivation at 30°C, each culture broth was found to contain L-leucine in amounts shown.

Separation of Leucine and Isoleucine

One conventional process for separating leucine from isoleucine is based on the fact that one of the two complexes which they form with a copper salt lending itself to complex formation has zero solubility, while the other shows low solubility in methanol. It is thus possible to isolate a pure, natural isoleucine by a series of successive crystallization and extraction operations. The number of repetitions of these operations is high and the yield very low.

An improved process for separating leucine and isoleucine from an aqueous solution of these two amino acids has been disclosed by *P. Hirsbrunner and R. Bertholet; U.S. Patent 3,960,942; June 1, 1976; assigned to Societe D'Assistance Technique pour Produits Nestle SA, Switzerland.* The process comprises the fractional crystallization of the copper or nickel salts of these amino acids at, alternatively, relatively low and relatively high pH values in the acid range.

It is desirable for the leucine and isoleucine to be present in the starting solution in a ratio by weight of at most 7:1, because separation becomes very difficult where the ratio is higher. The bases and acids, which may be used for pH-adjustment, are selected from those which do not form complexes with the metal ion used to form the leucinate or isoleucinate.

Preferably, copper sulfate or nickel sulfate is added to the starting solution and the pH-adjustments made by adding sulfuric acid or sodium hydroxide. Fractional crystallization may be carried out in two successive cycles of two crystallizations each for a starting solution in which the leucine and isoleucine are present in a ratio by weight of greater than 1:1. pH-values recommended for the so-called lower crystallization pH-values are in the range from 2 to 3.8.

Once actual separation has been obtained, all that remains is to free the amino acid molecules from the metal ions attached to them, which may be obtained by a high-temperature alkaline treatment for example.

Example: 96 g of $CuSO_4 \cdot 5H_2O$ are added to a solution of 80 g of leucine and 20 g of isoleucine in 1,000 g of water. The pH-value is adjusted to 1.5 with 50% sulfuric acid. 30% sodium hydroxide is added to the first, clear blue-colored solution obtained to adjust its pH-value to 2.5. 87 g of crystallized $Cu(Leu)_2$ are retained by filtering on an ordinary paper filter and collected.

The pH-value of the filtrate is adjusted to pH 7 by the addition of 30% NaOH. 33 g of a mixture of 50% of $Cu(Leu)_2$ and 50% of $Cu(Ileu)_2$ are retained by filtering on an ordinary paper filter. 20 g of water and 50% sulfuric acid are added to this mixture to obtain a second, clear blue-colored solution with a pH-value of 1.5. 30% sodium hydroxide is added to this second blue solution to adjust its pH-value to pH 3. 19 g of a mixture of 80% of $Cu(Leu)_2$ are retained by filtering on an ordinary paper filter and collected. The pH-value of the filtrate is adjusted to pH 7 by the addition of 30% NaOH. 12 g of a mixture of 15% of $Cu(Leu)_2$, and 85% of $Cu(Ileu)_2$, namely copper isoleucinate with a degree of purity of 85%, are collected by filtration on an ordinary paper filter.

The 87 g of $Cu(Leu)_2$ and the 12 g of $Cu(Ileu)_2$ are separately subjected to an identical treatment for separating the copper which is carried out as follows: 300 parts of $Cu(Leu)_2$ or $Cu(Ileu)_2$ mixed with 400 parts of 25% NaOH are held for 1 hour at 85°C. After stirring, the copper oxide is separated by filtration. The filtrate is neutralized with concentrated hydrochloric acid. The amino acids are collected by filtration, washed and dried.

In view of the fact that the 19 g of mixture of 80% of $Cu(Leu)_2$ and 20% of $Cu(Ileu)_2$ may be reused in a starting solution, i.e., recycled, the total yield of leucines is 95%, the losses totalling 5%. The yield of leucine is 98% and the yield of isoleucine is 78%.

ISOLEUCINE

L-isoleucine, or 2-amino-3-methylvaleric acid, one of the essential amino acids, has been used as an active ingredient of medicinal preparations and as a food additive. Known chemical synthesis of this amino acid is disadvantageous in that four stereoisomers of isoleucine are formed during reactions, and complicated procedures are required to separate biologically active L-isoleucine from the mixture of the isomers. It has been recognized, therefore, that a fermentative method is most advantageous and practical in the industrial preparation of L-isoleucine.

Fermentative Production Using *Serratia marcescens*

I. Chibata, M. Kisumi, S. Komatsubara and M. Sugiura; U.S. Patent 3,841,968; October 15, 1974; assigned to Tanabe Seiyaku Co., Ltd., Japan have found that a mutant of *Serratia marcescens* resistant to isoleucine hydroxamate or resistant to both isoleucine hydroxamate and α-aminobutyric acid has an excellent productivity of L-isoleucine from a normal source of carbon. Also, the L-isoleucine productivity of this mutant can be remarkably enhanced by carrying out the fermentation in the presence of L-threonine, L-homoserine or L-aspartic acid. According to the process, L-isoleucine is prepared by cultivating an isoleucine hydroxamate-resistant mutant or an isoleucine hydroxamate- and α-aminobutyric acid-resistant mutant of *S. marcescens* in a nutrient medium under aerobic conditions. This isoleucine hydroxamate-resistant mutant may be obtained by ultraviolet irradiation of a wild type strain of *S. marcescens* or by treating the wild type strain with a mutagen. For instance, a wild type strain of *S. marcescens* is treated with N-methyl-N'-nitro-N-nitrosoguanidine and then cultured at 30°C for 3 to 5 days on agar plates containing the following compositions: K_2HPO_4, 0.7 w/v percent; KH_2PO_4, 0.3 w/v percent; $MgSO_4 \cdot 7H_2O$, 0.01 w/v percent; $(NH_4)_2SO_4$, 0.1 w/v percent; glucose, 0.5 w/v percent; L-isoleucine hydroxamate, 1 mg/ml. The isoleucine hydroxamate-resistant mutant of *S. marcescens* may be isolated as large colonies.

Alternatively, the isoleucine hydroxamate- and α-aminobutyric acid-resistant mutant may be isolated as large colonies when the isoleucine hydroxamate-resistant mutant is cultured at 30°C for 2 to 3 days on the agar plates containing the same ingredients as mentioned above except that 10 mg/ml of α-aminobutyric acid is employed instead of isoleucine hydroxamate. A viable culture of an isoleucine hydroxamate-resistant mutant of *Serratia marcescens* has been deposited as ATCC No. 21741, and a viable culture of an isoleucine hydroxamate- and α-aminobutyric acid-resistant mutant of *Serratia marcescens* has also been deposited under ATCC No. 21740 and ATCC No. 21810.

The fermentation of the mutant of *S. marcescens* can be accomplished by either shaking cultivation or submerged fermentation under aerobic conditions. The fermentation may be preferably carried out at 25° to 37°C and at a pH of 6 to 9. Calcium carbonate and ammonia may be employed for adjustment of the pH of the medium.

L-isoleucine productivity of the above mutants may be further enhanced by addition of about 0.05 to 5 w/v of L-threonine, L-homoserine or L-aspartic acid. The fermentation can be accomplished in about 24 to 96 hours. L-isoleucine is accumulated in the fermentation broth.

This can be illustrated by culturing *Serratia marcescens* OUT 8259, the isoleucine hydroxamate-resistant mutant ATCC No. 21741 of *Serratia marcescens* and the isoleucine hydroxamate- and α-aminobutyric acid-resistant mutant ATCC No. 21740 of *Serratia marcescens* for 48 hours at 30°C in a nutrient medium (pH 7.0). When 1.0 w/v percent of L-threonine, L-homoserine or L-aspartic acid is added, the amounts of L-isoleucine accumulated in the medium are shown in the table below.

	Strains		
	OUT 8259	ATCC 21741	ATCC 21740
Amino Acid Added	mg/ml		
No addition	0	0.1	2.5
L-threonine	0	4.0	7.0
L-homoserine	0	2.0	5.0
L-aspartic acid	0	1.5	3.5

Example: An aqueous nutrient medium comprising the following ingredients is prepared: L-threonine, 2 w/v percent; glucose, 2 w/v percent; dextrin, 10 w/v percent; urea, 0.5 w/v percent; dibasic potassium phosphate, 0.1 w/v percent; magnesium sulfate, 0.05 w/v percent; and calcium carbonate, 2 w/v percent.

The above medium is adjusted to pH 7.0. 15 ml of the medium are charged into a 500 ml shaking flask and its contents are sterilized by autoclaving. A loopful of the isoleucine hydroxamate-resistant mutant ATCC No. 21741 of *Serratia marcescens* is inoculated aseptically into the medium. Then, the medium is cultivated for 48 hours at 30°C under shaking. The fermentation medium thus obtained contains 7 mg/ml of L-isoleucine.

1,000 ml of the fermentation medium are heated at 100°C for 10 minutes and then filtered. The filtrate is introduced into a column (3 cm x 45 cm) of strong cation exchange resin (H-form, Amberlite IR-120). After washing with water, the column is eluted with 5% aqueous ammonia. The fractions containing L-isoleucine are collected and concentrated under reduced pressure to about 80 ml. 100 ml of aqueous methanol are added to the solution. The precipitating crystals are collected by filtration and then recrystallized from aqueous methanol. 5 g of L-isoleucine are obtained, $[\alpha]_D^{20}$ = +40.1° (c = 1, 6 N HCl).

HYDROXYALIPHATIC AMINO ACIDS

Serine (2-amino-3-hydroxypropionic acid) and threonine (2-amino-3-hydroxy-butyric acid) are the two amino acids in this category. Although serine is classified as nonessential, threonine is one of the essential amino acids that is not synthesized in the human body. Of the four steroisomers of threonine only L-threonine is capable of supporting growth of the rat. Only processes regarding the production and use of serine have appeared in the recent U.S. patent literature.

SERINE

Production Using Culture Media Containing Threonine or Homoserine

It is known that L-serine may be produced with bacteria belonging to the genus Arthrobacter, the genus Brevibacterium or the genus Corynebacterium. It has been discovered by *K. Nakayama, H. Kase and H. Yoshida; U.S. Patent 3,795,583; March 5, 1974; assigned to Kyowa Hakko Kogyo KK, Japan* that the amount of L-serine accumulated in a culture medium can be increased with addition of L-threonine or L-homoserine to the medium. Concentrations of preferably 4 to 50 mg/ml of these compounds are used.

Either a synthetic culture medium or a natural nutrient medium is suitable for cultivation of the strains used in this process as long as it contains the essential nutrients for the growth of the strain employed. Such nutrients are well known and include a carbon source, a nitrogen source and inorganic compounds. In accordance with this process preferably more than 4 mg/ml of L-threonine or L-homoserine are further added to the culture medium to improve yields. Culturing is conducted under aerobic conditions, such as aerobic shaking of the culture or with aeration and agitation of a submerged culture, at a temperature of 20° to 40°C and at a pH of 4.0 to 9.5.

After the completion of culturing, the microorganism cells and undesirable precipitates are removed from the culture liquor and the L-serine is recovered from the liquor by conventional means, such as an ion exchange resin treatment, extraction with solvents, precipitation, adsorption, chromatography or concentration. An especially suitable method for recovery is the ion exchange resin treatment.

Example: L-serine-producing strain ATCC 21218 (isoleucine- and methionine-requiring) which belongs to *Arthrobacter paraffineus* was used as seed. The seed strain was inoculated into 20 ml of a sterilized culture medium (pH 7.4) in a 250 ml Ehrlenmeyer flask. The medium was composed of 2% sorbitol, 1% of beef extract, 1% of peptone, 0.5% of yeast extract and 0.3% of NaCl. The seed was cultured at 30°C for 24 hours with shaking, and the resultant seed culture (1 ml) was then inoculated into 10 ml of a sterilized fermentation medium in a large test tube. The seed was then further cultured at 30°C for 96 hours with shaking.

For the fermentation, a medium (pH 7.4) having a composition of 10% of n-alkane mixture ($C_{11}-C_{14}$), 2% of $(NH_4)_2SO_4$, 0.1% of K_2HPO_4, 0.1% KH_2PO_4, 0.1% of $MgSO_4 \cdot 7H_2O$, 2% of $CaCO_3$, 1 mg/ml of thiamine, 100 mg/l of L-methionine and 1 ml/l of a trace elements solution (composition: 83 mg/l of $Na_2B_4O_7 \cdot 10H_2O$, 37 mg/l of $(NH_4)_6Mo_7O_{24} \cdot 4H_2O$, 970 mg/l of $FeCl_3 \cdot 6H_2O$, 8.8 mg/l of $ZnSO_4 \cdot 7H_2O$, 20 mg/l of $CuSO_4 \cdot 5H_2O$ and 7.2 mg/l of $MnCl_2 \cdot 4H_2O$) was used as a control. L-threonine and L-homoserine in different amounts as shown in the table below were respectively added to the control medium. The amounts of L-serine produced using these media are shown below.

L-Threonine Added, mg/ml	L-Serine Produced, mg/ml
10	5.2
5	5.0
1	3.1
0 (control)	2.3
10	4.8
5	4.2
1	2.9
0 (control)	2.3

This example shows that the presence of L-threonine and L-homoserine enhanced the production of L-serine, with a range of about 4 to 10 mg/ml giving particularly good results.

Other cultures which may be used to produce L-serine by this method include: *Brevibacterium ketoglutamicum,* ATCC 21222; *Corynebacterium hydrocarboclastus,* ATCC 21221; and *Corynebacterium glutamicum,* ATCC 21586.

Production Using Artificially Induced Mutants

Artificially induced mutants of *Brevibacterium lactofermentum* and *Corynebacterium acetoacidophilum* are used by *K. Kubota, T. Onoda and S. Okumura; U.S. Patent 3,843,441; October 22, 1974; assigned to Ajinomoto Co., Inc., Japan* for the production of L-serine.

These sulfaguanidine-resistant mutants of parent strains of Brevibacterium and Corynebacterium produce serine from inexpensive carbohydrates, acetic acid or ethanol in an extremely high yield when they are cultured in an aqueous culture medium. The method of mutation is conventional, such as exposure of cells to ultraviolet light, x-rays, or gamma rays in mutagenic doses, or to sodium nitrite, nitrosoguanidine, or diethylsulfate solution in a conventional manner. The mutants resistant to sulfaguanidine are selected from the exposed parent strains by conventional methods. The exposed parent strains are cultured on an agar medium or in an aqueous medium containing enough sulfaguanidine to inhibit growth of the parent strains. Then, the mutants resistant to the sulfaguanidine are screened.

Effective serine-producing mutants include *B. lactofermentum* AJ 3360 (FERM-P 1371) and *C. acetoacidophilum* O-472 (FERM-P 1372). The respective parent strains are *B. lactofermentum* ATCC 13869 and *C. acetoacidophilum* ATCC 13870. They are incapable of producing serine in a culture medium. The nutrient media fermented by means of these mutants are conventional and contain at least one carbon source selected from carbohydrates, acetic acid and ethanol, assimilable sources of nitrogen, inorganic salts and other standard additives.

For a good yield of serine, the fermentation is carried out aerobically with aeration and/or agitation. Best yield requires pH control within 5 to 9. The desired pH may be maintained by adding gaseous or aqueous ammonia, calcium carbonate, alkali metal hydroxide, urea, organic or inorganic acids to the medium from time to time, some of which may also supply assimilable nitrogen. When the fermentation is carried out at 24° to 37°C, the concentration of serine in the broth reaches its maximum within 2 to 7 days.

The serine accumulated in the fermentation broth can be recovered by conventional methods, such as removing cells by filtration or centrifugation, passing the broth over an ion exchange resin and precipitation at the isoelectric point of serine.

Example: B. lactofermentum AJ-3360 was cultured at 31.5°C for 16 hours with shaking in an aqueous medium containing 3 g/dl glucose, 0.3 g/dl urea, 0.1 g/dl KH_2PO_4, 0.04 g/dl $MgSO_4 \cdot 7H_2O$, 0.2 mg/dl ferrous ion, 0.2 mg/dl manganese ion, 50 μg/l biotin, 200 μg/l thiamine·HCl, 1 ml/dl soy protein acid hydrolyzate and 0.2 g/dl yeast extract.

An aqueous fermentation medium was prepared to contain 10 g/dl glucose, 5 g/dl $(NH_4)_2SO_4$, 0.1 g/dl KH_2PO_4, 0.04 g/dl $MgSO_4 \cdot 7H_2O$, 0.2 mg/dl ferrous ion, 0.2 mg/dl manganese ion, 50 μg/l biotin, 100 μg/l thiamine·HCl, 0.3 g/dl soy protein acid hydrolyzate and 5 g/dl $CaCO_3$ and adjusted to pH 7.0. 300 ml batches of the fermentation medium were placed in 1,000 ml fermentation vessels, sterilized by steam, and inoculated with 15 ml of the cultured broth mentioned above. Fermentation was carried out aerobically at 31.5°C for 55 hours. After the cultivation, 0.20 g/dl serine was found in the fermentation broth.

Two liters of combined fermentation broth were passed through Amberlite IR-45 (OH-type) after separating cells by centrifuging. The eluate was concentrated, mixed with 2.5 volumes of methanol, and cooled. 3.5 g of crude crystals of serine were obtained.

C. acetoacidophilum O-472 was cultured by the above method and 0.22 g/dl serine was found in the fermentation broth.

Use of Mutants of *C. glycinophilum*

Artificially induced mutants of *Corynebacterium glycinophilum* ATCC 21341 are used by *K. Kageyama, I. Maeyashiki, K. Kubota, S. Konishi and S. Okumura; U.S. Patent 3,880,741; April 29, 1975; assigned to Ajinomoto Co., Inc., Japan* to produce L-serine from glycine. These mutants require leucine, isoleucine, methionine, tryptophan, serine or glycine for their growth.

The mutants are derived from *C. glycinophilum* by means of any conventional mutagenic agent, as by treating cells of the parent strain with 0.1 M diethyl sulfate solution at 30°C for 20 hours, screening the treated cells by the replication method, and examining them for their ability to produce L-serine. The five best strains of *C. glycinophilum* obtained so far have been deposited with accession numbers as listed below:

FERM-P 1685 (leucine)
FERM-P 1686 (tryptophan)
FERM-P 1687 (leucine + methionine)
FERM-P 1688 (leucine + isoleucine)
FERM-P 1689 (leucine + serine or glycine)

The culture media on which the mutant strains of this process are cultured may be entirely conventional aside from the required specific nutrients. The glycine concentration in the medium is preferably 0.5 to 4 g/dl, and glycine may be added to the medium before inoculation, early during culturing, or gradually.

Microbial fermentation of glycine is preferably carried out under aerobic conditions maintained by shaking, stirring, or by aeration. The pH should be held between 5.0 and 9.0 by adding acids, alkali hydroxide or carbonates of alkali or alkaline earth metals in a known manner. The preferred cultivation temperature is 24° to 37°C. Maximum serine concentration is reached within two to five days, depending on process variables. The cells may also be harvested from the broth and dispersed as an enzyme source in an aqueous glycine solution at pH 5.0 to 9.0 for 5 to 48 hours at 25° to 40°C to produce L-serine by enzyme action from the glycine and a source of carbon assimilable by the microorganisms.

L-serine is practically the only amino acid produced by the cultures or during enzymatic reaction so that L-serine may be recovered from a cell-free filtrate by means of a cation exchange resin of the H type, eluted, and crystallized from the eluate after partial evaporation of the water present.

Example: The five mutant strains identified above and the parent strain *C. glycinophilum* ATCC 21341 were cultured on respective 20 ml batches of a stock supplemented with the nutrients listed in the following table. The stock medium contained 5 g/dl glucose, 0.2 g/l ammonium sulfate, 0.3 g/dl ammonium nitrate, 0.1 g/dl KH_2PO_4, 0.05 g/dl $MgSO_4 \cdot 7H_2O$, 0.2 mg/l biotin, 1 mg/l thiamine·HCl, 1 mg/l nicotinamide, 1 mg/l pyridoxal·HCl, 1 mg/l folic acid, 2 mg/l riboflavin, 1 mg/l calcium pantothenate, 0.2 mg/l pyridoxin, 1 mg/l p-aminobenzoic acid, 0.05 g/dl DL-methionine, 0.05 g/dl L-glutamic acid, 0.05 g/dl L-valine, 2 g/dl glycine, and was adjusted to pH 6.5.

Each medium was sterilized in a 500 ml shaking flask with steam at 110°C for 5 minutes, and then inoculated with the microorganisms which had been cultured on bouillon agar slants at 30°C for 20 hours. The concentration of L-serine accumulated in each culture medium after 72 hours of shaking at 30°C is listed in the following table.

C. glycinophilum	Nutrient	Concentration (g/dl)	Serine (g/l)
FERM-P 1685	L-leucine	0.03	10.6
FERM-P 1686	L-tryptophan	0.02	9.7
FERM-P 1687	L-leucine	0.03	11.5
FERM-P 1688	L-leucine + L-isoleucine	0.03 0.03	10.2
FERM-P 1689	L-leucine	0.03	8.9
ATCC 21341	None	–	7.8

Enzymatic Production of Substituted Serines

It has been found by *H. Nakazawa, H. Enei, K. Kubota and S. Okumura; U.S. Patent 3,871,958; March 18, 1975; assigned to Ajinomoto Co., Inc., Japan* that an aldehyde having at least three carbons reacts with glycine or ethanolamine in the presence of an enzyme to form serine-derivatives of Formula (1) and serinol-derivatives of Formula (2).

$$(1) \quad RCH(OH)CH(NH_2)COOH \qquad\qquad (2) \quad RCH(OH)CH(NH_2)CH_2OH$$

R is the organic residue having at least two carbons derived from the aldehyde.

The enzyme source may be a broth in which a microorganism is cultivated, or a cell-free filtrate of the broth, intact cells of the microorganism recovered from the broth, an aqueous suspension of ground cells or a filtrate of the suspension. It is believed that the enzyme catalyzing the reaction between the aldehyde and glycine may be a threonine aldolase.

The microorganisms capable of producing the enzyme are distributed very widely; representative microorganisms belonging to the genera Escherichia, Citrobacter, Klebsiella, Aerobacter, Serratia, Proteus, Bacillus, Staphylococcus, Arthrobacter, Bacterium, Xanthomonas, Candida, Debaryomyces, Corynebacterium and Brevibacterium. The reaction can be carried out by culturing the microorganisms mentioned above in a conventional nutrient medium containing aldehyde and glycine and/or ethanolamine under aerobic conditions as usual, or by adding the enzyme source to a reaction mixture containing an aldehyde and glycine and/or ethanolamine. In certain cases the activity of the enzyme sources is improved by addition of threonine to the culture medium.

This reaction can be carried out by using saturated aliphatic aldehydes such as propionaldehyde, butyraldehyde, isobutyraldehyde, valeraldehyde, isovaleraldehyde, pivalic aldehyde, caproic aldehyde, heptaldehyde, caprylaldehyde, tridecylaldehyde, etc. and halogen substitution products thereof, aliphatic dialdehydes such as succinaldehyde, unsaturated aliphatic aldehydes such as acrylaldehyde, crotonaldehyde, propiolaldehyde, etc., heterocyclic aldehydes such as furaldehyde, aromatic aldehydes such as benzaldehyde and benzaldehyde-derivatives having one or more substituent groups such as p-nitro benzaldehyde, p-methyl benzaldehyde, p-chloro benzaldehyde, tolualdehyde, salicylaldehyde, cinnamaldehyde, naphthaldehyde, etc., formyl-carboxylic acids such as formylacetic acid, phthalaldehyde, and ketoaldehydes such as methylglyoxal, phenylglyoxal, etc.

The amount of aldehyde in the reaction mixture should be controlled so that the activity of the enzyme is not inhibited, and preferably having a range from 0.1 to 10% by weight of the reaction mixture.

Example: An aqueous culture medium was prepared containing 2.0 g/dl L-threonine, 0.2 g/dl KH_2PO_4, 0.1 g/dl $MgSO_4$ and 0.4 g/dl $(NH_4)_2SO_4$, and was adjusted to pH 5.0. 50 ml batches of the medium were placed in 500 ml shaking flasks and inoculated with the microorganisms shown on Table 2 which were previously cultured on bouillon agar slants at 31°C for 20 hours, and cultivation was carried out at 31°C for 24 hours with shaking. The microbial cells in 100 ml of each cultured broth were harvested by centrifuging, washed and suspended in 50 ml of reaction solutions (A), (B), (C) and (D) whose compositions are shown in Table 1 at pH 8.5 and at 31°C for 20 hours.

TABLE 1

Composition of the Reaction Solution	Reaction Solution, g/dl			
	A	B	C	D
p-Nitrobenzaldehyde	2.0	2.0	–	–
Isobutyraldehyde	–	–	2.0	2.0
Glycine	2.0	–	2.0	–
Ethanolamine	–	2.0	–	2.0
Sorpol W-200*	0.2	0.2	0.2	0.2

*Polyoxyethylene alkyl phenol ether surface active agent

β-p-Nitrophenyl-L-serine, β-p-nitrophenyl-L-serinol, β-isopropyl-L-serine and β-isopropyl-L-serinol were produced in solutions (A), (B), (C), and (D) respectively in the yields shown in Table 2.

TABLE 2

Microorganisms Employed	Yield in Each Reaction Solution, g/dl			
	A	B	C	D
Escherichia coli ATCC 15289	0.2	0.1	0.2	0.2
Citrobacter freundii ATCC 6750	0.8	0.7	0.5	0.6
Klebsiella pneumoniae ATCC 10031	0.5	0.4	0.3	0.3
Aerobacter aerogenes IFO 3319	0.1	0.1	0.2	0.1
Aerobacter cloacae IAM 1020	0.2	0.2	0.1	0.2
Serratia marcescens ATCC 14223	0.9	0.8	0.7	0.6
Proteus vulgaris ATCC 881	0.3	0.1	0.2	0.1
Bacillus mesentericus IAM 1026	0.2	0.1	0.3	0.2
Bacillus subtilis ATCC 13953	0.4	0.3	0.4	0.3
Staphylococcus aureus ATCC 4776	0.6	0.5	0.5	0.4
Arthrobacter simplex ATCC 13260	0.5	0.3	0.4	0.4
Arthrobacter globiformis ATCC 4336	0.6	0.6	0.2	0.5
Arthrobacter pascens ATCC 13346	0.6	0.6	0.5	0.4
Arthrobacter ureafaciens ATCC 7562	0.1	0.1	0.1	0.2
Arthrobacter aurescens ATCC 13344	0.7	0.6	0.6	0.6
Bacterium cadaveris ATCC 9760	0.8	0.7	0.8	0.6
Bacterium succinicum IAM 12090	0.4	0.3	0.4	0.3
Xanthomonas campestris IAM 1671	0.7	0.6	0.7	0.6
Candida rugosa ATCC 10571	0.6	0.5	0.5	0.4

(continued)

TABLE 2 (continued)

Microorganisms Employed	Yield in Each Reaction Solution, g/dl A	B	C	D
Candida humicola ATCC 14438	1.0	0.8	0.9	0.9
Candida utilis ATCC 9950	0.7	0.6	0.5	0.4
Candida guilliermondii ATCC 9058	0.5	0.4	0.3	0.2
Candida parapsilosis ATCC 7330	0.3	0.2	0.2	0.1
Debaryomyces hansenii ATCC 18110	0.1	0.1	0.2	0.1

Serine in Schizophrenia Treatment

It is known that approximately 40% of schizophrenics react adversely to the effects of L-methionine. That is, when chronic schizophrenics orally ingest methionine, approximately 40% of them develop an acute florid psychosis. Based on the foregoing observation, it is apparent that L-methionine, or a metabolic product thereof, acts in some manner in some schizophrenics to worsen symptons.

A method for temporarily alleviating symptoms of schizophrenia comprising administering to a schizophrenic an effective amount of L-serine has been disclosed by *J.R. Smythies; U.S. Patent 3,873,722; March 25, 1975; assigned to Nelson Research & Development Company.* An effective amount of L-serine varies with the type of patient, but generally ranges between 1 and 25 grams per day, and preferably 10 to 20 grams per day orally for a period of 1 to 10 weeks. Additional treatment may also be necessary. Generally, the L-serine (a nontoxic material) is mixed with the patient's food or swallowed by the patient in capsules.

Example: Rats given 250 mg/kg L-methionine evidenced behavioral disruptions similar to that induced by LSD and other hallucinogenic drugs. Simultaneous administration of L-serine (250 mg/kg) prevented behavioral disruption.

SULFUR-CONTAINING AMINO ACIDS

CYSTEINE AND DERIVATIVES

L-cysteine (2-amino-3-mercapto propionic acid) is a quasi-essential amino acid which is useful as a food additive and a pharmaceutical, e.g., as an antidote. In addition, L-cysteine is useful as an intermediate in the preparation of other pharmaceutical compounds. The derivatives of L-cysteine, e.g., the sulfoxides of S-methyl-L-cysteine and S-allyl-L-cysteine, are known to suppress increasing levels of cholesterol in blood and liver.

The known processes for the production of cysteine include: (1) extraction of cystine from a hydrolysate of natural products such as human hair, followed by reduction of cystine; (2) condensation of benzyl chloromethyl sulfide with diethyl phthalimidomalonate, followed by hydrolysis and reduction; (3) hydrolysis of the thiazoline derivative obtained by the reaction of N-thiobenzoylserine with thionyl chloride; and (4) enzymatic processes. Among these processes, organic syntheses are unfavorable because of high production costs due to the complicated nature of the reactions and the racemic resolutions required to obtain L-cysteine.

Enzymatic Production of L-Cysteine

A process for producing L-cysteine or a derivative in the presence of cysteine desulfhydrase is disclosed by *H. Yamada, H. Kumagai and H. Ohkishi; U.S. Patent 3,974,031; August 10, 1976; assigned to Mitsubishi Chemical Industries Ltd.* The process is illustrated by the following reaction sequence scheme:

$$\underset{\text{(II)}}{\underset{|\atop NH_2}{XCH_2CHCOOH}} + \underset{\text{(III)}}{RSH} \text{ or } \underset{\text{(IV)}}{RSY} \xrightarrow{\text{Cysteine desulfhydrase}} \underset{\text{(I)}}{\underset{|\atop NH_2}{RSCH_2CHCOOH}} + \underset{\text{(V)}}{HX} \text{ or } \underset{\text{(VI)}}{YX}$$

where R is hydrogen, C_1-C_{15} alkyl, C_2-C_{15} alkenyl, C_6-C_{15} aryl or C_7-C_{15} aralkyl; X is halogen, –OR' or –SR' where R' is hydrogen, allyl, C_1-C_{15} alkyl, C_6-C_{15} aryl or C_7-C_{15} aralkyl; and RSY is ammonium hydrosulfide or a metal hydrosulfide where R is hydrogen and Y is ammonium or a metal such as an alkali or alkaline earth metal. β-substituted L-alanines(II) suitable as the starting material include: β-halogeno-L-alanines; O-alkyl-L-serines, O-aryl-L-serines, O-aralkyl-L-serines, S-alkyl-L-cysteines, S-aryl-L-cysteines, S-aralkyl-L-cysteines, L-serine; L-cysteine; S-allyl-L-cysteine and O-allyl-L-serine. The preferred β-substituted alanines are β-chloro-L-alanine, L-serine and L-cysteine. It is preferred that the metal hydrosulfide used in the reaction be soluble in water. Suitable metal hydrosulfides include a hydrosulfide of an alkali metal or an alkaline earth metal. In place of the metal hydrosulfide, there may be used a metal sulfide which is capable of producing a water-soluble metal hydrosulfide and/or hydrogen sulfide in an aqueous medium.

The cysteine desulfhydrase used in the process is an enzyme known as a catalyst for the reaction where L-cysteine decomposes into pyruvic acid, ammonia and hydrogen sulfide. This enzyme is readily produced by a variety of microorganisms such as those shown in the following table.

Employed microorganism		Activity of the enzyme (μ)
Sarcina lutea	(IAM 1099)	3.04
Sarcina lutea	(IFO 3232)	3.92
Corynebacterium equi	(IAM 1038)	0.16
Arthrobacter simplex	(IFO 3530)	0.04
Brevibacterium ammoniagenes	(IFO 12071)	0.57
Pseudomonas fluorescens	(IFO 3081)	1.28
Proteus morganii	(IFO 3848)	1.25
Micrococcus roseus	(IFO 3764)	0.68
Citrobacter freundii	(IFO 12681)	18.59
Escherichia coli	(IFO 3301)	4.55
Serratia marcescens	(IFO 3054)	0.92
Alcaligenes faecalis	(IAM 1015)	4.27
Bacillus subtilis	(IFO 3009)	0.92
Agrobacterium tumefaciens	(IAM 1037)	3.86
Enterobacter aerogenes (*Aerobacter aerogenes*)	(IFO 3320)	5.59
Klebsiella pneumoniae	(IFO 3512)	3.18
Salmonella typhimurium	(IFO 12529)	3.28

Note: The activity of the enzyme is shown in micro units, expressing the activity of the enzyme in terms of the base activity wherein 1 micro mol of L-cysteine is decomposed in a period of 1 minute (1 micro unit) in a standard preparation.

Example 1: To 10 ml of 5 x 10^{-2} M ammonium buffer solution (pH 9.5) containing 270 mg of β-chloro-DL-alanine, β-chloro-L-alanine, L-serine or S-methyl-L-cysteine, and 150 mg of sodium sulfide was added 3 units of the purified enzyme obtained from *Citrobacter freundii* (IFO 12681), and the reaction was carried out at 30°C for 1 hour. L-cysteine was obtained in the amounts shown in the following table.

Substrate	Amount of L-cysteine obtained (mg)
β-chloro-DL-alanine	76.5
β-chloro-L-alanine	99.0
L-serine	4.7
S-methyl-L-cysteine	7.2

Example 2: To 10 ml of 5 x 10^{-2} M ammonium buffer solution (pH 9.5) containing 250 mg of β-chloro-L-alanine and 150 mg of sodium sulfide was added 2.5 units of the purified enzyme obtained from *Enterobacter aerogenes (Aerobacter aerogenes)* (IFO 3320), and the reaction was carried out at 30°C for 1 hour. 98 mg of L-cysteine was obtained.

Resolution of DL-Cysteine

The process disclosed by *P. Rambacher and S. Make; U.S. Patent 3,808,268; April 30, 1974; assigned to Diamalt AG, Germany* relates to the separation of DL-cysteine racemates. It has been discovered that DL-cysteine or its hydrochloride can be reacted in aqueous solution with dicyanodiamide to obtain DL-2-guanidine-1,3-thiazolinecarboxylic acid-4. The diastereomeric copper complex salts of this acid have substantial differences in solubility with optically active acids, preferably D- or L-aspartic acid. These differences can be used for the separation of the diastereomeric forms. Since on the other hand, such copper complex salts can be easily reduced again to cysteine, the separation of DL-cysteine into the optically opposite forms can be effected through them.

The conversion of cysteine by dicyanodiamide preferably takes place at room temperature in an acid or alkaline aqueous solution, and by ammonia cleavage yields DL-2-guanidine-1,3-thiazolinecarboxylic acid-4, which is tautomeric with the corresponding thiazolidinecarboxylic acid.

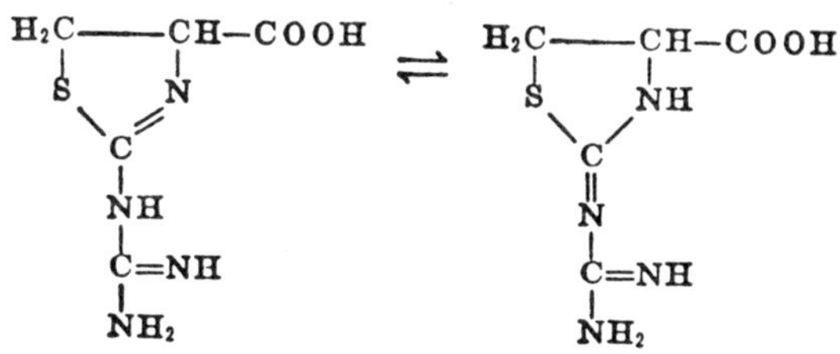

The DL-guanidinothiazolinecarboxylic acid is converted into diastereomeric copper complexes in an aqeuous solution preferably with copper(II)-L-aspartate or with copper(II) ions and L-aspartic acid in equimolar proportions, slightly heated if necessary.

The difficultly soluble L-guanidinothiazolinecarboxylic acid-L-aspartic acid-copper complex is dissociated and the D-L-Cu complex left in the solution can be split by the addition of cyanide. In this way, the D-guanidinothiazolinecarboxylic acid precipitates out and can be separated. It is converted, preferably by treatment in HCl acid solution, into S-guanidinocarbonyl-D-cysteine dihydrochloride, and this is split in partly concentrated aqueous alkali to yield D-cysteine which is oxidized in known manner to D-cystine, which due to its difficult solubility is easily isolated from weak acid solution. In like manner, the L-L-Cu complex can yield L-cystine.

It is also possible to obtain a direct cleavage of the D-guanidinothiazolinecarboxylic acid-L-aspartic acid-copper complex in acid solution without resorting to the isolation of the D-guanidinothiazoline, although an optically less pure cystine is obtained.

Example: Into a solution of 87.7 kg of DL-cysteine hydrochloride monohydrate in 500 liters of water, there are added 50 kg of dicyanodiamide. After 40 hours stirring at room temperature, 79.3 kg (equivalent to 84.5% of the theoretical quantity) of DL-2-guanidine-1,3-thiazolinecarboxylic acid-4 is precipitated. 33.9 kg DL-guanidinothiazolinecarboxylic acid (180 mols) and 49.6 kg copper-L-aspartate (180 mols) are added to 5,400 liters of water at 45°C. The solution is stirred for 4 days at 25° to 30°C. The solution is centrifuged after addition of 10 kg kieselguhr. From the residue there is obtained L-guanidinothiazolinecarboxylic acid-L-aspartic acid-copper complex, from which L-cystine can be obtained following the procedure given with respect to the D-L-Cu complex.

The centrifugate is treated at room temperature with 10 liters 25% caustic soda and approximately 23.5 kg potassium cyanide until decolorized. It is brought to pH 7.5 with 7 liters glacial acetic acid, cooled and restirred for at least 10 hours at 5° to 10°C. Precipitated D-guanidinothiazolinecarboxylic acid is discharged and dried. The result is 7.95 kg of acid which is then dissolved in 70 liters of water and 35 liters of concentrated hydrochloric acid. After standing overnight at room temperature, the solution is concentrated in vacuum to about ¼ of its volume, cooled, and the produced S-guanidinocarbonyl-D-cysteine dihydrochloride is sucked away, and washed and dried in the presence of air.

A resultant quantity of 10.2 kg of the salt is poured into 48 liters of 25% caustic soda at 35°C, stirred for 2 hours at 35° to 40°C, with about 25 liters concentrated hydrochloric acid to neutralize to a pH 7 to 8, and stirred again for 24 hours at 10° to 12°C. About 800 g of a by-product (S-guanidinocarbonyl-D-cysteine) is separated, the filtrate is treated with 1 liter of concentrated ammonia and oxidized while cooling with approximately 3 liters of 16.5% hydrogen peroxide solution until the SH reaction is negative. Concentrated hydrochloric acid and glacial acetic acid are added until the pH is brought to 5, and then it is restirred for 48 hours with pH control and the D-cystine is sucked off. After drying in air, about 4 kg of D-cystine is obtained having at least 98% purity. The specific rotation $[\alpha]_D^{20}$ amounts to +210° (c = 3, 1 N HCl).

Ammonium N-Acetyl-L-Cysteinate as Mucolytic Agent

N-acetyl-L-cysteine is a known pharmaceutical and has been used with considerable success as a mucolytic agent in ameliorating inflammatory conditions in mammals and in permanent deformation of human hair. For mucolytic purposes N-acetyl-L-cysteine has been administered by nebulization, direct application, or intratracheal instillation in the form of a solution of the sodium salt. The use of finely divided solid N-acetyl-L-cysteine for inhalation, i.e., insufflation, is via a dispenser in which the N-acetyl-L-cysteine is dispersed into an air stream by fluidization technique wherein the principal source of energy is the inspiratory action of the inhaler. However, administration of N-acetyl-L-cysteine by this method is not entirely satisfactory in that it produces severe coughing and irritation.

E.E. Hamlow and T.A. Martin; U.S. Patent 3,965,167; June 22, 1976; assigned to Mead Johnson & Johnson Company have discovered that the crystalline ammonium salt of N-acetyl L-cysteine which previously has not been prepared in pure pulverulent form is particularly suited for insufflation administration using only the energy of the inspiratory action of the inhaler. Pulverulent ammonium N-acetyl-L-cysteinate is also suited for inhalation administration by other means such as a pressurized aerosol container. Crystalline N-acetyl-L-cysteine ammonium salt can be readily micronized to provide ultrafine powders of a uniform particle size range. Moreover, unlike other salts of N-acetyl-L-cysteine such as the sodium, it is relatively nonhygroscopic. The preparation of crystalline N-acetyl-L-cysteine ammonium salt is more fully described below.

Example: A solution of concentrated ammonium hydroxide (52 ml, 0.78 mol) and 20 ml water is added in 10 minutes to a stirred mixture of N-acetyl-L-cysteine (114.2 g, 0.7 mol) in 1.2 liters of isopropanol under an atmosphere of nitrogen. During the addition the mixture thickens and the temperature increases to 35°C. The mixture is stirred overnight, collected and washed with isopropanol to yield 110 g (87%) of product having a melting point of 156° to 158°C (dec) (uncorr). Crystallization of this product from water-isopropanol with cooling to -10°C provides analytically pure N-acetyl-L-cysteine ammonium salt, MP 147.5° to 150°C (dec) (corr); $[\alpha]_D^{25} = +20.6°$ (c = 5, H_2O).

Powdered N-acetyl-L-cysteine ammonium salt was satisfactorily self-administered utilizing only the energy of the inspiratory action of the inhaler by employing the oral inhaler of U.S. Patent 3,518,992 without any appreciable incidence of coughing and irritation by an adult male. In contrast, oral inhalation of N-acetyl-L-cysteine produced severe coughing and irritation.

Cysteine Derivatives for Treating Seborrhea

The process disclosed by *G.Kalopissis and G. Manoussos; U.S. Patents 3,821,405; June 28, 1974; 3,950,542; April 13, 1976; and 3,629,452; December 21, 1971; all assigned to L'Oreal, France* relates to a method for treating a scalp characterized by an excessive secretion of sebum. This process comprises orally administering to a human a therapeutic composition comprising an ingestible carrier mixed with, as a nontoxic active compound, certain cysteine or cysteamine derivatives. The active compound used in the process is selected from compounds having the formula

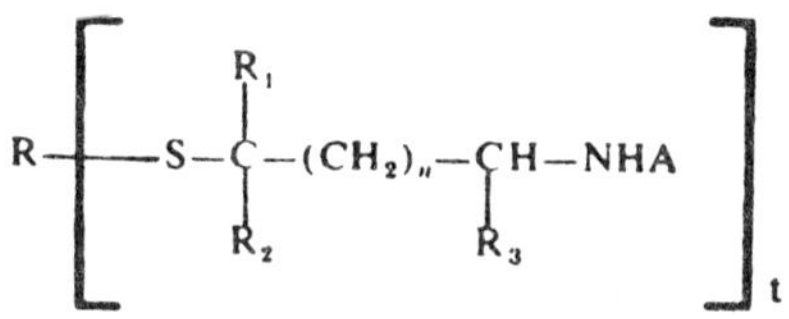

and its acid salts.

In the above formula:

n is 0 or 1; when n is 0, R_1 and R_2 are hydrogen or CH_3 and when n is 1, R_1 and R_2 are hydrogen.
R_3 is hydrogen or $-COR_7$ where R_7 is hydroxy, alkoxy containing 1 to 5 carbon atoms, glucosamine, $-NH-NH_2$, NHOH,

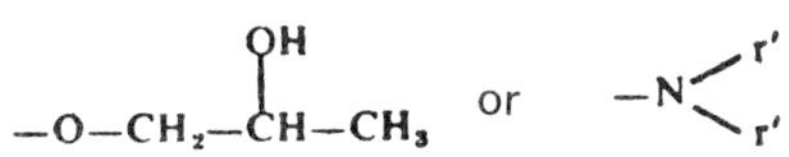

where r' is hydrogen or alkyl having 1 to 3 carbon atoms, or

$$-O-CH_2-CH_2-N\begin{matrix} r'' \\ r''' \end{matrix}$$

where r'' and r''' each independently represent C_{1-3} alkyl or together with an attached nitrogen form a ring selected from morpholine, piperidine, pyrrolidine or N'-methylpiperazine;

A is hydrogen, $-CONH_2$, nicotinoyl, glutamyl, $-COR_4$ or $-SO_2R_6$ where R_4 is hydrogen, C_{1-18} alkyls, C_{2-18} alkenyls,

$$-CH_2Cl,\quad -CF_3,\quad -CH(OH)-C(CH_3)_2-CH_2OH,$$

$$-CH_2-CH_2-NH-CO-CHOH-C(CH_3)_2-CH_2OH,$$

$-CH_2-CH_2-COOH$ or $-(CH_2)_p-$ 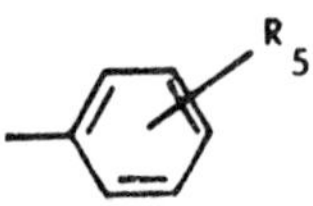

where p is 0 to 1 and R_5 is hydrogen, C_{1-4} alkyls, C_{1-4} alkoxys, halogen or acetamide, and where R_6 is C_{1-4} alkyls or,

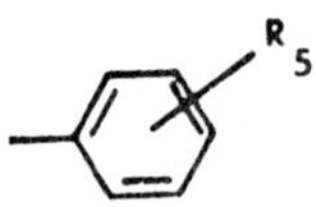

t is 1 or 2; when t is 1, R is linear or branched chain C_{1-18} alkenyls, propyne-2-yl, mono- or dihydroxyalkyl containing 2 to 4 carbon atoms, 1,2-dichlorovinyl, $-C(C_6H_5)_3$, $-C(CH_3)_3$, $-CH(C_6H_5)_2$, $-CH(C_6H_4\text{-p-}OCH_3)_2$, $-C(CH_3)_2(C_6H_4\text{-p-}OCH_3)$, $-CONH_2$,

$$-(CH_2)_m-CH\begin{matrix} OR''' \\ OR''' \end{matrix}$$

where R''' is C_{1-4} alkyl and m is 1 to 2,

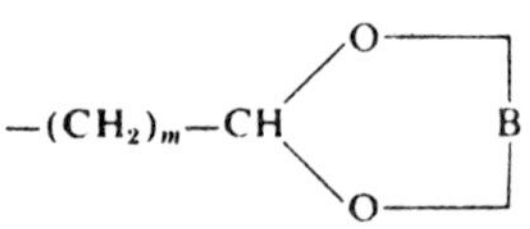

where m is 1 to 2 and B is $-CH_2-CH_2-$, or

$$-\underset{\displaystyle}{\overset{\displaystyle CH_3}{\overset{|}{C}}}H-CH_2-,\quad -CH_2-(CH_2)_n-\underset{\displaystyle NHA}{\underset{|}{C}}H-R_3,$$

where n, R_3 and A have the meanings given above, $-(CH_2)_s-COR_{10}$ where s is 1 to 5 and R_{10} is R_7, piperidino, morpholino, pyrrolidino, N-methylpiperazino or $-N(CH_2CH_2OH)_2$, $-(CH_2)_z-R_8$ where z is 0, 1 or 2 and when z is 0, R_8 is naphthyl-1, naphthyl-2, pyridyl-2, pyridyl-2 substituted with nitro or C_{1-4} alkyls, or pyridyl-2-N oxide when R_3 is hydrogen, and when z is 1, R_8 is naphthyl-1, naphthyl-2, thienyl-2, tetrahydrofuryl-2, furyl-2, pyridyl-2, pyridyl-2 substituted with nitro or C_{1-4} alkyls, or pyridyl-2-N-oxide, and when z is 2, R_8 is pyridyl-2, pyridyl-2 substituted with nitro or C_{1-4} alkyls and pyridyl-2-N-oxide, and

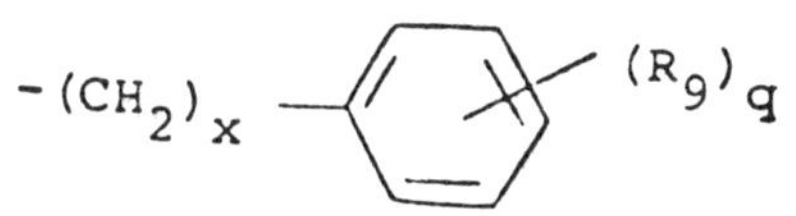

where x is 0 to 2 and when q is 1, 2 or 3, R_9 is hydrogen, halogen, C_{1-5} alkoxy and linear or branched chain C_{1-4} alkyl, and when q is 1, R_9 is acetamido, amino, phenoxy, cyclohexyl, methylenedioxy, trifluoromethyl, nitro, phenyl, dialkylamino where each alkyl moiety has 1 to 5 carbon atoms, alkylthio having 1 to 5 carbon atoms, alkylsulfinyl having 1 to 5 carbon atoms and alkyl-sulfonyl having 1 to 5 carbons,
when t is 2, R is C_{2-4} alkylene, C_{2-4} alkylene, C_{2-4} alkylene substituted with 1 to 2 hydroxy groups, butenylene and $-(CH_2)_2-SO_2-(CH_2)_2-$.

When the above active compounds contain at least one primary amine group, they are preferably used as a salt with an inorganic or organic acid selected from the group consisting of hydrochloric, hydrobromic, sulfuric, phosphoric, p-toluene sulfonic, dihydrocinnamic, cinnamic, mandelic, salicyclic, tropic, oxalic, malic, tartaric, succinic, acetic, lactic, 3-hydroxybutyric, fumaric, undecylenic, phenylacetic, monochloroacetic, trichloroacetic, glycolic, nicotinic, p-aminobenzoic, glutamic, aspartic, citric, pentothenic, crotonic and ascorbic acid. However, when the active compounds carry both an amine group and a carboxylic acid function an inorganic acid or a strong organic acid such as p-toluenesulfonic acid or trichloroacetic acid must be used in order to break the internal salt.

A total of 163 active derivatives of cysteine and 156 active derivatives of cysteamine are listed in the patents which may be prepared by standard known synthetic methods. Use of these compounds is illustrated in the following examples.

Example 1: The following composition is prepared: 3-allylthioalanine hydrochloride, 1 g; glycerin, 40 g; ethyl alcohol, 30 g; water, qsp, 100 g; and lemon tincture (qsp to impart a pleasing aroma). It is usefully employed to reduce excessive secretion of sebum on the scalp. It is orally administered in the form of drops.

The oral administration of this composition at the rate of 10 drops each day for 15 days by a person having greasy hair due to excessive secretion of sebum substantially improves the condition of the scalp and the appearance of the hair. Essentially similar effective results are obtained when the 3-allyl thio alanine hydrochloride is replaced with the following compounds: 3-(β-hydroxyethyl thio) alanine, 2-amino-4-thio dodecanoic acid, 2-(2-naphthyl thio) ethyl ammonium disuccinate, 2-(pyridyl-2-methyl thio) ethylamine dihydrochloride, 2-cetylthio ethylamine hydrochloride, 2-(3,4-dimethoxy benzylthio) ethylammonium malate and 2-amino 4-tetradecylthio butyric acid.

Example 2: The following composition is prepared:

3-methylthio alanine hydrochloride	50 mg
Glucose	300 mg
Water, qsp	5 ml
Orange juice (qsp to impart a pleasant aroma)	

The oral administration of this composition, at the rate of 2 ampules each day for 15 days by a person having greasy hair due to excessive secretion of sebum substantially improves the condition of the scalp and the appearance of the hair. Essentially similar effective results are achieved by replacing the 3-methylthio analine hydrochloride with the following compounds: 3-(2,4-dichlorobenzylthio)alanine, 3,3'-(2-butene 1,4 diyldithio)dialanine, 2-(2,2-dimethoxy ethylthio)-ethylamine hydrochloride, S-(3-aminopropyl)mercapto acetic acid, 2-(2,4-dichlorocetylthio)ethylamine and 5-dodecanamido 3-thio hexanedioic acid. Lozenges, chewable pellets and tablets may also be prepared from these compounds for oral administration in the treatment of seborrhea.

S-Carboxymethyl Cysteine for Respiratory Disorders

S-carboxymethyl cysteine has been previously used for administration in nasal sprays, nasal drops or aerosols in otolaryngology and in pneumology. *M.L. Joullie, G. Maillard, L. Lakah and P. Muller; U.S. Patent 3,891,749; June 24, 1975; assigned to Recherche Pharmaceutiques et Scientifiques, France* have found that when administered by the digestive route, this compound has the unexpected effect of reducing the viscosity of secretions of the respiratory tract. Administration to patients may be at a level of 1.5 to 3 g per day in the form of a syrup, gelatin-coated pills or capsules, tablets or suppositories, in which other active materials, e.g., antibiotics, antihistamine drugs, corticoids or bronchodilators, may also be present.

When administered by inhalation in the form of a 10% aqueous solution to patients suffering from respiratory insufficiency, S-carboxymethyl cysteine considerably increases the volume of secretions but their viscosity (measured by the extent to which the two arms of forceps can be moved apart before the seromucous thread breaks) is not modified.

When studying 26 patients who received S-carboxymethyl cysteine by oral route, on the other hand, it was found that the reduction in the viscosity of the secretions became statistically significant from the third day, the gradient p, which is 0.005 on the second day, falling to 0.001 on the third day and staying at that value for the whole duration of the treatment. This lowering in the viscosity of the secretions seems to be associated with the appearance of a much richer secretion of salivary acid and salivary mucine. A specific cellular localization of transferases responsible for the synthesis of salivary glycosamine glycans has been demonstrated at the level of the mucous cells of the bronchial tree, and that of sulfotransferases in the serous acini of the mixed glands and in the caliciform cells. One may therefore assume that S-carboxymethyl cysteine acts selectively on the metabolism of the mucous elements. In bronchial disease, these mucous tubules are hypertrophied but generally only produce a mucine which is low in salivary acid. The action of S-carboxymethyl cysteine on the mucous producing hyperplasia would therefore be due to a process of activation of the salivary transferases, which would explain why this compound is only active when administered orally and does not have the same action when applied locally.

Example:

S-carboxymethyl cysteine	2 g
Sucrose	70 g
Methyl-p-hydroxybenzoate (preservative)	0.15 g
Flavoring	
Strawberry concentrate	1 ml
Cherry concentrate	0.5 ml
Vanillin	0.05 g
Red coloring	0.005 g
Sodium hydroxide, qs for pH 6.2	
Water, qs 100 ml	

In view of the insolubility of S-carboxymethyl cysteine, preparation of the syrup is started by neutralizing S-carboxymethyl cysteine with sodium hydroxide to pH = 6.2. The other constituents are then added and the mixture is stirred and filtered.

Syrup for adults:	
S-carboxymethyl cysteine	5 g
Sugared and flavored excipient qsp	100 ml

The compound may also be formulated as capsules, pills, tablets, etc. and additives such as antibiotics, antihistamine drugs, corticoids, and bronchodilators may be added to the formulations.

S-Derivatives of Cysteine for Pharmaceutical Applications

Three series of S-substituted cysteine compounds have also been developed at Recherches Pharmaceutiques et Scientifiques for use in treatment of various medical disorders. In the first process *M. Joullie, L. Lakah, G. Maillard and P. Muller; U.S. Patent 3,892,852; July 1, 1975; assigned to Recherches Pharmaceutiques et Scientifiques, France* found that L forms of certain S-substituted cysteines are active in reducing atheromatous deposits and hypercholesterolemia in test animals and they are proposed for use in treatment of similar conditions in human beings.

The following cysteines are used: S-(3-hydroxypropyl)-cysteine, S-allylcysteine, S-allylcysteine sulfoxide, S-allyl-N-formylcysteine, S-allyl-N-acetyl-cysteine, S-(propen-1-yl)cysteine, S-(buten-2-yl)cysteine, S-propargylcysteine, S-(buten-2-yl)-cysteine, S-benzylcysteine and S-(parachlorobenzyl)cysteine. The S-substituted derivatives of cysteine may be used alone or together with substances which are themselves active against hypercholesterolemia, such as derivatives of 2-(p-chlorophenoxy)-2,2-dimethylacetic acid and 3-benzylidenebutyric acid or nicotinic acid.

The following examples illustrate the production of these compounds.

Example 1: L-S-(3-Hydroxypropyl)Cysteine (LJ 554) — To 500 ml of liquid ammonia cooled to -80°C are added 17.55 g (0.1 mol) of cysteine hydrochloride monohydrate and 6.9 g (0.3 g atom) of sodium. After the metal has dissolved, 9.5 g of redistilled propane-1,3-diol monochlorohydrin is added. The reaction mixture shows a negative thiol reaction only after 12 hours.

The ammonia is removed by evaporation, ultimately under reduced pressure. The residue is dissolved in 100 ml of iced water, is thrice extracted with diethyl ether and the pH of the aqueous solution is then brought to 6.1 using 4 N hydrochloric acid. The solution obtained is passed through a column of 400 ml of "Dowex 50" which has previously been activated with 10% hydrochloric acid.

When no chloride ions are present in the effluent, the product is eluted with 4 N ammonium hydroxide, the solution is evaporated to dryness under reduced pressure and the residue dissolved in the mininum amount of water and treated with "Norit" decolorizing carbon. After filtering the solution is again evaporated to dryness. 12.30 g (yield = 70%) of product are obtained. MP = 203°C (Buchi). Rotation: $[\alpha]_D^{22} = -20.7°$ (5% solution in water).

Example 2: 43.9 g (0.25 mol) of cysteine hydrochloride monohydrate is dissolved in 362 ml of distilled water. This mixture is cooled to 0°C and a solution containing 20.3 g (0.5075 mol, 1.5% excess) of sodium hydroxide in 100 ml of water is added, followed by 35 g (0.292 mol, 16.8% excess) of freshly distilled allyl bromide dissolved in 25 ml of distilled acetonitrile. After 15 to 20 minutes, the reaction mixture gives a negative reaction to nitroprusside. The solution is brought to pH 6.4 with concentrated hydrochloric acid and the subsequent procedure is as described in the preceding example. 24 g (yield 60%) of product are obtained. MP = 234°C. Isoelectric point = 6.4. Rotation: $[\alpha]_D^{20} = 11.04°$ (5% solution in water). The mother liquors still contain some product which may be recovered.

These cysteine derivatives, formulated in conventional pharmacologically acceptable ways, with added excepients and/or lubricants, depending on the manner in which they are to be administered, may be given to humans in doses of from 200 mg to 3 g per day, preferably 800 mg per day.

A second series of basic derivatives of cysteines have been disclosed by *M. Joullie, L. Lakah, G.F. Maillard and P. Muller; U.S. Patent 3,896,121; July 22, 1975; assigned to Recherches Pharmaceutiques et Scientifiques, France*. These sulfur containing compounds and their acid addition salts have a favorable action on epithelia as well as on their secretions. The compounds are characterized in that they carry a tertiary amine radical at the end of a carbon chain bonded to the sulfur atom of cysteine, specifically a tertiary amino-ethyl group, the tertiary amino group being the radical of dimethylamine, diethylamine, pyrrolidine, piperidine, morpholine or 4-methylpiperazine, or a tertiary amino-benzyl group, the tertiary amino group being the radical of pyrrolidine, piperidine, morpholine or 4-methyl-piperazine.

In order to prepare these tertiary-amine-type compounds, cysteine in the form of its hydrochloride is reacted with a tertiary amino-ethyl or tertiary amino-benzyl halide in the form of the hydrohalide, the halogen being chlorine or bromine, in the presence of an alkali metal hydroxide, in an aqueous medium.

Example: S-(β-Dimethylamino-Ethyl)-Cysteine (LJ 222) and its Monohydrochloride (LJ 388) —

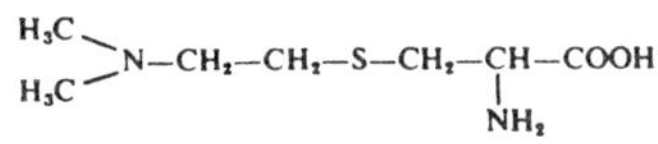

70 g (0.4 mol) of cysteine hydrochloride monohydrate in 100 ml of freshly distilled water are placed in a flask equipped with a stirrer and a nitrogen inlet. 48 g (1.2 mols) of sodium hydroxide dissolved in 100 ml of freshly distilled water are added gradually to the solution obtained. The temperature is increased to about 60° and 120 g (approximately 30% excess) of β-dimethylaminobromoethane hydrobromide are introduced over the course of 20 minutes. The solution obtained is passed over a Dowex 50 resin and eluted with ammonia solution containing 33% by volume of NH_4OH.

The elution product is distilled in vacuo to dryness, the residue is dried in vacuo in the presence of a dehydrating agent and the product is finally recrystallized from absolute ethanol. 49 g (64% yield) of it are obtained. The melting point is 212° to 213°; isoelectric point is 6.32, and R_f is 0.489. In order to obtain the hydrochloride from the base, 0.1 mol of the latter, e.g., is suspended in 60 ml of water, the theoretical amount of normal hydrochloric acid is added, the solution is brought to dryness under 15 mm of mercury and the residue is dried azeotropically as a suspension in benzene and then the product is dried in vacuo.

S-(β-diethylamino-ethyl)-cysteine (LJ 427), S-(β-pyrrolidino-ethyl)-cysteine (LJ 428), S-(β-morpholino-ethyl)-cysteine (LJ 223) and its monohydrochloride (LJ 394), S-(β-piperidino-ethyl)-cysteine (LJ 225) and its monohydrochloride (LJ 393), and S-[β-4-methyl-piperazino)-ethyl]-cysteine (LJ 541) may also be prepared by the same technique from their component starting materials.

These derivatives of cysteine were found to have a favorable action on the secretion of bronchial epithelium, bringing the amount of mucus produced back to the normal level and restoring biochemical equilibrium to a perturbed bronchial wall. These new derivatives can consequently be used as active principles of medicines, particularly medicines for oral administration in pneumology (bronchial-pneumopathy and acute and chronic bronchitis) and in oto-rhino-laryngology (acute sinusitis, acute otitis and rhino-pharyngitis).

These compounds can also be used in dermatology (illnesses of exposed parts) of the body, excess perspiration and seborrhea), in gynecology (trophic disorders of the vaginal mucous membrane, vaginitis and cervicitis) and in cosmetology, by virtue of the trophic action of these derivatives, especially the restoration of the SH groups of the damaged mucous membrane, on epithelia such as those of the vagina and of the skin. The following example illustrates the process in these applications.

Dermatological Cream for the Skin	Grams
LJ 223	2 (or 5)
Glyceryl monostearate	8
Diethylene glycol monostearate	6
Paraffin oil	5

(continued)

Dermatological Cream for the Skin	Grams
Sweet almond oil	3
Isopropyl myristate	5
Polyoxyethylene sorbitan oleate	3
Polyoxyethylene sorbitan stearate	1
Sorbic acid	0.10
Water + scent qsp	100

In order to prepare the cream, the fatty materials are melted at 65°C, the sorbic acid, followed by the LJ 223, is dissolved in hot water and the sorbitan oleate and stearate are added; the melted fatty materials are added to the solution, at 60°C, and the mixture is stirred until cold; then the perfume is added.

Derivatives of cysteine and glutathione have also been developed by *M. Joullie, L. Lakah, G. Maillard and P. Muller; U.S. Patent 3,950,387; April 13, 1976; assigned to Recherches Pharmaceutiques et Scientifiques, France* which may be used for treating numerous complaints such as elastosis, elastorexia, collagenosis, degenerative arthropathy, arthrosis, arthritis, atheromatosis and arteriosclerosis. Furthermore, the reduction in the level of blood cholesterol found in certain animals treated with these cysteine derivatives indicates activity against hypercholesterolemia. The L-cysteine or glutathione derivative has one of the formulas:

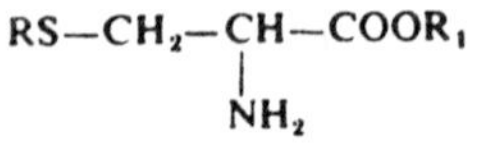

or

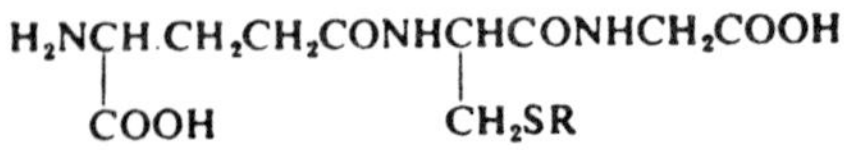

in which R is an allyl, geranyl, cinnamyl, benzylidine-3-butyl, 1-(para-chlorophenyl)ethyl or (pyridyl-3)methyl group and R_1 is a hydrogen atom or an alkyl group having 1 to 4 carbon atoms provided that when R is allyl, R_1 is an alkyl group having 1 to 4 carbon atoms, together with salts thereof with pharmacologically acceptable organic or mineral acids. The following compounds of the L series are included by this process:

Ethyl β-(propen-2-ylthio)-α-aminopropanoate hydrochloride or ethyl S-allylcysteinate,
Isopropyl β-(propen-2-ylthio)-α-aminopropanoate hydrochloride or isopropyl S-allylcysteinate,
S-allyl glutathione or γ-glutamyl-S-allyl cysteinylglycine,
S-geranylcysteine,
β-(3-phenyl-propen-2-ylthio)-α-aminopropanoic acid or S-cinnamyl cysteine,
β-(4-phenyl-3-methylbuten-3-ylthio)-α-aminopropanoic acid or S-(3-benzylidenebutyl)-cysteine,
β-[1-(p-chlorophenyl)ethylthio] α-aminopropanoic acid or S-[1-(p-chlorophenyl)ethyl] -cysteine and
β-[(pyridyl-3)methylthio] α-aminopropanoic acid or S-[(pyridyl-3)-methyl] cysteine.

These compounds are produced by (a) reacting cysteine or glutathione with a halide having the general formula RX in which R is as defined above and X is a halogen or (b) esterifying a cysteine of the general formula shown on the following page:

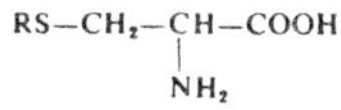

with an alkanol having 1 to 4 carbon atoms.

Example 1: Ethyl L-β-(Propen-2-ylthio)-α-Aminopropanoate Hydrochloride or Ethyl S-Allyl Cysteinate (LJ 565) —

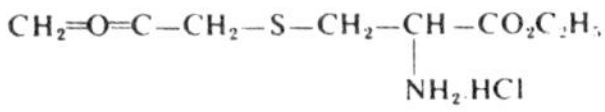

9.66 g (0.06 mol) of S-allylcysteine is added to 150 ml of anhydrous ethyl alcohol. After saturation with hydrochloric acid gas and solution of the S-allylcysteine, the mixture is refluxed for 5 hours and evaporated to dryness. The residue is dissolved in diethyl ether, filtered, and recrystallized from a mixture of 35 ml of cyclohexane and 35 ml of ethyl acetate. 9 g of product is obtained (yield 67%). MP = 121°C. Thin layer chromatography on silica gel shows that the product forms only a single spot. R_f = 0.70. (Solvent is a mixture of 15 parts of n-butanol, 15 of acetone, 10 of acetic acid and 20 of water.)

Example 2: L-S-Allyl Glutathione or L-γ-Glutamyl-S-Allylcysteinyl Glycine (LJ 545) —

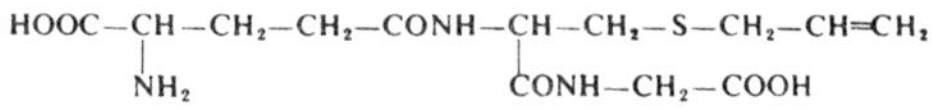

To 500 ml of liquid ammonia which has been cooled in a bath of carbon dioxide snow and acetone, is added in small portions 30.7 g (0.1 mol) of glutathione and 6.9 g (0.3 gram atom) of sodium. To this mixture is added 13.3 g (0.11 mol) of freshly distilled allyl bromide diluted in 25 ml of diethyl ether. When this reaction mixture gives a negative reaction to nitroprusside the ammonia is removed. The residue is dissolved in 200 ml of iced water containing 10% acetic acid.

This solution is extracted with diethyl ether and concentrated under a pressure of 0.5 mm of mercury. A colorless syrup is obtained which solidifies in ethanol. The powdery solid is washed a number of times with 90% ethanol until free from bromide ions. The product crystallizes with two molecules of water. 33 g of product are obtained (yield 86%). MP = 148° to 149°C (after softening at 108°C). Thin layer chromatography reveals that the product forms only a single spot. R_f = 0.32 (using the same solvent as in Example 1).

It was found that these compounds may be administered in doses of from 200 mg to 3 g per day, preferably 800 mg per day. The compounds may be admixed with conventional solid and liquid pharmacologically acceptable diluents and/or excipients as well as other additives such as lubricants.

Preparation of DL-S-Benzyl-N-Acetyl Cysteine

P. Rambacher and S. Make; U.S. Patent 3,798,263; March 19, 1974 disclose a process for making racemic cysteine, inactive cystine and DL-S-benzyl-N-acetyl cysteine by reacting serine and acetic anhydride in the presence of a thioacid such as thioacetic acid. The thio-compound which is attached to the azlactone or acetamino acrylic acid in statu nascendi, is then hydrolyzed by an alkaline or acid hydrolyzing agent, or by both successively, thereby forming racemic cysteine. This product may be used as such, or it may be oxidized to form inactive cystine, or it may be benzylated and acetylated to form DL-S-benzyl-N-acetyl cysteine.

Intially, synthesizing these compounds was tried, using acetaminoacrylic acid or its azlactone as the starting material, by splitting water from serine or by reaction of acetyl serine with acetic anhydride. This method, however, did not succeed, probably because a methylene compound is formed during the reaction which is polymerized and thus no longer useful as a basis for the complete reaction to produce cystine.

It was then discovered that a highly efficient and useful synthesis of cysteine and cystine may be accomplished by providing for the reaction to occur in the presence of a thioacid, such as, e.g., thioacetic acid which is attached to the methylene compound in statu nascendi. It was also discovered that a successful synthesis may also be obtained by carrying out the reaction in the presence of thiourea, which is attached in a similar manner, forming an acetylated isothiourea. In both instances the presence of acetic anhydride is critical and is required for the successful reaction to occur. This process is further illustrated by the production of cystine and DL-S-benzyl-N-acetyl cysteine in the following examples.

Example 1: 21 g of DL-serine are suspended in 50 ml of acetic anhydride and 15.2 g of thioacetic acid, to which are added a few drops of pyridine. During this addition, the temperature rises from 35° to 40°C. The suspension is agitated for 2 hours and then heated to 100°C. The resulting reaction is exothermic and the temperature rises further to 122° to 125°C. During refluxing the acetic anhydride and the formed glacial acetic acid are brought to a boil and after a short while a complete solution is obtained, having a reddish-yellow color. The solution is maintained at 120° to 125°C for 15 minutes, then allowed to cool and poured over a bed of ice.

Concentrated hydrochloric acid is added in an amount to form a 20% solution of hydrochloric acid, which is refluxed for 1 hour, settled with an active coal filter and evaporated under vacuum until a slurry is formed having the consistency of syrup. This syrup is absorbed in 200 ml of water, treated with a concentrated aqueous solution of ammonia until it becomes slightly ammoniacal and then is gradually oxidized with a 3% solution of hydrogen peroxide, while now and then small amounts of ice are being added, until a negative reaction with respect to SH-groups is obtained with sodium nitroprusside.

During this reaction at least some cystine is precipitated having a shape of fine needles, most of which form spherically shaped agglomerations. This precipitation is completed by adjusting to a pH of 4.5 and allowing to stand overnight. The product is then filtered and washed with water and acetone. The product is slightly colored but substantially pure. The yield is 14.9 g, which is 62% of theoretical.

Example 2: 315 g of DL-serine are suspended in 240 g of thioacetic acid and 100 ml of acetic anhydride. 5 g of pyridine are added to the suspension and within 30 minutes the temperature rises to 45°C. After heating to 75°C, the beginning exothermic reaction further raises the temperature, which is maintained at a constant 80°C by intermittent cooling. After ½ hour complete dissolution is obtained. The solution is further agitated for 4 hours while continuously being held at 80°C. It has a honey-yellow color. Then 330 ml of a concentrated ammonia solution are added drop by drop and at the same time the solution is cooled and agitated. After adding 35 ml of benzyl chloride the solution is continuously stirred overnight.

The next day the small amount of resinous by-products is decanted and the pH of the remaining solution is adjusted to pH 6 by adjusting the required amounts of hydrochloric acid. This solution is then evaporated under vacuum until ½ of its original volume is left. During the addition of hydrochloric acid and the subsequent evaporation a few additional by-products are precipitated, which are now removed by a conventional treatment with active charcoal. The clear filtrate is adjusted to a pH of 1 by adding concentrated hydrochloric acid. A yield of 7.2 g of DL-S-benzyl-N-acetyl cysteine is found, which is 9.5% of theoretical yield.

S-Substituted α-Alkylcysteine Compounds

S-aryl-α-methylcysteines have been disclosed by *J.T. Suh; U.S. Patent 3,849,574; November 19, 1974; assigned to Colgate-Palmolive Company* for use as heart stimulants. These compounds have the fomula

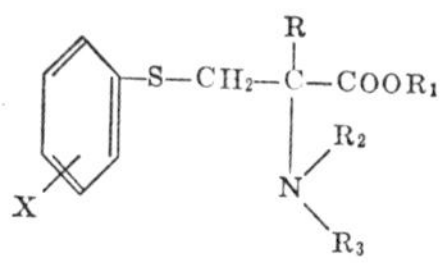

in which X is hydrogen, a halogen such as chloro or bromo, trifluoromethyl, a lower alkoxy such as methoxy or ethoxy or nitro, R is a lower alkyl of 1 to 4 carbon atoms, phenyl, a cycloalkyl of 3 to 7 carbon atoms, a cycloalkyl-lower alkyl in which the cycloalkyl contains 3 to 7 carbon atoms, and phenyl-lower alkyl of 7 to 13 carbon atoms, R_1 is hydrogen or a lower alkyl of 1 to 4 carbon atoms, and R_2 and R_3 are hydrogen, a lower alkyl of 1 to 4 carbons, a cycloalkyl of 3 to 7 carbon atoms, a cycloalkyl-lower alkyl in which the cycloalkyl contains 3 to 7 carbon atoms or phenyl-lower alkyl of 7 to 13 carbon atoms.

These compounds are preferably prepared by reacting a thiophenol such as p-chlorothiophenol with a chloroketone such as chloroacetone in a sodium hydroxide solution to form 1-(p-chlorophenylthio)-2-propanone. A dispersion of the 1-(p-chlorophenylthio)-α-propanone is then treated with a mixture of ammonium carbonate and potassium cyanide in aqueous ethanol at reflux to prepare 4-(p-chlorophenylthiomethyl)-4-methylhydantoin. The latter compound is then treated with barium hydroxide octahydrate in water at reflux until the reaction is complete and acidified to yield the β-(p-chlorophenylthio)-α-methylalanine.

Representative of the final compounds which may be prepared by the described process are:

β-(p-Chlorophenylthio)-α-methylalanine,
β-(p-Chlorophenylthio)-α-ethylalanine,
β-(p-Methoxyphenylthio)-α-methylalanine,

β-(p-Trifluoromethylphenylthio)-α-ethylalanine,
β-(m-Trifluoromethylphenylthio)-α-isopropylalanine,
β-(o-Bromophenylthio)-α-phenylalanine, and
β-(p-Benzyloxyphenylthio)-α-ethylalanine.

Example: 1-(p-Chlorophenylthio)-2-Propanone — To a solution of 120 g (3 mols) of NaOH in 2 liters of H_2O is added with stirring 435 g (3 mols) of p-chlorothiophenol and the resulting mixture is heated to 50°C for 0.5 hour. To the mixture is added dropwise with cooling and stirring 300 g (3.24 mols) of chloroacetone so as to maintain the reaction temperature at 21° to 24°C. The addition requires 45 minutes. The reaction mixture is stirred at 21° to 24°C for an additional two hours and is extracted with 1.5 liters of benzene. The benzene layer is dried over anhydrous Na_2SO_4 and the solvent removed in vacuo. The residue is distilled to give 1-(p-chlorophenylthio)-2-propanone, BP 104° to 105°C/0.4 mm.

The compounds disclosed are pharmacologically active as cardiac stimulants which increase the rate of the heart. The compound β-(p-chlorophenylthio)-α-methylalanine when administered intravenously in doses of 1, 3 and 10 mg/kg of body weight to dogs which were vagotomized and anesthetized, increased the heart rate significantly without effecting significant changes in the EKG responses. When used as pharmaceutical agents the active compounds are preferably combined with conventional pharmaceutical diluents, flavoring agents, disintegrating agents and lubricants and then formed into conventional dosage forms for oral or parenteral administration such as capsules, tablets and parenteral solutions. Generally the dosage forms will contain from 2 mg to 50 mg of the active ingredient. Representative of suitable pharmaceutical compositions which may be prepared is the following:

Tablets	Mg
(1) β-(p-chlorophenylthio)-α-methylalanine	10
(2) Starch USP	57
(3) Lactose USP	73
(4) Talc USP	9
(5) Stearic acid	6

Powders 1, 2 and 3 are slugged, then granulated, mixed with 4 and 5 and tableted. The daily dose of the active ingredient will normally range from 0.01 to 10 mg/kg per day. The exact amount to be prescribed will, of course, depend upon many factors, including the patient's age, weight and condition.

CYSTINE AND DERIVATIVES

Also known as dicysteine or more correctly as 3,3'-dithiobis(2-aminopropionic acid), cystine is readily obtained by the oxidation of cysteine.

Cystine or Cysteine in Treatment of Opthalmological Disorders

Cystine and/or cysteine or their derivatives have found a certain number of uses as anticollagenasic agents. It is also known that pyridoxine (vitamin B_6) in the form of the base or a salt has been utilized in numerous medicaments. It has been found by *A. Bigou; U.S. Patent 4,002,758; January 11, 1977; assigned to Societe B.F.B., France* that the combination, in the same orally administrable medicament, of cystine and/or cysteine with pyridoxine has unexpected beneficial results in ophthalmology and, in particular, upon disorders of the cornea in which there is lysis of collagen by the action of collagenase. This process thus relates to an orally administrable medicament which comprises at the same time pyridoxine and from 1 to 10 parts by weight, with respect to the pyridoxine, of cystine and/or cysteine.

Utilizable doses of the medicament according to the process depend upon the disorder to be treated, but they are preferably such that the patient should be able to absorb 200 to 1,000 mg of cystine and/or cysteine and from 50 to 300 mg of pyridoxine per dose, such dose being repeated 2 to 4 times per day if necessary. The medicament can be made up in the form of a coated tablet or pill or cachet or, in general, any form which can be taken orally. Clearly such pills can contain the ordinary excipients used, such as sugars, carboxymethylcellulose and talc, for instance.

The applications of the medicament according to the process essentially relate to treatments of keratoconus, corneal injury or damage, corneal herpes, subepithelial viral keratitis and epidemic epithelial keratitis punctata. They also relate to the treatment of the difficulties caused by cicatrization of the cornea and those due to nontolerance by the cornea of contact lenses.

Example: Various Applications – Coated tablets of cystine and pyridoxine corresponding to the following formula were made:

Cystine	500 mg
Pyridoxine (vitamin B_6)	100 mg
Lactose Carboxymethylcellulose Talc Magnesium stearate Water	Sufficient quantities to give a tablet weighing about 1 g

These tablets were coated with a known coating compound (Eudragit E, talc, gum arabic, sugar, tartrazine and yellow wax). These tablets were administered to various patients at the rate of four tablets per day on average, as follows:

(1) Three cases of trophic troubles of the cornea had been able to recover owing to a sufficiently long treatment;
(2) Two cases of subepithelial viral keratitis had recovered owing to treatment which was carried out for 15 days;
(3) Sixteen cases of epidemic epithelial keratitis punctata had recovered by treatment for 1 week approximately, though with the medicaments known the healing could only be obtained after several weeks or even several months;
(4) Fourteen cases of intolerance of corneal contact lenses had been treated. The patients had been able to continue to wear lenses throughout the treatment and the lenses had not needed to undergo any modification. At the end of the treatment, the patients could wear their lenses without inconvenience.

Preparation of Cystine Peptides

Two processes have been developed at Ciba-Geigy for the production of cystine peptides from cysteine peptides having protected mercapto groups. In the disclosure by *B. Kamber and W. Rittel; U.S. Patent 3,794,633; assigned to Ciba-Geigy Corporation*, the protective group is CH_2NHCOR, in which COR represents the acyl radical of a carboxylic acid. The S-protected cysteine-containing sequences are directly oxidized with iodine to yield the cystine disulfide bond.

The reaction takes place at room temperature but can, depending on the nature of the peptide, also be carried out at temperatures of 0° to 60°C. The reaction is preferably carried out in a solvent in which iodine and the peptide are at least partially soluble, preferably in an alcohol such as a lower alkanol, e.g., ethanol, or especially methanol, or in a mixture of an alcohol with an organic solvent, in which the peptide is soluble, such as ethyl acetate, dimethylformamide, methylene chloride or in acetic acid, e.g., glacial acetic acid or aqueous acetic acid.

In the process, peptides in which the amino groups are free or protected are used as starting substances. Free hydroxyl and carboxyl groups can also, if desired, be present in a protected form. As amino protective groups there should, e.g., be mentioned trifluoracetyl, phthaloyl, p-toluenesulfonyl, or above all groups derived from carbonic acid such as carbobenzoxy groups which are optionally substituted in the aromatic residue by halogen atoms, or lower alkyl, lower alkoxy or lower carbalkoxy groups, tolyloxycarbonyl, 2-phenylisopropoxycarbonyl, 2-tolylisopropoxycarbonyl and above all 2-p-diphenylisopropoxycarbonyl (under the abovementioned conditions) and also aliphatic oxycarbonyl groups, such as, e.g., cyclopentyloxycarbonyl, tert-amyloxycarbonyl, adamantyloxycarbonyl and primarily tert-butoxycarbonyl.

The carboxyl groups can, if desired, e.g., be protected by amidation or esterification. As esters, those of methanol, ethanol, benzyl alcohol, p-methoxybenzyl alcohol, 2,4,5-trichlorophenol, N-hydroxysuccinimide, N-hydroxyphthalimide or above all of tert-butanol should, e.g., be mentioned. Hydroxyl groups, e.g., of serine or tyrosine residues, can, e.g., be protected by etherification, e.g., with benzyl alcohol or preferably with tert-butanol. In arginine residues, the guanidino group can, e.g., be protected by the tosyl group. The disulfide-peptides obtained in the process and possessing protective groups can be directly used for the synthesis of peptides with a longer amino acid chain or, if desired, the protective groups can be split off in a known manner by hydrolysis and/or reduction.

The process is described in the examples which follow. The following abbreviations are used: BOC, tert-butoxycarbonyl; But, tert-butyl; Acm, acetylaminomethyl; OtBu, tert-butyl ester; Asn, Asparagine; and Trt, trityl.

Example 1: A solution of 463 mg of BOC–Cys(Acm)–Asn–OtBu in 5 ml of methanol is added to 500 mg of iodine in 5 ml of methanol and the mixture left to stand for 1 hour at 20°C. Excess iodine is then removed at 0°C by dropwise addition of aqueous 1 N thiosulfate solution and the product is precipitated by adding water. The dried precipitate is recrystallized from methanol. Crystals of melting point 194° to 196°C are obtained.

The starting material BOC–Cys(Acm)–Asn–OtBu can be manufactured as follows. 585 mg of BOC–Cys(Acm)–OH and 377 mg of H–Asn–OtBu in 10 ml of ethyl acetate and 5 ml of dimethylformamide are mixed at 0°C with 452 mg of dicyclohexylcarbodiimide and allowed to react for 24 hours at this temperature. The mixture is filtered off, the filtrate is evaporated to dryness, the residue is taken up in ethyl acetate and the solution is washed with 1 N citric acid, 1 N sodium bicarbonate and water. The organic phase is dried with sodium sulfate and evaporated. Crystallization from ethyl acetate and n-hexane yields BOC-Cys(Acm)–Asn–OtBu of melting point 97° to 100°C.

Example 2: This example describes the preparation of the following peptide:

BOC–Cys–Asn–OH
|
Trt–Cys–Gly–Glu(OtBu)$_2$

BOC–Cys(Acm–Asn–OtBu (1.18 g) and Trt–Cys(Acm)–Gly–Glu(OtBu)$_2$ (1.83 g) in 30 ml of methanol are added to a stirred mixture of 3.2 g of iodine and 1.38 g of pyridine in 75 ml of methanol. After 2 hours at 20°C the excess iodine is removed with 1 N thiosulfate solution, and the mixture is concentrated to about 20 ml and poured into 200 ml of ice water. The product is filtered off and dried over sodium hydroxide. The unsymmetrical cystine derivative is isolated from this product by countercurrent distribution in the system chloroform-carbon tetrachloride-methanol-0.1 N ammonium acetate (1:8:5:1). After 500 distribution steps the product is present in elements 50 to 75 (K = approximately 0.15).

Another process for preparing cystine-peptides is disclosed by *B. Kamber; U.S. Patent 3,875,136; April 1, 1975; assigned to Ciba-Geigy Corporation.* In this process a cysteine peptide whose mercapto group is substituted by trityl or acylamidomethyl is first reacted with Hal–S–CO–OR$_3$ where R$_3$ is an optionally substituted hydrocarbon residue. Then the resulting –S–COOR$_3$– protected cysteine peptide is reacted with a cysteine peptide with free mercapto group to yield the corresponding cystine peptide.

The process is described in the example which follows. The R$_f$ values relate to thin layer chromatography on silica gel plates; System: 100 = ethyl acetate-n-butanol-pyridine-glacial acetic acid-water (42:21:21:6:10). The following abbreviations are used: OtBu, tert-butyl ester; BOC, tert-butyloxycarbonyl; Trt, trityl; Acm, acetylaminomethyl; SCM, methoxycarbonylsulphenyl (–S–CO–OCH$_3$).

Example: 3.2 g of BOC–Cys(Trt)–Asn–OtBu in 15 ml of CHCl$_3$ and 5 ml of MeOH are added to 700 mg of Cl–S–CO–OCH$_3$ in 2 ml of CHCl$_3$ at 0°C. After 45 minutes at 0°C, 150 ml of CHCl$_3$ are added and the solution is extracted with water. The solution is dried over sodium sulfate and evaporated in a water pump vacuum. The oil thereby produced is a mixture of BOC–Cys(SCM)–Asn–OtBu and triphenylmethyl methyl ether. The latter does not have to be separated off since it does not cause any side-reactions in the subsequent reaction.

BOC–Cys(SCM)–Asn–OtBu (1.45 g) and BOC–Val–Cys–Gly–Glu(OtBu)$_2$ (1.85 g) are dissolved in 30 ml of CHCl$_3$–MeOH (1:1) and the solution is kept for 45 minutes at room temperature. It is then evaporated in a water pump vacuum at 35°C and the resulting foam is twice triturated with 5 ml of petroleum ether at a time. 3.0 g of a product of the formula

BOC–Cys–Asn–OtBu
|
BOC–Val–Cys–Gly–Glu(OtBu)$_2$

which is pure according to thin layer chromatography, are obtained as a colorless powder. R_f = 0.47 in chloroform-methanol (9:1). The BOC–Val–Cys–Gly–Glu(OtBu)$_2$ used as the starting material can be manufactured as follows: 700 mg of mercury(II) acetate are added to 1.72 of BOC–Val–Cys(Trt)–Gly–Glu(OtBu)$_2$ in 20 ml of ethyl acetate and 10 ml of methanol and the mixture is left to stand for 3 hours at room temperature. A stream of H_2S is then passed through the clear solution for 15 minutes, the mercury sulfide which has separated out is filtered off and the filtrate is evaporated. The resulting foam consists of BOC–Val–Cys–Cly–Glu(OtBu)$_2$ [R_f = 0.55 in chloroform-methanol (1:1)] and triphenylmethyl methyl ether. The latter does not interfere with the subsequent reaction.

Cystine Derivatives as Antibacterial Agents

A series of amino acids are disclosed by *B.V. Alexiev, S.B. Stoev, A.V. Spasov, L.S. Maneva, E.A. Emanuilov, and E.V. Golovinski; U.S. Patent 3,897,494; July 29, 1975; assigned to Institute po Biochimia pri Ban, Bulgaria* which have pharmaceutical utility, particularly against bacterial infections. These amino acids have the fomula

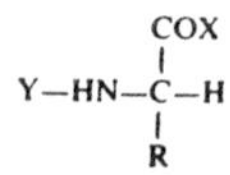

where

X is OH, OC_2H_5, $NHN(CH_2CH_2Cl)_2$;
Y is $-OCOCH_2C_6H_5$, H·HBr;
R is $-CH_2SO_2N(CH_2CH_2Cl)_2$,
$-CH_2-S-S-CH_2-CH(NHOCOCH_2C_6H_5)CONHN(CH_2CH_2Cl)_2$,
$-CH_2-S-S-CH_2-CH(NH_2{\cdot}HBr)CONHN(CH_2CH_2Cl)_2$.

These amino acid derivatives show an emphasized antibacterial activity towards *Sarcina lutea* and the staphylococcal mutants YF-2 and YF-3. For example, the product L-cystine-bis(N,N-β-chlorethylhydrazine) inhibits the growth of *Sarcina lutea* and the staphylococcal mutant YF-2 97% and 100%, respectively, in a concentration of 1 μm/ml and inhibits the growth of the staphylococcal mutant YF-3 100% in a concentration of 0.5 μm/ml. The amino acid derivatives suppress the development of some of the transplantable tumors in experimental animals. The preparation of these amino acid derivatives proceeds through the interaction of a sulfochloride or an acid chloride of the amino acid with a bis-(β-chlorethyl) amine in an acetone medium or with an asymmetical bis-(β-chlorethyl)-hydrazine, respectively in an ethyl acetate medium.

A solution of 5 g N-benzyloxicarbonylcystilchloride in 50 ml anhydrous ethyl acetate is added to the asymmetrical bis-(β-chlorethyl)-hydrazine, obtained from 8 g hydrazine hycrochloride. After standing for 12 hours at the ambient temperature a crystalline product is formed, which is filtrated and recrystallized from an ethanol-water mixture. Yield 5.7g of L-bis(N-α-benzyloxicarbonyl)-cystine-N,N-bis-(β-chlorethyl)-dihydrazide(I) (72.4%). MP = 191° to 192°C.

Example 2: 5.0 g of the substance (I) are tested at the ambient temperature with 35 ml 30% solution of hydrogen bromide in glacial acetic acid. The product is first dissolved under intensive liberation of carbon dioxide, after which the residue is deposited. After full deposition by means of anhydrous ether the solid product is filtrated and washed multifold with dry ether. Yield 3.9 g (91.2%) of L-cystine-bis-(N,N-β-chlorethylhydrazide). MP = 167° to 169°C. The substance is highly hygroscopic.

METHIONINE AND DERIVATIVES

Methionine or 2-amino-4-(methylthio)-butyric acid is essential to human nutrition and is not synthesized by the human body. It therefore finds use as a food additive. In this amino acid, both the D- and the L-forms are effective.

Preparation of DL-Methionine

Various routes have been used to prepare methionine from acrolein, methyl mercaptan and hydrogen cyanide to form 5-(β-methylthioethyl)hydantoin which is subsequently hydrolyzed. Previous processes had not always produced high yields of pure product. *S. Ouchi and C. Shibuya; U.S. Patent 3,917,683; Nov. 4, 1975; assigned to Deutsche Gold- und Silber-Scheideanstalt vormals Roessler, Germany* have developed an improved low cost process whereby high purity DL-methionine can be obtained in high yield.

The process comprises reacting acrolein cyanohydrin (AC) with methyl mercaptan, an ammonia or ammonium supplying agent and a carbonic acid or carbonate supplying agent, in the presence of an organic or aqueous organic solvent. This accomplishes an addition reaction of the methyl mercaptan to a double bond of the acrolein cyanohydrin (AC) simultaneously with a hydantoin ring forming reaction followed by hydrolysis of the resulting product.

The principal feature of the process resides in that the addition reaction of methyl mercaptan to a double bond of AC and the hydantoin ring forming reaction are simultaneously effected in one-stage and the hydrolysis of the resulting hydantoin type intermediate gives high purity DL-methionine in high yield. In addition, the process requires no catalyst and can be easily carried out, leading to the economical production of DL-methionine.

The first step (i.e., the hydantoin ring forming reaction simultaneously with addition reaction) is effected in the presence of an organic solvent or an aqueous organic solvent. As the organic solvents, there may be used any of those that do not decompose and would be inert to the reaction.

Examples of preferable solvents are methanol, ethanol, isopropanol, n-propanol, tert-butanol, n-butanol, ethylene glycol, acetonitrile, propionitrile, malonitrile, succinonitrile, ethylene glycol monomethyl ether, nitromethane, nitroethane, dimethylformamide, dimethylacetamide, N-methylpyrrolidone, dimethyl sulfoxide, hexamethyl phosphoramide, pyridine and the like. The organic solvent may be employed in the form of aqueous organic solvent. The water content of such aqueous organic solvent is preferably not greater than 50 wt %.

The reaction of this step is effected preferably at 50° to 150°C in view of rate of reaction, selectivity in the reaction and other factors. The reaction period in this step is not critical, but the reaction may be usually completed in 0.5 to 3 hours. The methyl mercaptan used in this step may be gaseous or liquid. The amount of methyl mercaptan used is theoretically equivalent to AC but preferably of a somewhat excess. The reaction may be conducted under atmospheric pressure while blowing gaseous methyl mercaptan into the reaction system, or under superatmospheric pressure while maintaining methyl mercaptan in the liquid state. After completion of the reaction in the first step, the reaction mixture is subjected to the second step either as such or, if necessary, after distilling off the solvent to obtain a concentrate.

The operation of hydrolysis (second step) may be conducted according to the method commonly used in the hydrolysis of hydantoin, e.g., heating in the presence of a suitable base or acid. Suitable bases include alkali metal hydroxides, e.g., sodium or potassium hydroxide; alkaline earth metal hydroxides, e.g., calcium or barium hydroxide; alkali metal carbonates, e.g., sodium or potassium carbonate; and the like. Suitable acids include inorganic acids, e.g., hydrochloric or sulfuric acid; and organic acids, e.g., acetic or propionic acid. After completion of the second step, the desired DL-methionine can be easily recovered as pure crystals from the reaction mixture through isolation and purification by a conventional method, for instance, by the treatment with a suitable ion exchange resin or by neutralization.

Example: A series of reactions using a variety of solvents were carried out as follows. Into a stainless steel pressure reactor having a capacity of 100 ml were charged 3.32 g of acrolein cyanohydrin and 9.5 g of ammonium bicarbonate together with 40 ml of the organic solvent indicated below. The reactor was cooled to -30°C externally with dry ice-methanol. Separately, 2.0 g of cold liquid methyl mercaptan were withdrawn from a bomb and introduced into the reactor. The reactor was then sealed and shaken in an oil bath at 80°C for 1.5 hours.

Thereafter, the solvent was removed from the reaction mixture and 40 ml of 2 N NaOH were added to the residue, which was then diluted with water to make up the whole amount to 50 ml. The resulting mixture was heated at 160°C for 1.5 hours. The reaction mixture was passed through Amberite-200 (ion exchange resin), which was then eluted with a 5% aqueous ammonia. The eluate so obtained was treated with active carbon and concentrated to give DL-methione at the yield indicated below.

Organic Solvent	Yield of DL-methionine (Percent)
Methanol	87.9
Ethanol	84.0
Isopropanol	82.0
tert-Butanol	80.0
Ethylene glycol	83.5
Acetonitrile	85.2
Nitromethane	76.5
Dimethylacetamide	83.3
Dimethylformamide	83.0
Tetrahydrofuran	72.1
Pyridine	84.5
Dimethylsulfoxide	80.5
Acetone	74.2
Chloroform	72.2
Benzene	70.5
Ethylene glycol monomethyl ether	75.5

S. Ouchi and C. Shibuya; U.S. Patent 3,833,651; September 3, 1974; assigned to Deutsche Gold- und Silber-Scheideanstalt, vormals Roessler, Germany have also developed a one-stage method to give a hydantoin type intermediate, which is then hydrolyzed. The process can be simply and easily conducted to obtain high purity DL-methionine in high yield at low cost.

In operation, acrolein, methyl mercaptan, hydrocyanic acid, a source of ammonium ion and a source of carbonate ion are reacted in the presence of a reaction medium at 30° to 200°C to accomplish an addition reaction of the methyl mercaptan and hydrocyanic acid to acrolein simultaneously with a hydantoin ring forming reaction. The hydantoin type intermediate thus obtained is then subjected to the second step, i.e., hydrolysis. Before hydrolysis, it is desirable to preliminarily distill the reaction medium or solvent from the reaction mixture comprising the hydantoin type intermediate. The operation of hydrolysis may be conducted according to the method commonly employed in the hydrolysis of hydantoin. Solvents used in the process are the same as those given in U.S. Patent 3,917,683.

Example: A series of reactions using a variety of organic solvents were carried out as follows. Into a stainless steel pressure reactor having a capacity of 100 ml were charged 2.24 g of acrolein containing 0.1 wt % of methoxyhydroquinone and 9.5 g of ammonium bicarbonate together with 40 ml of an organic solvent indicated below. The reactor was cooled to -30°C externally with dry ice-methanol. Separately, 2.0 g of cold liquid methyl mercaptan were withdrawn from a bomb and introduced into the reactor.

1.1 g of liquid hydrocyanic acid were also added to the reactor. The reactor was then sealed and shaken in an oil bath at 70°C for 2 hours. Thereafter, the solvent and the reactants remaining unreacted were removed from the reaction mixture. To the residue, 40 ml of 2 N NaOH were added and the mixture was hydrolyzed at 160°C for 1.5 hours. The reaction product was treated in the same manner as described in the example given in U.S. Patent 3,917,683, to give DL-methionine at the yield indicated below (based on acrolein).

Organic Solvent	Yield of DL-methionine (Percent)
Methanol	89.2
Ethanol	88.5
Ethylene glycol	80.6
Acetonitrile	86.5
Dimethylformamide	85.5
Tetrahydrofuran	75.6
Dimethylsulfoxide	82.1
Chloroform	76.2
Benzene	74.8
Acetone	73.0
Ethylene glycol monomethyl ether	76.3

Preparation of N-Acetyl-L-Methionine

Previous processes for the preparation of N-acetyl-L-methionine gave low yields, often accompanied by partial racemization. *H. Wagner and A. Maierhofer; U.S. Patent 4,011,263; March 8, 1977; assigned to Deutsche Gold- und Silber-Scheideanstalt vormals Roessler, Germany* have now developed a process whereby optically-pure N-acetyl-L-methionine is produced with yields of at least 90%. According to the process, the acetylation is undertaken suitably at temperatures of 20° to 60°C, preferably at 30° to 50°C, with the pH maintained, preferably between 7.0 and 9.5. It is advantageous to use the acetic anhydride in at least equimolar amounts based on the L-methionine. Preferably, there are used 1.05 to 1.70 mols, especially 1.10 to 1.30 mols, of acetic anhydride per mol of methionine.

In order to establish and maintain the preferable pH of 7.0 to 9.5, there is added alkali, e.g., an aqueous solution of sodium hydroxide or potassium hydroxide, preferably sodium hydroxide. The amount of alkali is dependent to a certain extent upon the amount of acetic anhydride and the type of alkali. In the case of the use of alkali hydroxide, there is generally added 1.6 to 2.1 mols of alkali hydroxide per mol of acetic anhydride. Advantageously, there is present in the reaction mixture about 15 to 30 mols of water per mol of L-methionine. However, the amount of water can vary from 5 to 200 mols per mol of L-methionine.

To carry out the process, preferably L-methionine is present in water and then the acetic anhydride is gradually added. The mixture is always held to a pH of 7.0 to 9.5 before and during the feeding of the acetic anhydride by addition of alkali. The N-acetyl-L-methionine is advantageously recovered from the reaction mixture by acidifying the reaction mixture, e.g., with sulfuric acid, and then extracting with a solvent such as ethyl acetate and separating the N-acetyl-L-methionine from the solvent. If necessary, the N-acetyl-L-methionine is recrystallized and, for this purpose, it is suitable to use water as the solvent. In the following example, the rotary power listed of the N-acetyl-L-methionine is given as the specific rotation $[\alpha]_D^{20}$ in degrees x cm^3/dm x g. Unless otherwise specified, all parts and percentages are by weight.

Example: 149 g (1.00 mol) of L-methionine were present in 390 ml of water. In the course of 45 minutes, there were simultaneously added 119 ml of a 50% solution of sodium hydroxide (2.27 mols of sodium hydroxide) and 114 ml (1.21 mols) of acetic anhydride. The procedure was such that the pH of the mixture was always kept between 7.0 and 8.0. The temperature of the mixture meanwhile, and for a further 30 minutes, was held at 40°C. Then 70 ml of concentrated sulfuric acid (1.26 mols of sulfuric acid) were added so that the pH fell to 1.8. Thereby, there resulted two phases. The lighter phase which consisted of N-acetyl-L-methionine was separated off. The heavier, aqueous phase was successfully extracted with 300 ml, 200 ml and 100 ml of ethyl acetate. The extracts were mixed with the N-acetyl-L-methionine phase.

This mixture was washed with 50 ml of water, then concentrated under reduced pressure to 300 ml, treated with 2.5 g of activated carbon, heated for a short time to 80°C, and subsequently filtered over a pressure filter. Upon cooling the filtrate to 5°C, the N-acetyl-L-methionine crystallized out. It was filtered off under suction, washed with 30 ml of ice cold water, and then dried at 50°C under reduced pressure. After concentration of the mother liquor to half its volume in a corresponding manner, there was recovered a further portion of N-acetyl-L-methionine. In all, there were obtained 132 g of N-acetyl-L-methionine, corresponding to a yield of 90% based on the L-methionine added. The product was 99.9% pure, had a melting point of 104° to 105°C and, in water at a concentration of 4 g/100 ml of water, had a specific rotation of –21.0°C.

Enzymatic Resolution of N-Acryl-DL-Methionine Esters

Proteolytic enzymes such as sulfhydryl proteinases or microbially derived serine proteinases are used by *C.E. Stauffer; U.S. Patent 3,963,573; June 15, 1976; assigned to The Procter & Gamble Company* to resolve D,L-methionine derivatives. It has been found that sulfhydryl proteinases and microbially derived serine proteinases exhibit high esterase activity for N-acyl-L-methionine esters and very low esterase activity for N-acyl-D-methionine esters. Furthermore, it has been found that the high esterase activity exhibited for the L-isomer is not exhibited by the presence of the D-isomer.

Subjecting N-acyl-D,L-methionine ester to the action of such a proteinase provides a mixture of N-acyl-D-methionine ester and N-acyl-L-methionine. The N-acyl-L-methionine can be readily separated from the mixture by conventional means, e.g., by adjusting the pH of an aqueous mixture and extracting with an organic solvent such as chloroform, ethyl acetate or butyl acetate. A variety of specific N-acyl-D,L-methionine esters can be used in this process. Preferably, the acyl group is derived from fatty acids containing from 1 to 9 carbon atoms. The ester group can be derived from a variety of alcohols containing from 1 to 10, preferably 1 to 6, carbon atoms. Racemic N-acetyl-D,L-methionine methyl ester is most preferred for use in this process.

Suitable sulfhydryl proteinases are, e.g., papain, ficin and bromelain. Serine proteinases suitable for use in this process are derived from microorganisms such as bacteria, fungi and mold. Microbially derived serine proteinases are preferred such as those derived from the bacteria *Bacillus subtilis* and termed subtilisins. A preferred subtilisin is the *Bacillus subtilis*-derived Carlsberg strain.

Example: (A) Preparation of N-Acetyl-D,L-Methionine Methyl Ester – One hundred grams of D,L-methionine were mixed with 500 ml of methanol containing 30 g NaOH to form a stirrable slurry. To this mixture there was slowly added with stirring 1.5 mol equivalents (based on D,L-methionine) of acetic anhydride. This mixture was continuously stirred until the slurry turned to a clear solution. The resulting clear solution contained N-acetyl-D,L-methionine dissolved in methanol.

To this solution was added with stirring sufficient anhydrous H_2SO_4 to neutralize the NaOH and render the solution slightly acidic. Under these conditions the N-acetyl-D,L-methionine was converted to N-acetyl-D,L-methionine methyl ester. The remaining methanol and methyl acetate (formed from the reaction of residual acetic acid and acetic anhydride with NaOH) was removed by warming the mixture under vacuum. The remaining mixture was slurried with chloroform and filtered to remove sodium sulfate salts. The ester-chloroform solution was then water washed to remove any remaining salts. The resulting ester-chloroform solution was then warmed under vacuum to remove the chloroform. The resulting material was N-acetyl-D,L-methionine methyl ester.

(B) Preparation of N-Acetyl-L-Methionine – An aqueous solution containing 20 wt % N-acetyl-D,L-methionine methyl ester was formed. The pH was adjusted to 7.5 by the addition of 1 N NaOH. To this solution there was slowly added 0.04 wt % of the aqueous solution of crystalline subtilisin Carlsberg, a serine proteinase (Alcalase). During the addition of the proteinase, the solution

was continuously stirred, and the pH was kept constant by the automatic addition of 1 N NaOH from an automatic Radiometer pH-stat type titrator. When no further addition of NaOH is needed to maintain pH, the reaction is complete (here about 30 minutes). The resulting mixture of N-acetyl-D-methionine methyl ester and N-acetyl-L-methionine were separated in the following manner. The aqueous mixture was raised to a pH of 10 to 11 by the addition of NaOH. This aqueous mixture was then extracted three times, with an equal volume of chloroform.

The separated chloroform layers containing N-acetyl-D-methionine methyl ester were combined. The chloroform was removed by warming the solution under vacuum. The resulting N-acetyl-D-methionine methyl ester can be added to water or methanol containing NaOH, and heated for a time sufficient to racemize the ester material. The resulting D,L-material can then be recycled in the process.

The water layer, containing N-acetyl-L-methionine, was acidified by the addition of H_2SO_4 to lower the pH to 2 or less. This acidified water was then extracted three times with equal volumes of chloroform. The separated chloroform layers were combined. The chloroform solution was then warmed gently under vacuum to remove the chloroform. The product was N-acetyl-L-methionine having an optical purity of better than 95%.

Resolution of N-Acyl-DL-Methionine by Crystallization

R. Fahnenstich and J. Heese; U.S. Patent 3,845,110; October 29, 1974; assigned to Deutsche Gold- und Silber-Scheideanstalt vormals Roessler, Germany have developed a process for the separation of N-acyl-D-methionine or N-acyl-L-methionine from mixtures of these two materials by fractional crystallization of their diastereomer salts with the optically active bases D-lysine or L-lysine.

D-lysine and L-lysine are optically active bases which are easily available. Surprisingly they form with N-acyl-D-methionine and N-acyl-L-methionine diastereomer salts which have such strongly different physical properties that there can be obtained a high degree of separation. By a single fractional crystallization from a mixture containing equal amounts of N-acyl-D-methionine and N-acyl-L-methionine there can be obtained about 50% yield of one of these two materials in an optical purity of 90 to 95%.

From the same solution there can be selectively separated one component by crystallization at higher temperature and the other by crystallization at lower temperature. Thus using D-lysine at higher temperatures there preferentially crystallizes D-lysine-N-acyl-D-methionine and at lower temperatures there preferentially crystallizes D-lysine-N-acyl-L-methionine. Conversely using L-lysine at higher temperature there is preferentially separated L-lysine-N-acyl-L-methionine and at lower temperatures there is preferentially separated L-lysine-N-acyl-D-methionine.

The acyl derivatives of D- or L-methionine which can be recovered by the process are chiefly derivatives of aliphatic carboxylic acids, especially containing up to 4 carbon atoms or the derivatives of aromatic carboxylic acids, especially benzoic acid, p-nitrobenzoic acid, benzene sulfonic acid and p-toluene sulfonic acid. Preferably the process is used for the recovery of N-formyl or N-acetyl derivatives of D-methionine or L-methionine.

The fractional crystallization of the diastereomeric salts is advantageously carried out in alcoholic solutions. As alcohols there are mainly used lower alkanols such as methanol, ethanol, isopropanol, propanol, butanol, and sec-butanol. Mixtures of alcohols can also be used. In many cases it is suitable to add alcohols in admixture with water. The amount of water, however, generally does not exceed 20%. An example of a suitable alcohol-water mixture is 95% isopropyl alcohol. For each mol of the methionine isomeric mixture there is suitably added about 0.5 to about 1.5 mols, preferably about 1.0 mol of the D-lysine or L-lysine. Details of the process are shown in the following examples.

Example 1: A mixture of 5.85 g of optically pure L-lysine and 7.64 g of the racemate N-acetyl-D,L-methionine were dissolved in a mixture of 8 g of water and 132 g of isopropanol at 60°C. The solution was inoculated with 0.1 g of L-lysine-N-acetyl-L-methionine and then held for 2 hours at 45°C. The crystallized product consisting essentially of L-lysine-N-acetyl-L-methionine, was filtered off, washed with 17 ml of isopropanol and dried. The yield was 5.9 g, corresponding to 88% based on the N-acetyl-L-methionine added. The product had a specific rotation of +4.6 (concentration of the measuring solution 4.0 g/100 ml, solvent 5 N hydrochloric acid), corresponding to an optical purity of 60%.

The mother liquor was cooled to 4°C, inoculated with 0.1 g of L-lysine-N-acetyl-D-methionine and held for 2 hours at that temperature. The product which crystallized out consisting essentially of L-lysine-N-acetyl-D-methionine was filtered off, washed with 17 ml of isopropanol and dried. The yield was 5.8 g, corresponding to 85% based on the N-acetyl-D-methionine added. The product had a specific rotation of +18.7 (concentration of the measuring solution 4.0 g/100 ml, solvent 5 N hydrochloric acid), corresponding to an optical purity of 67%.

Example 2: A mixture of 1.46 g of optically pure D-lysine and 1.96 g of the racemate N-acetyl-D,L-methionine was dissolved in 3.7 ml of water at 50°C. The solution was added to 66 ml of isopropanol and 0.1 g of D-lysine-N-acetyl-D-methionine heated to 50°C. The mixture was held a half hour at 50°C. The crystallizate was filtered off and washed with 50 ml of isopropanol. The yield of D-lysine-N-acetyl-D-methionine was 2.1 g, corresponding to a yield of 124% based on the N-acetyl-D-methionine added. The product had a specific rotation of -5.3, corresponding to an optical purity of 53%.

Storage-Stable Methionine

Methionine is generally produced by hydrolysis of the corresponding aminonitrile (α-amino γ-methylmercapto butyronitrile) or hydantoin (5-β-methylmercaptoethyl) hydantoin. The hydrolysis is carried out in acid or alkaline medium. The methionine is separated from the solutions resulting from the hydrolysis by neutralization of the solutions to a pH value near the isoelectric point of pH 5.6.

The resulting methionine is contaminated by inorganic salts and by-products occurring in the hydrolysis. Because of these impurities the methionine is impaired in its storability, even if there is a relatively small content of impurities. It has an unpleasant odor, discolors after some time, becomes lumpy, can no longer be sprinkled or poured freely and for most purposes is only still of limited utility. *G. Eikelmann, R. Fahnenstich, T. Lussling, S.-P. Mannsfeld, G. Pohl, H. Tanner and H. Wagner; U.S. Patent 3,931,307; January 6, 1976; assigned to Deutsche Gold- und Silber-Scheideanstalt vormals Roessler, Germany* have found that there can be obtained a storage stable methionine if the pH of the methionine is adjusted to between 3 and 5 in the presence of water. By the use of this process methionine which is contaminated with by-products occurring in the hydrolysis and by the inorganic salts formed in the separation from solution can be rendered storage stable. It is unnecessary to completely purify the methionine from these contaminations.

To carry out this process the methionine is adjusted to a pH of preferably 4.3 to 4.8. The necessary amount of acid is added to the methionine to accomplish this. There can be used any acid which supplies the required pH and is inert to methionine. These include especially mineral acids such as hydrochloric acid, sulfuric acid, phosphoric acid, and hydrobromic acid. Instead of acids there can also be added acid salts of the acids as, e.g., alkali hydrogen sulfates, e.g., sodium hydrogen sulfate and potassium hydrogen sulfate, etc. The preferred acidic material used is sulfuric acid, phosphoric acid, sodium hydrogen sulfate and sodium dihydrogen phosphate.

Example: Methionine was produced by the hydrolysis with potassium carbonate of the corresponding hydantoin, specifically 5-(β-methylmercaptoethyl) hydantoin, precipitated by saturation of the reaction mixture with carbon dioxide, filtering and washing with water. The wet methionine recovered had 24% of adhering water and had a methionine content of 99.3% based on the dry material.

The pH value measured on a 2 g sample of the dry material in 50 ml of water was 7.8. 500 g of the wet methionine were mixed with 29 ml of 5% sulfuric acid. The methionine was then dried at 70°C. A 2 g sample in 50 ml of water had a pH of 4.7. After storing for 5 months at 30°C it had a normal odor, light color and a good pourability. A portion of the wet methionine was not treated with acid but dried directly at 70°C. After the same period of storing this portion had a strong odor, changed color and poor pourability.

Separation of Methionine from Alkali Metal Salts

Earlier processes for the separation of free methionine from alkali metal salts have used oximinocyanoacetic acid in ethanol or cation exchange resins which are expensive. According to the process described by *T. Horisawa, T. Kojima and R. Itoh; U.S. Patent 3,920,737; November 18, 1975; assigned to Kanegafuchi Kagaku Kogyo KK, Japan* an aqeuous solution of alkali metal salt of methionine is neutralized with an inexpensive mineral acid and the resulting crystals of free methionine are recovered. The mother liquor is acidified with the mineral acid and then concentrated.

To the concentrate the mineral acid is again added to precipitate the alkali metal salt of mineral acid, and the salt is recovered by filtration or centrifugation. The resulting filtrate containing the mineral acid, methionine in the form of the mineral acid salt and the inorganic salt is recycled to the neutralization step of the alkali metal salt of the mineral acid, and the above procedure is repeated.

More particularly, an aqueous solution of alkali metal salt of methionine such as sodium or potassium methioninate is neutralized with a mineral acid such as hydrochloric acid or sulfuric acid to neutralize the alkali metal salt into free methionine and the mineral acid salt. The free methionine is remarkably less soluble in water than the mineral acid salt. For example, the solubility of sodium methionate in 100 g of water at 20°C is 3.3 g whereas those of sodium chloride and sodium sulfate are 35.85 g and 19.08 g, respectively. Major parts of the free methionine thus formed may be easily separated out by this difference in solubility. After recovering free methionine by filtration or centrifugation of the neutralized mixture, the filtrate (referred to as "methionine filtrate") is acidified by adding the mineral acid. The quantity of the mineral acid used is sufficient to convert all or substantial parts of the free methionine into the mineral acid salt.

The use of the mineral acid in large excess is undesirable as it causes corrosion of the apparatus and is conveyed to the distillate. The acidified filtrate is then concentrated, preferably under reduced pressure. If the methionine filtrate is concentrated without acidification, precipitation of free methionine will take place so that the concentration to a large extent becomes difficult on account of bubbling or spattering. Besides, impurities, e.g., alkali metal formate formed from the starting alkali metal cyanide during the synthesis of methionine, are accumulated in the cycle of the process to give an impure final product.

In the process, however, precipitation of methionine can be prevented due to the greater solubility of its mineral acid salt and the accumulation of the formate in the cycle can also be avoided as it becomes formic acid and is distilled off along with water during the concentration step. For example, when 860 parts by weight of methionine filtrate containing 2.5 wt % of methionine and 0.7 wt % of sodium formate was acidified with 26 parts by weight of concentrated hydrochloric acid and was concentrated, the distillate contained formic acid at a concentration of 2,600 ppm whereas the control wherein no hydrochloric acid was added gave a distillate containing no trace of formic acid.

The concentration may preferably be completed just before or after the mineral salt begins to precipitate. To the concentrate thus obtained, the mineral acid as used in the neutralizing step is added. The alkali metal salt with the mineral acid will then precipitate out as the solubility decreases, while methionine remains in the concentrate in the form of the mineral acid salt. The resulting inorganic salt is recovered by filtration or centrifugation. The filtrate containing methionine in the form of the mineral acid, the inorganic salt and the mineral acid is then recycled to the neutralizing step of the subsequent cycle of the operation to neutralize the alkali metal salt of methionine, and the procedure as described above is repeated.

Example: (A) – 4,000 parts of an aqueous solution containing 785 parts of sodium methionate, 268 parts of sodium carbonate, 294 parts of sodium hydrogen carbonate and 17 parts of sodium formate were neutralized with theoretical amounts of concentrated hydrochloric acid. The resulting crystals of free methionine were filtered off and washed with water whereby 465 parts of free methionine were obtained. The purity of the product was 99.8% after washing with water whereas that of the crude product before washing was 97.5%. The filtrate and the washing were mixed and adjusted to pH 3.3 with hydrochloric acid. The distillate contained 401 ppm of formic acid. To the concentrated mixture were added 1,150 parts of concentrated hydrochloric acid and the resulting crystals were recovered by filtering whereby 248 parts of sodium chloride having a purity of 99.7% were obtained. The filtrate was returned to the subsequent cycle of the operation.

(B) – 4,000 parts of an aqueous solution containing 785 parts of sodium methionate, 268 parts of sodium carbonate, 294 parts of sodium hydrogen carbonate and 17 parts of sodium formate were neutralized with the filtrate obtained in the preceding cycle. The resulting crystals of free methionine were filtered off and washed with water, whereby 615 parts of free methionine having a purity of 99.9% were obtained. The purity of the product before washing was 97.6%.

The filtrate and the washing were mixed and adjusted to pH 3.3 with hydrochloric acid. The mixture was then concentrated in vacuo. The distillate contained 2,080 ppm of formic acid. To the concentrated mixture were added 1,150 parts of concentrated hydrochloric acid and the resulting crystals were filtered off whereby 612 parts of sodium chloride having a purity of 99.7% were recovered. The resulting filtrate was returned to the next cycle of the operation and the procedure was repeated.

Separating Ammonium Sulfate from Methionine Solutions

In the preparation of methionine by hydrolyzing methionine nitrile with sulfuric acid and neutralizing the hydrolysis mixture with ammonia, ammonium sulfate is produced in a fairly large amount compared with the amount of methionine. It is important, in the production of methionine by the above process, to have a method for a good separation between the methionine and the ammonium sulfate. *G.J.J. Maass and H.A. Korpel; U.S. Patent 3,817,718; June 18, 1974; assigned to Stamicarbon N.V., Netherlands* have developed a process which overcomes the difficulties of previously known methods of using excess amounts to precipitate ammonium sulfate or the use of liquid ammonia to extract the methionine from a dried reaction mixture.

It was found that from an aqueous solution containing ammonium sulfate and methionine with weight ratio of methionine to ammonium sulfate of 0.01:0.3, and at a pH of 0.2 to 6, ammonium sulfate with less than 1 wt % of methionine can be recovered by crystallization. The crystallization temperature should be taken above a minimum value which depends on the weight ratio of methionine to ammonium sulfate and the pH. Accordingly, within the above-mentioned limits, the higher the weight ratio of methionine to ammonium sulfate and the pH, the higher the minimum crystallization temperature would be so that ammonium sulfate with a methionine content of below 1 wt % can be recovered from the solution.

Generally, the crystallization temperature will be from 30° to 110°C. Clearly, the water content of the solution must be such that crystallization will take place at the required temperature. Generally, the solution will contain too much water and excess water must then be removed by evaporation. It is possible, however, that in the preparation of methionine an aqueous solution containing methionine and ammonium sulfate is produced wherein the water content is of the required value. Although the required water content varies with the weight ratio of methionine/ammonium sulfate, the pH, and the crystallization temperature, in general the water content will be within 35 to 60 wt % water in the aqueous solution.

The process provides for the recovery of ammonium sulfate from an aqueous ammonium sulfate solution containing 1 to 30 g of methionine per 100 g of ammonium sulfate, and which is characterized in that at a pH of between 0.2 and

6 ammonium sulfate with a methionine content of below 1 wt % is removed from the solution by crystallization. Preference is given to a pH of 1.0 to 3.5 because it is possible then, under otherwise identical conditions, to crystallize a larger amount of ammonium sulfate of the same low methionine content.

Example 1: 161 g of an aqueous solution containing 1.8 g of methionine and 59.3 g of ammonium sulfate and having a pH of approximately 5, is evaporated under atmospheric pressure. After evaporation of 60 g of water, the precipitate is filtered off at 100°C. The solid so obtained is washed with 30 ml of a saturated aqueous ammonium sulfate solution and subsequently dried at 70°C and 50 mm Hg. This yields 27.8 g of ammonium sulfate with 0.6 wt % of methionine.

Example 2: 2 g of concentrated sulfuric acid (96 wt %) are added to 200 g of an aqueous solution containing 5.4 g of methionine and 62.4 g of ammonium sulfate and having a pH equal to 3.8. After the addition, the pH of the solution is 2.0. Subsequently, 65 g of water are removed from the solution by evaporation at atmospheric pressure. Then, the solution is cooled to 50°C and the crystallized solid material filtered off at that temperature. Washing the isolated solid with 20 ml of a saturated aqueous ammonium sulfate solution and drying it under the same conditions as in Example 1, yields 14.3 g of ammonium sulfate with a methionine content of 0.4 wt %.

Production of S-Adenosylmethionine or Methylthioadenosine by Yeast

A method has been developed by *T. Tsuchida, F. Yoshinaga and S. Okumura; U.S. Patent 3,962,034; June 8, 1976; assigned to Ajinomoto Co., Inc., Japan* for producing S-adenozylmethionine (S-5'-desoxyadenosine-5'-yl methionine) or methylthioadenosine (1-adenyl-5-methylthioribose). S-Adenosylmethionine and methylthioadenosine will be referred to simply as "SAM" and "MTA," respectively.

SAM is known as an active methionine, and has an antifatty liver activity. It is also known that a small amount of SAM is contained in a culture medium and/or cells of bakers' yeast, *Saccharomyces cerevisiae* and that small amounts of both S-adenosylmethionine and methylthioadenosine are contained in cells of *Torulopsis utilis*, when the microorganisms are cultured in the presence of small amounts of methionine. MTA is known as vitamin L_2 and promotes secretion of milk, and has been produced from SAM by hydrolysis.

It has now been found that high amounts of SAM and/or MTA are accumulated in the culture broth and/or in cells of various yeasts when the yeasts are cultured in otherwise conventional medium containing as much as 0.5 to 2.0 g/dl L-methionine with maintaining a pH of the medium at 2 to 5 and preferably 3 to 5. It is known that L-methionine is necessary for the accumulation of SAM by bakers' yeast, but it is surprising that extremely large amounts of SAM together with extremely large amounts of MTA are produced, when the same yeast of genus Saccharomyces is cultured by the specific culture conditions of pH and the amount of methionine in the medium.

Yeasts which produce SAM or MTA, found so far, belong to genus Candida, Pichia, Rhodotorula, Cryptococcus, Hansenula, Tricosporon, Kloeckeria, Torulopsis, Hanseniaspora, Sporoboromyces, Lipomyces or Debaryomyces. Most of the yeasts produce SAM and MTA at the same time in the culture broth and in the cells. The culture media of this process are conventional except for containing 0.5 to 2.0 g/dl L-methionine. The media also contain a carbon source, nitrogen source, and inorganic ions. For a certain yeast, minor organic nutrients such as vitamins or amino acids are necessary for promoting its growth. Cultivation of the yeasts is carried out maintaining the pH at preferably 3 to 5 with alkali or acid, and temperature to 25° to 40°C. Normally, the fermentation period is from 2 to 10 days, although some variation is possible, to accumulate SAM or MTA.

Example: An aqueous culture medium was prepared to contain, per deciliter, 5 g glucose, the amount of methionine shown in the following table, 3 ml soy protein hydrolyzate, 0.5 g KH_2PO_4, and 0.04 g $MgSO_4 \cdot 7H_2O$, and adjusted to pH 4 with ammonia. 20 ml batches of the aqueous culture medium were placed in 500 ml flasks, heated with steam, and inoculated with *Candida lipolytica* IFO 0746. Cultivation was carried out at 31.5°C with shaking for 72 hours. The medium was adjusted to a pH of 4 with ammonia water. SAM and MTA in the culture broth and in the homogenate of the cells were determined by cutting the spots of SAM and MTA on paper chromatography and measuring optical densities each at 260 mμ. The results are shown in the table.

Methionine Added (g/dl)	SAM (mg/dl)	MTA (mg/dl)
0.0	0	0
0.2	45	38
0.5	100	96
1.0	110	100
1.5	180	175
2.0	130	120
2.5	30	5

Other yeasts which may be used in the process include: *Saccharomyces marxianus* IFO 0272; *Hansenula polymorpha* IFO 0799; *Candida tropicalis* IFO 0618; *Pichia mogii* IFO 0607; and *Candida psuedotropicalis* IFO 1065.

Selenomethionine-Se^{75}

The process disclosed by *P.P.H.L. Otto; U.S. Patent 3,898,276; August 5, 1975; assigned to U.S. Philips Corporation* relates to a chemical synthesis of the radioactive selenomethionine-Se^{75} [$CH_3SeCH_2CH(NH_2)COOH$] starting from radioactive labelled selenium. The radioactive selenomethionine-Se^{75} is used as a tracer for tumors especially for the scanning of the pancreas.

The process for obtaining the known selenomethionine-Se^{75} is characterized in that selenium that is labelled with radioisotope Se^{75} is converted, in an oxygen-free solution of tetrahydrofuran at a temperature between +10° and –15°C, into lithium methylselenide-Se^{75} (CH_3SeLi). This selenide is then converted by means of an approximately equivalent amount of an acid into methylselenol-Se^{75} (CH_3SeH). Then this selenol compound by means of sodium methylate and/or means of lithium methylate, is converted into sodium methylselenide (CH_3SeNa) and/or lithium methylselenide (CH_3SeLi), which then is reacted with dihaloethyldiketopiperazine, whereupon the reaction product is cleaved by means of an acid and selenomethionine-Se^{75} is isolated, together with ordinary selenomethionine.

Example: Methyllithium was prepared in the usual way from lithium and methyl iodide in dry diethyl ether. The ether was distilled in vacuo and replaced by oxygen-free dry tetrahydrofuran. The content was approximately 1 mmol methyllithium per ml solvent (determined titrimetrically).

Thus to 3 mmol of powdered selenium, labelled with radioactive Se^{75} in an amount such that the activity was 1 mCi/mmol of selenium there was slowly added, while stirring and while passing over a stream of nitrogen, a solution of about 3 mmol of methyllithium in tetrahydrofuran under oxygen-free and anhydrous conditions and while cooling in a bath to -5°C. So much methyllithium was added, that the initially dark red colored solution was entirely discolored. The lithium methyl selenide Se^{75} formed was decomposed by means of an equivalent amount of diluted sulfuric acid (50 wt %) and the methylselenol was distilled over in high vacuum. All these operations were carried out under absolute exclusion of oxygen and moisture. The yield was approximately 90%.

To the methylselenol distilled over, an equivalent amount of sodium methylate in methanol (2 ml) was added, whereupon all volatile components were distilled off in high vacuum. To the remaining sodium methylselenide a suspension of an equivalent amount of dichloroethyldiketopiperazine (approximately 1.4 mmol) in 8 ml dry oxygen-free ethanol was added. This mixture was heated at 50°C for 2 hours while being stirred, evaporated to dryness in a high vacuum and the residue was hydrolyzed by heating with concentrated hydrochloric acid. After evaporating to dryness it was taken up in 2.5 N HCl and purified by means of column chromatography over an ion exchanger (Dowex 50 x 8, 100 to 200 mesh, eluate with 2.5 N HCl). Yield 50% (calculated on selenium). It is also possible to use dibromoethyldiketopiperazine in an optically pure L-form in place of the dichloroethyldiketopiperazine.

METHIONINE AS FOOD SUPPLEMENT

Methionine as Feed Supplement for Hoofed Animals

The process developed by *J.G. Forest and E.J. Czarnetzky; U.S. Patent 3,826,827; July 30, 1974; assigned to International Stock Food Corporation* relates to a feed supplement adapted to improve the hoofs of hoofed animals and notably of horses and cattle. In particular, it relates to the use of ingestible sulfur compounds for this purpose. These compositions are applicable to treatment of all hoofed animals, but for simplicity, the process will be described in terms of its applicability to horses.

In the horse, the hoof grows downward from the coronary band from just under the hairline at the top of the hoof. Any injury to the coronary band affects the growth of the hoof. For example, a quarter crack usually begins at the coronary band and often runs all the way to the bottom of the hoof. A quarter crack is one in that portion of the wall of the hoof lying between the heel and the toe, and usually is more spread at the lower end of the crack. Often, the cracked hoof has become dry, hard and brittle. In general, drying out of the hoof, improper shoeing, improper trimming all contribute to hoof trouble. Certainly, an important aspect of proper care for the horse is maintenance of the outside horny shell of the hoof in a healthy condition.

Heretofore, no adequate method of alleviating hoof problems and, in particular quarter crack, have been known. Rest or corrective shoeing is all that could be done for the horse. This process, however, provides a nutritional supplement, specifically adapted to assure rapid and healthy growth of the keratinaceous horny shell of the hoof. Accordingly, in its broadest sense, the process contemplates feeding 0.1 to 5.0 oz/day of either methionine or a methionine equivalent. Since many forms of sulfur are assimilable at least in part, cysteine or cystine may replace the methionine in whole or in part as may sulfur itself or sulfate salts, or even high sulfur content foods such as egg yolks.

In a preferred composition, a sulfur-containing amino acid, notably methionine, hydroxy methionine, cysteine or cystine or a mixture of two or all three of these should be present to an extent such that 0.5 oz/day thereof is fed to the animal. A separate point of preference is inclusion of DL-alanine in the supplement in amounts so that at least 0.05 oz/day thereof is fed to the animal. A preferred composition is as follows: DL-alanine, 1 to 100 g; DL-methionine, 1 to 100 g; sodium sulfate, 0.1 to 100 g; sulfur, 1 to 100 g; and powdered egg yolk, 1 oz to 1 lb.

Example: 10 g of DL-methionine were fed daily to each of 4 horses for a period of 50 days. A similar control group of 4 horses was maintained on the same feed rations (excepting, of course, the methionine). The hoofs of all the horses were measured to the nearest millimeter from the coronary band to the toe and they were all shod throughout the test period. The hoofs grew an average of 11.5 mm in the control group. In the experimental group, the average growth over the test period was 14.0 mm.

A visual examination of the new growth in the experimental animals showed striking changes, particularly in the older horses, whereas the new hoof growth in the control animals was about the same as older hoof growth. In the horses fed the methionine, new hoof growth was translucent in appearance and was covered with loosely adhering, fine silvery flakes of keratin. Vigorous brushing with a stiff brush exposed a clear line of demarcation between old growth and new growth. The new growth was dense and free of cracks.

Zinc Methionine Complex Salts as Diet Supplement

Complex salts have been disclosed by *M.M. Abdel-Monem; U.S. Patents 4,021,569; May 3, 1977; and 3,941,818; March 2, 1976; both assigned to Zinpro Corporation* which function as a readily-available source of zinc and methionine for dietary supplementation. These zinc compounds have the formula: $[CH_3SCH_2CH_2CH(NH_2)COO^-{\cdot}Zn^{++}]_wX^-$, where X is an anion and w is an integer equal to the anionic charge of X. These compounds all contain a 1:1 ratio complex of zinc and methionine as the cation. Such compounds, it is believed because of a complex formed between the zinc and methionine, are in a form which can be readily absorbed, distributed, and utilized within the biochemical system of animals and humans.

The selection of an anion is not critical. The anion can be inorganic, organic, monovalent, divalent, or polyvalent. Preferably, where the anion is an inorganic anion, it is selected from monovalent anions, such as halides, hydrogen sulfate, and dihydrogen phosphate because of the ready solubility of the zinc methionine complexed salts. Where organic acid sources are utilized as the source of the anion for the salts of this process, it is preferred that the source be a monobasic carboxylic acid and that the acid be either acetic, propionic, or benzoic acid. The method will be described in connection with the preparation of zinc methionine acid sulfate.

In accord with this process, zinc sulfate, either in an anhydrous or hydrated form is reacted with methionine at a pH of 7 or less. It is important that the pH be controlled to 6 to 7.5 either neutral or acid conditions because if the pH is allowed to become basic (greater than 7.5), the resulting product will be a 1:2 salt of methionine and zinc which is insoluble in water. The reactants, zinc sulfate and methionine are used in equimolar quantities. Use of equimolar quantities insures complete reaction with a minimum of side reactions, or excess ingredients. It is preferred that the zinc sulfate be hydrated zinc sulfate, in that little water is needed. Where anhydrous zinc sulfate is used, it is preferred that a quantity of water be added to the reaction ingredients about equal to the weight of zinc sulfate and methionine.

The product, zinc methionine acid sulfate, can be conveniently obtained from the reaction solution by the addition of an organic solvent such as isopropyl alcohol, acetone or ethyl alcohol. Alternatively, the reaction solution could be dried using conventional drying methods such as hot air oven, spray drying, or freeze drying to provide a white, solid, uniformly free-flowing powdered form of zinc methionine acid sulfate. The reaction can be carried out under similar conditions of pH, to prepare zinc methionine chloride by substituting zinc chloride as an initial reaction ingredient for the zinc sulfate previously described herein. However, the resulting product is zinc methionine chloride, which is very soluble in water and is obtained as a syrupy concentrated aqueous solution.

Example: The effect of 1:1 zinc methionine complexes on the degree of incorporation of radio-labeled methionine in protein was studied in rats. In one experiment 42 male Sprague-Dawley rats, average weight 174.3 g, were self-fed on a diet of Purina Chow and water. Nineteen animals were randomly selected and each was put under light ether anesthesia and administered a single oral dose of 40 mg/kg of body weight 2-[^{14}C]-DL-methionine acid sulfate which had been prepared from 2-[^{14}C]-DL-methionine, and was put into a metabolism cage.

The remaining 23 rats were put under light ether anesthesia and administered a single oral dose of a mixture of 19.0 mg/kg of 2-[^{14}C]-DL-methionine and 10.5 mg/kg zinc oxide, which are amounts of methionine and zinc equivalent to those present in the dose of zinc methionine acid sulfate administered to the first group of rats. 24 hours later the rats were sacrificed and the livers were immediately removed. The weights of the livers were determined and portions of the livers were treated and their radioactivity measured with a liquid scintillation counter. The data obtained from the tests are summarized in the table on the following page.

The data indicate that rats which received radio-labeled methionine in the form of zinc methionine sulfate complex converted more of the amino acid methionine to protein in the liver than the rats which received an equivalent amount of the amino acid alone. Thus 1:1 zinc methionine complexes provide a significantly greater body conversion of methionine to protein than is obtained when feeding methionine alone.

	Zinc [^{14}C]-DL-Methionine Acid Sulfate Group	[^{14}C]-DL-Methionine Group
Number of animals	19	23
Average weight of rats, g (S.E.)	171.5 (3.8)	177.0 (3.0)
Average weight of livers, g (S.E.)	7.8 (0.3)	8.1 (0.3)
Average ratio of liver weight to rat weight, % (S.E.)	4.56 (0.14)	4.54 (0.12)
Average cpm*/g liver (S.E.)	7.12×10^4 (4.68×10^3)	6.12×10^4 (3.65×10^3)
Average cpm/g liver (S.E.)	5.52×10^5 (2.6×10^4)	4.85×10^5 (2.7×10^4)

*Counts per minute.

N-Acyl Methionines as Food Additives

Fortification of proteinaceous foodstuffs with sulfur-containing amino acids, and particularly with respect to methionine, has typically involved direct addition of DL-methionine as the free acid. The DL-racemic mixture was used even though only the L stereoisomeric form is found incorporated in human protein because methionine is one of the few α-amino acids in which the D-form experiences the metabolic conversion to the L-form. However, it was discovered that foodstuffs so fortified with DL-methionine developed off-flavors and became discolored.

This effect was particularly noticeable for foodstuffs that were either heated prior to eating, or were stored for an indefinite period in a hydrated condition, but the effect was also noticed after dry storage. The chemical basis for the generation of these off-flavors and unsightly discolorations was, in part, attributable to the well-known Strecker degradation, or more commonly, the Maillard reaction (chemical browning).

Unexpectedly and contrary to the prior view, it has been discovered that of the racemic mixture, N-acyl DL-methionine, only the L-isomer is nutritionally available; the D-isomer does not serve as a source of methionine in animal metabolism. Accordingly, *R.A. Damico and R.W. Boggs; U.S. Patent 3,878,305; April 15, 1975; assigned to The Procter & Gamble Company* have developed a proteinaceous foodstuff comprising a sulfur-containing amino acid deficient protein and a nutritionally supplemental amount of an L-amino acid derivative selected from N-acyl-L-methionine, N,N'-diacyl-L-cystine, or N-acyl-L-cysteine. The foodstuffs are essentially free of the corresponding D-isomer of these amino acid derivatives.

The following table shows examples of proteinaceous foodstuffs fortified with N-acryl derivatives of L-methionine, L-cystine and L-cysteine. These examples are given to indicate the range of products which can be prepared. The percent addition figures in the table are calculated so as to benefit the indicated protein in sulfur amino acids to a level comparable to that of whole egg protein. In the table, equivalent benefication is achieved when the N-acetyl-L-methionine, N,N'-diacetyl L-cystine, and N-acetyl-L-cysteine, respectively, is replaced by a nutritionally equivalent amount of the corresponding derivatives having as acyl substituent: formyl, propionoyl, butyroyl, valeroyl, caproyl, enanthoyl, caprylyl, pelargonoyl, carpryl, undecanoyi, lauroyl, tridecanoyl, myristoyl, pentadecanoyl, palmitoyl, margaroyl, stearoyl, nonadecanoyl, arachidoyl, heneicosanoyl, behenoyl, tricosanoyl, tetracosanoyl, palmitoleoyl, oleoyl, elaidoyl, vaccenoyl, linoleoyl, linolenoyl, γ-linolenoyl, eleostearoyl, arachidonoyl, and clupanodonoyl, respectively.

Percent Addition of Total Protein

Protein Sources	N-Acetyl-L-Methionine a	b	N,N′-Diacetyl-L-Cystine c	d	N-Acetyl-L-Cysteine e	f
Animal protein						
Collagen (gelatin)	3.8	9.0	2.8	3.8	2.8	3.8
Plant protein						
Leaf (alfalfa)	0.8	5.9	0.8	1.8	0.8	1.8
Oilseeds						
Soybean						
Meal	1.2	6.3	1.4	2.3	1.4	2.3
Concentrate	1.3	6.4	1.5	2.4	1.5	2.4
Isolate	1.5	6.7	1.6	2.6	1.6	2.6
Cottonseed	0.3	5.4	0.8	1.8	0.8	1.8
Peanut	1.3	6.4	1.1	2.0	1.1	2.0
Sunflower	3.0	4.7	0.8	1.8	0.8	1.8
Flax meal (linseed meal)	0.1	5.2	0.9	1.9	0.9	1.9
Cereal grains						
Wheat (gluten)	2.0	4.6	0.3	1.2	0.3	1.2
Wheat	2.0	3.5	–	–	–	–
Corn	3.0	4.9	0.8	1.8	0.8	1.8
Barley	2.0	4.2	0.4	0.7	0.4	0.7
Oats	0.6	5.8	1.4	2.3	1.4	2.3
Rice	0.4	5.5	1.5	2.4	1.5	–
Milk protein						
Casein	2.0	4.5	2.4	3.4	2.4	3.4
Nonfat milk solids	3.0	5.0	1.8	2.7	1.8	2.7
Whey protein	2.0	4.0	–	–	–	–
Lactalbumin	2.0	4.6	0.1	0.3	0.1	0.3
Marine protein						
Fish protein concentrate	1.0	3.8	0.1	1.1	0.1	1.1

Example: A meat analog product was made according to the following formulation and procedure.

Component	Parts by Weight
Soybean concentrate (70% protein)	60.0
Water	40.0
N-acetyl-L-methionine	1.8
Fat	10.0
Starch	10.0
Beef extract flavoring	5.0
Food coloring and salt	

The components were blended to form a dough, which was passed through a three-fourth inch Brabender extruder using a temperature in the extruder of 150°C. A pressure of approximately 450 psi developed within the extruder and forced a cylindrical product from the 4 mm exit die. The extruded product was dried and ground to particle sizes having an average diameter of one-fourth inch. When the product was hydrated and shaped into hamburger-like patties and fried in an oiled pan, no off-flavors attributable to the degradation of products of methionine were observed. Fortified peanut butter, canned pet food or cattle feed could also be produced by addition of less than 1 part by weight of the N-acylated methionine or other S-containing amino acid.

N-Hydroxymethylmethionine as Food Additive

H. Bertram, R. Fahnenstich and H. Tanner; U.S. Patent 3,968,253; July 6, 1976; assigned to Deutsche Gold- und Silber-Scheideanstalt, Germany have found that N-hydroxymethylmethionine of the formula

$$CH_3{-}S{-}CH_2{-}CH_2{-}\underset{\displaystyle NH{-}CH_2OH}{\underset{|}{CH}}{-}COOH$$

is suited for feeding ruminants. Surprisingly there results by the addition of this compound as a feed additive a considerable improvement, for example, in wool growth in sheep. It is advantageous to add the N-hydroxymethylmethionine as a salt, for example, as the magnesium salt or especially the calcium salt. There can also be used other nontoxic salts, e.g., the potassium and sodium salts.

The N-hydroxymethylmethionine can be prepared, for example, by reaction of methionine with formaldehyde. The salts are obtained if there are added compounds of the cations involved, especially the hydroxide; for example, to recover the magnesium or calcium salt there is used magnesium or calcium hydroxide. The N-hydroxymethylmethionine and its salts can be fed directly. Generally, however, they are used in conventional feeds or with premixes such as vitamin and mineral mixtures or with feedgrade urea. The amount of N-hydroxymethylmethionine or salt added to the feed based on the entire amount of feed on a dry basis supplied in one day should be about 0.01 to 5%, especially about 0.05 to 0.5% by weight.

Example: The research animals used were twelve merino sheep. As the basic feed there were added per animal per day depending on the age of the animals, 800 to 1,000 g of a mixture of 10.8% grass meal, 44.9% oat hull meal, 42.9% tapioca and 1.4% of a mineral mixture for cattle as well as 100 g of wheat straw. In a preliminary period of 5 weeks all of the animals were fed this basic feed exclusively. In a subsequent main period of 5 weeks the animals were fed separately in 3 groups of 4 animals each.

Group 1, animals 1 to 4, only basic feed;
Group 2, animals 5 to 8, 4.0 g of methionine were added daily to the basic feed; and
Group 3, animals 9 to 12, 5.3 g of the calcium salt of N-hydroxymethylmethionine were added to the basic feed.

At the end of the preliminary period and of the main period on each animal there was ascertained the wool yield compared to the cuttings before the preliminary period. The results are collected in the following table.

	. . . Preliminary Period. . .		. . . . Main Period.	
	Wool (g)	Yield Group 1 (%)	Wool (g)	Yield Group 1 (%)
Group 1				
1	73.3		64.4	
2	62.8		72.9	
3	66.2		64.3	
4	63.9		67.1	
Total	266.2	100	268.7	100

(continued)

	. . . Preliminary Period. . .		. . . . Main Period	
	Wool (g)	Yield Group 1 (%)	Wool (g)	Yield Group 1 (%)
Group 2				
5	64.2		68.3	
6	67.3		69.5	
7	71.4		73.4	
8	68.7		69.1	
Total	271.6	102	280.3	104
Group 3				
9	66.8		84.8	
10	59.5		78.4	
11	68.3		83.7	
12	65.9		80.8	
Total	260.7	98	327.7	122

Dipeptides of Glycine with Methionine as Food Additives

Supplementation of poultry foods with low levels of methionine results in marked increases of meat production and feed efficiency. Supplementation of feeds for poultry, pigs, sheep, beef and dairy animals should result in improved meat and milk production. Methionine, however, suffers from a serious shortcoming in that it develops on standing an unpleasant odor, which can adversely affect the palatability of foods or feeds fortified therewith.

According to this disclosure by *R.S. Johnson; U.S. Patent 3,959,519; May 25, 1976; assigned to E.I. du Pont de Nemours and Company* various food materials can be fortified with one or more methionine sources selected from glycyl-D,L-methionine, glycyl-L-methionine, D,L-methionylglycine, and L-methionylglycine. Such fortified food materials are stable under normal processing and storage conditions without development of bad odor. There are many ways to incorporate a methionine source into food materials. The exact method is not critical as regards this process, but low odor will be important in any of them.

Glycyl-D,L-methionine, glycyl-L-methionine, D,L-methionylglycine, and L-methyionylglycine are well-known, readily available chemicals. L-methionylglycine has a specific optical rotation of +86.5±0.5°, and glycyl-L-methionine has a specific optical rotation of –10.0°, each in water solution at a concentration of 2 g/100 g of water.

Example 1: Samples of virtually odorless, recrystallized D,L-methionine, glycyl-D,L-methionine, and D,L-alanyl-D,L-methionine were stored in sealed containers for 29 days at 45°C. The sample of glycyl-D,L-methionine was still virtually odor-free, while the other two samples developed a typical sulfurous odor.

Example 2: For this test, cockerels are produced from a cross of Barred Rock males and Rhode Island females. All the 150 cockerels are selected on the basis that their weight differs no more than 2 grams. The birds are maintained on an all vegetable protein corn-soy basal diet for a 2-week preliminary period. At the end of this period, a median group of 90 birds is selected on the basis of weight, and they are then evenly distributed into three experimental groups, which in turn are placed on the experimental rations as shown in the following table. The level of glycyl-D,L-methionine in Diet No. 3 is equivalent to the level of D,L-methionine in Diet No. 2, on the basis of L-methionine content. The birds are weighed weekly and their feed consumption also determined weekly, to permit weekly feed efficiency evaluations.

	..... Experimental Rations......		
	1*	2	3
Ground corn, lb	37	37	37
Soybean meal, lb	29	29	29
Wheat bran, lb	15	15	15
Wheat middlings, lb	10	10	10
Alfalfa leaf meal, lb	5	5	5
Bone meal, lb	1.5	1.5	1.5
Calcium carbonate, lb	1.0	1.0	1.0
Iodized salt, lb	0.5	0.5	0.5
Riboflavin supplement (550 mcg/g), lb	0.3	0.3	0.3
Vitamin A and D oil, lb	0.5	0.5	0.5
$MnSO_4 \cdot 4H_2O$, g	7.85	7.85	7.85
D,L-Methionine, g	–	68	–
Glycyl-D,L-methionine, g	–	–	92
Crude protein, %	21.70	21.62	21.65

*Control.

After 4 weeks, the average weight of the chicks in the group on Control Ration 1 increases from 37 g to 272 g. The average weight of chicks receiving Rations 2 and 3 is about 11% greater than those of the control group. After 8 weeks, the chicks on Ration 1 average 837 g in weight. The average weight of chicks receiving Rations 2 and 3 for 8 weeks is about 14% greater than those on Ration 1. Feed efficiencies in terms of grams of feed per gram of body weight gain after 8 weeks are 2.75 for Rations 2 and 3 and 3.22 for the unfortified Ration 1.

Methionyl-Methionine Dipeptide as Food Additive

DL-methionyl-DL-methionine is prepared and used as a food and fodder additive in the process disclosed by *H. Bertram, R. Fahnestich and J. Hesse; U.S. Patent 4,056,658; November 1, 1977; assigned to Deutsche Gold- und Silber-Scheideanstalt vormals Roessler, Germany.* This dipeptide of DL-methionine can be prepared in customary manner for making peptides, for example, by reaction of N-carbobenzoxy-methionine with methionine ethyl ester and dicyclohexylcarbodiimide to form N-carbobenzoxymethionyl-methionine and removal of the N-carbobenzoxy group with an alkali metal, e.g., sodium or potassium, in liquid ammonia. This procedure, however, is expensive.

In this process, the 3,6-bis(2'methylmercaptoethyl)-2,5-diketopiperazine is saponified at a pH between 7 and 12. 3,6-bis(2'methylmercaptoethyl)-2,5-diketopiperazine is produced by heating esters of methionine, for example, methionine isopropyl ester to the boiling temperature. If the saponification of the bis-methylmercaptoethyl-diketopiperazine is carried out under customary conditions, by mixing the bis-methylmercaptoethyl-diketopiperazine with an equivalent amount of alkali hydroxide, e.g., sodium hydroxide or potassium hydroxide, and then heating the mixture, the bis-methylmercaptoethyl-diketopiperazine is saponified in only small amounts of methionyl-methionine, but instead is saponified essentially to methionine.

In contrast by keeping the pH between 7 and 12, preferably between 9.0 and 11.5, especially between 10.0 and 10.8 there is obtained methionyl-methionine in high yields. Surprisingly the methionyl-methionine produced in this manner is of excellent purity and in distinction from pure methionine is completely odorless and tasteless. As saponification agents there can be used substances which are suitable for producing the desired pH value in the saponification medium. Thus, useful materials include ammonia, alkaline salts such as sodium carbonate and potassium carbonate, and alkali hydroxides, preferably potassium hydroxide and sodium hydroxide. The saponification agent is preferably added as an aqueous or alcoholic solution, that is, the solvent is water, methyl alcohol, ethyl alcohol, or isopropyl alcohol.

Example: 131 g (0.5 mol) of 3,6-bis(2'-methylmercaptoethyl)-2,5-diketopiperazine was first stirred into a paste with 250 ml of water and then mixed with a further 750 ml of water. The suspension thus formed was heated and held at the boiling temperature under reflux. Then in the course of 6 hours there were gradually added dropwise 0.5 mol of sodium hydroxide in 10% aqueous solution at such a slow rate that the pH did not exceed 10.8. The mixture was subsequently held for 30 minutes at the boiling temperature, then cooled and clarified by the addition of 10 g of activated carbon, subsequently concentrated to a volume of 600 ml by evaporation under reduced pressure, adjusted to a pH of 10.5 by addition of sulfuric acid and filtered with suction.

There were recovered 13.6 g of unreacted 3,6-bis(2'-methylmercaptoethyl)-2,5-diketopiperazine. The filtrate was adjusted to a pH of 5.6 by the addition of sulfuric acid and then kept at room temperature with stirring for 2 hours. The DL-methionyl-methionine which separated was filtered with suction and then washed with 100 ml of water. There was recovered 90.0 g corresponding to a yield of 65% based on the 3,6-bis(2'-methylmercaptoethyl)-2,5-diketopiperazine added. The DL-methionyl-DL-methionine was identified mass spectroscopically. As was established by thin layer chromatography the product was homogeneous, it melted between 221° and 255°C with decomposition and was odorless and tasteless.

The DL-methionyl-methionine is used as a food or fodder additive, e.g., for cattle or sheep. It can be used in any place in which there is required a sparingly water-soluble, but easily alkali soluble derivative of methionine which besides is odorless and tasteless. After taking up the methionyl-methionine with the food the methionine is soon set free and is then available unhindered for protein synthesis in the organism.

Methionine Dipeptides as Food Additives

Various methods for improving the flavor problems coincident with the addition of methionine and other amino acids have received considerable attention. Some success has been achieved with the use of DL-methionyl-DL-methionine (U.S. Patent 4,056,658) and N-acyl derivatives of sulfur-containing L-amino acids (in U.S. Patent 3,878,305). Although each of these derivatives shows some promise in the fortification of proteinaceous foodstuffs, each has drawbacks. DL-methionyl-DL-methionine is almost totally insoluble in water and dilute acid and the N-acyl derivatives of methionine have an inherent characteristic sour, acid flavor which limits its use for palatably fortifying foodstuffs.

The process that is disclosed by *D.A. Cornelius and M.M. Sternberg; U.S. Patent 4,024,286; May 17, 1977; assigned to Miles Laboratories, Inc.* provides palatable and nutritious proteinaceous foodstuffs. These proteinaceous foodstuffs comprise an edible, methionine deficient protein and a supplemental amount of a bland and water-soluble C-terminal amino acid substituted methionine dipeptide where the C-terminal amino acid substituent is a nonsulfur containing amino acid selected from glycine, alanine, valine and glutamic acid.

The methionyl dipeptides of this process are prepared by known methods. One such method is based on the conversion of the DL- or L-form of methionine, which carries the reactive carbonyl group, into its N-carboxyanhydride (NCA). The methionine-NCA derivative is obtained by treatment of methionine with phosgene in a dry inert solvent such as dioxane, ether, toluene and the like. The NCA derivative of methionine is then treated with the selected amino acid substituent to produce the bis-anion of the dipeptide carbamate which is decarboxylated by acidification to give the unprotected dipeptide as illustrated by the following example.

Examples of raw materials which are improved by the process include various protein fractions of the leguminous and other proteins, as characterized by their protein content and method of preparation such as meals, flours, concentrate and isolate. Such leguminous seed proteins include soybeans, peanuts, cottonseeds, sunflower seeds and the like. The most preferred leguminous protein is soybean. Other raw materials include casein, corn gluten, fish protein concentrate, nonfat milk solids, wheat germ, wheat gluten, whey protein, yeast, lactalbumin, microbial protein and the like.

Example: (A) Preparation of DL-Methionine-N-Carboxyanhydride – A 500 ml reaction vessel was charged with 14.0 g (0.094 mol) of DL-methionine and 200 ml of dioxane (twice distilled from sodium). The vessel was purged with nitrogen and the mixture warmed to 40° to 45°C. Phosgene was then slowly passed into the mixture with stirring for a time sufficient to completely dissolve the solids (ca 30 min). Excess phosgene was removed by nitrogen purging. The reaction mixture was then filtered and the filtrate was concentrated by evaporation to yield an oil which solidified after standing under a high vacuum overnight. A portion of the crude title compound was further purified by recrystallization from 1,2-dichloroethane. The title compound had a melting point of 67° to 69°C.

(B) Preparation of DL-Methionylglycine – A solution containing 1.50 g (0.02 mol) of glycine in 175 ml of cold (0°C) sodium borate buffer (pH 10.2) was prepared. The above crude DL-methionine-N-carboxyanhydride was diluted to 40 ml with anhydrous acetone. Ten ml (ca 0.023 mol) of this acetone-NCA solution was then added to the glycine-borate buffer solution with rapid mixing. The pH of the resulting solution was adjusted and maintained at 10.2 by the addition of 4 N NaOH. The reaction was completed in 1 to 2 minutes.

The pH of the mixture was then adjusted to pH 7.0 by addition of concentrated HCl. The reaction mixture was filtered and the filtrate was then desalted on a 4 x 40 cm anion-exchange column of Dowex 2X-8 (OH^- form). The column was washed with water until the pH of the elute was neutral, then the column was eluted with 2,000 ml of 1.0 N acetic acid. Fractions, 100 ml in volume, were collected and assayed for positive ninhydrin reaction. All ninhydrin positive fractions were combined and concentrated to ca 50 ml by evaporation. The desalted filtrate was lyophilized to give a white solid.

One gram of the crude, desalted product above was dissolved in a minimum amount of 1.0 N acetic acid and applied to a 1.8 x 35 cm cation exchange column of Amberlite CG-120 (200 to 400 mesh) resin that had been equilibrated with 0.2 N ammonium acetate buffer (pH 4.25). The column was eluted with about 1,000 ml of the ammonium acetate buffer at a flow rate of 15 ml/hr to remove free glycine and methionine. The column was then eluted with 0.2 N ammonium acetate buffer (pH 7.00). Fractions were collected as before, combined, concentrated and lyophilized to yield the title compound as a white solid, MP 210° to 212° (d). The compound was found to be water-soluble.

Hydrolysis of 10 mg of this solid in 1.0 ml of 6 N HCl for 24 hours at 100°C gave two spots by TLC corresponding to glycine and methionine. A taste test of this derivative revealed that it had a slightly saline taste with no noticeable sulfur aftertaste.

A portion of DL-methionylglycine was added as a dry powder to meat analog sausage patties corresponding to 1.0 and 2.0 wt % of methionine equivalents based on the total protein. The fortified samples were autoclaved at 100°C for 1 hour and subsequently fried at 180°C for 10 minutes. No unusual odors or color changes were observed. Samples of the analog containing the two levels of dipeptide and an unfortified sample (control), but similarly processed, were evaluated for flavor by a 24-member trained sensory panel. A one-way analysis of the taste panel data revealed that there was a nonsignificant difference in the flavor evaluation of the three samples. These data indicate the methionyl dipeptides do not adversely affect the taste of a finished foodstuff.

DIAMINO ACIDS

ORNITHINE

The chemical names for this nonessential amino acid are 2,5-diaminopentanoic acid or α,α-diaminovaleric acid.

Preparation of Nω-Acyl Dibasic Amino Acids

Because of the presence of two amino groups, it has been difficult to produce a monoacyl derivative of basic amino acids with the acyl group on the omega-amino group.

K. Takizawa and R. Yoshida; U.S. Patent 3,897,466; July 29, 1975; assigned to Ajinomoto Co., Inc., Japan have now disclosed a process for producing these compounds in high yields.

In this process the amino acids are first converted to the salts of the carboxylic acids from which the acyl group is to be derived, and the salts so obtained are held at 100° to 250°C until the desired result is achieved and a corresponding amount of water is released.

The basic amino acids readily form salts with carboxylic acids in aqueous media, and the salts may be recovered by conventional methods. The basic amino acids in which the amino groups outnumber the carboxyl group, and which are suitable for this process include the α,ω-diaminomonocarboxylic acids such as lysine, ornithine, and α,γ-diaminobutyric acid, and also their Nα-lower alkyl derivatives, of which those having one to four carbon atoms in the lower alkyl group have greatest practical significance. Nα-methyllysine, Nα,Nα-dimethyllysine, and Nα,Nα-dimethylornithine are typical of this group.

A wide variety of carboxylic acids free from amino groups including the mono- and polycarboxylic alkanoic acids having one to 22 carbon atoms, such as acetic acid, propionic acid, capric acid, lauric acid, myristic acid, stearic acid, behenic acid, succinic acid, and adipic acid will form salts with these basic amino acids. Analogous salts capable of conversion to the analogous N^{ω}-acylamino acids are formed with alkenoic acids such as acrylic, methacrylic, crotonic, oleic, linolic, maleic, and itaconic acid. Benzoic acid, phthalic acid and terephthalic acid are representative of suitable aromatic acids. Polycarboxylic acids of high molecular weight, such as polymethacrylic acid, also form suitable salts. Details of the process are shown in the following examples.

Example 1: 25.3 g ornithine benzoate was held in an autoclave at 180°C for 4 hours. The reaction mixture was dissolved in 300 ml 2 N aqueous sodium hydroxide solution, and a crystalline precipitate was formed when the solution was adjusted to pH 2 with 6 N hydrochloric acid. It was recovered by filtering, thoroughly washed with water, and dried. It weighed 21.0 g.

Benzoic acid was removed from the crude crystals in a Soxhlet extractor by means of low-boiling gasoline. The residue consisted of 19.4 g pure, crystalline N^{ϵ}-benzoylornithine (82.6% yield) melting at 258° to 260°C and identified by elementary analysis and infrared spectrum.

Example 2: 34.6 g lysine laurate was suspended in 300 ml xylene, the suspension was kept at its boiling point, and the water produced was distilled from the reaction system as an azeotropic mixture with the solvent. The calculated amount of water was collected after 3 hours, and heating was continued for 2 additional hours. The solution then was pale brown, and was permitted to cool to ambient temperature.

A crystalline precipitate formed, was filtered off, washed with 100 ml 50% ethanol and dried. It weighed 29.4 g. It was dissolved in 500 ml 3 N sulfuric acid and reprecipitated in purified form by partly neutralizing the solution.

27.6 g pure, crystalline N^{ϵ}-lauroyllysine melting at 230° to 232°C was obtained (84.2% yield) and identified by elementary analysis and infrared spectrum.

Al or Mg Salts as Antiulcer Agents

Therapeutically useful aluminium and/or magnesium salts of dibasic amino acids have been prepared by the process disclosed by *A.H. Passedouet and R. Pipon; U.S. Patent 4,013,699; March 22, 1977; assigned to Rhone-Poulenc SA, France.*

These metal salts are compounds of the formula:

$$(1)\qquad \left[CH_3CONH{-}(CH_2)_n{-}\underset{\displaystyle NHCOCH_3}{CH}{-}COO\right]_a Al_b Mg_c (OH)_d$$

In the above formula the amino acid moiety is of the D and/or L form, n is 3 or 4, a is an integer from 1 to 10, b is zero or an integer from 1 to 9, c is zero or 1, at least one of the indices b and c is an integer, and d is zero or an integer from 1 to 19, the numbers represented by the indices a, b, c and d having the relationship a + d = 3b + 2c. When b is zero, preferably d also is zero when c is zero preferably d is an integer from 1 to 19.

Preferred salts are those of formula (1) wherein:

c is zero, and a, b and d represent respectively 10, 9 and 17, or 5, 3 and 4, or 2, 7 and 19 when n represents 3, and 6, 7 and 15 or 1, 1 and 2 when n represents 4,

b is zero and a, c and d are respectively 2, 1 and zero when n is 3, and

b is an integer equal to or higher than 1 and

c is 1, and a, b c and d are respectively 4, 2, 1 and 4 when n is 3 or 4.

The metal salts of formula (1) have useful pharmacological properties combined with low toxicity. They are particularly valuable as antiulcer agents and as agents for protecting the gastrointestinal mucous membrane. They also possess good cicatrizant activity.

The salts of formula (1) are prepared by the process which comprises (i) when b is zero and c is 1, reacting an aqueous solution of an amino acid of the formula:

(2) $$CH_3CONH-(CH_2)_n-\underset{NHCOCH_3}{\underset{|}{CH}}-COOH$$

(where n is 3 or 4), i.e., N,N'-diacetyl ornithine or N,N'-diacetyl lysine, with a reactive magnesium compound, or (ii) when b is an integer from 1 to 9 and c is zero, reacting an aqueous solution of an amino acid of formula (2) with a reactive aluminum compound, or (iii) when b is an integer from 1 to 9 and c is 1, carrying out the reaction mentioned in (i) above and then reacting the resulting magnesium salt conforming to formula (1) with a reactive aluminum compound, the amount of reactive magnesium compound and reactive aluminum compound used in each reaction being such as to give the required number of gram atoms of magnesium and/or aluminum in the desired salt of formula (1).

In the preparation of magnesium salts and mixed magnesium aluminum salts of formula (1) the reactive magnesium compound usually is the oxide, hydroxide or an alkoxide. Preferably an aqueous suspension of magnesium hydroxide is reacted with the aqueous solution of the amino acid of formula (2) at 50° to 80°C.

Aluminum compounds used as reactant are preferably alkoxides, obtained from an aliphatic alcohol containing 1 to 3 carbon atoms. For example, a suspension of aluminum isopropoxide in isopropanol is reacted with an aqueous solution of the amino acid of formula (2), or with the above magnesium salt at 50° to 80°C.

Example 1: Aluminum isopropoxide (183.6 g; 0.9 mol) suspended in isopropanol (150 cc) is added to N,N'-diacetyl ornithine (216 g; 1 mol) dissolved in distilled water (900 cc) at 60°C. After the end of the addition, the reaction mixture is stirred and heated at 60°C for 3 hours. The isopropanol is then removed by distillation under reduced pressure. After filtering the reaction mixture in order to remove the insoluble impurities, the clear and colorless filtrate is concentrated under reduced pressure so as to obtain a gel after removal of the water. The gel is taken up several times in acetone before a precipitate is obtained which can be filtered off. After drying the precipitate, an aluminum salt of N,N'-diacetyl ornithine (254 g) is obtained in the form of a white powder corresponding to the formula:

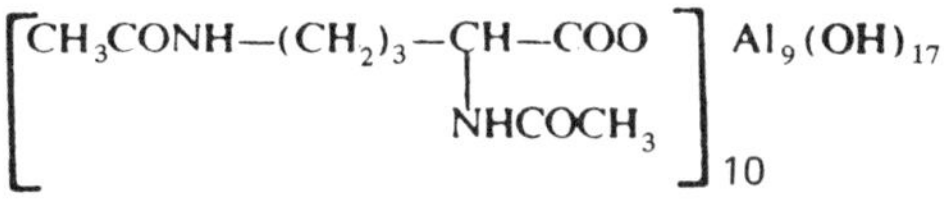

Example 2: Magnesium hydroxide (3.5 g) in water (6 cc) is added to a solution, heated to 60°C, of N,N'-diacetyl-L-lysine (55.2 g) and water (360 cc), and then the reaction mixture is kept at 60°C for 30 minutes with stirring. A suspension of aluminum isopropoxide (24.5 g) in isopropanol (30 cc) is then added with stirring and while keeping the temperature at 60°C. The isopropanol is removed by distillation under reduced pressure (25 mm Hg). The reaction mixture is then filtered in order to remove insoluble material and the filtrate concentrated to dryness under reduced pressure (25 mm Hg) at 50°C. The paste-like residue obtained is triturated three times in acetone and then dried under reduced pressure (25 mm Hg). A solid is thus obtained which is washed by filtration using acetone (3 x 150 cc). After drying, a mixed aluminum magnesium salt of N,N'-diacetyl-L-lysine (60.35 g)

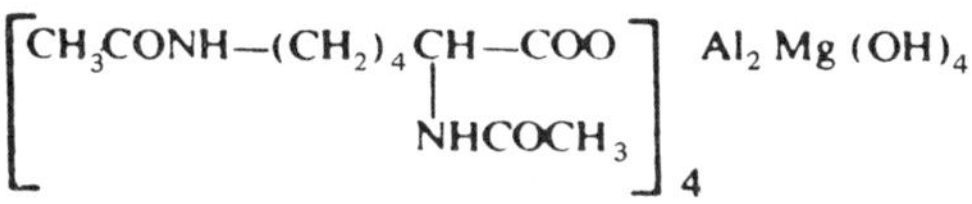

is obtained.

LYSINE—PRODUCTION BY FERMENTATION METHODS

This essential amino acid can also be named as 2,6-diaminohexanoic acid or α,ϵ-diaminocaproic acid. It is used as a food additive for human nutrition.

L-lysine previously has been produced by fermentation relying on three types of bacteria. The first type requires for its growth amino acids related to the biosynthesis of L-lysine and is exemplified by a homoserine-requiring mutant of *Micrococcus glutamicus* disclosed in U.S. Patent 2,979,439. The second type includes threonine- or methionine-sensitive mutants and mutants whose growth is inhibited by low concentrations of threonine or methionine, but which require threonine for their growth (French Patent 1,533,688). The third type is characterized by resistance to S-(2-aminoethyl)-L-cysteine (AEC) which is a sulfur analogue of L-lysine, as disclosed in U.S. Patent 3,707,441.

Use of Mutants with Minor Organic Nutrient Requirements

K. Kubota, Y. Yoshihara, H. Hirakawa, H. Kamijo, S. Nosaki, F. Yoshinaga, S. Okumura and H. Okada; U.S. Patent 3,825,472; July 23, 1974; assigned to Ajinomoto Co., Inc., Japan have found that L-lysine is produced at higher concentrations than were available heretofore under comparable conditions by mutants which combine resistance to feedback inhibition by lysine or lysine analogues. These mutants also require at least one specific nutrient which is serine, proline, alanine, nicotinamide, nicotinic acid, pantothenic acid, thiamine, guanidine, adenine, hypoxanthine, or vitamin B_{12} when these mutants are cultured on otherwise conventional media including sources of assimilable carbon and nitrogen, inorganic ions, and unspecific organic growth-promoting agents.

These mutants are derived by means of conventional mutagenic agents from parent strains of the genera Brevibacterium and Corynebacterium, and by screening of the mutants so produced for the necessary resistance to feedback inhibition by lysine and its analogues, such as S-(2-aminoethyl)-L-cysteine (AEC) in concentrations of 1 mg/ml or more, and a requirement for one of the nutrients enumerated above. Because of their resistance to lysine analogues, these mutants produce lysine at a predictable rate and in predictable amounts without being seriously affected by variations in the amounts of threonine and other organic nutrients in the fermentation medium employed.

Example 1: Cells of *Brevibacterium lactofermentum* ATCC 13869 were irradiated with x-rays to induce mutation, and cultured thereafter in a complete agar flat plate at 31°C for 4 to 10 days. Four mutant strains were isolated by the replica method from colonies appearing on the plate according to specific nutrient requirements for serine, pantothenic acid, thiamine, and guanine or adenine, which were determined by standard methods known in the art.

The four strains so obtained were each treated with 200 μg/ml nitrosoguanidine at 30°C for 30 minutes, and inoculated on agar flat plates containing 2 g/dl glucose, 0.3 g/dl urea, 0.5 g/dl threonine, 0.1 g/dl KH_2PO_4, 0.04 g/dl $MgSO_4{\cdot}7H_2O$, 2 ppm Fe^{++}, 2 ppm Mn^{++}, 50 μg/l biotin, 100 μg/l thiamine, 500 μg/ml AEC, 2 g/dl agar, and, respectively, 30 mg/dl serine, 10 mg/l calcium pantothenate, and 100 mg/l adenine for the three strains requiring these nutrients.

The four cultures were adjusted to pH 7.0 and cultured at 31°C for 4 to 10 days. The colonies appearing on the plates were tested for their lysine-producing ability and the following four strains were isolated:

B. lactofermentum Y-108 (ATCC 21798; FERM P-1443)
B. lactofermentum AC-73 (ATCC 21799; FERM P-1444)
B. lactofermentum AD-5 (ATCC 21800; FERM P-1445)
B. lactofermentum AD-162 (ATCC 21801; FERM P-1446)

The parent strain *B. lactofermentum* ATCC 13869 and the four mutants were cultured 24 hours at 31°C with shaking on a minimal medium of 2 g/dl glucose, 1 g/dl ammonium sulfate, 0.3 g/d! urea, 0.1 g/dl KH_2PO_4, 0.04 g/dl $MgSO_4{\cdot}7H_2O$,

2 ppm Fe^{++}, 2 ppm Mn^{++}, 50 μg/l biotin, and 200 μg/l thiamine hydrochloride. In parallel control tests, the four mutants were grown on respective modified media differing from the minimal medium by an addition of 30 mg/dl serine (Y-108), an addition of 10 mg/l calcium pantothenate (AC-73), the absence of thiamine hydrochloride (AD-5), and the addition of 100 mg/l adenine (AD-162).

Each cultured broth was diluted with 25 volumes of water, and its optical density was determined as a measure of growth by measuring light absorbancy at 562 mμ. The results are listed in the following table.

	Optical density	
Brevibacterium lactofermentum	Minimal medium	Modified medium
ATCC 13869	0.400	--------------------
Y-108	0.065	0.410
AC-73	0.059	0.405
AD-5	0.380	0.075
AD-162	0.052	0.375

Example 2: The four mutant strains of *B. lactofermentum* referred to in the table were inoculated on respective seed culture media each containing 15 g/dl glucose, 0.3 g/dl ammonium acetate, 0.1 g/dl KH_2PO_4, 0.04 g/dl $MgSO_4·7H_2O$, 2 ppm Fe^{++}, 2 ppm Mn^{++}, 50 μg/l biotin, 200 μg/l thiamine hydrochloride, 1.5 ml/dl soybean protein hydrolyzate, and 0.5 g/dl yeast extract, and having a pH of 8.0. The seed cultures were held at 31°C for 16 hours with stirring and aeration.

Respective 15 ml inocula of the seed cultures were added to 300 ml batches of fermentation media containing 3 g/dl glucose, 0.5 g/dl ammonium acetate, 0.2 g/dl urea, 0.1 g/dl KH_2PO_4, 0.04 g/dl $MgSO_4·7H_2O$, 2 ppm Fe^{++}, 2 ppm Mn^{++}, 50 μg/l biotin, 50 μg/l thiamine HCl, 3 ml/dl soybean protein hydrolyzate, at pH 7.5. The media for *B. lactofermentum* AC-73 additionally contained 5 mg/l calcium pantothenate, and those for *B. lactofermentum* AD-162 100 mg/l each of adenine and guanine.

The fermentation mixtures were stirred at 1,500 rpm at 31.5°C and aerated with one volume of air per minute and volume of broth. The pH of each medium was held automatically between 7.2 and 8.0 by additions of a 60% solution of acetic acid containing 0.25 mol ammonium acetate per mol of acetic acid. After 48 hours, it was found that strain Y-108 has produced 6.4 g/dl lysine, strain AC-73 4.8 g/dl, strain AD-5 5.7 g/dl, and strain AD-162 6.0 g/dl.

Use of Specific Corynebacterium Mutants

Mutants with a sensitivity to methionine are used by *K. Kubota, Y. Yoshihara and H. Okada; U.S. Patent 3,871,960; March 18, 1975; assigned to Ajinomoto Co., Inc., Japan* to produce higher yields of L-lysine. It was found that mutants of Corynebacterium combining resistance to feedback inhibition by lysine and its analogues such as S-(2-aminoethyl)-L-cysteine (AEC) in concentrations of 1 mg/ml or more with methionine sensitivity or methionine sensitivity and a nutrient requirement for proline and/or arginine, produce large amounts of L-lysine in a culture medium.

The methionine sensitivity of the mutants used in the process is different from that of the threonine- or methionine-sensitive mutant raised in French Patent 1,533,688. That is, the growth of the mutant in the French patent is inhibited by threonine or methionine, in the medium under conditions in which the growth of the wild strain is not inhibited. The growth inhibition by threonine is overcome by addition of methionine, and the inhibition by methionine is overcome by addition of threonine. On the other hand, the methionine-sensitive strains of the process are inhibited by methionine but not by threonine and the growth inhibition by methionine is not overcome by addition of threonine.

Cells of *Micrococcus glutamicus* ATCC 13032 (now named *Corynebacterium glutamicum*) were treated with 250 μg/ml nitrosoguanidine at 30°C for 30 minutes, and then inoculated on agar flat plates and cultured at 31°C for 4 to 10 days.

Many mutant strains which have an ability to produce L-lysine will appear among the colonies on the plates after cultivation. Strain AJ3400 was isolated as a good lysine producer and was found to be resistant to AEC and sensitive to methionine.

The strain AJ 3400 could not grow in a medium containing more than 0.1 mM/l (14.9 mg/l) of L-methionine during 24 hours' cultivation.

However, the parent strain *M. glutamicus* ATCC 13032 could grow well in a medium containing more than 5 mM (7,450 mg/l) of L-methionine during 24 hours' cultivation.

The strain AJ3400 was again treated with 250 μg/ml of nitrosoguanidine at 30°C for 30 minutes. After treatment, mutant strains having some nutrient requirement were isolated by the known replication method. Mutant strains AJ3609, which combines an arginine requirement with resistance to AEC and sensitivity to methionine, and AJ3610 which requires proline were obtained as good L-lysine producers.

Based on obtained results, a methionine-sensitive microorganism for the process is defined as a strain whose growth during 24 hours' cultivation on a minimal medium supplemented with 0.1 mmol of L-methionine is inhibited severely and whose growth inhibition by L-methionine is not overcome by addition of L-threonine.

Example 1: 20 ml batches of a medium containing 10% glucose, 4.5% $(NH_4)_2SO_4$, 0.1% KH_2PO_4, 0.04% $MgSO_4 \cdot 7H_2O$, 2 ppm Mn^{++}, 2 ppm Fe^{++}, 50 μg/l biotin, 200 μg/l thiamine HCl, and 5% calcium carbonate (sterilized separately), and of pH 7.0 were placed in separate 500 ml shaking flasks, and sterilized.

C. glutamicum AJ 3400 (FERM P-1638) previously cultured on bouillon agar slants was introduced into the flasks, and cultured at 31°C with aeration and agitation for 72 hours. The cultured broth was found to contain 3.1 g/dl lysine and centrifuged. One liter of the supernatant liquid was passed through a column packed with an ion exchange resin, Amberlite IR-120 (H^+ type), and lysine was eluted with 3% aqueous ammonia. The eluate was concentrated under reduced pressure.

Hydrochloric acid was added to the concentrated solution which was then cooled in an icebox to precipitate L-lysine, and 20.7 g of crude crystalline L-lysine hydrochloride dihydrate was obtained.

Example 2: C. glutamicum AJ3400, AJ3609 and AJ3610 were cultured in media respectively containing beet molasses or cane molasses as carbon source, under the conditions of Example 1. The composition of the culture medium is as follows.

Beet molasses or cane molasses (calculated as glucose), %	10
$(NH_4)_2SO_4$, %	4.5
KH_2PO_4, %	0.1
$MgSO_4 \cdot 7H_2O$, %	0.04
Fe^{++}, ppm	2
Mn^{++}, ppm	2
Biotin, μg/l	50
Thiamine HCl, μg/l	200
Calcium carbonate, % (sterilized separately)	5
Soy protein hydrolyzate, % (total nitrogen: 7%)	1.5
pH	7.0

The amount of L-lysine (as hydrochloride) accumulated in each medium was as follows:

	. . . . . L-Lysine Accumulated (g/dl).	
	Beet Molasses	Cane Molasses
AJ3400	3.4	3.3
AJ3609	4.3	4.0
AJ3610	4.5	4.2

Use of Hydrocarbon-Assimilating Microorganisms

Three processes have been developed by the staff at Kanegafuchi on the production of L-lysine using culture media containing hydrocarbons.

In the first process *K. Watanabe, T. Tanaka, T. Hirakawa, H. Kinoshita, M. Sasaki and K. Obayashi; U. S. Patent 3,905,866; September 16, 1975; assigned to Kanegafuchi Chemical Industries, Japan* have developed mutant L-lysine-producing strains which are resistant to at least one amino acid. These amino acids include L-valine, α-aminobutyric acid, norvaline, β-hydroxynorvaline, α-amino-β-chlorobutyric acid and L-threonine.

The microorganisms used in this process are newly isolated from the soil and upon comparison of their microbiological properties are considered to be new strains designated as *Pseudomonas brevis* No. 22 (ATCC 21940) and *Achromobacter coagulans* No. 42 (ATCC 21934).

The mutational treatment may be carried out by the conventional means, for example, by using mutagenic agents such as N-methyl-N'-nitro-N-nitrosoguanidine, or the physical treatment such as irradiation with ultraviolet rays, radioactive rays or other procedures.

A typical procedure for producing an L-valine-resistant strain comprises culturing a mutant obtained by a usual mutational treatment from a strain of the genus Pseudomonas or Achromobacter in a minimum culture medium containing L-valine and ethanol (preferably 0.5 to 1.0% by weight of ethanol based on the total weight of the medium) as a carbon source for a period of 3 to 5 days to obtain a culture enriched with L-valine-resistant strain.

Alternatively, it is also possible to replace L-valine by the various analogues or L-threonine at a concentration more than 0.5 mg/ml for the enrichment culture. It is generally important to increase the L-valine resistance of the mutant stepwise by subjecting the mutant to repeated mutational treatments. In producing a resistant strain it is generally necessary to predetermine the optimum culture conditions in order to impart to the microorganism the strongest sensitivity to amino acids or amino acid analogues used for the mutational treatment. For this purpose, the selection of a carbon source to be used in the culture medium and the combination of various amino acids or amino acid analogues (two or more agents) should carefully be investigated. To this effect the most preferable carbon source for L-valine-resistant strain is found to be ethanol.

In carrying out the L-lysine production, an L-valine-resistant strain is aerobically cultured in a medium containing a hydrocarbon as a main carbon source, a nitrogen source, inorganic salts and other additives. Hydrocarbons which can preferably be used in the culture medium are n-paraffins containing 10 to 20, preferably 13 to 18, carbon atoms or kerosene. By such methods high-lysine producing strains such as *Pseudomonas brevis* No. 56 (ATCC 21941) were obtained from *P. brevis* No. 22 (ATCC 21940), and *Achromobacter coagulans* ATCC 21935 and ATCC 21936 were obtained from *A. coagulans* ATCC 21934.

Example: Valine-resistant strain, *Pseudomonas brevis* No 56 (ATCC 21941) was cultivated on a bouillon agar slant medium at 33°C overnight and then inoculated in a 2-liter shake-flask containing 600 ml of the previously sterilized seed culture medium having the following composition followed by being shake-cultured at 33°C for 24 hours.

Seed Culture Medium

75% Phosphoric acid, ml	12
$(NH_4)_2SO_4$, g	6
NaCl, g	1
$MgSO_4 \cdot 7H_2O$, g	0.2
$CaCl_2 \cdot 2H_2O$, g	0.1
$FeSO_4 \cdot 7H_2O$, g	0.1
$ZnSO_4 \cdot 7H_2O$, g	0.03
$MnSO_4 \cdot 7H_2O$, g	0.002
KOH, g	14
Tap water, ml	1,000
n-Paraffin (C_{11}-C_{18}), ml	7

A 600 ml portion of the above seed culture was inoculated in a jar fermentor containing 20 liters of a sterilized fermentation medium having the following composition and the inoculated culture was cultivated at 35°C by agitating at a rate of 800 rpm with aeration at a rate of 27 l/min.

Fermentation Medium

n-Paraffin (C_{11}-C_{18}), % (w/v)	10
K_2HPO_4, %	0.1
KH_2PO_4, %	0.1
$MgSO_4 \cdot 7H_2O$, %	0.1
$FeSO_4 \cdot 7H_2O$, %	0.002
$ZnSO_4 \cdot 7H_2O$, %	0.001
$MnSO_4 \cdot 4H_2O$, %	0.005
Tween 85, %	0.05
$(NH_4)_2SO_4$, %	1.0
$CaCO_3$, %	1.0
pH	7.0

During the cultivation, the medium was adjusted to pH of 6 to 8 with an aqueous ammonia solution. The amount of L-lysine hydrochloride produced in the medium after 90 hours' cultivation was found to be 40.5 g/l (yield: about 40% based on the n-paraffin). The microbial cells were removed by centrifugation, and the L-lysine produced was adsorbed onto an ion-exchange resin in a usual manner. After elution by ammonia, the eluate was concentrated, and hydrochloric acid and then an alcohol were added thereto to obtain crystals weighing 35.3 g containing L-lysine hydrochloride having a purity of more than 98% from 1 liter of the culture broth.

In a modification of the previous process *K. Watanabe, T. Hirakawa, K. Takahara, Y. Nakamura, S. Iwasaki and T. Tanaka; U.S. Patent 3,888,737; June 10, 1975; assigned to Kanegafuchi Chemical Industries Co., Ltd., Japan* have used a mixture of microorganisms for producing lysine. The process comprises aerobically culturing a mixture of (1) a hydrocarbon-assimilating and L-lysine-producing bacterium, and (2) at least one hydrocarbon-nonassimilating microorganism in a nutrient medium in which a hydrocarbon or a mixture of hydrocarbons is the predominant carbon source.

Microbial strains which have a high capability for producing L-lysine from hydrocarbons are often contaminated with various microorganisms in a fermentor. The yield of L-lysine is greatly reduced by such contaminating microorganisms.

An important advantage of the process is that the growth of these contaminating microorganisms is greatly inhibited and then adverse effects minimized, and thus a reproducible and high yield of L-lysine is obtained in a shorter period of time.

A wide range of microorganisms which have a hydrocarbon-assimilating and L-lysine-producing ability can be used. The L-lysine-producible strains can be obtained from the hydrocarbon-assimilating microorganisms using known artificial mutation techniques. Included are strains derived from bacteria belonging to the genera of Pseudomonas, Arthrobacter, Achromobacter, Micrococcus, Corynebacterium, Brevibacterium, Acinetobacter, Alkaligenes and Nocardia which can assimilate hydrocarbons.

Examples of preferred mutants are *Pseudomonas brevis* ATCC 21941 (valine-resistant strain), *Achromobacter coagulans* ATCC 21936 (α-aminobutyrate and norvaline-resistant strain), *Arthrobacter alkanicus* M 1554 (threonine-resistant strain), which were derived using an artificial mutation technique and have a high productivity for L-lysine.

The hydrocarbon nonassimilating microorganisms used are found among various bacteria, fungi, yeast or actinomyces which cannot assimilate hydrocarbons and preferably do not decompose the L-lysine accumulated in the culture broth during the fermentation process. These include bacteria belonging to the genera Bacillus, Pseudomonas, Corynebacterium, Microbacterium, Micrococcus, Brevibacterium, Aerobacter, Aeromonas, Escherichia coli, Proteus, Flavobacterium and the like.

As representative examples of bacterial strains, *Bacillus subtilis* (IAM-1130), *B. megatherium* (IAM-1145), *B. polymyxa* (IAM-1189), *Pseudomonas xanthe* (IAM-1310), *Microbacterium flavum* (ATCC-10340), *Micrococcus candidus* (ATCC-14852); *M. lysodeikticus* (IAM-1313), *Corynebacterium faciens* (IAM-1079), *Brevibacterium ammoniagenes* (IAM-1641), *Acromonas formicans* (ATCC-13137), *Escherichia coli* (K-13), *Proteus vulgaris* (IAM-1025) can be employed.

Also included are fungi belonging to the genera Penicillium and Aspergillus, with representative strains being *Aspergillus niger* (IAM-3008) and *Penicillium citrinum* (ATCC-9849); Actinomycetes belonging to the genera Streptomyces and Actinomyces with representative strains being *Streptomyces mitakaensis* (ATCC-15297), *S. hygroscopicus* (ATCC-10976), *S. griseus* (ATCC-23345), *S. virginiae* (ATCC-19817), and *S. albus* (ATCC-3381); and yeast belonging to the genera Rhodotorula, Zygosaccharomyces, Debaryomyces and Trichosporon.

In carrying out the L-lysine production in accordance with the process, the two kinds of strains described hereinbefore (i.e., the hydrocarbon-assimilating and hydrocarbon-nonassimilating) are aerobically cultured in the culture medium containing hydrocarbons as a main carbon source, nitrogen sources, inorganic salts and other additives.

Hydrocarbons which can preferably be used in the culture medium for the production of L-lysine are n-paraffins containing 13 to 18 carbon atoms. The concentration of hydrocarbons in the medium is usually 5 to 20% w/v.

Example: Pseudomonas brevis (ATCC 21941) as a hydrocarbon-assimilating and L-lysine-producing microorganism was cultured on a bouillon agar slant at 33°C for 24 hours, and then was used to inoculate the following seed culture medium and was then cultured at 33°C.

The composition of the seed culture medium was as follows:

n-Paraffin (C_{11}-C_{18}), g/l	5
75% H_3PO_4, ml/l	12
$(NH_4)_2SO_4$, g/l	6
NaCl, g/l	1
$MgSO_4 \cdot 7H_2O$, g/l	0.2
$CaCl_2 \cdot 2H_2O$, g/l	0.1
$FeSO_4 \cdot 7H_2O$, g/l	0.1
$ZnSO_4 \cdot 7H_2O$, g/l	0.03
$MnSO_4 \cdot 4H_2O$, g/l	0.0002

The pH was adjusted to 7.0 with KOH. After 24 hours, 1 ml of the above seed culture (inoculum ratio ~3%) was used to inoculate 30 ml of a fermentation medium in shaking flasks which was sterilized at 120°C for 30 minutes and cultured with shaking at 33°C.

The composition of the fermentation medium was as follows.

	Percent, w/v
n-Paraffin (C_{11}-C_{18})	11.6
K_2HPO_4	0.1
KH_2PO_4	0.1
$MgSO_4 \cdot 7H_2O$	0.1
$FeSO_4 \cdot 7H_2O$	0.001
$ZnSO_4 \cdot 7H_2O$	0.001
$MnSO_4 \cdot 4H_2O$	0.005
$CaCO_3$	3.0
$(NH_4)_2SO_4$	3.5
Yeast extracts	0.1
Tween 85	0.1

(pH 7.0)

After 24 hours from the beginning of the cultivation, a platinum loopful amount of each of the hydrocarbon-nonassimilating bacteria as shown in the table below was used to inoculate the culture to provide a mixed culture and the culturing was continued for 9 days. A control culture was also conducted using only the *Pseudomonas brevis* for the purposes of comparison. The concentration of L-lysine (as the hydrochloride) produced in the broth of each of the fermentations was measured at 7 and 9 days. The results obtained are shown in the following table.

Strains of Mixed Bacteria	L-Lysine (HCl) . . . Accumulated. . . . 7 Days	9 Days
	. . . . . .mg/ml.	
Pseudomonas brevis		
Control*	20.5	26.0
Pseudomonas brevis plus		
Bacillus megatherium IAM-1030	25.0	30.0
Bacillus subtilis IAM-1145	24.0	30.0
Bacillus polymyxa IAM-1189	30.0	32.8
Microbacterium flavum ATCC-10340	30.5	34.5
Microbacterium lacticum IAM-1640	28.0	30.0
Micrococcus candidus ATCC-14852	32.8	32.8
Brevibacterium ammoniagenes IAM-1641	28.8	30.8
Aeromonas formicans ATCC-13137	35.0	34.4
Aerobacter clocae IAM-1020	27.6	29.0
Escherichia coli K-12	34.0	36.8
Corynebacterium faciens IAM-1079	31.2	25.6
Corynebacterium rathayi ATCC-13659	28.0	30.8
Proteus vulgaris IAM-1025	31.2	29.5

*Single culture using *Pseudomonas brevis* only.

As shown in this table, the production of L-lysine was greatly improved when a mixed culture was employed.

The time required to reach the maximum yield was 7 to 9 days in the case of a mixed culture, while it was 10 to 11 days in the case of a single culture.

It was further found by *T. Tanaka, T. Hirakawa and K. Takahara; U.S. Patent 3,920,520; November 18, 1975; assigned to Kanegafuchi Chemical Industries Co., Ltd, Japan* that some mutant strains of a microorganism belonging to the genus Acinetobacter have an ability to produce L-lysine in high yield from hydrocarbons.

A typical example of the microorganism belonging to the genus Acinetobacter is species No. 38 (ATCC 31023). This strain was found to produce only a small amount of L-lysine in a culture medium containing n-paraffins, but it was found that various mutant strains having an improved L-lysine-producing ability can be obtained when subjected to a mutational treatment such as ultraviolet radiation or a treatment with mutagenic agents such as nitrosoguanidine, nitrogen mustard and the like. These mutant strains were designated as No. 38-15 (ATCC 31024), No. 38-19 (ATCC 31025 and No. 38-20 (ATCC 31038).

As with other lysine-producing mutants, strains are produced which are resistant to one or a combination of the amino acids such as threonine, valine, methionine, lysine, serine and the like, and amino acid analogues such as norvaline β-hydroxynorvaline, α-aminobutyric acid, α-amino-β-chlorobutyric acid, S-aminoethylcysteine, lysine hydroxide, serine hydroxide, methionine hydroxide, valine hydroxide and the like.

From tests conducted on the above strains, it was found that strain No. 38-15 is resistant to threonine and valine, and strains No. 38-19 and 38-20 are resistant to valine.

Example: A platinum loopful amount of each inoculum of Acinetobacter species No. 38 (ATCC 31023) and the mutant strains derived from this parent strain, i.e, Acinetobacter species No. 38-15 (ATCC 31024), Acinetobacter species No. 38-19 (ATCC 31025) and Acinetobacter species No. 38-20 (ATCC 31038) was inoculated into a large size test tube containing 10 ml of a culture medium comprising 1% peptone, 1% meat extract and 0.5% yeast extract. The culture medium had previously been sterilized at 120°C for 15 minutes. The inoculated culture was then cultivated for 24 hours at 33°C to prepare a seed culture.

Each of the above 4 strains was inoculated into an n-paraffin inorganic salt slant medium having the following composition at a temperature of 33°C for 2 days.

	Percent
n-Paraffin (C_{13}-C_{18})	0.8
Ammonium sulfate	0.6
K_2HPO_4	1.0
KH_2PO_4	1.0
$MgSO_4 \cdot 7H_2O$	0.02
NaCl	0.1
$FeSO_4 \cdot 7H_2O$	0.01
$CaCl_2 \cdot 2H_2O$	0.01
$ZnSO_4 \cdot 7H_2O$	0.003
$MnSO_4 \cdot 4H_2O$	0.0003
Agar	2

(pH, 7.0; sterilized at 120°C for 15 minutes)

A platinum loopful amount of the resulting culture was then inoculated into 20 ml of the following L-lysine-producing medium contained in a 500 ml Sakaguchi flask followed by cultivation with shaking at a temperature of 33°C for 10 days.

	Percent
n-Paraffin (C_{12}-C_{18})	10
K_2HPO_4	0.05
KH_2PO_4	0.05
$MgSO_4 \cdot 7H_2O$	0.05
$FeSO_4 \cdot 7H_2O$	0.001
$ZnSO_4 \cdot 7H_2O$	0.001
$MnSO_4 \cdot 4H_2O$	0.001
Tween 85	0.1
$CaCO_3$	3
$(NH_4)_2SO_4$	3.5

(sterilized at 120°C for 15 minutes)

Upon completion of the cultivation, the resulting L-lysine hydrochloride in the broth was measured using a microbioassay method to obtain the following results.

Strains Used	Amount of L-Lysine HCl, g/l
Acinetobacter species No. 38	0.4
Acinetobacter species No. 38-15	20.2
Acinetobacter species No. 38-19	21.4
Acinetobacter species No. 38-20	1.1

Use of *Candida lipolytica* Mutants

Strains of Candida yeast are sensitive to all lysine analogues, called "toxic analogues" by specialists. Examples of such toxic analogues include: amino-cyclohexyl alanine, trans-dehydrolysine (TDL), S-(β-amino-ethyl) cysteine, 2,6-diaminoheptanoic acid, oxalysine, amino-3-methyl cyclohexane glycine, aza-4-lysine, 5-methyl lysine, α-amino-ϵ-hydroxycaproic acid and δ-hydroxylysine. The sensitivity is such that if a lysine analogue is present in the culture medium for a Candida yeast, the growth of the yeast is completely stopped.

Mutants of *Candida lipolytica* have been developed by *P. Maldonado, C. Gaillaridin, G. Sylvestre and G. Glikmans; U.S. Patent 3,986,933; October 19, 1976; assigned to Institut Francais du Petrol, des Carburants et Lubrifiants and Entreprise de Recherches et d'Activities Petrolieres Elf, France* which produce both L-lysine and citric acid. This process comprises subjecting a yeast of the species *Candida lipolytica* to the action of a mutagenic agent (either a chemical agent, e.g., nitrosomethyl urethane or nitrosomethyl guanidine, or radiation, e.g., x-rays or ultraviolet rays), cultivating the treated yeast in a culture medium containing a lysine analogue in a concentration 50 to 100 times greater than that to which the cells of the original yeast are sensitive to obtain a mutated resistant yeast, and cultivating the resistant yeast on a medium containing at least one hydrocarbon.

Example: A typical gelose containing *Candida lipolytica* Mg-5 cells (ATCC 20-462) made resistant to concentrations of 10^{-3} M of trans-4-dehydrolysine was used for seeding a 100 ml flask containing 20 ml of a liquid medium A having the following composition.

KH_2PO_4, g/l	3.4
$Na_2HPO_4 \cdot 12H_2O$, g/l	1.5
$MgSO_4 \cdot 7H_2O$, g/l	0.7
$(NH_4)_2SO_4$, g/l	4
$CaCl_2$, g/l	0.1
$FeSO_4 \cdot 7H_2O$, mg/l	2

(continued)

$CuSO_4 \cdot 5H_2O$, μg/l	5
H_3BO_3, μg/l	10
$MnSO_4 \cdot 7H_2O$, μg/l	10
$ZnSO_4 \cdot 7H_2O$, μg/l	10
$(NH_4)_6Mo_7O_{24} \cdot 4H_2O$, μg/l	100
$Co(NO_3)_2 \cdot 6H_2O$, μg/l	10
Yeast extract, mg/l	100
Tap water	to make up to 1 liter
C_{12}-C_{19} n-paraffin fraction, g/l	15

The Candida cells were incubated at 30°C after the flask containing the inoculum had been secured on an agitation table rotating at 140 rpm. After incubation for 36 hours, 10 ml of the inoculum were used to seed a 1.5-liter Fernbach flask containing 200 ml of nutrient medium A. After incubation for 36 hours at 30°C on the agitation table, the 200 ml of medium containing the growing Mg-5 cells were used to seed a fermentor having a capacity of 4 liters and containing two liters of nutrient medium A, the pH being adjusted by the addition of 4 N aqueous ammonia solution.

The medium was agitated at 2,000 rpm at 30°C and the pH was maintained at 5, the n-paraffin layer being injected continuously into the fermentor at an average rate of 2.5 ml/hr.

After cultivation for 150 hours, the total quantity of hydrocarbons introduced was 140 g/l, whereas the concentration of excreted products in the fermentation medium was as follows.

Citric acid, g/l	135
Isocitric acid, g/l	40.5
Glutamic acid, g/l	1.8

The biological substance collected was found to contain 14 to 15% L-lysine with respect to the proteins, measured on hydrolyzed cells using a JEOL automatic amino acid analyzer.

By way of comparison, Candida cells which had not been made resistant to a toxic analogue of L-lysine were cultivated under the same conditions. The quantities of citric, isocitric and glutamic acids excreted into the fermentation medium were nil or less than 1 gram, whereas the proportion of lysine with respect to the cellular proteins was 8 to 9%.

Use of Protaminobacter Mutants

The mutants used in the process disclosed by *K. Nakayama, M. Kohata, Y. Tanaka, T. Nomura and R. Katsumata; U.S. Patent 3,907,637; September 23, 1975; assigned to Kyowa Hakko Kogyo Co., Ltd., Japan* are capable of utilizing methanol and have a resistance to S-2-aminoethyl-L-cysteine and L-threonine. A particularly preferred mutant is *Protaminobacter thiaminophagus* SLR-77 ATCC 21926, derived from a methanol-utilizing strain of *P. thiaminophagus* ATCC 21371 disclosed in U.S. Patent 3,663,370, but any microorganism of the genus Protaminobacter capable of utilizing methanol may be mutated to possess a resistance to relatively high concentrations of S-2-amino-ethyl-L-cysteine and L-threonine.

Any culture medium usually used for the culturing of microorganisms of the genus Protaminobacter is suitable for the process as long as it contains methanol as a carbon source, a nitrogen source, inorganic materials and other growth-promoting factors which may be required by the specific strain employed.

Example 1: *P. thiaminophagus* SLR-77 (FERM-P 2020) ATCC 21926 is inoculated in 10 ml of a seed medium having the following composition in a test tube and cultured at 30°C for 20 hours with shaking.

Methanol, ml	20
$(NH_4)_2SO_4$, g	10
KH_2PO_4, g	2
K_2HPO_4, g	7
$MgSO_4 \cdot 7H_2O$, g	0.5
$FeSO_4 \cdot 7H_2O$, mg	10
$MnSO_4 \cdot 4H_2O$, mg	8
Thiamine hydrochloride, mg	1
Biotin, μg	10
Yeast extract, g	0.1
$CaCO_3$, g	20
Water to make up the total volume, l	1
(pH 7.2)	

The resulting seed culture is inoculated into 10 ml portions of a fermentation medium having the same composition as the seed medium in test tubes in a ratio of 1 ml. Fermentation is carried out with shaking at 30°C for 72 hours, while feeding 1% of methanol after 16 hours of culturing and 2% of methanol after 25 and 40 hours of culturing, respectively. After the completion of fermentation, 1.5 mg/ml (as hydrochloride) of L-lysine is produced in the culture liquor.

The culture liquors in the test tubes are combined to make the total volume 2 liters and the microbial cells are removed from the culture liquor. The cell-free culture liquor is subjected to adsorption on a strongly acidic cation exchange resin, Amberlite IR-120 B. Elution is carried out with an aqueous ammonia. The eluate is concentrated and neutralized with hydrochloric acid. After crystallization, 2.4 g of crystals of L-lysine hydrochloride is obtained.

Using Microorganisms Resistant to L-Lysine Analogues

Mutants having resistance to a series of L-lysine analogues have been used by *Y. Kurimura, Y. Furutani, N. Makiguchi and K. Souda; U.S. Patent 3,905,867; September 16, 1975; assigned to Mitsui Toatsu Chemicals, Incorporated, Japan* in the production of L-lysine. These L-lysine analogues include:

(a) L-lysine hydroxamate,
(b) S-(β-aminoethyl)-L-cysteine hydroxamate,
(c) 3-(β-aminoethyl-sulfinyl)-L-alanine (S-dl),
(d) 3-[S-methyl-S-(β-aminoethyl)-thio]-L-alanine,
(e) S-β-(2-pyridyl)ethyl-L-cysteine, and
(f) S-β-(4-pyridyl)ethyl-L-cysteine.

When a mutant which has a resistance to one of these metabolic antagonists is cultured in a suitable culturing medium, L-lysine is produced and accumulated in a large amount. The L-lysine produced is easily recovered from the medium.

The mutants used in this process are obtained from parent strains of Arthrobacter sp. B-1, Corynebacterium sp. 18S and Brevibacterium sp. B-4 treated with N-methyl-N-nitro-N'-nitrosoguanidine.

After testing for L-lysine-producing ability in the presence of the L-lysine analogues, the following mutants were selected.

Resistant Mutant	FERM-P No.	ATCC No.	Resistant to Compound	L-Lysine HCl Produced, g/l
Arthrobacter sp. B-1 1772-193	1295	21868	a	23.8
501				17.3
1002				21.5
1798				12.1
Brevibacterium sp. B-4 1443-54	1299	21864	a	28.5
1401				21.8
1405				10.2
1477				15.6
Arthrobacter sp. B-1 874-75	1302	21861	b	41.5
349				23.4
1008				38.2
1067				37.6
301				15.7
Corynebacterium sp. 18S 351-23	1301	21862	b	39.8
391				18.9
1068				26.1
1100				35.3
1306				32.9
Brevibacterium sp. B-4 1304-31	1303	21860	b	42.7
1509				26.4
1513				37.4
1301				38.9
1306				11.2
Arthrobacter sp. B-1 55-8	1296	21867	c	15.3
67				13.8
80				14.9
33				12.2
72				14.7
Brevibacterium sp. B-4 1899-31	1298	21865	c	18.1
1580				14.8
1560				13.1
1922				16.9
1955				18.0
Arthrobacter sp. B-1 1162-166	1304	21859	d	39.4
417				38.4
1155				24.3
1179				27.1
820				18.8

(continued)

Resistant Mutant	FERM-P No.	ATCC No.	Resistant to Compound	L-Lysine HCl Produced, g/l
Corynebacterium SO 18S 1202-138	1306	21857	d	42.4
1137				40.7
1217				29.8
1227				31.0
1206				27.5
Brevibacterium sp. B-4 1506-13	1297	21866	d	32.8
1501				29.7
1829				20.3
1834				18.6
1839				21.4
Arthrobacter sp. B-1 269-51	1305	21858	e	23.0
6				16.5
127				12.1
362				10.7
493				13.5
Corynebacterium sp. 18S 738-26	1300	21863	f	19.0
225				12.4
341				17.2
755				13.3
791				13.1

Example: A fermentation medium was prepared of the folllowing composition.

Glucose, g/l	160
Ammonium sulfate, g/l	20
K_2HPO_4, g/l	1
$MgSO_4 \cdot 7H_2O$, g/l	0.4
$FeSO_4 \cdot 7H_2O$, g/l	0.01
$MnSO_4 \cdot 7H_2O$, g/l	0.01
Casaminoic acid, g/l	0.2
Biotin, mg/l	300
Thiamine hydrochloride, mg/l	200
$CaCO_3$, g/l	25
Urea, g/l	8
pH 7.5	

Ten liters were introduced into a 201 Jar Fermentor, and were then sterilized at 110°C for 30 minutes. The medium was inoculated with the microorganisms shown in the table below and incubated at pH 7.5 to 7.8, agitation rate 600 to 650 rpm, aeration rate 1 l/l/min, and temperature 30° to 32°C. After completion of the fermentation, L-lysine produced was determined by Warburg method and recovered by ion-exchange procedure. The results are listed in the following table.

Strain	Culturing Duration (hr)	L-Lysine HCl Produced (g/l)	L-Lysine HCl Recovered (g)
Arthrobacter sp. B-1 874-75 FERM-P 1302 ATCC 21861	89	43.8	385

(continued)

Strain	Culturing Duration (hr)	L-Lysine HCl Produced (g/l)	L-Lysine HCl Recovered (g)
Arthrobacter sp. B-1 1162-166 FERM-P 1304 ATCC 21859	76	41.9	366
Brevibacterium sp. B-4 1506-13 FERM-P 1297 ATCC 21866	76	39.7	–
Arthrobacter sp. B-1 269-51 FERM-P 1305 ATCC 21858	72	35.5	288
Corynebacterium sp. 18S 738-26 FERM-P 1300 ATCC 21863	69	25.7	–

Using Culture Media Containing Benzoic Acid

The microorganisms used in the process developed by *K. Tanno and Y. Ikeno; U.S. Patent 3,816,255; June 11, 1974* are mutant strains free from amino acid requirement which belong to the genus Brevibacterium or Corynebacterium. These mutant strains are free from amino acid requirements and grow in a medium containing benzoic acid.

Brevibacterium divaricatum HC-615 (FERM P-928: ATCC 21792), a mutant strain derived from *B. divaricatum* NRRL B-2312, and *Corynebacterium lilium* HC-B020 (FERM P-973: ATCC 21793), a mutant strain derived from *C. lilium* NRRL B-2243, are used in the process.

The above mutants were obtained with a mutating agent, N-methyl-N'-nitro-N-nitrosoguanidine. These mutants have none of the amino acid requirement for accumulating L-lysine. Further, these mutant strains can accumulate a considerable amount of L-lysine, even when cultivated in a medium containing a large amount of L-threonine or other amino acids.

Example: A medium having a composition of 1.0% ammonium benzoate, 0.1% K_2HPO_4, 0.05% $MgSO_4 \cdot 7H_2O$, 0.0005% $FeSO_4 \cdot 7H_2O$, 0.0005% $MnSO_4 \cdot 4H_2O$, 100 μg/l of d-biotin, 100 μg/l of thiamine monohydrochloride, and 0.5% corn steep liquor was adjusted to pH 7.0 to 7.5 with ammonia water, and 20 ml of the medium was placed in each shaking flask having a capacity of 500 ml, and sterilized at 121°C for 10 minutes. The medium in the flasks was inoculated with 5% bouillon preculture broth of *Corynebacterium lilium* HC-B020 (FERM-P 973; ATCC 21793) obtained through shaking cultivation at 30°C for 24 hours, and subjected to shaking cultivation with 120 reciprocations per minute.

Benzoic acid as a carbon source was added to the medium in the form of ammonium salt successively at 1% concentration, and after 60 hours from the start of cultivation total 10% (w/v) thereof as benzoic acid was added. Then, the cultivation was completed.

Determination of the amount of L-lysine accumulated in the culture broth after the completion of fermentation by microbioassay method revealed that 40.6 g/l of L-lysine monohydrochloride was formed. After the removal of cells from the culture broth by centrifuge, crystals were collected by the ordinary ion exchange resin method, whereby 37.3 g/l of L-lysine monohydrochloride was obtained.

Additives for Improved Culture Media

The addition of antibiotics, surface active agents and/or antioxidants to the culture media has been found by *K. Kubota, Y. Yoshihara and Y. Hirose; U.S. Patent 3,929,571; December 30, 1975; assigned to Ajinomoto Co., Inc., Japan* to improve the production of L-lysine.

The microorganisms used in the process may be of any L-lysine-producing strain and may be derived by means of conventional mutagenic agents from parent strains of the genera Brevibacterium, Corynebacterium, Microbacterium, and Arthrobacter, and by screening of the mutants so produced for the necessary resistance to S-(2-aminoethyl)-L-cysteine (AEC) and/or a nutrient requirement.

The addition agents suitable for the process include the following substances:

(1) Antibiotics: chloramphenicol, erythromycin, oleandomycin, kanamycin, streptomycin, kasugamycin, tetracycline, oxytetracycline, mytomycin C, actinomycin D, cyloserine, members of the penicillin and cephalosporin groups, polymyxin and azaserine. Chloramphenicol, erythromycin, oleandomycin, streptomycin and kasugamycin which are thought to react with the protein synthesizing part of the microorganism were found to be especially effective.

(2) Surface active agents: anionic surface active agents such as higher alcohol sulfates, alkylbenzenesulfonates, alkyl phosphates and dialkyl sulfosuccinates, cationic surface active agents such as alkylamine salts and quaternary ammonium salts, nonionic surface active agents such as polyoxyethylene alkyl ether, polyoxyethylene sorbitan monoalkyl ether and sorbitan monolaurate, amphoteric surface active agents such as imidazoline and betaine.

(3) Antioxidants: 4,4'-dihydroxy-3,3'dimethyldiphenyl, 2,6-di-3-butylphenol, catechol, butylcatechol, protocatechuic acid, α-tocopherol, pyrogallol, gallic acid, esters of gallic acid, naphthol, phenolic compounds such as aminophenol, amines such as naphthylamine, diphenylamine, di-2-butyl-p-phenylenediamine, 6-ethoxy-2,3,4-trimethyl-1,2-dihydroquinoline, semicarbazide, phenothiazine and tetraphenylhydrazine, and sulfur-containing compounds such as thiodipropionic acid and thiodiglycolic acid.

Example 1: A basic culture medium containing 10% glucose, 4.5% ammonium sulfate, 0.1% KH_2PO_4, 0.04% $MgSO_4 \cdot 7H_2O$, 2 ppm Fe^{++}, 2 ppm Mn^{++}, 50 μg/l biotin, 200 μg/l thiamine hydrochloride, 1% soybean protein hydrolyzate (total nitrogen 7%), and 5% calcium carbonate (sterilized separately), of pH 8.0, was divided into 20 ml batches. Each batch was further mixed with chloramphenicol or Miranol 2MCA (cationic surface active agent produced by Miranol Chemical

Co., Inc.), as indicated below, placed in a 500 ml shaking flask, and sterilized. The batches were inoculated with *Brevibacterium lactofermentum* FERM P-1944 (AEC-resistant strain induced from *B. lactofermentum* ATCC 13869) or *Corynebacterium glutamicum* FERM P-1987 (AEC-resistant strain induced from *Micrococcus glutamicus* ATCC 13032) which had previously been cultured on bouillon-glucose slants, and cultivation was carried out at 31°C with shaking. The batches were divided into two groups.

To one group, chloramphenicol was added initially and/or 15 hours after inoculation in an amount shown in Table 1. The fermentation was permitted to proceed for a total of 72 hours. The amount of L-lysine, as hydrochloride, accumulated in the broths was as shown in Table 1.

From the fermented broth of each of the two strains which were cultured in media without antibiotics, more than 3,000 colonies were examined for their ability of producing L-lysine. All showed resistance to AEC and were able to produce L-lysine. This shows that there was no back mutation even after the 72 hours' cultivation.

To another group of batches, Miranol 2MCA was added initially or 7 hours after inoculation in an amount shown in Table 2. After a total of 72 hours' cultivation, L-lysine accumulated in each broth as shown in Table 2.

TABLE 1

Strain	Chloramphenicol (μg/ml) 0 hr	15 hrs	Growth (OD)	Residual glucose (g/dl)	L-lysine (g/dl)	Yield (%)
FERM P-1944	1	0	1.18	0	2.0	20
	2	0	1.18	0.5	2.0	21
	1	5	0.91	0.2	3.1	32
	1	50	0.76	3.5	2.0	31
	0	0	1.25	0	1.8	18
FERM P-1987	1	5	0.95	0.5	3.0	30
	0	0	1.25	0	1.8	18

(Note)

Growth: Specific optical density (OD) was determined at 562 mμ in a sample diluted to 26 times its initial volume.

Yield: Based on consumed glucose.

TABLE 2

Strain	Miranol 2MCA (%) 0 hr	7 hrs	L-lysine (g/dl)
FERM P-1944	0.01	0	2.0
	0.05	0	2.3
	0	0.05	2.3
	0	0.1	2.4
	0	0	1.8
FERM P-1987	0.05	0	2.1
	0	0.1	2.5
	0	0	1.8

Example 2: L-lysine fermentation was carried out in the manner described in Example 1, but employing *B. lactofermentum* FERM P-1944, FERM P-1711, *C. glutamicum* FERM P-1982, and *Micrococcus glutamicus* ATCC 13286, adding required nutrients as shown in Table 3 and the antioxidants shown in the table. After 72 hours' cultivation, L-lysine accumulated in the broth as shown in Table 3.

TABLE 3

Strain	Nutrient	Antioxidant (mg/ml)	L-lysine (g/dl)
FERM P-1944	—	Gallic acid Initially: 0.5	2.5
		None	1.8
FERM P-1711	—	L-Aspartic acid & Laurylgallate After 12 hrs: 0.05 each	4.6
		None	4.2
FERM P-1982	—	BHA After 12 hrs:0.02 & Thio-diglycolic acid After 12 hrs: 0.1	3.8
		None	3.0
ATCC 13286	DL-Methionine 15 mg/dl	Ethyl gallate After 12 hrs:0.1	4.4
		None	3.8

LYSINE–PRODUCTION BY CHEMICAL SYNTHESIS

See also U.S. Patent 3,897,466, page 160 for another method of producing lysine.

Production from 2-Nitrocyclohexanone

Lysine is produced in the four-step process disclosed by *J. Pisanchyn, R. Fuhrmann, S. Sifniades and A.A. Tunick; U.S. Patents 3,876,685; April 8, 1975 and 3,681,460; August 1, 1972; both assigned to Allied Chemical Corporation* from 2-nitrocyclohexanone. These steps comprise:

(1) nitrosation of 2-nitrocyclohexanone to form 2-nitro-6-oximinocyclohexanone, (NOC);

(2) ring opening of 2-nitro-6-oximinocyclohexanone in alcoholic base to form an alkyl ester of 2-oximino-6-nitrocaproic acid;

(3) hydrogenating 2-oximino-6-nitrocaproic acid alkyl ester to form an alkyl ester of lysine; and

(4) saponifying lysine alkyl ester to afford lysine or lysine hydrochloride.

The starting material, 2-nitrocyclohexanone is formed by the known process of nitrating cyclohexanone. The preparation of the starting material 2-nitrocyclohexanone, also known as α-nitrocyclohexanone, is described, for example, in *J. Org. Chem.,* 31, 357 (1966).

Step 1, the nitrosation is ordinarily carried out by dissolving the 2-nitrocyclohexanone in a solvent (e.g., ether, 1,2-dimethoxyethane), cooling to about -15°C, and then adding 1.0-1.5 mols of the nitrosating agent (e.g., methyl nitrite, nitrosyl chloride) per mol of 2-nitrocyclohexanone. The presence of strong acid catalyst (e.g., HCl) aids the reaction.

Step 2, the ring opening is suitably effected by heating the 2-nitro-6-oximinocyclohexanone at 50° to 90°C with an alcoholic solution of strong base. Suitable alcohols are many of the C_1-C_4 normal alkanols, preferably methanol. Suitable bases include the alkali metal hydroxides or the alkali metal alkoxide of the alcohol used as solvent. Preferably 0.05-0.10 mol of base per mol of 2-nitro-6-oximinocyclohexanone is used. The ring opening is rapid, and heating for about 1 hour is ordinarily sufficient to effect complete formation of the ester of 2-oximino-6-nitrocaproic acid corresponding to the alcohol used as the solvent.

Reduction of the nitro and oximino groups of the ester to afford the ester of lysine (Step 3) is best effected by catalytic hydrogenation. Suitable catalysts include palladium on a support such as carbon, alumina, calcium carbonate, or barium sulfate. The concentration of palladium on the support is suitable from about 1 to 10% by weight.

The ester may be converted to the free acid, (Step 4) by known methods. Details of the reactions are illustrated in the following examples, where all parts are by weight unless noted otherwise.

Example 1: Nitrosation of Nitrocyclohexanone with Methyl Nitrite in Glyme — A 10.0-gram (0.070 mol) sample of 2-nitrocyclohexanone was dissolved in 350 ml of 10 N HCl in dry 1,2-dimethoxyethane (glyme) and the resulting solution then treated with 6.0 ml (6.0 grams, 0.098 mol) of methyl nitrite at -20°C. After 3.5 hours at -13°C, a 50-ml aliquot was removed for work-up and analysis. Nearly quantitative conversion of the nitrocyclohexanone to the desired 1-nitro-6-oximinocyclohexanone was indicated by gas chromatographic analysis of the aliquot.

The remaining solution was worked up after an additional standing of one-half hour as follows: the volatile components were stripped on a rotary evaporator at -10°C and the residue triturated with cold methanol. The resulting yellow crystals of NOC were collected by filtration and washed with a few milliliters of cold methanol. The 50-ml aliquot was worked up in a similar fashion. Yield from both portions: 7.74 grams (64.5%).

Example 2: To a solution of 10.0 grams (0.058 mol) of NOC in 400 ml absolute methanol was added at 50°C a solution of 2.40 grams (0.058 mol) of sodium hydroxide in 400 ml absolute methanol. A voluminous yellow precipitate appeared, which dissolved after several minutes. After standing at 50°C for 5½ hours, the reaction mixture was cooled to room temperature and then treated with a solution containing 5 ml 12 M HCl and 7 grams urea in 500 ml water. The resulting solution was concentrated to 450 ml by stripping under reduced pressure, and it was exhaustively extracted with ether. After drying of the ether extracts and evaporation of the ether, 11.6 grams of crude methyl-2-oximino-6-nitrohexanoate was obtained. The material was of sufficient purity for hydrogenation without additional purification.

Example 3: A 515-mg (2.5 mmol) sample of methyl-6-nitro-2-oximinohexanoate was dissolved in 20 ml of methanol containing 1 ml concentrated HCl and 500 mg 5% palladium on carbon. The mixture was hydrogenated at 60°C and 50 to 70 psig hydrogen pressure in a Parr bomb. The uptake of hydrogen was 97% of theoretical in 69 hours. The mixture was filtered, evaporated to dryness on a rotary evaporator, and the residue refluxed with 50 ml 6 N HCl for 3 hours. The resulting solution was concentrated on a rotary evaporator to a thick syrup, which was dissolved in 10 ml 95% EtOH at reflux.

The solution was treated with 0.5 ml pyridine in 1 ml 95% ethanol, cooled slowly, and left standing for 25 hours at room temperature followed by 16 hours in the refrigerator. The lysine hydrochloride which precipitated was collected by suction filtration, washed with ethanol, and dried in air at 120°C. A total of 275 mg was obtained. Analysis of the mother liquors indicated the presence of an additional 35 mg of lysine hydrochloride, for a total yield of 310 mg (68%). A small amount (30 mg) of pipecolinic acid was also present in the mother liquor.

An improved process for the ring opening of 2-nitro-6-oximinocyclohexanone (NOC) has been disclosed by *R. Fuhrmann, F.W. Koff and J. Pisanchyn; U.S. Patent 3,803,231; April 9, 1974; assigned to Allied Chemical Corporation*. This is Step 2 of the previous process.

The transformation of NOC to 2-oximino-6-nitrocaproamide (ONC) is effected by contacting the former with a sufficient quantity of anhydrous ammonia in the presence of a solvent selected from C_1 to C_4 alkanols and C_4 to C_8 aliphatic and alicyclic mono- or di-ethers. Contacting NOC with anhydrous ammonia in the absence of solvent affords substantial amounts of by-product. Reaction with ammonia in the presence of water affords an intractable complex mixture of products, possibly including ONC.

Examples of the suitable solvents include methanol, ethanol, isopropanol, n-butanol, t-butanol, diethyl ether, diisopropyl ether, glyme, diglyme, diethyl carbitol, tetrahydrofuran and dioxan. The preferred solvents are glyme (dimethoxyethane), t-butanol, and dioxan. Mixtures of any of the above solvents can also be used. Further details are given in the following example.

Example: To a 6 oz Fischer & Porter aerosol compatability tube was charged 8.3 g of 2-oximino-6-nitrocyclohexanone and 80 ml of dioxane. Stirring was effected with a magnetic stirring bar at room temperature to achieve solution. The reaction vessel (tube) and charge was purged of air by five pressurizations with N_2 followed by venting to atmospheric pressure. The reaction vessel was then placed in a heated silicone oil bath and the temperature brought up to 62°C. At this temperature, the reactor was pressurized with ammonia to 100 psig. Immediate precipitation of the ammonium salt of 2-oximino-6-nitrocyclohexanone occurred. Stirring was continued together with repressurizations with ammonia until complete solution of the precipitated salt occurred.

After 5.75 hours of reaction, heating was terminated, the reaction mixture cooled to room temperature, and the ammonia vented. The solution was then flash evaporated at reduced pressure in a Buchler rotating flash evaporator to obtain 9.10 g of solids (theoretical is 9.08 g). The dry crude product was recrystallized from ethylene dichloride affording 8.00 g of the pure product, 2-oximino-6-nitrohexanamide (88.7% yield). The recrystallized product was shown to be free of the starting material and any other impurity by thin-layer chromatography.

R. Fuhrmann, F.W. Koff and J. Pisanchyn; U.S. Patent 3,849,481; November 19, 1974; assigned to Allied Chemical Corporation have developed improved methods for the hydrogenation of 2-oximino-6-nitrohexanamide to form lysine amide bases or salts. This is Step 3 of the overall process.

The two hydrogenation methods that may be used are:

(a) A catalytic method of hydrogenating this precursor (2-oximino-6-nitrohexanamide) to produce lysine amide base in good yield using anhydrous ammonia as a solvent, and Raney nickel or Raney cobalt as the catalyst, and

(b) A catalytic method of hydrogenating this precursor (2-oximino-6-nitrohexanamide) to produce D,L,-lysine amide salts in good yield using a carboxylic acid, such as acetic acid, as a solvent, and one or more of the noble metals as the catalyst. The precursor, 2-oximino-6-nitrohexanamide, is obtainable from low-cost raw materials and is itself an intermediate in the production of lysine from cyclohexanone.

Example 1: The reactor consists of a 45 ml Parr general purpose screw cap bomb fabricated of T-303 stainless steel, Model 4712. Within the bomb is a Teflon-coated, serrated piston-type agitator, magnetically activated externally, and controlled by a Wilson repeat-cycle timer to provide any required degree of agitation. Heat is supplied by a constant temperature silicone oil bath.

The reactor is charged with 0.5263 g of 2-oximino-6-nitrohexanamide and 0.66 g of methanol-wet No. 27 Raney active cobalt catalyst (dry weight, 0.56 g) and after attaching the screw cap, the reactor is purged of air. This is done by pressurizing the bomb with nitrogen to 15 psig, venting, and repeating the procedure to a total of five pressurizations. The reactor is then cooled in a Dry-Ice-acetone bath to -77°C and charged with a slight excess of anhydrous ammonia over the intended charge. This excess is vented, thus serving as a further purge of the system and leaving 8.8 g of anhydrous ammonia as the charge. While still cold, the reactor is pressurized to 1,650 psig with pure hydrogen gas.

The reactor is next brought to 100°C by immersion in the constant temperature bath, agitation is initiated, and the reduction of the 2-oximino-6-nitrohexanamide is allowed to proceed at 100°C, with agitation, for 7 hours. The pressure during the hydrogenation remains at about 2,100 psig.

After the hydrogenation the reaction product is cooled to ambient temperature. The reactor is vented, the ammonia is allowed to evaporate, and a small amount of a suitable solvent such as methanol is added. The quantity of solvent is not critical. The solution is filtered to remove the catalyst, which is rinsed with the solvent used. The filtrate and wash then is made up to a standard volume such as 100 cc, and an aliquot portion taken for analysis by gas chromatography. The recovered catalyst is held for reuse in subsequent runs.

The aliquot of crude product is dried over P_2O_5, then charged into a 3 oz Fisher & Porter aerosol compatibility tube, and treated with 5 cc of a 4 N solution of anhydrous HCl in methanol. A Teflon closure and Teflon valve are used, and the magnetically controlled agitator used while the tube is heated in an oil bath for 1 hour at 100°C. The methanol is then flash-evaporated directly from the tube. 8 cc methylene chloride and 2 cc trifluoroacetic anhydride are added to the contents of the tube, which is closed as before, and agitated for 15 minutes until solution is substantially complete. Now the tube is heated to 150°C for 10 minutes, then cooled, and to it is added from a pipette a solution consisting of 3 cc of a 3% solution of octadecane in toluene, which serves as an internal standard in the gas chromatograph.

The crude product is found to contain 81.3 mol % lysine amide, 12.7 mol % pipecolinamide as by-product, and traces of three other unidentified by-products.

Example 2: The reactor (same as Example 1) is charged with 1.003 g of 2-oximino-6-nitrohexanamide, 0.303 g of a catalyst consisting of 5% rhodium on carbon and 10 cc of glacial acetic acid. The reactor cap is then fixed in place, and the reactor purged with nitrogen as described in Example 1. It is then pressurized with hydrogen to 1,200 psig and heated with agitation in the constant temperature bath at 75°C for 5 hours. The reactor is next cooled to ambient temperature, vented, and the contents filtered to remove the catalyst.

The acetic acid solution is flash-evaporated at reduced pressure and at a maximum temperature of 60°C using a Buchler Rotary Flash Evaporator to remove the acetic acid and obtain the crude product. This crude product is hydrolyzed, esterified, and trifluoroacetylated as in Example 1, then subjected to gas-liquid chromatography. The crude product is found to contain 83.3 mol % lysine amide and 17.3 mol % pipecolinamide.

From 2-Alkoxy-3-Oximinocyclohexenes

M.M. Rogic and F. Mares; U.S. Patent 3,928,445; December 23, 1975; assigned to Allied Chemical Corporation have developed an inexpensive and efficient route for the production of lysine using 2-alkoxy-3-oximinocyclohexene as the starting material. This process comprises the synthetic steps of:

(a) halogenating a 2-alkoxy-3-oximinocyclohexene or the mineral acid salt thereof with at least one equivalent of chlorine, bromine or iodine;

(b) treating the 1,2-dihalo-2-alkoxy-3-oximinocyclohexane intermediate thus formed with an alcohol having 1 to 12 carbon atoms, more preferably with ethanol, methanol, propanol, butanol or isobutanol to form the corresponding 2-halo-6-oximinocyclohexanone dialkyl ketal or its mineral acid salt;

(c) ring opening of the ketal using a Beckmann cleavage reaction to form an alkyl 2-halo-5-cyanovalerate;

(d) converting the alkyl 2-halo-5-cyanovalerate to form an alkyl ester of lysine by reductive amination;

(e) hydrolyzing the lysine alkyl ester to form lysine dihydrochloride, and

(f) neutralizing the lysine dihydrochloride to form lysine hydrochloride.

Moreover, since 2-alkoxy-3-oximinocyclohexene can be formed from cyclohexanone in situ, this lysine synthesis may be carried out in two steps involving conversion of the cyclohexanone to the alkyl 2-halo-5-cyanovalerate followed by reductive amination and hydrolysis to produce the lysine hydrochloride.

Example 1: A mixture of 42.1 g (0.237 mol) 2-methoxy-3-oximinocyclohexene hydrochloride in 150 ml SO_2 was cooled to -60°C and 17.05 g (0.24 mol) Cl_2 were gradually added. The reaction occurred immediately as the temperature of the mixture rose to -40°C. After completion of the addition of Cl_2, the reaction mixture was slowly warmed to -20°C and refluxed for 1½ hours, after which time it was cooled to -40°C and 25 ml of MeOH were added and allowed to react. The reaction was completed in about 30 minutes and the SO_2 and excess HCl were evaporated at a temperature lower than 0°C.

To isolate the product, an additional 100 ml MeOH were added and the mixture neutralized with Na_2CO_3. The mixture was filtered to remove the salts and the methanol was evaporated at room temperature to produce 38 g crystalline 2-chloro-6-oximinocyclohexanone dimethyl ketal (A).

Example 2: Eight grams (0.041 mol) of (A) were combined with 60 ml of glyme and mixed at 0°C with 16 g (0.134 mol) $SOCl_2$ in 20 ml of ether. The mixture was warmed to 20°C under N_2 for one hour until the oxime crystals disappeared. For the purposes of yield determination, the solvents and excess $SOCl_2$ were then evaporated in vacuum. The residue was diluted with chloroform and extracted with $NaHCO_3$ and water. The chloroform layer was dried and chloroform was stripped off in vacuo. The residue was distilled giving 6.56 g of the methyl 2-chloro-5-cyanovalerate.

Example 3: Into a mixture of 6 g MeOH and 150 ml SO_2, 8.5 g HCl were absorbed while maintaining the temperature at -60°C, followed by condensation of 15.5 g NOCl into the reaction flask. In a separate operation, 23.5 g cyclohexanone and 4 g methanol were combined and then added dropwise to the reaction mixture. During the addition of cyclohexanone (about 20 minutes) the temperature increased from -60° to -30°C and it was maintained at 30°C for another 30 minutes until the red color of the NOCl disappeared. The mixture was then cooled to -60°C and 16.6 g Cl_2 was added. The chlorine was consumed as it came into contact with the reaction mixture and the temperature increased to -40°C.

After completion of the addition of Cl_2, the reaction mixture was slowly warmed to reflux SO_2 for about 1½ hours. The mixture was then cooled to -40°C and then 25 ml methanol added and allowed to react. After 30 minutes, 73.8 g $SOCl_2$ were added. The reaction mixture was refluxed under N_2 for 2 hours. The SO_2 and excess HCl and $SOCl_2$ were evaporated in vacuum, the residue diluted with chloroform and extracted with $NaHCO_3$ and water. The chloroform layer was dried and chloroform was stripped off in vacuo. The residue was distilled giving 31.5 g methyl 2-chloro-5-cyanovalerate. The 31.5 g methyl 2-chloro-5-cyanovalerate was dissolved in 1 liter of a 1:1 mixture of methanol and ammonia containing 1 g Raney nickel.

The mixture was hydrogenated at 60°C and 50 to 70 psig hydrogen pressure in a Parr bomb. The uptake of hydrogen was 97% of theoretical in 4 to 6 hours.

The mixture was filtered, evaporated to dryness on a rotary evaporator, and the residue refluxed with 50 ml 6 N HCl for 3 hours. The resulting solution was concentrated on a rotary evaporator to a thick syrup, which was dissolved in 10 ml 95% EtOH at reflux. The solution was treated with 0.5 ml pyridine in 1 ml 95% ethanol, cooled slowly, and left standing for 25 hours at room temperature followed by 16 hours in the refrigerator. The lysine hydrochloride which precipitated was collected by suction filtration, washed with ethanol, and dried in air at 120°C.

By Acylation of 2,6-Dioximinocyclohexanone

In the process developed by *T.T. Yee; U.S. Patent 3,794,684; February 26, 1974; assigned to Atlantic Richfield Company* acylation of the 2,6-dioximinocyclohexanone is carried out easily and simply by the use of ketenes or carboxylic acid anhydrides in the absence of any catalyst.

Among the ketenes, ketene itself, $CH_2=C=O$, is preferred because of its ease of generation from acetone or acetic anhydride. The alkyl-substituted ketenes such as the dimethyl derivative, $(CH_3)_2C=C=O$, or diethyl derivative,

$$(CH_3CH_2)_2C=C=O$$

are also suitable. Higher derivatives are likewise suitable.

The carboxylic acid anhydrides which can be used include the alkyl compounds such as acetic anhydride, propionic anhydride, and the like in which the alkyl group may vary from 1 to 10 carbon atoms. Likewise, mixed alkyl anhydrides can be employed, for example, such as that obtained from acetic acid and propionic acid or the like. The diacylated derivatives may be converted to lysine by known methods of ring cleavage and reduction.

The acylation reaction can be carried out at temperatures ranging from 0°-100°C. Low temperatures of from 0°-10°C are preferred when using ketene because of its high reactivity, whereas temperatures in the range of 60°-100°C are preferred with the carboxylic acid anhydrides.

In carrying out the reaction since the 2,6-dioximinocyclohexanone is a solid it is necessary to use a solvent unless the acylating agent itself is a liquid and can act as the solvent. Thus, for example, when ketene which is a gas is used as the acylation agent a solvent such as tetrahydrofuran or acetone can be used. If acetic anhydride is used as the acylation agent since it is liquid it is unnecessary to use an additional solvent, although one can be employed if desired. If acylating with benzoic acid anhydride, for example, since it is a solid, a solvent such as acetone or tetrahydrofuran can be used. The choice of solvents is not critical although it is preferred, of course, that they be polar and nonreactive with the acylation agent.

The progress of the reaction can be followed by the formation of a yellow precipitate characteristic of the acylated derivative. When no further formation of a precipitate is observed the reaction is completed. In general, extremely high yields of the acylated derivatives are obtained by this method, i.e., of the order of 90 to 100% and commonly from 94 to 98%.

Example 1: To a suspension of 10 g of 2,6-dioximinocyclohexanone in 200 ml of dry acetone was passed a stream of ketene (and methane) generated by pyrolysis of anhydrous acetone. This was continued for approximately 3 hours (about 0.3 mol ketene per hour), while the temperature was maintained between 0° and 10°C with external cooling utilizing an ice bath. The reaction mixture became homogeneous and then a bright yellow solid precipitated. Filtration gave 2.9 g of solid and further concentration of the filtrate gave an additional 4.8 g of the yellow solid. These solids were examined by infrared and the spectra were found to be identical with that of 2,6-diacetoximinocyclohexanone.

The remaining filtrate was evaporated under vacuum to dryness and 5.7 g of the yellow solid, 2,6-diacetoximinocyclohexanone, was obtained. The yield of the acylated derivative was approximately 96% of theoretical.

Example 2: A flask containing 30 g of 2,6-dioximinocyclohexanone and 100 ml of acetic anhydride was heated on a steam bath with occasional shaking for about 15 minutes until the solid had disappeared. The mixture was then allowed to cool and a bright yellow solid precipitated which was collected by filtration. After washing with ethyl ether, 35.4 g of the 2,6-diacetoximinocyclohexanone was obtained. Concentration of the filtrate gave an additional 7.1 g of the diacetoximinocyclohexanone product. The total yield was about 93%. A similar run was carried out utilizing acetic anhydride and acetone and a yield of 84% by weight was obtained.

From 2-Cyano Cyclic Amines

It is known that α,ω-diaminocarboxylic acids can be produced by hydrolyzing corresponding 5-(ω-aminoalkyl)hydantoins in the presence of an acid or alkali. However, these hydantoins are generally synthesized by extremely complicated procedures which involve various reaction steps.

R. Kikumoto, M. Ito, A. Nakamura and T. Nomura; U.S. Patent 3,886,209; May 27, 1975; assigned to Mitsubishi Chemical Industries Ltd., Japan have disclosed an improved process for preparing α,ω-diaminocarboxylic acids. It was found that a 5-(ω-aminoalkyl)hydantoin can be obtained in one step by reacting a 2-cyano saturated cyclic amine as a starting material with ammonia, carbon dioxide and water. The product obtained by the above reaction not only contains the desired hydantoin, but also a precursor capable of being converted easily into an α,ω-diaminocarboxylic acid by hydrolysis with an acid or alkali. Therefore, the α,ω-diaminocarboxylic acid could easily be obtained in high yields by hydrolyzing the above reaction product.

This process is easily accomplished by contacting a 2-cyano saturated cyclic amine with a reactant selected from: (a) H_2O, NH_3 and CO_2; (b) a compound which is capable of providing NH_3 and CO_2 together with an aqueous medium; and (c) a mixture of the compound and at least one of H_2O, NH_3 and CO_2 to obtain a reaction product containing 5-(ω-aminoalkyl)hydantoin, and hydrolyzing the reaction product to obtain an α,ω-diaminocarboxylic acid. Details of the preparative methods are illustrated in the following examples.

Example 1: 22 g (0.2 mol) of 2-cyanopiperidine, 145 g (1.5 mols) of ammonium carbonate and 200 cc of water were charged into a 500 cc autoclave, heated to 100°C and maintained at that temperature for 2 hours at a pressure of about 6 kg/cm^2. Excess ammonium carbonate and water were then distilled off from the reaction mixture to give a residue. Analysis of the residue by heat-detecting liquid chromatography indicated that the residue contained 0.095 mol of 5-(4-aminobutyl)hydantoin.

To this residue was added 400 g of 25% sulfuric acid and the mixture was heated at 180°C for 1 hour at a pressure of 10 kg/cm^2. After cooling, the mixture was neutralized with $Ca(OH)_2$ to form a precipitate, which was then filtered off and, after driving off ammonia, 30 cc of hydrochloric acid was added to the filtrate followed by evaporation to dryness. The dried residue was then dissolved in 95% ethanol and 20 g of pyridine was added to the ethanol solution to obtain 30.3 g of a white solid having a melting point of 254°C. This solid substance was identified as DL-lysine hydrochloride by elementary analysis and the infrared spectrum thereof. The yield based on the mols of 2-cyano piperidine was 83%.

The results indicated that the above-described residue after the hydantoin-forming reaction was mainly 5-(4-aminobutyl)hydantoin, but also contained lysine precursor.

Example 2: 19.2 g (0.2 mol) of 2-cyanopyrrolidine, 145 g (1.5 mols) of ammonium carbonate and 200 cc of water were charged into a 500 cc autoclave, heated to 100°C and maintained at that temperature for 2 hours at a pressure of 6 kg/cm^2. Distillation of the ammonium carbonate and water from the reaction mixture gave a residue. Quantitative analysis of the residue using heat-detecting liquid chromatography indicated that the residue thus obtained contained 0.093 mol of 5-(3-aminopropyl)hydantoin.

To the above residue was then added 400 g of 25% sulfuric acid and the mixture was heated at 180°C for 1 hour at a pressure of 10 kg/cm^2. After cooling, the mixture was neutralized with $Ca(OH)_2$ and the precipitate which formed was filtered. After driving off ammonia from the filtrate, 30 cc of hydrochloric acid was added thereto and the mixture was then evaporated to dryness. The dried residue was dissolved in 95% ethanol and 20 g of pyridine was added to obtain 21.9 g of a white solid having a melting point of 232° to 233°C. This solid was identified as DL-ornithine hydrochloride by elementary analysis and infrared spectrum. The yield based on mols of 2-cyanopyrrolidine was 65%.

The results indicated that the above residue was mainly composed of 5-(4-aminobutyl)hydantoin, but also contained other ornithine precursors.

Example 3: 22 g (0.2 mol) of 2-cyano-2-methylpyrrolidine, 145 g (1.5 mols) of ammonium carbonate and 200 c of water were charged into a 500 cc autoclave, heated to 150°C and maintained at that temperature for 30 minutes at a pressure of 15 kg/cm^2. The ammonium carbonate and water were distilled off from the reaction mixture to give a residue, and 63.1 g of barium hydroxide [$Ba(OH)_2 \cdot 8H_2O$] and 200 cc of water were added to the residue followed by heating at 180°C for 1 hour in an autoclave at a pressure of 4 kg/cm^2.

Carbon dioxide gas was then bubbled through the reaction mixture and the precipitated barium carbonate was then separated by filtration. The filtrate was distilled off under reduced pressure to leave 2-methylornithine carbonate as a residue. The resulting product was dissolved in 20 cc of water and 300 cc of ethanol was added to the solution to form a white precipitate which was then filtered to yield 27.4 g of 2-methylornithine carbonate. The monopicrate was formed by known methods and analyzed to confirm the formation of 2-methylornithine.

From Mixtures of 5-(3-Cyanopropyl)-Hydantoin and 2-Ureido-5-Cyanovaleramide

It is known from U.S. Patent 2,688,023 that 4-cyanobutyraldehyde can be converted into 5-(3-cyanopropyl)-hydantoin by reacting it with a cyanide reactant, an ammonium reactant and a carbonate reactant. In this known process, the resulting reaction mixture is rendered slightly acidic and heated in order to improve product purity and yield.

It has been found by *G.H. Suverkropp and W. Reichrath; U.S. Patents 3,911,001; October 7, 1975 and 3,891,702; June 24, 1975; both assigned to Stamicarbon NV, Netherlands* that by maintaining the pH above 7, the reaction of 4-cyanobutyraldehyde will not only form 5-(3-cyanopropyl)-hydantoin, but also 2-ureido-5-cyanovaleramide. In an acid medium, however, this last compound is converted into 5-(3-cyanopropyl)-hydantoin when it is heated. Although the production of the valeramide compound as a by-product would appear to be an inconvenience, the formation of such undesirable by-products can be prevented through the use of a pH above 7.

It has further been found that 2-ureido-5-cyano-valeramide can be hydrogenated to 2-ureido-5-aminocapronamide in the same manner as the 5-(3-cyanopropyl)-hydantoin. Moreover, 2-ureido-6-aminocapronamide can be hydrolyzed to form lysine using hydrolysis conditions which are essentially equivalent to the hydrolysis conditions normally used for the formation of lysine from 5-(4-aminobutyl)-hydantoin.

Example 1: Preparation of a Mixture of 5-(3-Cyanopropyl)-Hydantoin and 2-Ureido-5-Cyanovaleramide — To a reaction flask provided with a stirrer and a reflux cooler, 200 ml of ammonia water (25% by weight of NH_3), 31 g of ammonium carbonate and 9 ml of liquid hydrogen cyanide were added. At approximately 25°C and with continuous stirring, 19.8 g of 4-cyanobutyraldehyde were added over a period of 15 minutes. The mixture was stirred for 30 minutes at 25°C after which the mixture temperature was raised to 60°C and stirred for an additional 4 hours at the higher temperature.

The resulting reaction mixture was dried by evaporation under reduced pressure (18 mm Hg), and a residue of 35 g of white crystalline product was obtained. From the nuclear magnetic resonance spectrum of this product, it was determined that 5-(3-cyanopropyl)-hydantoin and 2-ureido-5-cyanovaleramide were present in a molar ratio of 85:15. From the thin layer chromatogram, it was further determined that, except for the compounds mentioned, there were practically no other substances present.

By titration of a pyridine solution of the resulting product with tetrabutylammonium hydroxide, it was found that the product contained 83.4% by weight of 5-(3-cyanopropyl)-hydantoin.

The above analysis establishes that the conversion of the 4-cyanobutyraldehyde into a mixture of 5-(3-cyanopropyl)-hydantoin and 2-ureido-5-cyanovaleramide was practically quantitative.

Example 2: Preparation of Lysine Dihydrochloride – To a 1-liter autoclave provided with a stirrer, 1 g of Raney nickel, 100 ml of methanol and 70 g of ammonia were added. After the autoclave had been closed, hydrogen was forced into the autoclave until a pressure of 20 atm had been reached. Subsequently, and with stirring, the temperature was raised to 110°C, the pressure being increased to 62 atm by supply of hydrogen. While stirring the catalyst suspension so obtained, a solution of 12 g of 5-(3-cyanopropyl)-hydantoin, 2.4 g of 2-ureido-5-cyanovaleramide and 20 g of ammonia in 200 ml of methanol was added, whereupon stirring was continued until hydrogen was no longer taken up by the reaction mixture.

After cooling and release of pressure, the reaction mixture was filtered and the filtrate evaporated till dry under reduced pressure (20 mm Hg). The residue obtained, together with 100 ml of hydrochloric acid (36% by weight of HCl), was heated for 4 hours to 170° to 180°C in an autoclave, whereupon the resulting solution was evaporated till dry. The residue obtained contained 16.2 g of lysine dihydrochloride, which corresponds to conversion of 87.3% based on the original quantity of 5-(3-cyanopropyl)-hydantoin and 2-ureido-5-cyanovaleramide provided.

By Resolution and Racemization of α-Aminocaprolactam

It is known that D,L-α-aminocaprolactam can be optically resolved and the L-α-aminocaprolactam ring can be opened to form L-lysine. It is an object of the process developed by *A.M. Kubanek, S. Sifniades and R. Fuhrmann; U.S. Patent 3,824,231; July 16, 1974; assigned to Allied Chemical Corporation* to provide a rapid and effective method of resolving D,L-α-aminocaprolactam and recovering the L-isomer. In addition, the process provides a method of racemization and recycle of D-α-aminocaprolactam. This process comprises the steps of:

(a) Forming a complex of D,L-α-aminocaprolactam with a metal ion selected from Ni^{++}, Co^{++}, Cu^{++}, Fe^{++}, Fe^{+++} and Zn^{++};

(b) Optically resolving the D,L-α-aminocaprolactam metal complex whereby separate phases rich in D- and L-α-aminocaprolactam metal complex, respectively, are obtained;

(c) Recovering and decomposing L-α-aminocaprolactam metal complex to give L-α-aminocaprolactam plus metal ion; and

(d) Racemizing D-α-aminocaprolactam metal complex to form D,L-aminocaprolactam metal complex.

The complexing of the D,L-α-aminocaprolactam mixture with the metal ion of step (a) is accomplished by contacting the α-aminocaprolactam with the metal ion. The mol ratio of metal ion to D,L-α-aminocaprolactam in the combined solution is preferably about 1:3. Preferably, the chloride or nitrate salts of the above metals are used. The preferred metal ion is Ni^{++}. The reaction temperature for complex formation is preferably from 35°C up to the boiling point at atmospheric pressure of the solvent used.

Any solvent which will dissolve both the metal salt and the α-aminocaprolactam without reaction is suitable. The preferred solvents are methanol, ethanol, n- or isopropanol and mixtures thereof, with each other and/or with up to about 50% by weight water. More than 50% water is undesirable in that it makes it difficult to crystallize the L-α-aminocaprolactam metal complex in step (b).

In step (b), one obtains by preferential crystallization, the metal complex of either the D or L isomer of α-aminocaprolactam. Crystallization of the L-isomer is achieved by seeding the solution containing D,L-α-aminocaprolactam metal complex with crystals of L-α-aminocaprolactam metal complex. On addition of the seed crystals the L-isomer in solution will preferentially crystallize out around such seed crystals and precipitate leaving the solution mother liquor rich in the D-isomer.

To recover uncomplexed L-α-aminocaprolactam, the L-α-aminocaprolactam metal complex is dissolved in solvent, preferably a C_1 to C_3 aqueous alcohol, and treated with any strong nonoxidizing mineral acid, preferably hydrochloric, sulfuric, or phosphoric at 0°C. The acid effects decomposition of the complex and the simultaneous precipitation of the L-α-aminocaprolactam as the acid salt.

Step (d) entails racemization of the D-α-aminocaprolactam metal-complex-rich phase to form D,L-α-aminocaprolactam metal complex which is then recycled to step (b). The racemization is effected by heating a solution of the D-isomer in the presence of at least a catalytic amount of base at a temperature of at least 100°C. It should be noted that such temperatures will usually require superatmospheric pressures, but this does not otherwise affect the course of the racemization. The preferred solvent is a C_1 to C_3 alkanol which may contain up to 50% by weight water, i.e., the preferred solvent of the previous steps.

The preferred catalytic base is D,L-α-aminocaprolactam. When using this latter base to effect racemization, the preferred procedure is to add D,L-α-aminocaprolactam to a solution of the D-isomer metal complex and heat this mixture, e.g., at 120°C for 1 hour. Details of the first three steps are given in the following examples. In these examples aminocaprolactam is abbreviated as ACL.

Example 1: Resolution of $(DL\text{-}ACL)_3NiCl_2$ by Seeding – 0.476 part (2 mmol) $NiCl_2 \cdot 6H_2O$, 0.770 part (6 mmol) DL-ACL, and 1.70 parts EtOH were placed in a sealed glass vessel containing a magnetic bar. The vessel was immersed in a water bath at 80° to 90°C until no solids remained in the dark blue solution (2 to 3 minutes). Then 0.089 part of $(L\text{-}ACL)_3NiCl_2$, equivalent weight 280 g/Cl^-, was added and the vial was placed in a thermostated bath at 50°C. The contents were magnetically stirred for 10 seconds at the beginning, and thereafter for 10 seconds every 10 minutes. After 20 minutes, the vessel was shaken vigorously 2 to 3 times to loosen up its contents, the mixture was filtered by suction through medium frit and the blue crystals were washed with 4 parts EtOH and dried in vacuo at 50°C. Yield 0.160 part, equivalent weight 280 g/Cl^-, $[\alpha]_D^{25} = -22.6°$ (c = 3, 1 N HCl) or optical purity of L-complex 97%.

Total L-complex = 0.160 x 0.97 = 0.155 part
Seed L-complex = 0.089 part
Gain L-complex = 0.066 part

This represents a gain of 74% with respect to seed and 11.8% with respect to L-ACL available in the DL-ACL charged.

Resolution of $(DL\text{-}ACL)_3NiCl_2$ by Seeding

Temperature: 50°C
Solvent: 1.82 g of $8EtOH/1H_2O$ (by weight)
DL-Complex (EW = 280 g/Cl^-): 1.12 g formed in situ

Run No.	Seed L (mg)	Time (min)	. . . . .Crop. . . . . Mg	Optical Purity (%)	. . .Resolution. . . Seed (%)	Charge* (%)
1	81	10	99	100	22	3.2
2	88	13	148	99	50	8.0
3	89	20	172	97	88	13.9
4	90	20	164	92	68	10.9
5	89	22	160	97	74	11.8
6	91	30	179	94	81	13.5
7	88	40	181	96	93	14.9
8	89	50	195	96	112	17.6
9	208	15	294	94	32	12.0
10	91**	5	106	96	13	2.1
11	93**	10	136	99	45	7.5
12	89**	15	152	91	55	8.8

*Resolution as percent of L-ACL available in the DL-ACL charged.
**Seeds aged for 10 minutes in EtOH.

Example 2: Decomposition of $(L\text{-}ACL)_3NiCl_2$ – L-aminocaprolactam nickel chloride complex 2.05 parts (3.67 mols) of 91% optical purity, was stirred for 30 minutes at 70° to 80°C with 3 parts ethanol, 1 part water and 0.40 part (11 mols) hydrogen chloride. The mixture was allowed to cool to 25°C, then it was filtered and the colorless crystals were washed with ethanol and dried in vacuo at 50°C. Yield 1.56 parts L-ACL·HCl (86% of total ACL, or 95% of L-ACL charged), $[\alpha]_D^{25} = -26.7°$ (c = 2, 1 N HCl) or 100% optically pure L-ACL·HCl. A 10% aqueous solution of the product gave a faint pink color with dimethylglyoxime, indicating the presence of traces of Ni^{++}, estimated colorimetrically as ~150 ppm Ni^{++}.

Example 3: Elimination of Ni^{++} Traces from ACL·HCl – DL-ACL·HCl (1 g) was dissolved in 10 ml of an aqueous solution containing 50 ppm Ni^{++}. The solution was passed within 5 minutes through 15 ml of chelating resin Chelex-100 in the H^+ form, at 60°C. The resin was subsequently washed with 10 ml H_2O in 5 minutes.

At this point, only traces of ACL remained in the resin. The combined liquors were analyzed for Ni^{++} colorimetrically: 5 drops bromine/H_2O were added to the liquors, followed by 0.5 ml 14 N NH_4OH and 0.5 ml of 1% ethanolic dimethylglyoxime and made up to 25 ml with H_2O. The optical density was zero. A standard containing 1 g DL-ACL·HCl and 0.5 ppm Ni^{++} with respect to DL-ACL·HCl had OD = 4.5 after similar treatment with dimethylglyoxime.

An improved process of simultaneous resolution and racemization of α-aminocaprolactam (ACL) has also been disclosed by *S. Sifniades, W.J. Boyle, Jr. and*

J.F. Van Peppen; U.S. Patent 3,941,776; March 2, 1976; assigned to Allied Chemical Corporation. It has been found that solutions of aminocaprolactam complexes or salts can be rapidly racemized at relatively low temperatures, and that optically active crystals suspended in a supersaturated solution of the corresponding racemic compound effect resolution by preferential crystallization of one isomer even under conditions of relatively fast racemization in the solution. The process comprises:

(a) forming a supersaturated feed solution of an aminocaprolactam compound and maintaining the solution below 120°C and under conditions such that the racemization rate constant of aminocaprolactam is at least 0.001 min^{-1};

(b) contacting the solution with seed crystals of the desired isomer of the compound; and

(c) removing the grown crystals of the compound.

Three types of salts may be used in the process. These comprise:

(a) the D,L-aminocaprolactam complexes, tris(aminocaprolactam)-nickel(II) salts and tris(aminocaprolactam)cobalt(II) salts,

(b) the D,L-aminocaprolactam salts of the acids hydrogen chloride, hydrogen bromide, β-naphthalenesulfonic acid and 2-naphthylamine-1-sulfonic acid, and

(c) the D,L-aminocaprolactam diastereomeric salts of the acids L-pyrrolidonecarboxylic acid, D-N-carbamoylvaline, L-N-p-nitrobenzoylglutamic acid, L-N-benzoylglutamic acid, L-N-p-toluylglutamic acid and L-N-p-chlorobenzoylglutamic acid.

When the D,L-aminocaprolactam compound is a salt (b) or diastereomeric salt (c), a racemization catalyst of the formula 1 or 2

(1) 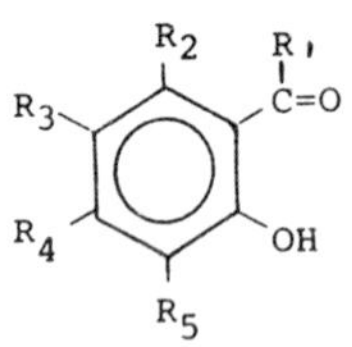

(2) 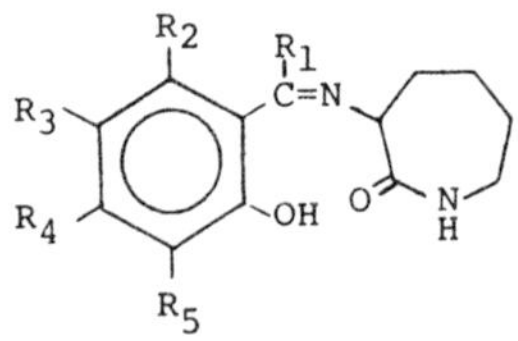

where R_1 is hydrogen or methyl and R_2 through R_5 are hydrogen, nitro, halogen, cyano or carboxylic substituents, and a metal ion selected from the group consisting of Fe^{++}, Fe^{+++}, Al^{+++}, Cu^{++}, Co^{++}, Zn^{++} and Ni^{++} is used in the reaction.

The following examples illustrate the batch simultaneous resolution/racemization of DL-ACL using the three types of salts.

Example 1: $(D,L\text{-}ACL)_3NiCl_2$ – A 0.556 g sample of $NiCl_2$, anhydrous (4.3 mmol) in 6.1 g ethanol was heated at reflux, then 2.05 g (16 mmol) solid D,L-ACL was added and the mixture heated at reflux for 2 hours to dissolve the nickel chloride. The solution was filtered to remove 10 mg of undissolved solids. A 0.60 ml sample of 0.99 M KOH in ethanol was added, the solution stirred 10 minutes, then filtered to remove KCl. The solution was again heated and 3.2 ml ethanol distilled out, leaving a solution containing about 33% complex by weight.

A 200 mg sample of $(L\text{-}ACL)_3NiCl_2$ (100% optically pure) was added and the flask heated at 85° to 90°C for 1.5 hours, with occasional stirring. The crystals were collected by filtration, washed with ethanol and dried in vacuo at 60°C. Yield: 646 mg $[\alpha]_D = -22.3°$, (c = 4, 1 N HCl) 95% optically pure. This represents 20% resolution with respect to the available complex. The mother liquors from the filtration were diluted to 25 ml with 3 ml 6 N HCl and the remainder 1 N HCl. The observed optical rotation, +0.01°, corresponds to a 10 mg excess of $(D\text{-}ACL)_3NiCl_2$ in solution and indicates essentially complete racemization. Analysis of the mother liquor by TLC showed the absence of lysine.

Example 2: ACL·β-Naphthalenesulfonate – D,L-ACL, 1.51 g, and D,L-ACL·β-naphthalenesulfonate, 3.90 g, were dissolved in 25 ml methanol at reflux. Then 10 ml methanol were distilled off and 1 ml methanol solution containing 15.8 mg $FeCl_3 \cdot 6H_2O$ and 58.6 mg 5-nitrosalicylaldehyde was added. Then 0.50 g of L-ACL·β-naphthalenesulfonate seed crystals were added and methanol was distilled off at the rate of about 0.1 ml/min. After 1 hour, 6 ml methanol had been distilled off. Heating was continued for an additional period of 10 minutes. The grown crystals were filtered, washed with a little methanol and dried in vacuo at 60°C.

Yield of L-ACL·β-naphthalenesulfonate was 0.97 g of optical purity 93%. This corresponds to 0.40 g gain in L-salt. Examination of the mother liquor showed the presence of less than 0.10 g D-salt excess in solution, indicating the occurrence of racemization simultaneously with the resolution.

Example 3: Diastereomeric Salt with L-Pyrrolidonecarboxylic Acid (L-PCA) – D,L-ACL, 2.56 g (20 mmol) was dissolved in 31 ml of refluxing glyme, then a racemization catalyst solution of 16.6 mg (0.1 mmol) $FeCl_3$ and 5.31 mg (0.3 mmol) 5-nitrosalicylaldehyde in 5 ml glyme was added. Next a solution of 1.28 g (9.9 mmol) L-PCA in 75 ml of hot glyme was added with stirring while maintaining reflux. A crystalline precipitate formed. Next the mixture was refluxed and stirred for 1 hour, filtered hot, the crystals were washed with glyme and dried in vacuo at 60°C. Yield of L-ACL·L-PCA salt was 2.61 g (theoretical is 2.56 g).

A 1.30 g aliquot of the salt was slurried with 15 ml isopropanol and decomposed with 1 ml of 12 N hydrochloric acid; L-ACL·HCl, 0.824 g (99% of theory) was obtained, $[\alpha]_D = -20.1°$ (c = 4, 1 N HCl). This corresponds to 76% optical purity of L-ACL.

The mother liquor from the diastereomer salt filtration was treated with 1 ml of 12 N HCl in glyme; ACL HCl, 1.48 g (87% of theory) was obtained, $[\alpha]_D = +0.7$ (c = 4, 1 N HCl). This indicates that the ACL in the mother liquor was about 96% racemic.

Conversion of Lysine·2HCl to Lysine·HCl

Anion exchange resins have previously been used to convert lysine·HCl produced by the hydrolysis of aminocaprolactam (ACL) to lysine·HCl. However, this process requires bulky, costly and time-consuming anion exchange resin equipment. Moreover, the lysine monohydrochloride is obtained as a relatively dilute aqueous solution which must be further processed to recover the lysine monohydrochloride in the desired form.

A process has been developed by *S. Sifniades; U.S. Patent 3,931,308; January 6, 1976; assigned to Allied Chemical Corporation* to convert lysine·2HCl solution to crystalline lysine·HCl by neutralization with α-amino-ϵ-caprolactam or a transition metal complex thereof and crystallization from the solution with an organic solvent. After removal of the precipitated lysine monohydrochloride, the remaining solution is fractionated and that fraction consisting essentially of the by-product of the neutralization may be recycled to ensure recovery of substantially all the desired product and utilization of all processing agents.

The following example illustrates the process whereby the neutralization is carried out using L-ACL, the solution is fractionated, the organic solvent recycled to the neutralization and the fractionation residue containing primarily L-ACL·HCl is recycled to the hydrolysis step without further isolation.

Example: L-ACL·HCl, 197 g (1.2 mols) and 6 N hydrochloric acid, 400 ml (2.4 mols) were placed in a 1 liter autoclave and heated at 142°C for 80 minutes. At the end of the period all L-ACL·HCl had been converted to L-lysine·2HCl. No racemization took place.

The reaction mixture was subjected to distillation at atmospheric pressure. During distillation, the amount of water that distilled off together with HCl was partly replaced by periodic additions of fresh water. 51.2 g (1.4 mols) HCl distilled. This HCl was then recycled to the hydrolysis step.

The residue contained 219 g (1 mol) L-lysine·2HCl, 36.5 g (0.2 mol) L-lysine·HCl and 360 ml water. To this was added a solution of L-ACL, 128 g (1 mol) in 1,000 ml methanol. Crystals of L-lysine·HCl·$2H_2O$ started forming immediately. The crystallization was completed after 30 minutes at room temperature, the crystals were filtered and washed with MeOH:H_2O (4:1) and dried at 120°C. The yield of L-lysine·HCl was 182.5 g (1 mol).

The combined mother liquors and wash liquors were fractionated. Almost pure MeOH, 1,000 ml, was obtained. The residue contained 164.5 g (1 mol) L-ACL·HCl and 36.5 g (0.2 mol) L-lysine·HCl in water. This solution, after adequate concentration, was treated as described in the above hydrolysis step, with HCl in an autoclave to produce more L-lysine·2HCl.

If the neutralization is carried out using other than the optically pure L enantiomer, e.g., a racemic modification of D,L-ACL or D-ACL, the solution is fractionated to separate the organic solvent which is recycled to the neutralization and a portion of the water, the fractionation residue is filtered to remove the precipitated crystalline D,L-ACL·HCl or D-ACL·HCl which is recycled to a resolution operation and the filtrate is recycled to the neutralization step.

If the neutralization is carried out using L-ACL-transition metal complex, e.g., with $NiCl_2$ or $CoCl_2$ or an optically impure L-ACL-transition metal complex, the solution remaining after removal of the precipitated lysine monohydrochloride is dried, and the organic solvent recycled. The remaining crystalline residue is then extracted with an alcohol to separate the crystalline L-ACL·HCl from the alcohol solution containing the transition metal salt; the extraction residue containing primarily crystalline L-ACL·HCl is recycled to the hydrolysis step and the alcohol extract containing primarily the transition metal salt and any D,L-ACL·HCl, if present is recycled to a resolution operation.

S. Sifniades and A.A. Tunick; U.S Patent 3,917,684; November 4, 1975; assigned to Allied Chemical Corporation have also disclosed a modification of the previous process whereby the lysine·2HCl is neutralized with ammonia in an organic solvent.

After removal of the precipitated lysine·HCl, the solvent mixture containing NH_4Cl, some lysine·HCl and organic solvent is fractionated to recycle the solvent and then acidified with hydrochloric acid to salt out a major portion of the NH_4Cl. The remaining solution is then combined with a lysine precursor to produce lysine·2HCl. This method ensures virtually complete recovery of lysine from the processing system.

Example: 165 g L-α-amino-ε-caprolactam·HCl (L-ACL·HCl) (1 mol) were introduced into a 1 liter autoclave together with a recycle solution containing 26 g (0.12 mol) L-lysine·2HCl, 21 g (0.40 mol) NH_4Cl, H_2O and 2.0 molar equivalents hydrochloric acid. The overall volume of the reaction mixture was about 500 ml. The mixture was heated at 142°C for 80 minutes, after which all the L-ACl·HCl had been converted to L-lysine·2HCl. No racemization occurred.

The reaction mixture was subjected to distillation at atmospheric pressure. During distillation, the amount of water that distilled off together with HCl was partly replaced by periodic additions of fresh water. The 44 g HCl (1.2 mols), which distilled, was recycled downstream to the acidification step.

To the residue containing 201 g (0.92 mol) L-lysine·2HCl, 36.9 g (0.2 mol) L-lysine·HCl, NH_4Cl and about 300 ml water was added about 1.0 mol anhydrous ammonia in 1,100 ml methanol at room temperature to cause precipitation, over a period of 30 minutes, of the L-lysine·HCl·$2H_2O$. These lysine crystals were then filtered and washed with fresh aqueous methanol and air dried at 120°C. The yield of L-lysine·HCl was 182 g (1.0 mol), an amount which represented about 89% of the L-lysine·HCl present.

The combined mother liquors and wash liquors were fractionated by distillation. The methanol and excess ammonia were recycled to the neutralization/crystallization step. The resulting mixture was concentrated to remove some of the water and the small amount of remaining ammonia.

The residue containing the remaining 11% (22 g) L-lysine·HCl plus 72 g (1.35 mols) NH_4Cl and water was acidified at about 30°C by the addition of 28 g hydrochloric acid combined with the 43 g hydrochloric acid from the first distillation operation to supply a total of 71 g (2 mols) HCl in sufficient water to make a final volume of the reaction mixture equal to about 350 ml. This resulted in salting out NH_4Cl. The crystals were washed with a small amount of concentrated hydrochloric acid and the combined liquors, containing 25 g (0.11 mol) L-lysine·2HCl, 21 g (0.4 mol) NH_4Cl and water were returned to the autoclave and treated as described in the above hydrolysis step to produce additional L-lysine·2HCl. The overall L-lysine·HCl yield was 99%.

LYSINE AND DERIVATIVES—USES

Preparation of Effervescent KCl-Lysine Compositions

Normally, in the treatment of potassium depletion (hypokalemia), it is customary to administer potassium chloride in tablets containing up to 500 mg of the salt. Where larger amounts of potassium chloride are to be administered, it is generally done intravenously using dilute solutions of the salt so as to avoid a sudden increase in blood plasma potassium, which could be fatal. Intravenous administration is, of course, much less desirable than oral administration; but tablets containing large amounts of potassium chloride are difficult to administer because of the size and unpleasant taste, which also renders multiple administration objectionable. Any effort made to improve the taste of such tablets in the usual manner with flavoring and sweeteners results in tablets even larger and more difficult to administer conveniently.

E.W. Brockemeyer and M.B. Rodell; U.S. Patent 3,970,750; July 20, 1976; assigned to Sandoz, Inc. have developed a stable effervescent composition containing potassium chloride which forms, on addition to water, a pleasant-tasting drink in which the ratio of potassium ion to chloride ion is 1 to 1.

The desired taste characteristics and balance between potassium ion and chloride ion can be provided by including in the potassium chloride effervescent composition a sufficient quantity of a basic amino acid hydrochloride to provide an amount of chloride ion exactly equivalent to the amount of potassium ion liberated in the reaction by the potassium carbonates during the effervescent reaction. The stoichiometric amount of basic amino acid hydrochloride, i.e., hydrochloride salt of a basic amino acid, required is readily calculated from the amount of potassium carbonates present in the potassium chloride effervescent composition. The basic amino acid hydrochlorides include lysine hydrochloride, arginine hydrochloride, and histidine hydrochloride, but the preferred amino acid hydrochloride is L-lysine hydrochloride.

Normally, the compositions of the process will be made up in units (e.g., tablets or cachets) containing about 1,125 g of potassium chloride. A suitable composition and range of ingredients according to the process is a tablet comprising the following.

	Grams
Amino acid hydrochloride	0.800-3.000
Potassium chloride, USP	0.500-1.500
Potassium bicarbonate, USP	0.500-2.000
Citric acid, USP, anhydrous	0.500-2.000
Saccharin, USP	0.005-0.100
Polyethylene glycol-6000	0.101-0.150
Color and flavor, q.s.	

Example: Tablets containing the following amounts of ingredients are prepared in the manner described below.

	Milligrams
L-lysine hydrochloride	913
Potassium chloride	1,125
Citric acid	550
Potassium bicarbonate	500
Saccharin	35
Polyethylene glycol-6000	60
Lemon-lime Spralene (Kohnstamm L-1236)	20
Punch flavor (Firmenich 59.722/AP)	10
Pineapple Aromalok (Fritzsche 28992)	5

Granular potassium bicarbonate for the desired number of tablets is ground through a No. 12 screen and then dried at 220°F for about 16 hours to remove any moisture present. It is then mixed with the remaining ingredient using standard techniques and apparatus for about 20 minutes. The granulation is tabletted in a conventional manner using 13/16 inch punches.

When the tablet is added to a glass of water, it disintegrates to yield a pleasant-tasting effervescent drink, which contains potassium ion and chloride ion in a solution in which the ratio of potassium ion to chloride ion is 1 to 1.

Similar results are obtained when the L-lysine hydrochloride above is replaced by an equivalent amount of L-arginine hydrochloride or L-histidine hydrochloride.

Lysine for Resolution of D,L-Penicillamine

The amino acid D-penicillamine is known to be an important medicine for the treatment of Morbus Wilson, defective schizophrenia, scleroderma, cystinuria and chronic aggressive hepatitis as well as basic therapy of primary chronic polyarthritis. D-penicillamine is also useful as an antidote for heavy metal intoxications. Therapeutic uses are only found for D-pencillamine since the L-isomer is much more toxic.

Optically active bases such as d-pseudoephedrine or l-ephedrine have been used to resolve synthetic D,L-penicillamine derivatives such as N-acylated or S-benzyl as well as the acylation products of the D,L-penicillamine with carbonyl compounds. However, the antipode which precipitates is the L-isomer.

R. Fahnenstich, J. Heese and H. Offermanns; U.S. Patent 3,980,665; September 14, 1976; assigned to Deutsche Gold- und Silber-Scheideanstalt, Germany have found that it is advantageous to use L-lysine as the optically active base for the recovery of D-penicillamine from D,L-penicillamine. The splitting of the racemate of D,L-penicillamine proceeds with the help of this optically active base with very high yields. The D-penicillamine is obtained in high purity since the desired salt of the D-acid and the L-base is the very much more difficultly soluble diastereomeric salt and precipitates.

The D,L-penicillamine must first be converted into a suitable derivative for resolving before the reaction with the L-lysine can take place. At least one of the hydrogen atoms of the amino group must first be protected. At the same time the hydrogen atom of the mercapto group can be protected.

Such protection can be obtained, for example, if D,L-penicillamine is converted in known manner to a compound having the formula:

(1)

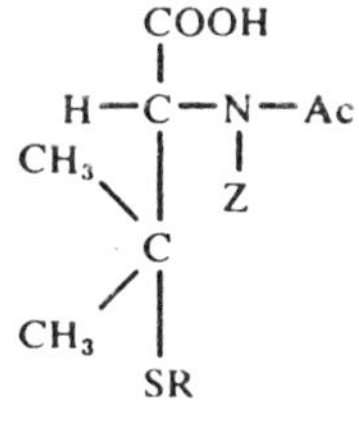

in which Ac is an acyl group, especially benzoyl, tosyl, nitrophenyl sulfenyl, lower alkanoyl, e.g., acetyl or preferably formyl, Z is hydrogen or together with Ac can be a diacyl group, especially phthaloyl and R is hydrogen or benzyl or can have the same meaning as Ac.

Preferably, however, the protection of such groups is provided by converting the D,L-penicillamine in known manner into a thiazoline-4-carboxylic acid of the formula:

(2)

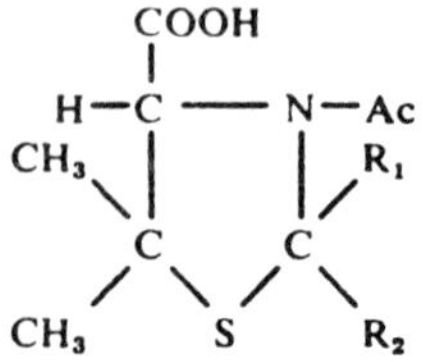

in which R_1 and R_2 are the same or different and are hydrogen, alkyl of 1 to 8 carbon atoms, cycloalkyl with 5 to 8 carbon atoms or aryl, i.e., for example, benzyl or R_1 and R_2 together with the adjacent carbon atom form a cycloalkylene ring, e.g., of 5, 6, 7 or 8 carbon atoms and Ac is an acyl group, especially benzoyl, tosyl, nitrophenylsulfenyl, lower alkanoyl, e.g., acetyl or preferably formyl.

Example: 32.1 g of L-lysine and 43.8 g of D,L-3-formyl-2,2,5,5-tetramethyl-thiazolidine-4-carboxylic acid were dissolved in 240 ml of methanol at 25°C. The solution was cooled to 0°C with stirring and inoculated with 3.6 g of an optically pure salt of L-lysine and D-3-formyl-2,2,5,5-tetramethyl-thiazolidine-4-carboxylic acid. After 3 hours the precipitated crystals were filtered off with suction; it was washed twice with 30 ml (each time) of cold methanol and then dried at 40°C under reduced pressure. The yield of salt was 23.7 g, corresponding to 59%. The salt had an optical rotation of +50°, which corresponds to an optical purity of 94%. (A salt produced from pure components had an optical rotation of +53.5°.

14.5 g of the salt of L-lysine and D-3-formyl-2,2,5,5-tetramethyl-thiazolidine-4-carboxylic acid recovered was dissolved in 50 ml of water and treated with 8 ml of 6 normal hydrochloric acid. The precipitate formed was filtered off with suction, washed with water and dried. There were obtained 7.45 g of D-3-formyl-2,2,5,5-tetramethyl-thiazolidine-4-carboxylic acid. This had an optical rotation of +54° (concentration = 1 g/100 ml in methanol). The D-3-formyl-2,2,5,5-tetramethyl-thiazolidine-4-carboxylic acid was mixed with 45 ml of 15% hydrochloric acid which was heated to 70°C. The mixture was held for 2 hours at 70°C with distillation of acetone and then brought to dryness in a rotary evaporator. There were recovered 5.04 g of crude D-penicillamine hydrochloride.

The D-penicillamine hydrochloride was dissolved in 40 ml of 96% ethanol, the solution treated with 29.5 g of triethylamine, whereupon the free D-penicillamine precipitated. It was filtered off with suction, washed with 96% ethanol and dried under reduced pressure at 50°C. The D-pencillamine had an optical rotation of -62° and a melting point of 182° to 184°C. The yield was 3.6 g.

By evaporation of the mother liquor remaining in the crystallization of the salt of L-lysine and D-3-formyl-2,2,5,5-tetramethyl-thiazolidine-4-carboxylic acid there were recovered a salt of L-lysine and substantially L-3-formyl-2,2,5,5-tetramethyl-thiazolidine-4-carboxylic acid. From this there was separated L-lysine by treatment with dilute hydrochloric acid. The L-3-formyl-2,2,5,5-tetramethyl-thiazolidine-4-carboxylic acid was racemized by heating in toluene in the presence of acetic anhydride.

Lysine Pantothenate for Treatment of Leukopenia

Pharmaceutical preparations have been developed by *G. Tixier; U.S. Patent 3,876,801; April 8, 1975* for the treatment of leukopenia. It was found that lysine pantothenate stimulates the production of leukocytes while having no acute or lingering toxicity.

The administration of lysine pantothenate to a rat made leukopenic through injection of cyclophosphamide (Endoxan) 40 mg/kg caused an alleviation of the poisonous effects. While the leukocyte numerations of the reference batch remain constant during the test, i.e., less than the initial value, the animals treated by the lysine derivatives have their number of leukocytes return to normal and even become higher than the original value. The following table shows an abstract of a pharmacological test.

Percentage of Variation in Relation to the Reference Leukocyte Numerations

Batch No.	4th Day	6th Day	11th Day	13th Day
Batch No. 1				
Reference	-32.35	-40.5	-16.9	-11.67
Batch No. 2				
Lysine pantothenate	-49.6	-26.5	+26.2	+16.5

Formulations designed for the treatment of syndromes under consideration are as follows:

(a) an ampule containing 0.60 g lysine pantothenate, aromatized excipient and distilled water, qs 5 ml (dosage is 1 to 3 ampules per day);

(b) a coated tablet containing 0.25 g lysine pantothenate and excipient enough to form a coated tablet (dosage is 2 to 5 tablets per day);

(c) a suppository containing 0.40 g of lysine pantothenate and enough excipient to form the suppository (dosage is 1 to 3 suppositories per day); or

(d) an injectable preparation containing 0.20 g lysine pantothenate and twice-distilled water, qs 5 ml (dosage is 1 to 2 ampules per day injected intramuscularly).

The administration of lysine pantothenate at suitable doses to patients having leukopenia of various etiologies, causes in a relatively short time, a return to normal of the leukocytary formula of the patients. The following is an example case study: Female–68 years old. Physical and psychical asthenia with anorexia, loss of weight, poor general condition. Treatment–lysine pantothenate

0.90 g/day for 15 days. Improvement of the general condition, psychical and physical.

	Leukocytes	
	Before Treatment	After Treatment
Numeration	3,000	8,000
Polynuclears, %	38	59
Eosinophils, %	2	3
Lymphocytes, %	57	34
Monocytes, %	3	4

ϵ-Acyllysines as Pro-Lysine Food Additives

Cereal products are deficient in the required amino acid lysine. However, on cooking, some added lysine reacts with reducing sugars in a "Maillard" reaction and becomes inactivated. Consequently, as too much lysine could produce a nutritional imbalance, the addition of just the right amount of lysine to cereals is a complex operation.

The process developed by *J. Mauron, P.-A. Finot and F. Mottu; U.S. Patent 3,993,795; November 23, 1976; assigned to Société d'Assistance Technique pour Produits Nestle SA, Switzerland* provides the means of adding to the foodstuff an effective amount of at least one metabolic substitute of L-lysine. The expression "metabolic substitute of L-lysine" denotes a substance which plays the same role as L-lysine in human and/or animal metabolism, being converted on ingestion into free L-lysine or directly assimilable bound L-lysine. For convenience, such a metabolic substitute will be referred to as "pro-lysine." Moreover, the term lysine will be used for the natural isomer, i.e., L-lysine and, likewise, the residue lysine designates L-lysine.

It has been found that ϵ-acyl derivatives of lysine, especially ϵ-aminoacyl derivatives, are particularly effective as pro-lysines. Various derivatives of lysine were assessed in in vivo growth tests in rats. In each test, one group of rats was fed a diet based on proteins deficient in lysine supplemented with a lysine derivative, and the second group was fed a similar diet enriched with lysine itself, in varying proportions. The assimilation of the lysine and of its derivatives is assessed by determining the gain in weight of the rats in a group. The relative assimilation values of these derivatives, by comparison to lysine, are the following.

α-Acyl Derivatives	Percent Assimilation
Formyl	1-26
Acetyl	0-20
Propionyl	0
Palmityl	0
Oleyl	0
Linoleyl	0
ϵ-Acyl Derivatives	**Percent Assimilation**
Formyl	43-87
Acetyl	49-83
Propionyl	0
Lauryl	0

(continued)

ε-Acyl Derivatives	Percent Assimilation
Palmityl	0
Oleyl	0
Linoleyl	0

ε-Aminoacyl Derivatives	Percent Assimilation
Glycyl	55-92
α-L-Glutamyl	80-100
γ-L-Glutamyl	75-100
α-L-Aspartyl	80-100
β-L-Aspartyl	0-20
L-Alanyl	80-100
L-Phenylalanyl	68-93
L-Leucyl	80-100
β-Alanyl	0
L-Cysteinyl	0
α-L-Pyroglutamyl	0
L-Arginyl	80-100

α,ε-Diaminoacyl Derivatives	Percent Assimilation
Diglycyl	80-100
Di-L-alanyl	80-100

ε,ε'-Diaminoacyl Derivatives	Percent Assimilation
Bis-L-cystinyl	0

Schiff Bases	Percent Assimilation
ε-Benzylidene	69-100

Lysine Fixed On	Percent Assimilation
Polyacrolein (13.5%)	0-22
Peroxidized starch (34.3%)	11-43

The values listed in this table are ranges of which the limits correspond to the heaviest and lightest rats as compared with the lysine-fed control.

Thus, among the acylated derivatives only ε-formyl-lysine and ε-acetyl-lysine are of interest. Among the aminoacylated derivatives or peptides particularly useful are ε-glycyl-lysine, ε-(α-L-glutamyl)-lysine, ε-(γ-L-glutamyl)-lysine, ε-(α-L-aspartyl)-lysine, ε-(L-alanyl)-lysine, ε-(L-phenylalanyl)-lysine, ε-(L-leucyl)-lysine, ε-(L-arginyl)-lysine, α,ε-diglycyl-lysine and α,ε-di-L-alanyl-lysine. There does not appear to be a particular relationship between assimilability and the aminoacyl radical linked to the lysine.

Example: One group of 6 weaned rats is fed a diet based on wheat and gluten zein, both being proteins having a particularly low lysine content. The diet is supplemented with 0.82 mmol of lysine hydrochloride per 100 g of the diet (equivalent to 4.55 mmol per 100 g protein). A partial lysine balance is thus restored.

A second group of 6 weaned rats is fed an analogous diet which, however, in place of lysine hydrochloride is supplemented with 0.98 mmol of ε-(α-L-glutamyl)-lysine per 100 g of feed (equivalent to 5.44 mmol per 100 g protein). After 14 days the average weight gain of the rats of the second group is similar to that of the first, namely 24.9 g as against 24.8 g. 0.98 mmol of ε-(α-L-glutamyl)-lysine are thus equivalent to 0.82 mmol of lysine hydrochloride; the relative assimilation of this peptide is 83%, with a range of 80 to 100% (lighest rat-heaviest rat).

ARGININE—PRODUCTION METHODS

Arginine or 1-amino-4-guanidinovaleric acid, is an important amino acid widely used for medicine and feed, and has been prepared on a commercial scale from protein hydrolyzates at relatively high cost in a complex isolation procedure. It is known to be produced by certain wild strains of hydrocarbon-assimilating Corynebacterium and Brevibacterium in a very low concentration.

Using a *Microbacterium ammoniaphilum* Mutant

It was found by *K. Kubota, T. Onoda, H. Kamijo, F. Yoshinaga and S. Okumura; U.S. Patent 3,833,473; September 3, 1974; assigned to Ajinomoto Co., Inc., Japan* that certain mutant strains of Microbacterium produce significant amounts of arginine in an aqueous medium.

The mutant strains used in this process can be derived by known mutation methods from parent strains of Microbacterium which are incapable of producing extracellular arginine. Usually, an effective arginine-producing mutant strain is found to be resistant to a growth-inhibiting agent for the parent strain, such as 2-thienylserine, D-serine, ethionine, 2-thiazolealanine, α-amino-β-hydroxyvaleric acid, 6-chloropurine and sulfa drugs (sulfaguanidine, sulfamerazine, sulfisoxazole, etc.).

A mutant strain capable of growing on a medium containing more than 1,000 μg/ml 2-thiazolealanine is most effective. Sometimes the addition of nutrients required for growth of the resistant strain improves the yield of arginine.

The most effective arginine-producing mutant found so far is *Microbacterium ammoniaphilum* AJ 3353 (FERM-P976). The culture media used in the process are conventional in themselves, and contain assimilable carbon sources, assimilable nitrogen sources, inorganic ions and, if necessary, minor organic nutrients.

For good yield of arginine, the fermentation is carried out aerobically with aeration and agitation. Best yields require pH of 5 to 9, maintained by means of gaseous or aqueous ammonia, calcium carbonate, alkali metal hydroxide, urea, organic or inorganic acids.

When fermentation is carried out at 24° to 37°C, the maximum concentration of arginine in the broth is reached within 2 to 7 days. The arginine accumulated in the fermentation broth can be recovered by conventional methods, such as by using ion exchange resin in combination with precipitation. The arginine was identified by its ninhydrin reaction on a paper chromatogram, the R_f value on the paper chromatogram, a positive Sakaguchi reaction and growth curves of arginine requiring mutants of lactic acid bacteria. The arginine in the broth was determined by bioassay employing *Leuconostoc mesenteroides* ATCC 8042.

Example: 300 ml batches of a medium containing 10 g/dl glucose, 0.2 g/dl KH_2PO_4, 1.5 g/dl $(NH_4)_2SO_4$, 0.04 g/dl $MgSO_4 \cdot 7H_2O$, 50 μg/l biotin, 50 μg/l thiamine HCl, 1 mg/dl $FeSO_4 \cdot 7H_2O$, 1 mg/dl $MnSO_4 \cdot 4H_2O$ and 1 ml/dl soy protein hydrolyzate, and of pH 7.5 were placed in 1,000 ml fermentation vessels, and sterilized by heat.

Microbacterium ammoniaphilum AJ 3353 (FERM-P976) previously cultured on bouillon agar slants was inoculated in each medium, and cultured at 31°C with aeration and agitation for 48 hours, maintaining pH at 7.5 with gaseous ammonia. The cultured broth was found to contain 0.41 g/dl arginine. Two liters of the cultured broth was centrifuged to remove cells. The supernatant was passed through a column packed with an ion exchange resin (Amberlite C-50, NH_4^- type), and arginine was eluted with 2 N NH_4OH. The eluate was concentrated to precipitate crude crystalline arginine; 4.9 g of the crystalline material was obtained.

Using Mutants of *Corynebacterium glutamicum*

K. Nakayama and H. Yoshida; U.S. Patent 3,849,250; November 19, 1974; assigned to Kyowa Hakko Kogyo KK, Japan have found that a substantial amount of L-arginine can be accumulated by fermentation using mutant strains derived from the coryneform glutamic-acid-producing bacteria (such as, for example, *Corynebacterium glutamicum*). The coryneform glutamic-acid-producing bacteria are sometimes classified into Brevibacterium, Corynebacterium, Micrococcus, Microbacterium, Arthrobacter or Azotobacter, which form a group closely related taxonomically to bacteria described by Abe et al in *J. General and Applied Microbiology,* Vol. 13, pp 279-301 (1967).

These mutant strains capable of producing significant amounts of L-arginine are generally characterized as those being resistant to L-arginine-analogues (e.g., canavanine, homoarginine, etc.) so that the latter can be regarded as typical strains for the process. However, there are some mutant strains capable of producing large amounts of L-arginine although their gene mutations, except their L-arginine productivity, have not been clarified.

The preferred microorganisms are mutants of *Corynebacterium glutamicum,* particularly *C. glutamicum* ATCC 21659 and *C. glutamicum* ATCC 21831. Either synthetic or natural medium can be used for the process so long as it contains preferable amounts of carbon sources, nitrogen sources, inorganic salts and small amounts of nutrients required for the growth of the strains used, as is shown in the example.

Example 1: An L-arginine-producing strain (ATCC 21659) (canavanine-resistant) obtained from *C. glutamicum* (synonym of *Micrococcus glutamicus*) ATCC 13032 by mutation treatment was used. This strain was cultured with shaking in a seed medium having a composition of 2% glucose, 1% peptone, 1% yeast extract and 0.3% NaCl for 24 hours at 30°C. One milliliter of the resultant seed culture was inoculated into 10 ml of a fermentation medium in a 250 ml Erlenmeyer flask, and cultured with shaking at 30°C for 72 hours to obtain 5.1 mg/ml of L-arginine in the culture liquor. The composition of the fermentation medium used is shown on the following page.

Glucose, %	10
KH_2PO_4, %	0.05
$MgSO_4 \cdot 7H_2O$, %	0.025
$MnSO_4 \cdot 4H_2O$, %	0.001
Biotin, g/l	50
$CaCO_3$, %	2
K_2HPO_4, %	0.05
$(NH_4)_2SO_4$, %	3
$FeSO_4 \cdot 7H_2O$, %	0.001
NZ-amine,* %	0.5
Thiamine hydrochloride, mg/l	2

*Casein hydrolyzate

After the removal of microbial cells and $CaCO_3$ from the fermented liquor, L-arginine contained in the liquor was recovered as hydrochloride by using weak cation exchange resin. The yield was 3.8 g (hydrochloride) per one liter of the fermented liquor.

Example 2: As the seed culture, an L-arginine-producing strain (canavanine-resistant) of *C. glutamicum* ATCC 21831 was used. This strain was cultured with shaking in a seed medium (composition: 2% gluclose, 2% peptone, 0.15% K_2HPO_4, 0.05% KH_2PO_4, 0.05% $MgSO_4 \cdot 7H_2O$, 50 g/l biotin and 1% yeast extract) at 30°C for 24 hours. One milliliter of the resultant seed culture was inoculated into 10 ml of a fermentation medium in a 250 ml Erlenmeyer flask and was cultured with shaking at 30°C for 72 hours. The composition of the used fermentation medium was as follows: cane molasses, 15% (calculated as glucose); soybean meal hydrolyzate (decomposed with 6 N H_2SO_4 and neutralized with aqueous ammonia), 2% (calculated as soybean meal); urea, 0.3%; K_2HPO_4; 0.07%; $MgSO_4 \cdot 7H_2O$, 0.05%; and $CaCO_3$, 3%.

The pH of the used fermentation medium was adjusted to 7.2. The amount of L-arginine produced in the cultured liquor was 17.7 mg/ml.

Using *Brevibacterium flavum* Mutant

In the process disclosed by *K. Kubota, T. Onoda, H. Kamijo and S. Okumura; U.S. Patent 3,878,044; April 15, 1975; assigned to Ajinomoto Co., Inc., Japan* an artificially induced mutant strain of *Brevibacterium flavum* which requires guanine for its growth and is resistant to 2-thiazolalanine or other arginine analogs produces L-arginine in high yields.

It was found that large amounts of arginine were produced in conventional culture media on which guanine requiring mutant strains of Brevibacterium resistant to arginine or arginine analogs had been cultured. Mutant strains which differ from the parent by resisting feedback inhibition and/or repression by arginine or arginine analogs are readily obtained by exposing the parent strain to mutagenic agents and screening the treated microorganisms for their ability of growing on culture media containing arginine or arginine analogs in amounts of 5 mg/ml or more.

Suitable arginine analogs include canavanine, homoarginine, α-methylarginine, 2-thienylserine, D-serine, ethionine, 2-thiazolalanine, α-amino-β-hydroxyvaleric acid, 6-chloropurine, and sulfa drugs such as sulfaguanidine, sulfamerazine, sulfisoxazole.

Mutants of Brevibacterium capable of growth of media containing the arginine analogs, such as *B. flavum* ATCC 21493, produce arginine to some extent even if they do not require specific organic nutrients. However, substantially enhanced arginine productivity is found in strains derived therefrom by further mutation and requiring guanine for their growth. The most effective arginine-producing mutant of genus Brevibacterium found so far is the guanine requiring strain *B. flavum* AJ 3401 (FERM-P 1639) which is resistant to 3-thiazolalanine and canavanine.

The following examples illustrate the methods employed in obtaining microorganisms for use when applying the process, and the production of arginine by means of such microorganisms.

Example 1: Cells of *B. flavum* ATCC 14067 were exposed to mutagenic doses of x-rays and then cultured in a complete agar flat plate at 31°C for 4 to 10 days. From the colonies appearing on the plate, a mutant strain No. 33038 requiring guanine was isolated by replica plating. It was treated with 250 μg/ml nitrosoguanidine at 30°C for 30 minutes, and the treated microorganism was inoculated on agar flat plates and cultured as above.

The medium of the last-mentioned plates contained 2% glucose, 0.3% urea, 1% $(NH_4)_2SO_4$, 0.1% KH_2PO_4, 0.04% $MgSO_4 \cdot 7H_2O$, 2 ppm Fe^{++}, 2 ppm Mn^{++}, 100 μg/l biotin, 200 μg/l thiamine hydrochloride, 0.015% guanine, 5 mg/ml 2-thiazolalanine, and 2% agar, and had a pH of 7.0.

Screening of the colonies appearing on the culture plates yielded strain No. 112 which produces L-histidine. The strain was further treated with nitrosoguanidine as above, and inoculated on agar plates differing in composition from the last-mentioned plates by containing 10 mg/ml 2-thiazolalanine.

All strains growing on the medium of higher thiazolalanine concentration resisted 2-thiazolalanine and the other arginine analogs enumerated above and were capable of producing arginine. The best arginine producer was the strain AJ 3401 which was isolated on the basis of its requirement for guanine and its resistance to canavanine.

It produces arginine in culture media containing at least one assimilable carbon source, at least one assimilable nitrogen source, and the usual inorganic ions and minor organic nutrients including guanine. Carbon sources suitable for all arginine-producing strains include carbohydrates (glucose, sucrose, starch, starch hydrolyzate, molasses), organic acids (acetic acid, propionic acid, gluconic acid, succinic acid, citric acid, benzoic acid), and alcohols (ethanol), and some are capable of metabolizing n-alkanes. Suitable nitrogen sources include ammonium salts, nitrates, amino acids, urea, and ammonia.

Aerobic fermentation with aeration and agitation produces good yields, and best yields require pH of 5 to 9 maintained in the usual manner by additions of ammonia, calcium carbonate, alkali metal hydroxide, urea, or inorganic or organic acids. At fermentation temperatures between 24° and 37°C, maximum arginine concentration is reached within 2 to 7 days. The arginine concentration in the broth may be monitored colorimetrically according to the method of Voges-Proskauer.

Example 2: Twenty milliliter batches of an aqueous medium containing 10% glucose, 6% ammonium sulfate, 0.1% potassium dihydrogen phosphate, 0.04% magnesium sulfate heptahydrate, 2 ppm each of Fe and Mn ions, 50 μg/l biotin, 20 μg/l thiamine hydrochloride, 1% soy protein hydrolyzate (2.4% total nitrogen), 0.015% guanine were sterilized in respective 500 ml shaking flasks and adjusted to pH 7 with 5% separately sterilized calcium carbonate.

Inocula of *B. flavum* AJ 3401 (FERM-P 1639) prepared on bouillon agar slants were added to each flask which was thereafter held at 31°C with aeration and agitation for 72 hours. The combined broth contained 2.5 g/dl arginine and was centrifuged to remove the cells. One liter of the supernatant liquid was passed over a column packed with an ion exchange resin (Amberlite C-50, NH_4 type), and the arginine adsorbed by the resin was eluted with 2 N ammonium hydroxide solution. The eluate was partly evaporated to precipitate crude, crystalline arginine which, when dried, weighed 16.2 g.

Using Mutants of *Bacillus subtilis*

I. Chibata, M. Kisumi and J. Kato; U.S. Patent 3,902,967; September 2, 1975; assigned to Tanabe Seiyaku Co., Ltd., Japan disclose a process for producing L-arginine by cultivating a pyrimidine metabolic antagonist-resistant mutant capable of producing L-arginine.

The resistant mutants which are used in the process include the 6-azauracil-resistant mutant, 2-thiouracil-resistant mutant and 6-azauridine-resistant mutant. Such mutants may be derived from an arginine hydroxamate-resistant mutant of *Bacillus subtilis.* Specifically, they are *Bacillus subtilis* AHr.AUr-9 (arginine hydroxamate-resistant and 6-azauracil-resistant mutant: ATCC 31002, FERM-P 1998), *Bacillus subtilis* AHr.TUr-61 (arginine hydroxamate-resistant and 2-thiouracil-resistant mutant: ATCC 31003, FERM-P 1999 and *Bacillus subtilis* AHr-62 (arginine hydroxamate-resistant and 6-azauridine-resistant mutant: ATCC 31004, FERM-P 2000).

These mutants may be obtained by mutating an original strain by, for example, ultraviolet irradiation or a chemical mutating agent (e.g., N-methyl-N'-nitro-N-nitrosoguanidine, or ethyl methanesulfonate), cultivating the mutant thus obtained in a plate medium (e.g., Spizizen's minimal medium) containing 6-azauracil (0.5 mg/ml), 2-thiouracil (0.5 mg/ml) or 6-azauridine (2 mg/ml) for 3 to 4 days, and separating the large colonies thus produced.

The resistant mutants may also be obtained by any other mutating methods, for instance, by giving pyrimidine metabolic resistance to a microorganism being capable of producing L-arginine or reversely by giving at first pyrimidine metabolic resistance to a microorganism and then giving thereto a capacity of production of L-arginine, or by giving simultaneously both properties to a microorganism.

The medium used for the cultivation of the resistant mutants may include 5 to 15% by weight of saccharides (e.g., glucose or starch hydrolysates) as a carbon source, 0.5 to 4% by weight of an inorganic ammonium salt (e.g., ammonium chloride or ammonium sulfate), urea or the like as a nitrogen source and 0.02 to 2% by weight of peptone, yeast extract, corn steep liquor or the like as an organic nutrient, and optionally a small amount of inorganic salt (e.g., potassium phosphate, magnesium sulfate, manganese sulfate or ferric sulfate). It is also preferable to add calcium carbonate or the like to maintain the pH value of the medium at 6 to 9. Furthermore, it is preferable to add 1 to 5% by weight of glutamic acid and/or aspartic acid to the medium for enhancing the accumulation amount of the desired L-arginine.

According to the process, the resistant mutant may be inoculated into a medium containing the above components and it is cultivated under an aerobic condition such as by strongly shaking at 25° to 37°C for 1 to 4 days and thereby the desired L-arginine is accumulated in the broth. The L-arginine is accumulated largely in the broth and on the other hand, the by-product amino acids other than L-arginine are included in extremely slight amount. The L-arginine thus produced can be easily harvested by a conventional method, for instance, by using an ion exchange resin. In the following examples % means % by weight.

Example 1: To 500 ml shake flask is added a medium (pH 7.6; 30 ml) containing 8% of glucose, 2.5% of ammonium chloride, 3.5% of L-glutamic acid, 0.1% of peptone, 0.1% of yeast extract, 0.5% of potassium dihydrogen phosphate, 0.05% of magnesium sulfate and 2% of calcium carbonate, and the mixture is sterilized under pressure. The glucose and calcium carbonate added to the medium are previously sterilized. The medium is inoculated with one platinum loop of *B. subtilis* AHr.AUr-9 (ATCC 31002, FERM-P 1998) and subjected to shake culture at 30°C for 72 hours. L-arginine is produced in an amount of 28.6 mg/ml in the broth.

Example 2: To 500 ml shake flask is added a medium (pH 7.6; 30 ml) containing 8% of glucose, 2.5% of ammonium chloride, 3.5% of L-aspartic acid, 0.1% of peptone, 0.1% of yeast extract, 0.5% of potassium dihydrogen phosphate, 0.05% of magnesium sulfate and 2% of calcium carbonate and the mixture is sterilized under pressure. The glucose and calcium carbonate added to the medium are previously sterilized. The medium is inoculated with one platinum loop of *B. subtilis* AHr.AUr-9 (ATCC 31002, FERM-P 1998). It is treated in the same manner as described in Example 1. L-arginine is produced in an amount of 31.2 mg/ml in the broth.

The broth (one liter) thus obtained is harvested, heated and then filtered. The filtrate is passed through a column filled with Amberlite IR-120 (H type) and then the column is washed with water. The adsorbed L-arginine is eluted with 5% aqueous ammonia. The eluate is concentrated under a reduced pressure. To the residue are added hydrochloric acid and methanol, and the precipitated crystals are separated by filtration. The crystals are recrystallized from water-containing methanol to give L-arginine hydrochloride (29 g).

ARGININE AND DERIVATIVES – USES

Arginine-Potassium Phospho Citro Glutamate

An arginine-containing compound useful in treatment of physical and psychic asthenia has been developed by *D. Bocher and R. Faure; U.S. Patent 3,809,759; May 7, 1974.*

This compound has the empirical formula $C_{35}H_{76}O_{23}N_{17}P$–K and can be represented by the following formula:

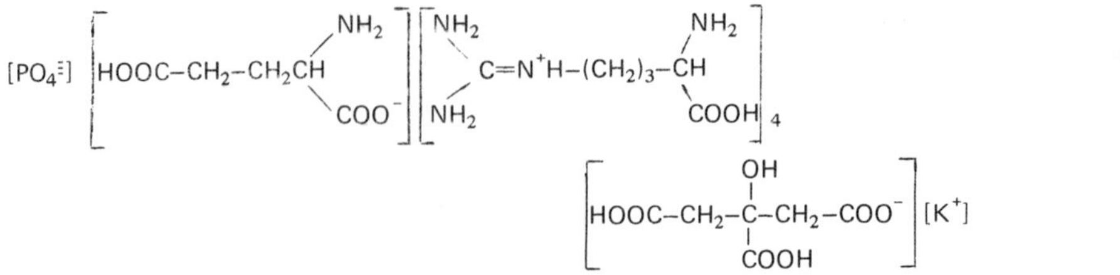

This compound is in the form of a white hygroscopic microcrystalline powder and has a melting point of 163°C. In 10% aqueous solution, it is in the form of a clear solution with a pH of 5.8. The compound is insoluble in ethyl alcohol, but is water-soluble at a rate of 59.2% by weight.

The process for preparing arginine-potassium phospho citro glutamate comprises reacting in a first-stage phosphoric acid and glutamic acid diluted in water with the base, L-arginine, to produce arginine phospho glutamate. Separately, potassium monobasic citrate solution is prepared by reacting citric acid with potash.

In the second stage, the potassium monobasic citrate solution is then poured slowly, with vigorous stirring, into the arginine phospho glutamate solution, the temperature of which has been brought to 60°-65°C. Stirring is continued for 20 minutes at this temperature and the mixture is then permitted to stand for an hour while the temperature returns to ambient temperature. The cooled solution is then filtered, and the filtrate is concentrated, thereby crystallizing the arginine-potassium phospho citro glutamate.

As a variation, the arginine-potassium phospho citro glutamate can be obtained by lyophilization of the above filtrate.

Preparation: To obtain 50 kg, or 43 mols of arginine-potassium phospho citro glutamate, an arginine phospho glutamate solution is prepared in a first stage by mixing together 29.928 kg of pure L-arginine, 6.321 kg of pure glutamic acid, 4.950 kg of 85% phosphoric acid and 150 liters of purified water.

To this resulting solution of arginine phospho glutamate there is added in a second stage with vigorous stirring and at 60°-65°C, a solution of potassium citrate which is prepared by dissolving 9.036 kg of monohydrated citric acid and 2.832 kg of 85% potash, in pellet form, in 18 liters of purified water.

After the potassium citrate has been added, stirring is continued for 20 minutes while maintaining the temperature at 60°-65°C. Then stirring is discontinued and the solution is allowed to stand for an hour so that the temperature returns to ambient temperature. The pH of the resulting solution is then about 6. Thereafter, the solution is filtered and the filtrate is concentrated, thereby crystallizing the arginine-potassium phospho citro glutamate.

The arginine-potassium phospho citro glutamate can advantageously be recrystallized in water and when obtained as described above it is present in the form of a white microcrystalline powder having a melting point of 162°-163°C.

The arginine-potassium phospho citro glutamate is recommended for its refreshing and energy action, both in human and animal therapeutics. Pharmacological studies made on rats and guinea pigs show a strong hepato-protective power, by an action on the nervous system, and by an action of resistance to muscular fatigue. The pharmaceutical compositions used generally contain from 0.05 to 80% by weight arginine-potassium phospho citro glutamate, the remainder being essentially the pharmaceutically acceptable or ingestible carrier. These compositions can be present in the form of a liquid or a solid as shown by the following examples.

Example 1: A syrup is made for children by preparing the following mixture.

Arginine-potassium phospho citro glutamate, g	10
Methyl para-hydroxybenzoate, g	0.125
Propyl para-hydroxybenzoate, g	0.125
Syrup, ml	70
Raspberry flavor, ml	4
Distilled water sufficient for, ml	100

Example 2: Ampoules having a volume of 10 cc are prepared for adults by making the following mixtures.

Arginine-potassium phospho citro glutamate, g	2
Sucrose, g	3
Honey scent, g	1.10
Caramel scent, g	0.30
Distilled water, ml	10

Example 3: Tablets are prepared by making the following mixture.

	Gram
Arginine-potassium phospho citro glutamate	0.500
Lactose sufficient for	0.660

N^2-Dansyl-L-Arginine Esters and Amides

In the past, there have been many attempts to obtain new and improved agents for the treatment of thrombosis. Of these, N^2-(p-tolylsulfonyl)-L-arginine esters are known to be effective in dissolving blood clots (U.S. Patent 3,622,615). A need continues to exist, however, for a highly specific inhibitor on thrombin for the control of thrombosis.

S. Okamoto, R. Kikumoto, Y. Tamao, S. Tonomura and K. Ohkubo; U.S. Patent 3,978,045; August 31, 1976; assigned to Mitsubishi Chemical Industries Ltd., Japan have provided a group of N^2-dansyl-L-arginine esters, amides and pharmaceutically acceptable acid addition salts which are useful in the diagnostic determination of thrombin in blood and in drug therapy as antithrombotic agents.

These N^2-dansyl-L-arginine esters and amides are compounds of formula (1) (shown on the following page) or their acid addition salts with a pharmaceutically acceptable acid.

(1) 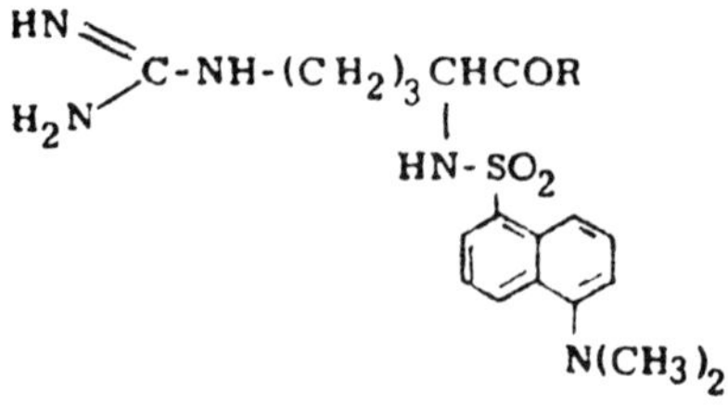

In formula (1), R is selected from alkoxy, alkenyloxy, alkynyloxy, or cycloalkoxy, containing not more than 10 carbon atoms, aralkyloxy of not more than 15 carbon atoms, tetrahydrofurfuryloxy, and alkoxy of not more than 10 carbon atoms substituted with an alkoxy group of not more than 10 carbon atoms, halogen or nitro;

(2) 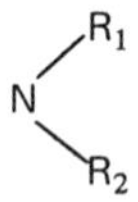

where R_1 and R_2 are hydrogen, alkyl, aryl, aralkyl, cycloalkyl, cycloalkylalkyl, alkenyl, or alkyl with no more than 10 carbon atoms, and where the alkyl may be substituted with alkoxy, alkoxycarbonyl, arylcarbamoyl, acyl, acyloxy, N,N-polymethylenecarbamoyl, respectively, containing not more than 10 carbon atoms, and carboxyl; and

(3) 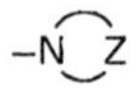

where Z is a divalent group containing up to 10 carbon atoms, which consists of more than one group selected from methylene $-CH_2$, monosubstituted methylene $-CR_3H-$, where R_3 is alkyl, acyl, alkoxy or alkoxycarbonyl, respectively, containing not more than 10 carbon atoms, and carbamoyl, and disubstituted methylene $-CR_4R_5-$, where R_4 and R_5 are alkyl groups of not more than 10 carbon atoms, and which may further contain at least one member selected from oxy, thio, cycloalkylene of not more than 10 carbon atoms, imino $-NH-$, alkyl substituted imino $-NR_6-$, where R_6 is an alkyl group of not more than 10 carbon atoms, acyl substituted imino

$$>NCOR_7$$

where R_7 is an alkyl group of not more than 10 carbon atoms, phenylene and carbonyl, which completes the NZ ring together with the methylene monosubstituted methylene or disubstituted methylene.

N^2-dansyl-L-arginine, which is the starting material for the preparation of N^2-dansyl-L-arginine esters, is most generally obtained by reacting L-arginine and dansyl chloride, that is, 5-dimethylamino-1-naphthalenesulfonyl chloride, in the presence of a base. However, N^2-dansyl-L-arginine may also be obtained by reacting ornithine, the ω-position of which is protected, with dansyl chloride in the presence of a base, removing the protective group at the ω-position of the product, and thereafter guanidylating the obtained N^2-dansyl ornithine by conventional procedures.

Some seventy-three esters and amides are described in the specifications prepared by known esterification and amide-forming processes including the methods shown in the following examples.

Example 1: To an ice-cooled suspension of 1.0 g of N^2-dansyl-L-arginine in 15 ml of methanol was added dropwise 0.5 ml of thionyl chloride with vigorous stirring. After being allowed to stand for 2 hours at room temperature, the reaction mixture was refluxed for 2 hours, and was evaporated to dryness (syrup). Treatment of the residual syrup with cold ethyl ether and a small amount of water gave crude crystals. After recrystallization from methanol-ethyl ether, colorless N^2-dansyl-L-arginine methyl ester dihydrochloride monohydrate was obtained in 92% yield; MP 147°-150°C.

Example 2: A 1.0 g amount of N^2-dansyl-L-arginine methyl ester dihydrochloride monohydrate was dissolved in 2 ml of n-butylamine with vigorous agitation. After the resulting solution was allowed to stand at room temperature for 2 days, the n-butylamine was removed by vacuum distillation and to the residual syrup was added 5 ml of water to obtain a crystalline deposit.

The precipitate was collected and recrystallized from 50% aqueous methanol to give N^2-dansyl-N-(n-butyl)-L-argininamide monohydrate in 90% yield; MP 150°-152°C. The antithrombotic activities of N^2-dansyl-L-arginine derivatives were compared with those of a known antithrombotic agent, (N^2-p-tolylsulfonyl)-L-arginine methyl ester, by determining the fibrinogen coagulation time.

The term "concentration required to prolong the coagulation time by a factor of two" is the concentration of an active ingredient required to prolong the normal coagulation time 50-55 seconds to 100-110 seconds. Addition salts of selected compounds of formula (1) were formed. Those salts which had a value of 2 or less for the concentration required to prolong the coagulation time by a factor of two are shown in the table below where Ts is p-tolylsulfonyl.

R	Addition Salt
$-O-nC_3H_7$	2HCl
$-O-nC_4H_9$	2TsOH
$-OCH_2\phi$	2TsOH
$-OCH_2CH{=}CHCH_3$	2TsOH
$-N(CH_3)(n\text{-}C_4H_9)$	$2CH_3COOH$
4-Methyl-1-piperidyl	$2CH_3COOH$

N^2-Alkoxynaphthalenesulfonyl-L-Argininamides

N^2-alkoxynaphthalenesulfonyl-L-argininamides and their pharmaceutically acceptable salts have been found by *S. Okamoto, R. Kikumoto, K. Ohkubo, T. Tezuka, S. Tonomura, Y. Tamao and A. Hijikata; U.S. Patent 4,018,915; April 19, 1977; assigned to Mitsubishi Chemical Industries Ltd., Japan* to be effective for the inhibition and suppression of thrombosis. They are prepared by reacting N^2-alkoxynaphthalenesulfony-L-arginyl halide with a secondary amine. These compounds can be represented by the formula:

(1)

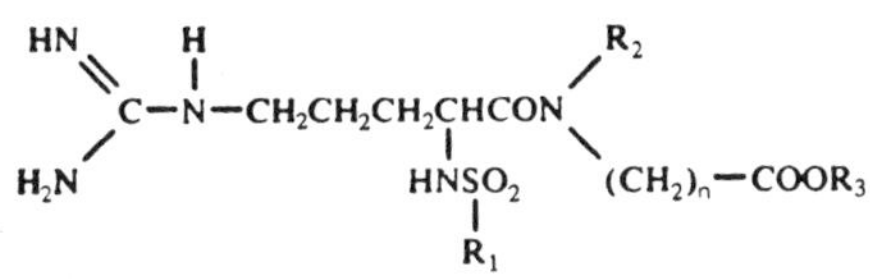

where R_1 is naphthyl substituted with at least one C_1-C_5 alkoxy; R_2 is C_7-C_{15} aralkyl; R_3 is hydrogen, C_1-C_{10} alkyl, C_6-C_{10} aryl or C_7-C_{12} aralkyl; and n is 1, 2 or 3.

Suitable illustrations of R_1 in the above formula (1) are 5-methoxy-1-naphthyl, 6-methoxy-2-naphthyl, 7-methoxy-2-naphthyl, 4,6-dimethoxy-2-naphthyl, 6,7-dimethoxy-2-naphthyl and 6,7-diethoxy-2-naphthyl. Suitable R_2 groups in the above formula (1) are benzyl, phenethyl and 3-phenylpropyl.

Suitable $-(CH_2)_n-COOR_3$ groups in the above formula (1) are carboxymethyl, 2-carboxyethyl, 3-carboxypropyl, ethoxycarbonylmethyl, tert-butoxycarbonylmethyl, phenoxycarbonylmethyl, benzyloxycarbonylmethyl, 2-ethoxycarbonylethyl, 2-tert-butoxycarbonylethyl, and 3-tert-butoxycarbonylpropyl. These compounds may be prepared by several synthetic routes such as:

(a) Condensation of an L-argininamide with an alkoxynaphthalenesulfonyl halide;

(b) Removal of the N^G-substituent from an N^G-substituted-N^2-alkoxynaphthalenesulfonyl-L-argininamide;

(c) Condensation of an N^2-alkoxynaphthalenesulfonyl-L-arginyl halide with an amine; or

(d) Guanidylation of an N^2-alkoxynaphthalenesulfonyl-L-ornithinamide or an acid addition salt thereof.

The following example shows the method (c) above in detail.

Example: (A) To 71.0 g of benzylamine in 100 ml of benzene was added with stirring 25.0 g of tert-butyl chloroacetate over a period of 30 minutes, while maintaining the temperature at 30°-70°C. The reaction mixture was held at 70°C for an additional hour. At the end of this period, the reaction mixture was taken up in 70 ml of 2 N NaOH solution and 80 ml of benzene, transferred into a separatory funnel and well shaken. The benzene solution was separated, washed with water, dried over anhydrous sodium sulfate and filtered. After evaporation of benzene, the residue was distilled under reduced pressure to give 32.7 g (89.0%) of N-benzylglycine tert-butyl ester; BP is 120°-125°C/2 mm Hg.

(B) A suspension of 2.00 g of N^2-(6,7-dimethoxy-2-naphthalenesulfonyl)-L-arginine in 20 ml of thionyl chloride was stirred for 2 hours at room temperature. Addition of cold dry diethyl ether resulted in a precipitate which was collected by filtration and washed several times with dry diethyl ether to give N^2-(6,7-dimethoxy-2-naphthalenesulfonyl)-L-arginyl chloride hydrochloride.

(C) To a stirred solution of 4.15 g of N-benzylglycine tert-butyl ester in 20 ml of chloroform was carefully added N^2-(6,7-dimethoxy-2-naphthalenesulfonyl)-L-arginyl chloride hydrochloride obtained above. The reaction mixture was allowed to stand at room temperature for 1 hour. At the end of this period, the reaction mixture was washed twice with 20 ml of saturated sodium chloride solution and evaporated to dryness.

The residue was triturated with a small amount of water to give a crystalline material. This was collected by filtration, washed with diethyl ether and recrystallized from ethanol-diethyl ether to give 2.67 g (85%) of N^2-(6,7-dimethoxy-2-naphthalenesulfonyl)-L-arginyl-N-benzylglycine tert-butyl ester hydrosulfite, MP 189°-191°C, IR (KBr): 3,360, 3,160, 1,730 cm^{-1}.

(D) To a solution of 2.00 g of N^2-(6,7-dimethoxy-2-naphthalenesulfonyl)-L-arginyl-N-benzylglycine tert-butyl ester hydrosulfite in 20 ml of chloroform was added 50 ml of 15% HCl-ethyl acetate. The reaction mixture was stirred for 5 hours at room temperature. At the end of this period, the reaction mixture was evaporated to dryness. The residue was washed several times with dry diethyl ether and chromatographed on 80 ml of Daiaion SK 102 ion exchange resin (200-300 mesh, H^+ form) packed in water, washed with water and eluted with 3% ammonium hydroxide soltuion.

The fraction eluted from 3% ammonium hydroxide solution was evaporated to dryness to give 1.20 g (70%) of N^2-(6,7-dimethoxy-2-naphthalenesulfonyl)-L-arginyl-N-benzylglycine as an amorphous solid, IR (KBr): 3,370, 3,200, 1,615 cm^{-1}.

Another series of N^2-alkoxynaphthalenesulfonyl-L-argininamides and their pharmaceutically acceptable salts have been disclosed by *S. Okamoto, R. Kikumoto, K. Ohkubo, T. Tezuka, S. Tonomura, Y. Tamao and H. Hijikata; U.S. Patent 4,018,913; April 19, 1977; assigned to Mitsubishi Chemical Industries Ltd., Japan.*

These compounds are also represented by formula (1) above, but here R_1 is naphthyl substituted with at least one C_1-C_5 alkoxy; R_2 is C_2-C_{10} alkylthioalkyl; R_3 is hydrogen, C_1-C_{10} alkyl, C_6-C_{10} aryl or C_7-C_{12} aralkyl; and n is 1, 2 or 3.

Suitable illustrations of R_1 in the above formula (1) are 5-methoxy-1-naphthyl, 6-methoxy-2-naphthyl, 7-methoxy-2-naphthyl, 4,6-dimethoxy-2-naphthyl, 6,7-dimethoxy-2-naphthyl and 6,7-diethoxy-2-naphthyl. Suitable R_2 groups in the above formula (1) are 2-methylthioethyl, 2-ethylthioethyl and 3-methylthiopropyl. Suitable $-(CH_2)_n-COOR_3$ groups in the above formula (1) are carboxymethyl, 2-carboxyethyl, 3-carboxypropyl, ethoxycarbonylmethyl, tert-butoxycarbonylmethyl, phenoxycarbonylmethyl, benzyloxycarbonylmethyl, 2-ethoxycarbonylethyl, 2-tert-butoxycarbonylethyl, and 3-tert-butoxycarbonylpropyl. These compounds may be prepared by the following synthetic routes:

(a) Condensation of an L-argininamide with an alkoxynaphthalenesulfonyl halide;

(b) Removal of N^G-substituent from an N^G-substituted-N^2-alkoxynapthalenesulfonyl-L-argininamide;

(c) Condensation of an N^2-alkoxynaphthalenesulfonyl-L-arginyl halide with an amine; or

(d) Guanidylation of an N^2-alkoxynaphthalenesulfonyl-L-ornithinamide or an acid addition salt thereof.

The details of method (c) are given in detail in the following example.

Example: (A) A suspension of 2.50 g of N^2-(6,7-dimethoxy-2-naphthalenesulfonyl)-L-arginine in 70 ml of thionyl chloride containing a few drops of dimethyl formamide was stirred for 2 hours at room temperature.

Addition of cold dry diethyl ether resulted in a precipitate which was collected by filtration and washed several times with dry diethyl ether to give N^2-(6,7-dimethoxy-2-naphthalenesulfonyl)-L-arginyl chloride hydrochloride.

(B) To a stirred solution of 4.80 g of N-(2-ethylthioethyl)-glycine tert-butyl ester in 40 ml of chloroform was carefully added N^2-(6,7-dimethoxy-2-naphthalenesulfonyl)-L-arginyl chloride hydrochloride obtained above. The reaction mixture was allowed to stand at room temperature overnight. The reaction mixture was washed twice with 20 ml of saturated sodium chloride solution and evaporated to dryness.

The residue was triturated with a small amount of water to give a powder. This was collected by filtration and reprecipitated with ethanol-diethyl ether to give 5.00 g of N^2-(6,7-dimethoxy-2-naphthalenesulfonyl)-L-arginyl-N-(2-ethylthioethyl)glycine tert-butyl ester, IR (KBr): 3,350, 1,745, 1,650, 1,360 cm^{-1}.

(C) A solution of 5.0 g of N^2-(6,7-dimethoxy-2-naphthalenesulfonyl)-L-arginyl-N-(2-ethylthioethyl)glycine tert-butyl ester in 10 ml of trifluoroacetic acid was stirred for 5 hours at room temperature. At the end of this period, the reaction mixture was evaporated to dryness. The residue was washed several times with dry diethyl ether and chromatographed on 80 ml of Daiaion SK 102 ion exchange resin (200-300 mesh, H^+ form) packed in water, washed with water and eluted with ethanol-H_2O-NH_4OH (5:4:1).

The main fraction eluted from the ethanol-H_2O-NH_4OH solution was evaporated to dryness to give a crystalline material. This was recrystallized from water to give 1.2 g of N^2-(6,7-dimethoxy-2-naphthalenesulfonyl)-N-(2-ethylthioethyl)glycine, MP 171°-172°C.

A third series of N^2-alkoxynaphthalenesulfonyl-L-argininamides and their pharmaceutically acceptable salts have been disclosed by *S. Okamoto, R. Kikumoto, K. Ohkubo, T. Tezuka, S. Tonomura, Y. Tamao and A. Hijikata; U.S. Patent 4,055,651; October 25, 1977; assigned to Mitsubishi Chemical Industries Ltd., Japan*. These compounds can be represented by the formula

(1)

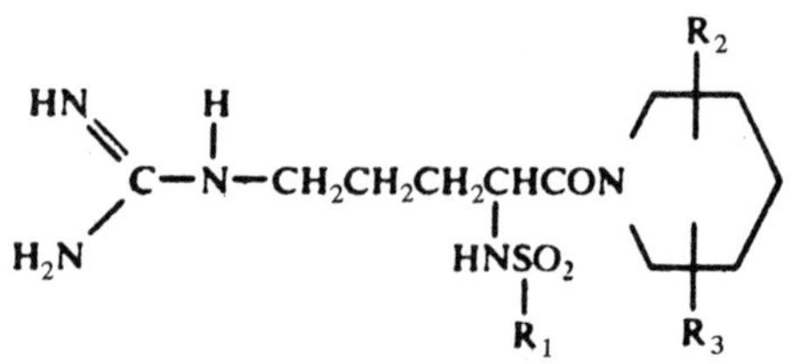

where R_1 is naphthyl substituted with at least one C_1-C_5 alkoxy; R_2 is hydrogen or C_1-C_{10} alkyl; R_3 is $-COOR_4$ where R_4 is hydrogen, C_1-C_{10}alkyl, C_6-C_{10} aryl or C_7-C_{12} aralkyl; R_2 can be substituted at the 2, 3, 4, 5 or 6 position; and R_3 is substituted at the 2 or 3 position.

Suitable illustrations of R_1 in the above formula (1) are 5-methoxy-1-naphthyl, 6-methoxy-2-naphthyl, 7-methoxy-2-naphthyl, 4,6-dimethoxy-2-naphthyl, 6,7-dimethoxy-2-naphthyl and 6,7-diethoxy-2-naphthyl. Suitable R_2 groups in the above formula (1) are hydrogen, methyl, ethyl, propyl, and isopropyl, and the suitable position of R_2 is 2, 4 or 6. Suitable R_4 groups in the above formula (1) are hydrogen, methyl and ethyl.

These compounds can be prepared by the following synthetic routes:

(a) Condensation of an L-argininamide with an alkoxynaphthalenesulfonyl halide;

(b) Removal of the N^G-substituent from an N^G-substituted-N^2-alkoxynaphthalenesulfonyl-L-argininamide;

(c) Condensation of an N^2-alkoxynaphthalenesulfonyl-L-arginyl halide with an amine; and

(d) Guanidylation of an N^2-alkoxynaphthalenesulfonyl-L-ornithinamide or an acid addition salt thereof.

Example: (A) To a well-stirred solution of 3.0 g of benzyl 1-(N^G-nitro-L-arginyl)-4-methyl-2-piperidinecarboxylate hydrochloride and 1.4 g of K_2CO_3 in 15 ml of water and 40 ml of dioxane, was added in portions 2.87 g of 6,7-dimethoxy-2-naphthalenesulfonyl chloride, while maintaining the temperature at 0°C. The reaction mixture was stirred overnight at room temperature. At the end of this period, the reaction mixture was evaporated to dryness. The residue was taken up to 50 ml of chloroform and the chloroform solution was washed consecutively with 10% citric acid, saturated NaCl, saturated $NaHCO_3$ and saturated NaCl solutions.

The chloroform solution was evaporated and the residue was chromatographed on silica gel packed in chloroform, and eluted from chloroform containing 3% ethanol. The main fraction was evaporated to dryness to give 1.0 g of benzyl 1-[N^G-nitro-N^2-(6,7-dimethoxy-2-naphthalenesulfonyl)-L-arginyl]-4-methyl-2-piperidinecarboxylate.

(B) To a solution of 0.7 g of benzyl 1-[N^G-nitro-N^2-(6,7-dimethoxy-2-naphthalenesulfonyl)-L-arginyl]-4-methyl-2-piperidinecarboxylate in 10 ml of tert-butanol, 5 ml of water and 2 ml of acetic acid, was added 0.2 g of palladium black and then the mixture was shaken in a hydrogen atmosphere for 15 hours at room temperature. The solution was filtered to remove the catalyst and evaporated to give an oily product. Reprecipitation with ethanol-diethyl ether gave 0.5 g of 1-[N^2-(6,7-dimethoxy-2-naphthalenesulfonyl)-L-arginyl]- 4-methyl-2-piperidinecarboxylic acid.

Enzymatic Preparation of L-Citrulline from Arginine

Various methods are also known for producing L-citrulline (α-amino-δ-ureidovaleric acid) by enzymic reaction of L-arginine deiminase with L-arginine. *I. Chibata, T. Tosa, T. Sato and K. Yamamoto; U.S. Patent 3,886,040; May 27, 1975; assigned to Tanabe Seiyaku Co., Ltd., Japan* have developed a process whereby the enzyme can be used for more than one reaction.

It was found that an L-arginine deiminase-producing microorganism immobilized with a semipermeable membrane can be used advantageously in producing L-citrulline from L-arginine or an inorganic acid addition salt thereof. A mixture of L-citrulline and D-arginine can also be obtained from DL-arginine or an inorganic acid addition salt thereof. The process can be carried out by polymerizing at least one acrylic monomer in an aqueous suspension containing an L-arginine deiminase-producing microorganism, and subjecting the resultant immobilized L-arginine deiminase-producing microorganism to enzymic reaction wtih L-arginine, DL-arginine or their inorganic acid addition salt.

The polymerization can be preferably carried out in the presence of a polymerization initiator and a polymerization accelerator. Typical of suitable initiators are potassium persulfate, ammonium persulfate, vitamin B_2 and Methylene Blue. β-(Dimethylamino)-propionitrile and N,N,N',N'-tetramethyl-ethylenediamine are among polymerization accelerations which can be used. It is preferred to carry out the reaction at 20° to 40°C. The reaction may be completed within 10 to 60 minutes. The acrylic monomers used include acrylamide, N,N'-lower alkylene-bis-acrylamide and bis-(acrylamidomethyl)ether. N,N'-methylene-bis-acrylamide and N,N'-propylene-bis-acrylamide are suitable as the N,N'-lower alkylene-bis-acrylamide.

Moreover, preferred examples of L-arginine deiminase-producing microorganisms which are used in the process include *Pseudomonas putida* ATCC 4359, *Pseudomonas fluorescens* IFO 3081, *Pediococcus cerevisiae* P-60 ATCC 8042, *Sarcina lutea* IAM 1099, *Mycobacterium avium* IFO 3154, *Streptomyces griseus* IFO 3122 and *Streptococcus faecalis* ATCC 11420.

The immobilized enzyme can be used as a suspension in a solution of the L- (or DL)-arginine, or a solution of the arginine can be passed over a column of the immobilized enzyme as shown in the following examples.

Example 1: An aqueous nutrient medium containing the following ingredients is prepared:

	Weight per Volume Percent
Glucose	1
Yeast extract	1
Polypeptone	0.5
L-arginine hydrochloride	0.5
Ammonium chloride	0.1
Dipotassium phosphate	0.1
Magnesium sulfate heptahydrate	0.05
Manganese sulfate tetrahydrate	0.01
Ferrous sulfate heptahydrate	0.0005
Sodium chloride	0.2

The aqueous nutrient medium is adjusted to pH 6.2. *Pseudomonas putida* ATCC 4359 is inoculated into 300 ml of the medium, and the medium is cultivated at 30°C for 40 hours under shaking. The medium is then filtered by centrifugation. The microbial cells thus collected are suspended in 12 ml of a physiological saline solution. 2.25 g of acrylamide, 0.12 g of N,N'-methylene-bis-acrylamide, 1.5 ml of 5% β-(dimethylamino)-propionitrile and 1.5 ml of 2.5% potassium persulfate are added to the suspension. The suspension is then allowed to stand at 25°C for 10 minutes. The insoluble product is ground and washed with a physiological saline solution. 25 g of an immobilized preparation of *P. putida* ATCC 4359 are obtained.

Example 2: 2.25 g of the immobilized preparation of the above *P. putida* ATCC 4359 are charged into a 1.6 cm x 15.5 cm column, and one liter of an aqueous 10% L-arginine hydrochloride solution is passed through the column at 37°C at a flow rate of 15 ml/hr. One liter of the effluent is concentrated at 60°C to bring the total volume to 250 ml. Then, the concentrated solution is allowed to stand at 5°C overnight. The crystalline precipitate is collected by filtration and washed with methanol. 74.0 g of L-citrulline are obtained $[\alpha]_D^{25} = +21.3°$ (c = 10, 1 N HCl).

HISTIDINE

L-histidine [α-amino-4(or -5)-imidazolepropionic acid or glyoxaline-5-alanine] is a basic amino acid which is essential in the nutrition of animals. Therefore, the establishment of a relatively inexpensive, industrial process for the production thereof has been intensively sought. Heretofore, limited methods have been available for the production of L-histidine. For example, there is a process for the production of L-histidine by optically resolving DL-histidine prepared by a synthetic method. It has also been known to produce L-histidine by hydrolyzing proteins. These processes, however, are not commercially feasible.

Production Using Culture Media with High Nitrogen Content

In U.S. Patent 3,713,977 there is described a process for producing L-histidine where certain strains of microorganisms belonging to the genera Brevibacterium, Corynebacterium, Arthrobacter, Microbacterium, Micrococcus, Nocardia, Bacillus and Azotobacter are cultured in a medium and the accumulated L-histidine is recovered therefrom. Such strains are mutants which are resistant to analogues of L-histidine. Although the abovementioned process produces high yields of L-histidine in an industrial manner, improvements thereof are still in demand.

The process by *K. Nakayama and K. Araki; U.S. Patent 3,791,925; February 12, 1974; assigned to Kyowa Hakko Kogyo Co., Ltd., Japan* provides a more efficient process for producing L-histidine in high yield by fermentation. It has been found that the concentration of the nitrogen source in the culture medium is a very important factor in the production of L-histidine. More particularly, it has been found that L-histidine is accumulated in high yield when the weight ratio of the nitrogen source in the medium to the carbon source is at least 5:100. The upper limit of the nitrogen to carbon ratio is determined by the amounts tolerated by the organism.

In practice, the nitrogen to carbon ratio will be between about 5:100 and 50:100. This excess concentration of the nitrogen source is more than is usually required for the growth of microorganisms. The effect of the concentration of the nitrogen source in the medium is observed in L-histidine-producing microorganisms in general.

The organisms found useful in the process belong to the genera Brevibacterium, Corynebacterium, Arthrobacter, Micrococcus, Bacillus, Nocardia and Microbacterium and are mutant strains of these genera which exhibit resistance to analogues of L-histidine, for example, α-methylhistidine, 1,2,4-triazole-3-alanine, 2-thiazolealanine, 3-amino-1,2,4-triazole and hydrazinoimidazole propionic acid.

Among the strains of L-histidine-producing microorganisms belonging to the above genera, the following are preferred: *Corynebacterium glutamicum* (ATCC

21604), *Corynebacterium glutamicum* (ATCC 21607), *Brevibacterium ammoniagenes, Brevibacterium flavum* (ATCC 21605), *Arthrobacter citreus* (ATCC 21600), *Bacillus megaterium* (ATCC 21603), *Nocardia globerula* (ATCC 21602), and *Microbacterium ammoniaphilum.*

Example: In this example, fermentation is carried out using an L-histidine-producing mutant strain (ATCC 21604) obtained from *C. glutamicum.* This microorganism is inoculated into a medium comprising: 15% glucose, 0.15% KH_2PO_4, 0.15% K_2HPO_4, 0.05% $MgSO_4 \cdot 7H_2O$, 0.001% $FeSO_4 \cdot 7H_2O$, 0.001% $MnSO_4 \cdot 4H_2O$ and 200 μg/l of biotin. A nitrogen source is added and the medium is maintained at a pH of 7.2. The amount of nitrogen source added, the growth of the microorganism, and the yield of L-histidine are set forth in the table below. As indicated in the table, ammonium sulfate and ammonium acetate are used in separate proportions as the nitrogen source.

As shown in the table, when the nitrogen concentration is less than 5:100 of the carbon concentration the amount of L-histidine produced is relatively small. On the other hand, when the nitrogen:carbon ratio exceeds 5:100, the increase of the growth of the microorganism is minimal, whereas the amount of L-histidine produced is considerable.

Nitrogen Source (%)	Nitrogen (%)	Carbon:Nitrogen	Growth (OD*)	L-Histidine (mg/ml)
Ammonium sulfate				
0.5	0.11	100:1.8	0.32	Negligible
1.0	0.21	100:3.5	0.48	3.1
2.0	0.43	100:7.2	0.62	4.9
3.0	0.64	100:10.7	0.65	5.5
4.0	0.85	100:14.2	0.65	6.0
5.0	1.05	100:17.5	0.68	6.6
6.0	1.26	100:24.0	0.68	5.8
7.0	1.47	100:24.5	0.68	6.3
8.0	1.68	100:28.0	0.65	5.1
9.0	1.89	100:31.5	0.68	6.4
10.0	2.10	100:35.0	0.63	5.4
Ammonium acetate				
1.0	0.18	100:2.9	0.42	2.5
2.0	0.36	100:5.5	0.49	3.9
3.0	0.55	100:7.9	0.49	4.7
4.0	0.73	100:10.0	0.65	4.8
5.0	0.91	100:12.0	0.47	5.3

*Optical density indicated by absorbancy (using solution diluted by 50 times).

Production Using Culture Media Containing Growth Inhibitor

The use of a growth-inhibiting agent has conventionally been applied to the production of L-glutamic acid by fermentation, but it has never been reported in the art in the case of the fermentation of L-histidine.

Y. Arai and F. Kato; U.S. Patent 3,816,258; June 11, 1974; assigned to Kyowa Hakko Kogyo KK, Japan have found that it is possible to produce L-histidine in better yield and at higher concentration by the addition of a growth-inhibiting agent to the culture medium so as to inhibit the growth of the microorganism used.

A seed culture of a microorganism capable of producing L-histidine is inoculated in a main culture medium containing suitable amounts of carbon source, nitrogen source, inorganic salts and, if desired, other nutrients, and is cultured under aerobic conditions at 25° to 35°C at a pH of 5 to 8 for not more than 24 hours. After this, a suitable amount of a growth-inhibiting agent is added to the medium and the cultivation is additionally carried out in similar conditions to those of the prior cultivation except the culturing time is from 48 to 72 hours so that a large amount of L-histidine is produced and accumulated in the medium.

L-histidine-producing microorganisms which may be used are those belonging to the genera Brevibacterium, Corynebacterium, Arthrobacter, Micrococcus, Bacillus, Nocardia and the like, although any and all strains capable of producing L-histidine may be used regardless of their taxonomical classification.

Growth-inhibiting agents which may be used include substances having antibiotic properties such as antibiotics, surface-active agents as well as various organic and inorganic inhibiting agents.

Preferable antibiotics are exemplified by penicillin, streptomycin, cephalosporin, cycloserine, chloramphenicol, erythromycin, tetracycline, bacitracin, kanamycin, spiramycin, viomycin, novobiocin, colistin, etc. Preferable surface-active agents are exemplified by anionic surface-active agents such as alkyl sulfates, alkylaryl sulfonates, alkyl phosphates, amide sulfonates, ester sulfonates and amide sulfates; cationic surface-active agents such as amines, and salts thereof, quaternary ammonium salts, pyridinium salts, and picolinium salts; amphoteric surface-active agents of betaine-type, sulfuric acid ester-type and sulfonic acid-type; and nonionic surface active agents, such as polyoxyethylene alkyl ether, etc.

Preferable organic and inorganic inhibiting agents are exemplified by mercuric chloride, acrinol, resorcinol, and cresol, etc. It has been found that among these agents penicillin and nonionic surface-active agents are particularly useful.

Example: An L-histidine-producing strain of *Corynebacterium glutamicum* FERM 900 (ATCC 21604) was used. The seed culture consisting of 20 ml of a medium containing 4% glucose, 2% peptone, 1% beef extract, 0.5% yeast extract, 0.05% $MgSO_4 \cdot 7H_2O$, 0.15% KH_2PO_4, 0.05% K_2HPO_4, 30 γ/l biotin and 0.3% urea and having an adjusted pH of 7.2 was put into a 250 ml Erlenmeyer flask and sterilized by heating. The strain was inoculated to the medium and cultured for 24 hours at 28°C with shaking.

The main culture consisting of 40 ml of a medium containing 20% sucrose, 0.2% KH_2PO_4, 0.1% K_2HPO_4, 0.05% $MgSO_4 \cdot 7H_2O$, 0.001% $FeSO_4 \cdot 7H_2O$, 0.001% $MnSO_4 \cdot 4H_2O$, 0.5% corn steep liquor, 200 γ/l biotin, 2% $(NH_4)_2SO_4$ and 0.2% urea having an adjusted pH of 7.2 was put into a Sakaguchi flask and sterilized by heating. Then 2 ml of the seed culture was inoculated. After culturing for 24 hours at 28°C with shaking, various fractions of antibiotics and organic and inorganic inhibitors in various concentrations which had been separately prepared under sterile conditions were added to the fractions of culture liquor, as shown in the following table. Each fraction was additionally cultured for 48 hours with shaking. The amounts of L-histidine accumulated in the respective culture liquor when using different inhibitors are shown in the table.

	Amount of Addition (γ/ml)	Dry Cell Weight (mg/ml)	L-Histidine (mg/ml)
Additives			
Pencillin G	2.5	17.4	6.94
Potassium salt	5	16.3	7.08
Potassium salt	10	15.2	7.98
Potassium salt	10	18.7	7.01
Antibiotics			
Cycloserine	20	15.9	6.82
Cycloserine	10	16.8	7.11
Cephalosporin	20	15.7	6.63
Cephalosporin	5	16.2	5.38
Streptomycin	10	15.1	5.55
Inorganic inhibitor			
Mercuric chloride	2	14.0	4.90
Organic inhibitor			
Acrinol	2	14.2	5.00
No addition	–	21.1	4.55

Production Using Bacterial Mutants Resistant to 2-Thiazolealanine

In U.S. Patent 3,716,453 it was disclosed that histidine can be produced by fermentation of conventional culture media by microorganisms of the genera Brevibacterium, Corynebacterium, and Arthrobacter which are artificially induced mutants of parent strains, incapable of producing histidine under the same conditions. These mutants were resistant to 2-thiazolealanine concentrations of more than 1 mg/ml of culture medium.

K. Kubota, H. Kamijo, O. Mihara, S. Okumura and H. Okada; U.S. Patent 3,875,001; April 1, 1975; assigned to Ajinomoto Co., Inc., Japan have disclosed an improvement of the above process.

It has been found that other mutants belonging to the genera Brevibacterium and Corynebacterium which combine resistance to 2-thiazolealanine with specific nutrient requirements are substantially superior to the microorganisms of the earlier patent in their ability of producing histidine by fermentation. The required nutrients are arginine, methionine, tryptophan, leucine, phenylalanine, lysine, threonine, xanthine, uracil, and/or shikimic acid.

The mutants of the process are produced from the respective parent strains by conventional methods, as by exposure of the parent strain to ionizing radiation (ultraviolet, x-rays, gamma rays) or to chemical mutagenic agents (sodium nitrite, nitrosoguanidine, diethyl sulfate). Typically exposure to 250 μg/ml nitrosoguanidine at 30°C for 30 minutes will have the desired effect.

The mutants produced are inoculated on an otherwise conventional culture medium containing enough 2-thiazolealanine for suppressing the growth of the parent strain. This requires generally more than 1 mg/ml 2-thiazolealanine, and the best histidine producing strains can grow on media containing 5 mg 2-thiazolealanine per mililiter. The resistant strains are collected and may be subjected to yet another treatment with mutagenic agents. The strains requiring the specific nutrients listed above are then selected from the first or second generation of mutants in a conventional manner.

The best histidine-producing strains available at this time are listed below. They were derived from *B. flavum* ATCC 14067, *B. lactofermentum* ATCC 13869, and *C. acetoacidophilum* ATCC 13870.

Brevibacterium flavum
- FERM-P 1561 (arginine)
- FERM-P 1562 (methionine)
- FERM-P 1563 (trypotophan)
- FERM-P 1564 (uracil)
- FERM-P 2168 (shikimic acid)
- FERM-P 2169 (xanthine)
- FERM-P 2170 (threonine)

Brevibacterium lactofermentum
- FERM-P 1565 (leucine)
- FERM-P 1566 (leucine and tryptophan)
- FERM-P 1567 (leucine and phenylalanine)
- FERM-P 1568 (phenylalanine)

Corynebacterium acetoacidophilum
- FERM-P 1569 (lysine)

For a good yield of histidine, the fermentation is carried out aerobically with aeration and agitation. Best yields require a pH controlled within 5 to 9, and maintained by means of gaseous or aqueous ammonia, calcium carbonate, alkali metal hydroxide, urea, organic or inorganic acids, and some of these addition agents may also provide the necessary nitrogen supply. When the fermentation is carried out at 25° to 37°C, the histidine concentration in the broth reaches a maximum within 2 to 7 days.

Example: *B. flavum* FERM-P 1561 was cultured on bouillon agar slants. An aqueous fermentation medium containing, per deciliter, 10 g glucose, 5 g $(NH_4)_2SO_4$, 0.1 g KH_2PO_4, 0.04 g $MgSO_4 \cdot 7H_2O$, 0.2 mg Fe^{++}, 0.2 mg Mn^{++}, 50 μg biotin, 5.0 μg thiamine HCl, 0.3 ml soy protein hydrolyzate, 50 mg arginine, and 5 g $CaCO_3$, was adjusted to pH 7.5, 20 ml batches of the medium were transferred to respective 500 ml shaking flasks, sterilized with steam, and thereafter inoculated with the microorganisms. Each culture medium was shaken for 72 hours at 31°C, and the combined broths then were found to contain 1.11 g/dl histidine.

The microbial cells were removed from one liter of the broth by centrifuging, and the supernatant was stripped of histidine by passage over a column of the strongly acidic ion exchange resin Amberlite IR-120 (H-type). The column was washed, and the histidine eluted with 3% ammonium hydroxide solution. The eluate was partly evaporated in a vacuum, and histidine crystallized from the concentrate was recovered in an amount of 7.3 g. In a comparison test, *B. flavum* ATCC 21406 was cultured on the same fermentation medium under the same conditions, and a histidine concentration of 0.40 g/dl was reached in good agreement with U.S. Patent 3,716,453.

Production Using Mutant of *Serratia marcescens*

No fermentative production of L-histidine by using a strain of the Serratia genus has been reported. *I. Chibata, M. Kisumi, M. Sugiura and N. Nakanishi; U.S. Patent 3,902,966; September 2, 1975; assigned to Tanabe Seiyaku Co., Ltd., Japan* have found that a mutant of *Serratia marcescens* exhibits excellent productivity of L-histidine in a nutrient medium. This mutant lacks L-histidine ammonia-lyase and is resistant to 2-methylhistidine.

The L-histidine ammonia-lyase-lacking and 2-methylhistidine-resistant mutant may be obtained by treating a wild strain of *S. marcescens* with a mutagen or ultraviolet rays to produce an L-histidine ammonia-lyase-lacking mutant of *S. marcescens* or a 2-methylhistidine-resistant mutant of *S. marcescens*, and further treating the L-histidine ammonia-lyase-lacking mutant or the 2-methylhistidine-resistant mutant with a mutagen or ultraviolet rays. For example, a wild strain of *S. marcescens* is treated with N-methyl-N'-nitro-N-nitrosoguanidine and then cultivated at 30°C for 2 to 3 days on agar plates containing the following compositions:

	Weight per Volume Percent
Glucose	0.2
K_2HPO_4	0.3
KH_2PO_4	0.7
$MgSO_4 \cdot 7H_2O$	0.01
$(NH_4)_2SO_4$	0.0005
L-histidine hydrochloride monohydrate	0.2
Sodium citrate trihydrate	0.05

The L-histidine ammonia-lyase-lacking mutant of *S. marcescens* may be isolated as small colonies. The L-histidine ammonia-lyase-lacking mutant of *S. marcescens* is treated with N-methyl-N'-nitro-N-nitrosoguanidine and then cultivated at 30°C for 3 to 4 days on agar plates containing the following compositions:

	Weight per Volume Percent
Glucose	0.2
K_2HPO_4	0.3
KH_2PO_4	0.7
$MgSO_4 \cdot 7H_2O$	0.01
$(NH_4)_2SO_4$	0.1
Sodium citrate trihydrate	0.05
2-Methylhistidine	0.1-1

The L-histidine ammonia-lyase-lacking and 2-methylhistidine-resistant mutant is isolated as large colonies. A viable culture of the mutant has been deposited as ATCC 31026. This mutant of *S. marcescens* may be accomplished by either shaking cultivation or submerged fermentation under aerobic conditions. Details of the fermentation for production of histidine are shown in the following example.

Example: An aqueous nutrient medium (pH 7.0) comprising the following ingredients is prepared:

	Weight per Volume Percent
Glucose	3
Dextrin	10
Urea	2
Ammonium phosphate	1
Dibasic potassium phosphate	0.1
Magnesium sulfate heptahydrate	0.05
Calcium carbonate	2

Fifteen milliliters of the medium are charged into a 500 ml shaking flask and its contents are sterilized by autoclaving. Glucose and dextrin are separately sterilized and added to the medium aseptically. A loopful of the L-histidine ammonia-lyase-lacking and 2-methylhistidine-resistant mutant (Hd-MHr) ATCC 31026 of *S. marcescens* is inoculated into the medium. Then the medium is cultivated for 72 hours at 30°C under shaking (140 rpm, 8 cm stroke). The fermentation medium thus obtained contains 5 mg/ml of L-histidine. 100 ml of the fermentation medium are heated at 100°C for 20 minutes and then filtered. The filtrate is passed through a column (1 cm x 10 cm) of strong cation exchange resin H-form, (Amberlite IR-120).

After washing with water, the column is eluted with 5% aqueous ammonia. The fractions containing L-histidine are collected and concentrated under reduced pressure, and aqueous methanol is added to the concentrated solution. The crystalline precipitate is collected by filtration, and recrystallized from aqueous methanol. 350 mg of L-histidine are obtained, $[\alpha]_D^{25}$ = –37.38° (c = 2, in H_2O).

Use as Anorexigenic Agents

Various therapeutic agents previously have been used in the treatment of obese individuals in order to decrease their appetite. The most widely used agent for this purpose has been d-amphetamine. This compound, however, being a central nervous system stimulant, produces some undesirable and unpleasant side effects, for example, nervous tension, insomnia, headache, palpitations and elevation of blood pressure.

R.I. Henkin; U.S. Patent 3,867,539; February 18, 1975; assigned to the U.S. Secretary of Health, Education, and Welfare has found that internal administration of the amino acid histidine has an anorexigenic or appetite-suppressing effect on humans. Histidine does not produce the undesirable side effects associated with the previously employed anorexigenic agents, and hence is highly effective in the treatment of obesity.

The histidine, which is preferably L-histidine but may also be D-histidine, DL-histidine or mixtures thereof, may be used either as the free base or in the form of various pharmaceutically acceptable, nontoxic, water-soluble salts thereof. Examples of such salts are the mono- or di-hydrochlorides, sulfates and phosphates, and such nontoxic alkaline metal salts as the potassium and sodium salts.

Anorexigenic pharmaceutical preparations may conveniently be produced by compounding the histidine or a pharmaceutically acceptable salt thereof with conventional inert pharmaceutical carriers into dosage forms which are suitable for oral or parenteral administration.

The daily dosage of histidine which is effective in producing anorexia may be modified within a relatively wide range and, to some extent, is dependent upon the needs and requirements of the individual patient. It has been found, however, that as a general rule, the daily administration of 1 to 32 g of histidine, calculated as free base, will prove to be effective. Preferably the histidine is orally administered to the patient in suitable dosage forms containing from 0.5 to 16 g of histidine per dosage unit one to three times a day as conditions demand, desirably before meals.

Under ordinary circumstances, a suitable oral dosage form containing about 8.1 grams of histidine, administered twice a day before meals for a total daily dose of about 16.2 g, will prove to be effective for producing anorexia without any undesirable side effects. Moreover, at this dosage level, normal appetite will generally return within 24 to 48 hours following discontinuation of the treatment.

Example: A patient was placed on a constant 9 meq Na^+ diet with 100 meq of Na^+ added in a salt shaker and a fixed amount of distilled water, the daily amount set by the patient. The patient was then studied for two control days during which two daily blood samples and total excretion of urine for all 24-hour periods were collected and zinc and copper measured. Mean control values for serum zinc and copper concentrations were 85 and 130 μg/100 ml, respectively; mean control values for urinary zinc and copper excretion were 950 and 35 μg/24 hours, respectively.

On the third and fourth days of the study, 8.1 gm L-histidine was administered orally, once daily, in pears, without any resultant change in either serum zinc or copper concentrations, but with a significant increase in urinary zinc excretion to 1,123 μg/24 hours. No change in any clinical parameters occurred at the end of these two days. On the fifth and sixth days of the study, 8.1 gm L-histidine was administered orally, twice daily, in pears, for a total daily dose of 16.2 gm, without any significant change in either serum zinc or copper concentrations, but with a further increase in urinary zinc excretion to 1,490 μg/24 hours.

On the morning of the sixth day of the study, at breakfast, the patient became aware of a marked loss of appetite resulting in his wish to avoid food. This symptom was very unusual for him. Following discontinuation of the L-histidine, urinary zinc excretion and appetite returned to normal within 24 to 48 hours. No other side effects were noted.

Histidine Derivatives as Antidepressants

Compounds derived from L-pyroglutamic acid and L-histidine have been prepared by *B.A. Morgan and D. J. Schafer; U.S. Patent 4,060,603; November 29, 1977; assigned to Reckitt & Colman Products Limited, England* and may be expected to be of use in the treatment of depression. The compounds, being chemically related to the hypothalamic thyrotropin-releasing hormone (TRH), L-pyroglutamyl-L-histidyl-L-prolineamide, may be expected to exhibit similar effects on the pituitary gland.

These histidine derivatives are compounds of the formula:

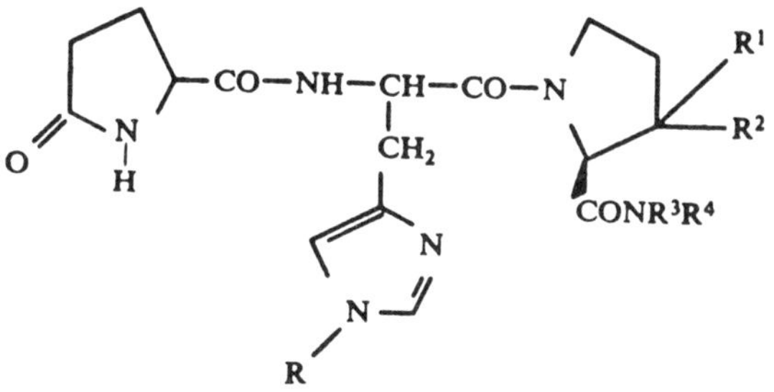

where R is hydrogen or alkyl C_{1-3}, R^1 is alkyl or alkoxy C_{1-3}; R^2 is hydrogen, alkyl or alkoxy C_{1-3}; R^3 is hydrogen, alkyl C_{1-6}, or cycloalkyl C_{3-6}; and R^4 is hydrogen or alkyl C_{1-6}. These compounds may be prepared by standard techniques of peptide chemistry such as shown in the following example.

Example: This example describes the preparation of L-pyroglutamyl-L-histidyl-trans-3-methyl-L-prolineamide (R = H, R^1 = trans-3-CH_3, R^2 = R^3 = R^4 = H).

(A) N-t-butoxycarbonyl-trans-3-Methyl-L-Proline — Triethylamine (3.5 ml) and t-butyl-2,4,5-trichlorophenyl carbonate (3.5 g) were added to a solution of trans-3-methyl-L-proline (1.25 g) in t-butanol (12 ml) and water (8 ml). The mixture was heated to 60° for 2 hours, evaporated, the residue dissolved in 1 N sodium bicarbonate solution (50 ml) and the solution washed with ether. Excess solid citric acid was added, the product extracted into ethyl acetate, the combined extracts dried (Na_2SO_4) and evaporated giving a crude crystalline product (1.6 g). Recrystallization from ethyl acetate/light petroleum (40°-60°) gave the desired product (1.33 g), MP 146°-150°C, $[\alpha]_D^{20}$ -60.1° (c, 1.0, chloroform).

(B) N-t-butoxycarbonyl-trans-3-Methyl-L-Prolineamide — N-Methylmorpholine (6.0 ml) was added to a solution of N-t-butoxycarbonyl-trans-3-methyl-L-proline (1.1 g) in tetrahydrofuran (50 ml) at -20°. Freshly distilled isobutylchloroformate (0.8 g) was added dropwise and the solution stirred for 10 minutes at -20°. Concentrated ammonia solution (5 ml) was added and the solution stirred for 2 hours, evaporated, the residue taken up in ethyl acetate (100 ml) and the solution washed with 10% aqueous sodium bicarbonate, brine and dried (Na_2SO_4). Evaporation gave the crude crystalline product (1.0 g). Recrystallization from ethyl acetate/light petroleum (40°-60°) gave the desired product (0.91 g) MP 100°-101°C, $[\alpha]_D^{20}$ -79.7° (c, 1.0, chloroform).

(C) trans-3-Methyl-L-Prolineamide Hydrochloride — A solution of hydrogen chloride in dioxan (7.5 N, 5 ml) was added to a solution of N-t-butoxycarbonyl-trans-3-methyl-L-prolineamide (0.65 g) in 10 ml dioxan. After 2 hours the crystalline solid was filtered and recrystallized from ethanol/ether to give the product (0.47 g), MP 235°-237°C $[\alpha]_D^{20}$ -14.8° (c, 1.0, water).

(D) L-Pyroglutamyl-L-Histidyl-trans-3-Methyl-L-Prolineamide — trans-3-Methyl-L-prolineamide hydrochloride (0.50 g) in dimethylformamide (5 ml) was added to a solution of L-pyroglutamyl-L-histidine azide, prepared from the hydrazide (0.9 g), in dimethylformamide (15 ml) at –20°, and triethylamine (0.45 ml) added. The mixture was left at room temperature for 12 hours, filtered, evaporated, the residue dissolved in 1% aqueous pyridine (10 ml) and applied to a column of sulfoethyl "Sephadex" C-25 resin (50 g weighed as moist resin; 2 cm column diameter) equilibrated with 1% aqueous pyridine.

The column was eluted by a gradient technique using 5% aqueous pyridine via a 1-liter spherical mixing vessel filled with 1% aqueous pyridine. The fractions containing product (as detected by optical rotation and TLC) were combined, evaporated, and the residue triturated with ether giving the product (as the monohydrate) as an amorphous solid (0.70 g), $[\alpha]_D^{20}$ –46.7° (c, 1.0, water), MP 145°-148°.

The compound of this example when tested in mice in the reversal of reserpine-induced hypothermia test was three to four times more potent than TRH when administered in doses of 1-30 mg/kg per os.

MONOAMINO DIBASIC ACIDS

These amino acids are straight chain α-amino acids with the ω-carbon atom forming the second carboxylic acid group. None of these amino acids is characterized as being "essential".

ASPARTIC ACID–PREPARATIVE METHODS

Also known as aminosuccinic acid, apartic acid can be used as: tonic, hepatic regulator, for eliminating the nitrogenous residues, metabolism regulator, myocardium regulator, regulator of neuro-vegetative disorders, food additive, sweetener for beverages, and basic intermediate for organic syntheses.

Use of Immobilized Aspartase Enzyme

Various methods for producing L-aspartic acid by the enzymatic reaction of aspartase with ammonium fumarate are known. In former processes, the enzyme could be used for only one conversion which makes this process uneconomical. Recent developments using immobilized aspartase have required the use of extracted enzyme.

In the process of *I. Chibata, T. Tosa and T. Sato; U.S. Patent 3,791,926; February 12, 1974; assigned to Tanabe Seiyaku Co., Ltd., Japan,* the aspartase-producing microorganisms have been immobilized with a semipermeable membrane and used for the production of aspartic acid.

According to this process, the membrane can be prepared by polymerizing at least one monomer selected from acrylamide, N,N'-lower alkylene-bis(acrylamide) and bis(acrylamidomethyl)ether in an aqueous suspension containing an aspartase-producing microorganism, and subjecting the resultant immobilized aspartase-producing microorganism to enzymatic reaction with ammonium fumarate or a mixture of fumaric acid or its salt and an inorganic ammonium salt.

The polymerization is preferably carried out in the presence of a polymerization initiator and a polymerization accelerator. Potassium persulfate, ammonium persulfate, vitamin B_2 and methylene blue are suitable as the polymerization initiator. On the other hand, β-(dimethylamino)propionitrile and N,N,N',N'-tetramethylethylenediamine are used as the polymerization accelerator. It is preferred to carry out the reaction at 20° to 40°C. The reaction may be completed within 10 to 60 minutes.

Examples of aspartase-producing microorganisms which may be used include *Escherichia coli* ATCC No. 11303, *Pseudomonas aeruginosa* OUT No. 8252, *Serratia marcescens* OUT No. 8259, *Proteus vulgaris* OUT No. 8226, *Bacterium succinium* AM No. 1017 and *Alcaligenes faecalis* OUT No. 8030.

L-aspartic acid can be prepared by contacting the resultant immobilized microorganism with ammonium fumarate or a mixture of fumaric acid or its salt and an inorganic ammonium salt. Suitable examples of the salt of fumaric acid include alkali metal salts thereof such as sodium fumarate and potassium fumarate. Ammonium chloride, ammonium sulfate and ammonium phosphate are preferred as the inorganic ammonium salt. Further details of the preparative process are included in the following examples.

Example 1: 4.8 grams of the microbial cells of *Escherichia coli* ATCC No. 11303 are suspended in 48 ml of physiological saline solution, 9 grams of acrylamide, 480 milligrams of N,N'-methylene-bis(acrylamide), 6 milliliters of 5% β-(dimethylamino)propionitrile and 6 milliters of 2.5% potassium persulfate are added to the suspension. Then, the suspension is allowed to stand at 37°C for 10 minutes. 120 milliliters of an immobilized preparation of *Escherichia coli* ATCC No. 11303 are obtained.

120 ml of the immobilized preparation of *Escherichia coli* ATCC No. 11303 are charged into a 2.1 x 34.8 cm column. 150 grams of ammonium fumarate are dissolved in 900 ml of water. The aqueous solution is adjusted to pH 8.5 with aqueous ammonia and diluted with water to bring the total volume to 1 liter. Then the aqueous solution is continuously passed through the column at 37°C at a flow rate of 20 ml/hr. The effluent is adjusted to pH 2.8 to 3.0 with concentrated sulfuric acid. The crystalline precipitate is collected by filtration and then washed with methanol. 126.3 grams of L-aspartic acid are obtained; $[\alpha]_D^{20}$ +24.8° (c = 1, 6 N HCl).

Example 2: An immobilized preparation of *Escherichia coli* ATCC No.11303 is prepared in the manner described in Example 1. 50 ml of the immobilized preparation are charged into a 2.1 x 14.5 cm column. An aqueous solution (pH 8.5) containing 1 M concentration of ammonium fumarate is continuously passed through the column at 37°C at a flow rate as shown in the following table. The concentration of L-aspartic acid in the effluent is assayed biologically by using *Leuconostoc Mesenteroides* P-60 as a sensitive microorganism, and the percentage conversion of ammonium fumarate to L-aspartic acid is calculated therefrom. The results are shown in the following table.

Flow Rate (ml/hr)	Conversion to L-Aspartic Acid (%)
25	45
12.5	92
6.5	100
3.5	100

Use of Pseudomonas PO 7111

The microorganisms considered as being the best for producing the amino-transferase necessary to the conversion of fumaric acid into L-aspartic acid have required a complex culture medium containing in addition to a carbon-providing material, organic nitrogen sources, which frequently are proteic hydrolyzates which are raw materials of variable composition and relatively high cost. As a consequence, the known methods using these microorganisms can be considered as being complex and not convenient for a large-scale production.

A new microorganism of the Pseudomonacea family which avoids the above drawbacks has been disclosed by *N.-C. Duc; U.S. Patent 3,933,586; January 20, 1976; assigned to Les Produits Organiques du Santerre Orsan, France.* This microorganism can perform, by using simple means, the bioconversion of non-saturated organic acids, such as fumaric acid, into L-aspartic acid with reaction rates considerably higher than those of prior methods. The obtained yields are very high and the product can be easily extracted with a very high purity grade.

This microorganism Pseudomonas PO 7111 (ATCC 21973) has been isolated from liquid effluents in the manufacture of beet sugar. It is considered new because it differs from other Pseudomonas bacteria by one or several reproducible and constant features according to the criteria defined in *Bergey's Manual* issued 1967.

Unlike the known microorganisms used for producing amino-transferase, this strain can grow in a common medium comprising essentially for example extraction residues of sugar, i.e., molasses of sugar, beet or cane, which residues are provided with a supplement of mineral phosphorus, for instance in the form of phosphate or phosphoric acid, and of nitrogen, for instance in the form of urea or ammonia, and some trace elements.

Example: The optimal bioconversion conditions which are specific to the PO 7111 strain have been applied on the industrial scale during operations carried out in an apparatus allowing obtainment of 1 m^3 of culture with a high concentration in amino-transferase.

In a fermentation apparatus fitted with a stirrer and heating means, 1,000 liters of a medium having the following composition is prepared: 40 g/l (expressed in total sugar quantity) molasses, 2.3 g/l phosphoric acid, 0.5 g/l magnesium sulfate, 1 g/l fumaric acid and distilled water to make one liter.

The pH is adjusted to 7 with ammonia prior to sterilization at 123°C for 30 min. The medium is inoculated with a 20 liter preculture of Pseudomonas (PO 7111). The culture is developed in a sterile medium and under controlled conditions of temperature (30°C) with stirring (100 rpm) and aeration at 0.8 volume per volume per minute. After a culture period of 11 hours, the bacteria population has reached 10^{10} bacteria/ml and the pH is 8.95. The enzymatic activity is then maximum.

At the end of the culture period, the fermentation apparatus is opened to air and the agitation is stopped. The temperature is progressively increased to 55° to 57°C, and the substrate is then progressively added by pouring crystallized fumaric acid directly into the tank, and this until a concentration of 100 g/l is obtained.

The pH of the medium is maintained between 8.9 and 9.1 by adding gaseous ammonia.

The conversion is observed by measurement of the formed L-aspartic acid. The addition of fumaric acid is adjusted so as to keep its concentration at about 10%. The operation goes on during 12 hours, then the addition of fumaric acid is decreased while maintaining the temperature and pH. The content is thus lowered until the 15th hour, and, at this time, the medium contains 180 g/l of L-aspartic acid and 30 g/l of nonconverted fumaric acid. The addition of fumaric acid is then stopped. The total amount of added substrate is equal to 160 kg. The stirring is continued while maintaining the temperature and pH until the 20th hour.

At this moment, the L-aspartic acid content in the medium is 21.1% and that of the residual fumaric acid 0.7%. Consequently, a fumaric acid amount of about 155 kg has been converted with a yield of 109%, corresponding to a molar yield of 95%.

The above solution is adjusted to pH 5.8 by concentrated hydrochloric acid; the temperature increased to 53° to 55°C and a 10 g/l solution of a flocculating agent of the sulfonated polystyrene type is added. After a 15-minute contact time, the insoluble parts and flocculated materials are removed by filtration under vacuum with a drum-filter coated with a layer of expanded perlite. The filtration rate is 500 $l/m^2/hr$. The filtered solution is fully clear. Then the solution is cooled to 30°C and, while stirring, made acid by adding dried concentrated hydrochloric acid to a pH of 2.8 to 3. After 8 hours, the crystals can be extracted from the mother liquor, and they are dried by centrifugation.

The purity degree of the water-washed crystals is better than 99%. 183 kg of L-aspartic acid are obtained.

Use of Mutants Resistant to 6-(Dimethylamino)Purine

The mutants used in the process developed by *T. Tsuchida, K. Kubota and Y. Hirose; U.S. Patent 4,000,040; December 28, 1976; assigned to Ajinomoto Co., Inc., Japan* are derived from parent strains belonging to the genus Brevibacterium and Corynebacterium.

The mutation is carried out by conventional techniques, for example by exposure of the parent strain to ultraviolet light, x-rays or gamma rays, or to chemical mutagenic agents. The mutant resistant to 6-(dimethylamino)purine can be selected from the exposed strains by culturing the exposed strains in a medium which contains an amount of 6-(dimethylamino)purine which suppresses the growth of the parent strain. In the culturing of the exposed strains, the mutant strain preferentially grows over the nonmutant strains and thereafter is isolated. The best L-aspartic-acid-producing mutants are listed below with their accession numbers.

	Mutant	Parent
Brevibacterium flavum	AJ 3859 (FERM-P 2799)	ATCC 14067
Brevibacterium lactofermentum	AJ 3860 (FERM-P 2800)	ATCC 13869
Corynebacterium acetoacidophilum	AJ 3877 (FERM-P 2803)	ATCC 13870
Corynebacterium glutamicum (Micrococcus geritamicus)	AJ 3876 (FERM-P 2801)	ATCC 13032

The following examples show the formation of the mutants and production of L-aspartic acid.

Example 1: Brevibacterium flavum ATCC 14067 was cultured in 5 ml of a medium consisting of 1 g/dl yeast extract, 1 g/dl peptone, 0.5 g/dl NaCl, and 0.5 g/dl glucose (pH 7.0), and cultured at 31°C for 16 hours.

Cells thus obtained were suspended in 5 ml of a phosphate buffer containing 250 μg/ml N-methyl-N'-nitro-N-nitrosoguanidine, and the suspension was held at 31°C for 30 minutes with stirring. Subsequently cells were recovered from the suspension and washed twice with the phosphate buffer, and spread on a plate-medium containing 1,000 μg/ml 6-(dimethylamino)purine, 2 g/dl glucose, 1 g/dl ammonium sulfate, 0.1 g/dl KH_2PO_4, 0.04 g/dl $MgSO_4 \cdot 7H_2O$, 0.2 mg/dl ferrous ions, 0.2 mg/dl manganese ions, 5.0 μg/dl biotin, 10.0 μg/dl thiamine·HCl and 0.2 g/dl urea, (pH 7.0).

Strains grown on the plate-medium were picked up, and the most effective L-aspartic acid-producer AJ 3859 was selected by culturing the strain in the same manner shown in Example 2, AJ 3859 and ATCC 14067 were cultured on Medium A at 31.5°C for 24 hours, and cells grown on Medium A were suspended in Basal-medium B.

The suspension (0.1 ml) was inoculated in 3 ml of Basal-medium B additionally containing the amount of 6-(dimethylamino)purine shown in the following table. Growth was determined after culturing at 31.5°C for 24 hours. Medium A contains 1 g/dl yeast extract, 1 g/dl peptone, 0.5 g/dl NaCl, and 0.5 g/dl glucose. Basal-medium B contains 2 g/dl glucose, 1 g/dl $(NH_4)_2SO_4$, 0.1 g/dl KH_2PO_4, 0.04 g/dl $MgSO_4 \cdot 7H_2O$, 0.2 mg/dl ferrous ions, 0.2 mg/dl manganese ions, 5.0 μg/dl biotin, 10.0 μg/dl thiamine·HCl and 3 g/dl $CaCO_3$. The pH is 7.0 (KOH).

6-(dimethylamino)purine μg/ml	Relative growth (%) ATCC 14067	AJ 3859
0	100	100
200	100	98
400	80	100
600	60	100
800	20	95
1,000	0	90

AJ 3860, AJ 3876 and AJ 3877 were obtained in a manner analogous to that mentioned above. Those strains were more resistant to 6-(dimethylamino)purine than the parent strain.

Example 2: An aqueous culture medium was prepared to contain, per deciliter, 10 grams glucose, 7 grams $(NH_4)_2SO_4$, 0.25 gram KH_2PO_4, 0.2 gram $MgSO_4 \cdot 7H_2O$, 0.4 μg biotin, 35.0 μg thiamine·HCl, 0.2 mg ferrous ions, 0.2 mg manganese ions, 1 ml soy-protein hydrolyzate ("Mieki") and 5 grams calcium carbonate.

300 ml batches of the aqueous culture medium were placed in 1 liter fermentors, and inoculated with each microorganism listed in the following table, which was previously cultured on bouillon-slant at 30°C for 24 hours. Cultivation was carried out aerobically at 31°C for 48 hours.

The amount of L-aspartic acid shown in the following table accumulated in the resultant culture broths. A one liter amount of the culture broth of AJ 3859 was passed through a cation exchange resin (H^+). The L-aspartic acid which adsorbed on the resin was eluted with ammonia water, and the eluate was adjusted to pH 2.8. Upon cooling, 6.05 grams of crystals of L-aspartic acid were obtained.

	L-Aspartic acid accumulated (g/dl)
ATCC 14067	0.01
AJ 3859	1.00
AJ 3860	0.73
AJ 3876	0.85
AJ 3877	0.95

Production in Hydrocarbon-Containing Medium Using Two Organisms

In the process developed by *V. Zangrandi and P. Peri; U.S. Patent 4,013,508; March 22, 1977; assigned to Liquichimica SpA, Italy* two types of microorganisms are used: one which assimilates hydrocarbons transforming them into fumaric acid and the other which transforms the fumaric acid into L-aspartic acid.

The microorganisms used in this process are of two kinds: the first presents morphologic and physiologic characteristics not dissimilar to those described for the species *Candida Hydrocarbofumarica* (ATCC 20473) (*Agricultural Biological Chemistry,* 34, 670, 1970), the second belongs to the genus Bacillus species ATCC 31177.

According to this process, the culture medium, containing paraffin, mineral salts, sources of assimilable nitrogen, is inoculated with an inoculum mass of the strain which metabolizes paraffin into fumaric acid and the culture is aerated directly under strong stirring to the almost complete transformation of the paraffin into dicarboxylic acid. At this point the direct recovery of fumaric acid or the direct inoculation of the inoculum mass of the strain which catalyzes the transformation into L-aspartic acid may be performed.

Example: Candida Hydrocarbofumarica ATCC 20473, cultivated on an agar culture medium containing paraffins at 30°C for 24 hours, was inoculated into a 3 liter flask, containing 500 ml of broth (A) (autoclaved at 115°C for 20 min) having the following composition.

Broth (A)

	Grams per Liter
n-Paraffins	30
NH_4Cl	5
KH_2PO_4	1
$MgSO_4 \cdot 7H_2O$	0.5
$FeSO_4 \cdot 7H_2O$	0.03
$MnSO_4 \cdot 4H_2O$	0.03
Yeast extract	0.5
Meat flour	1
Tween 80	0.5
H_2O	qs

The development of the inoculum occurred at 30°C on a reciprocating vibrator at 100 strokes per minute for 36 hours.

The entire culture of the inoculum flask was transferred into a jar containing 20 liters of fermention broth (B) and maintained in incubation at a constant 30°C with stirring at 800 rpm and an aeration of 1.5 v/v/min. The pH of the fermenting culture was adjusted to pH 6.5 by addition of ammonia.

Broth (B)

	Grams per Liter
n-Paraffins C_{13}-C_{18}	80
NH_4Cl	5
KH_2PO_4	1
$MgSO_4 \cdot 7H_2O$	0.5
$FeSO_4 \cdot 7H_2O$	0.03
$MnSO_4 \cdot 4H_2O$	0.03
$ZnSO_4 \cdot 7H_2O$	0.05
Yeast extract (paste)	0.5
Meat flour	2
Tween 80	0.5
H_2O	qs

The concentration of the fumaric acid after 72 hours of fermentation was 46 grams per liter. The yeast cells were separated by centrifugation and the clear broth recovered was inoculated into the same jar with 500 ml of an inoculum culture previously prepared, of the strain of Bacillus species (ATCC 31177) which transforms fumaric acid into L-aspartic acid.

The composition of the broth (C) for the preparation of the second inoculum is the following.

Broth (C)

	Grams per Liter
Ammonium fumarate	60
K_2HPO_4	2
$MgSO_4 \cdot 7H_2O$	0.5
Yeast extract (paste)	2
H_2O	qs

Broth (C), after adjustment of the pH to 7 was distributed into 3 liter flasks in the ratio of 500 ml each. The flasks were sterilized at 115°C in an autoclave for 20 minutes. Each flask was inoculated with a cellular suspension of the strain of Bacillus species (ATCC 31177), obtained from a slant of the strain cultivated on agar, containing also fumaric acid, K_2HPO_4, $MgSO_4$, yeast extract.

The development of the inoculum culture of ATCC 31177 was done at 30°C on a reciprocating vibrator at 100 strokes per minute for a period of 18 to 20 hr.

The concentration of L-aspartic acid in the broth culture, at this point, is only 0.5% w/v. From this moment on, the transformation of the fumaric acid continues rapidly, and is completed in 56 to 60 hours to a final concentration of L-aspartic acid of 41 g/l. Therefore the conversion yield, in the paraffins-L-aspartic acid fermentation is 51%.

Ten liters of broth culture was acidified to pH 4.5 with concentrated H_2SO_4. 20 grams of activated carbon was added to the acidified broth culture and the mixture then heated at 85° to 90°C for 20 minutes. The product was filtered through a Buchner filter after addition of 1% w/v of diatomaceous earth and then the cake was washed with water. The clear filtrate was concentrated under vacuum until it reached a concentration of L-aspartic acid of 18% w/v.

The clear concentrate, cooled in a water stream, was acidified under stirring to a pH 3, with H_2SO_4 and the stirring continued for 5 hours while the pH was maintained at 3. Because of the slight water solubility of L-aspartic acid, the acid crystallizes and precipitates. The precipitation was completed by bringing the cooling temperature to 5°C and maintaining the stirring for 10 more hours. The precipitate was recovered by washing the collected crystals with water at +5°C. The crystals obtained were dried under vacuum at 50°C. The weight of the recovered crystals was 342 grams (% recovery 83.75). The purity of the crystals was 92%.

ASPARTYL PEPTIDE AND AMIDE SWEETENING AGENTS—PREPARATION

Previous attempts to produce artificial sweeteners have involved the use of saccharins and/or cyclamates. In addition, problems have arisen with the use of these compounds, specifically, the saccharins which leave a somewhat bitter aftertaste in the mouth.

It has recently been found that certain dipeptide compounds possess an intense sweetness level. Examples of these compounds are set forth in U.S. Patents 3,475,403 and 3,492,131. Most suitable among these compounds are the lower alkyl esters of aspartylphenylalanine wherein the stereochemical configuration is L-L, DL-DL, DL-L or L-DL. Illustrative of the lower alkyl esters are methyl, ethyl, propyl, butyl, pentyl, hexyl, heptyl and the branched chain groups isomeric therewith, with the methyl ester being the most preferred embodiment.

These dipeptide materials would appear to have great commercial applications as sweetening agents since they range up to 200 times the sweetness level of sucrose, have no discernible unpleasant aftertaste, and can be produced from naturally occurring amino acids. Problems have arisen, however, with the use of these compounds in that their rate of solution into water is markedly slower than sucrose, and that the compounds tend to decompose, losing substantially all their sweetness upon heating at temperatures above about 80°C.

α-L-Aspartyl-L-Phenylalanine Methyl Ester

An improved process for the preparation of α-L-aspartyl-L-phenylalanine methyl ester is disclosed by *G.W. Anderson; U.S. Patent 3,901,871; August 26, 1975; assigned to American Cyanamid Company.*

Esters such as α-L-aspartyl-L-phenylalanine methyl ester have previously been prepared by the reaction of aspartic anhydride hydrochloride with methyl-L-phenylalaninate in the presence of ethylene dichloride and carbon dioxide and recovering the product. It has been found that when the filter cake from the above reaction is slurried with isopropanol, the product recovered and recrystallized from about equal portions of isopropanol and water, a pure product in high yields is obtained. The product is obtained in yields at least 5% greater than prior methods; less procedural steps are involved in obtaining a substantially pure product. The reaction of L-aspartic anhydride hydrochloride with methyl L-phenylalaninate in acetic acid and ethylene dichloride gives an α,β mixture of compounds as shown below which is separated through the hydrochloride salt which is then neutralized with sodium carbonate and recovered by subsequent column chromatography. This process involves many steps, a number of which are eliminated in this process.

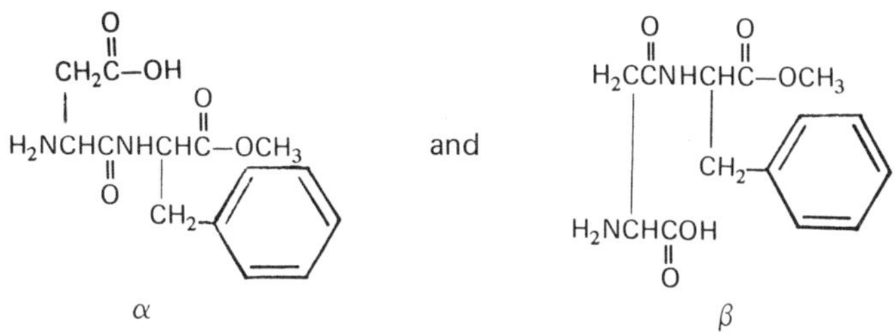

In carrying out this process, it is preferable that methyl-L-phenylalaninate be present in quantities at least four times that of aspartic anhydride hydrochloride. This is desirable in order to neutralize the hydrochloride salt and to obtain a greater quantity of the α-compound over the β-compound. The α-compound is the desired product and is usually obtained in quantities of about five times that of the β-compound.

In carrying out the process of reacting methyl-L-phenylalaninate with aspartic anhydride hydrochloride a by-product consisting essentially of methyl-L-phenylalaninate hydrochloride is obtained. Since the latter compound is the more expensive component of the reaction, it can be recycled to produce a further quantity of the desired end product.

When the reaction product of methyl-L-phenylalaninate and aspartic anhydride hydrochloride is diluted with a small amount of water, the product precipitates. On slurrying with isopropanol, the isopropanol removes undesirable by-products including starting materials. When the product is crystallized from isopropanol-water, substantially pure product is obtained. This process eliminates the necessity of converting the impure product to a salt, crystallization of the salt, neutralization, desalting and further crystallization as described in the prior art.

Example: Preparation of α-L-Aspartyl-L-Phenylalanine Methyl Ester — A mixture of 9.22 grams aspartic acid, 7.88 grams of phosphorus trichloride and 32.8 grams of glacial acetic acid is mechanically stirred at room temperature overnight. The mixture is protected from moisture during the operation. The mixture is filtered under nitrogen and the residue washed with 16 ml of acetic acid and then with 38 ml of ethylene dichloride. The residue is either suspended in 100 milliliters of ethylene dichloride to be added subsequently to methyl phenylalanine or stored in the dry state in a vacuum desiccator. The aspartic anhydride hydrochloride can be added in the dry state. Either mode of addition is satisfactory.

A portion of 54.2 grams of methyl phenylalaninate hydrochloride, 55 g of sodium carbonate in 150 ml water and 500 ml of ethylene dichloride are shaken together. The water layer is withdrawn and the ethylene dichloride layer mixed with 100 grams of sodium sulfate. A second extraction of the aqueous layer is made with 100 ml of ethylene dichloride. The two ethylene dichloride solutions are combined. The sodium sulfate is removed by filtration and the residue washed with 100 ml of ethylene dichloride. The washings and filtrate are combined (total volume 700 ml).

The solution containing methyl-L-phenylalaninate (prepared above) is cooled to -10°C, carbon dioxide supplied from solid carbon dioxide is passed into the stirred solution, and the aspartic anhydride hydrochloride (prepared above) added in one portion. The mixture becomes clear, then after about 15 minutes becomes opaque and increasingly viscous. Two hours after the addition of the anhydride, the cooling bath is removed, 6.0 ml of water added and the mixture stirred for an additional 15 minutes. The gelatinous solid is then isolated by filtration. The wet cake is slurried with 100 ml of isopropanol, filtered and washed with 50 ml of isopropanol (residue A).

The ethylene dichloride filtrates are shown to have H·Phe·OMe and moderate amounts of α- and β-aspartylphenylalanine methyl esters by thin layer chromatography (CMW system chloroform 40, methanol 10 and water 4). A small amount of precipitate is obtained by the addition of 2 ml of water and refrigeration. This is combined with the precipitate formed in the isopropanol washes (overnight at room temperature), washed with several small portions of isopropanol and dried in a steam oven (residue B) weight 3.21 grams.

Residue A is stirred vigorously in a beaker with 250 ml isopropanol, collected, washed with 500 ml isopropanol and dried, weight 12.0 grams. Thin layer chromatography shows both residues A and B to be principally the desired α-dipeptide with possibly a small amount of β, and small amounts of H·Phe·OMe, H·Asp·OH, and an unidentified material (R_f 0.1). The second isopropanol filtrate (250 + 50 ml) has a strong H·Phe·OMe spot at R_f 0.98 on TLC, with a lesser α-dipeptide (about 0.6).

Residues A and B are combined and recrystallized from 250 ml of 50-50 isopropanol-water, including wash, to give 8.36 grams (oven dried) of product, melting point 248° to 253°C (brown), softens about 190°C. An authentic sample has the same melting point but with less softening at 190°C. Thin layer chromatography in (CMW) showed only very faint H·Phe·OMe and R_f 0.1 impurities when very heavily spotted.

In the BAW system (40 n-butanol-4 HOAc-12 H_2O) a strong α-dipeptide spot at R_f 0.39 with a trace impurity at 0.25 (corresponding to the 0.1 in CMW) was obtained. The addition of 100 ml isopropanol to the recrystallization filtrate gives 0.44 gram of crystalline β-isomer on refrigeration for a week. Concentration of the filtrates to 50 ml and refrigerating yielded 1.80 grams of a mixture of α and β.

L-Aspartyl Dipeptides of Substituted Alanines and Lysines

Amino acids substituted with cycloalkenyl or phenyl groups have been used by *J.T. Sheehan and M.A. Ondetti; U.S. Patent 3,978,034; August 31, 1976; assigned to E.R. Squibb & Sons, Inc.* to form dipeptides with aspartic acid.

These artificial sweeteners have the formula:

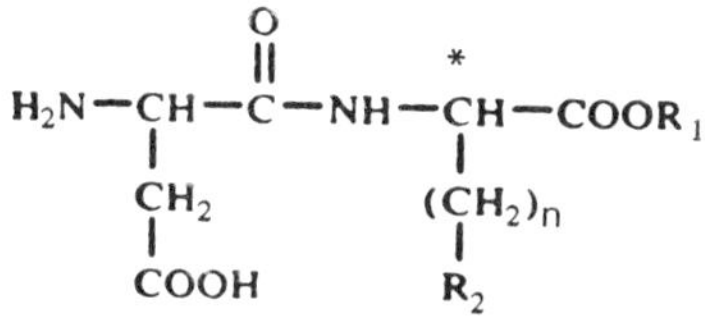

where n is an integer from 0 to 5, R_1 is an alkyl radical of up to 6 carbon atoms, an alkylaryl or alicyclic radical of up to 10 carbon atoms, and R_2 is a cycloalkenyl radical, and, in addition, may be phenyl when n is 0. The carbon marked with an asterisk is always of the L-configuration.

These dipeptide sweetening agents are stable, water-soluble substances which may be used under various physical forms, for example, as powders, tablets, syrups, and so forth. They may also be used in liquid or solid carriers, such as water, glycerin, sorbitol, starch, salt, citric acid and other appropriate nontoxic substances.

While it is not possible to predict whether a given dipeptide structure will taste sweet or even have a taste at all, it has been determined that the sweet tasting property is influenced by the stereochemistry of the amino acids constituting the structure of the dipeptide. In order to obtain the maximum sweetening effect, the carbon atom marked with an asterisk in the above formula is always of the L-configuration. The following examples illustrate the preparation of these dipeptides.

Example 1: 2,5-Dihydro-L-Phenylalanine Methyl Ester Hydrochloride — Thionyl chloride (15 ml) is added to methanol (150 ml) while stirring in a dry ice bath. The mixture is allowed to come to room temperature and 2,5-dihydro-L-phenylalanine (5.0 grams) is added. After refluxing for 30 minutes, the solvent is removed in vacuo and the residue is crystallized from ether and then recrystallized from acetonitrile. Yield: 4.95 grams.

Example 2: β-tert-Butyl-N^{α}-tert-Butyloxycarbonyl-L-Aspartyl-2,5-Dihydro-L-Phenylalanine Methyl Ester — β-tert-butyl-N^{α}-tert-butyloxycarbonyl-L-aspartic acid p-nitrophenyl ester (26.7 grams) is added to a mixture of 2,5-dihydro-L-phenylalanine methyl ester hydrochloride (11.8 grams), dimethylformamide (50 ml) and triethylamine (7.6 ml). After two hours at room temperature the reaction mixture is diluted with ethyl acetate and washed with 20% aqueous citric acid, water, saturated sodium bicarbonate and water. After drying over magnesium sulfate the solvent is removed in vacuo. Yield: 13.2 grams.

Example 3: L-Aspartyl-2,5-Dihydro-L-Phenylalanine Methyl Ester — β-tert-butyl-N^{α}-tert-butyloxycarbonyl-L-aspartyl-2,5-dihydro-L-phenylalanine methyl ester (13 grams) is dissolved in trifluoroacetic acid (80 ml) and the solution kept under nitrogen for one hour. The solvent is concentrated in vacuo and the residue is triturated with ether-hexane until it becomes solid. This material is dissolved in water (30 ml) neutralized with DEAE Sephadex (OH cycle), filtered and the filtrate concentrated in vacuo to a small volume (5 ml). The dipeptide ester is crystallized on standing in the refrigerator.

In a similar manner: L-aspartyl-L-(1-cyclohexene-1)-alanine methyl ester was prepared from the protected dipeptide β-tert-butyl-Nα-tert-butyloxycarbonyl-L-aspartyl-L-(1-cyclohexene-1)-alanine methyl ester in turn prepared from 1-cyclohexene-1-L-alanine methyl ester hydrochloride; L-aspartyl-L-phenylglycine methyl ester was prepared from β-benzyl-Nα-tert-butyloxycarbonyl-L-aspartyl-L-phenylglycine; and L-aspartyl-L-2,5-dihydrophenylglycine methyl ester was prepared from β-tert-butyl-Nα-tert-butyloxycarbonyl-L-aspartyl-L-2,5-dihydrophenylglycine methyl ester in turn prepared from L-2,5-dihydrophenylglycine methyl ester hydrochloride.

L-Aspartyl Amides of Branched Chain Amino Acids

Branched chain amino acids have been incorporated into aspartyl dipeptides in the process disclosed by *R.H. Mazur; U.S. Patent 3,799,918; March 26, 1974; assigned to G.D. Searle & Co.* These dipeptides are represented by the formula:

```
                        R'
                        |
(1)       H2NCHCONHCCOOR
              |         |
              CH2       R"
              |
              COOH
```

where R and R' are each a methyl or ethyl radical and R" is an alkyl radical containing 4 to 7 carbon atoms, e.g., butyl, pentyl, hexyl and heptyl and the branched-chain groups isomeric therewith.

An especially desirable group of compounds includes those compounds in which R and R' are methyl and R" is a branched-chain alkyl group of 5 or 6 carbon atoms, e.g., L-α-aspartyl-DL-2-amino-2,5-dimethylhexanoic acid methyl ester, advantageously used in the form of its hemihydrate, or L-α-aspartyl-L-2-amino-2,5-dimethylhexanoic acid methyl ester.

The compounds of Formula (1) are surprisingly superior, in terms of stability and sweetness potency, to the structurally related aspartyl dipeptide esters with phenylalanine.

It has been determined that the property of sweetness is affected by the stereochemistry of the individual amino acid units comprising the dipeptide. The α-aspartyl portion of the compounds of Formula (1) must have the L configuration in order for sweetness to be present. It is obvious, however, that isomeric mixtures which contain DL-α-aspartyl compounds, are also sweet and that such isomeric mixtures are within the scope of the process.

With respect to the aliphatic amino acid portion of the compounds of Formula (1), the L configuration has been found necessary for optimum sweetness. Isomeric mixtures containing, e.g., DL-aliphatic amino acid compounds, are consequently also sweet and such mixtures are also within the scope of this process. The compounds of Formula (1) can be prepared by the known methods for peptide formation such as shown in the following series of experiments.

Example 1: 57.09 parts of 5-methyl-2-hexanone is mixed with 42.3 parts of potassium cyanide and 159 parts of ammonium carbonate. That mixture and 570 parts by volume of a 1:1 by volume mixture of ethanol and water are placed in a stainless steel reaction vessel, sealed, heated to 55°C with shaking and maintained at that temperature for 24 hours. The crystalline material is removed, rinsed with methanol-denatured ethanol, filtered and dried to afford DL-5 isopentyl-5-methylhydantion, MP 154° to 159°C. That compound possesses antiulcer and antihypertensive activity.

Example 2: 55 parts of DL-5-isopentyl-5-methylhydantoin, 283 parts of barium hydroxide octahydrate and 1,000 parts of water are placed in a stainless steel reaction vessel. The vessel is sealed, heated to 100°C and maintained at that temperature for 38 hours. The reaction vessel is cooled and the contents are removed. The foamy mixture thus obtained is heated to near the boiling point and carbon dioxide is bubbled through the mixture for 1 hour. The resulting mixture is filtered to remove precipitated barium carbonate.

The filtrate is stripped and the residual slurry is shaken with methanol-denatured ethanol, then stripped of solvent. That residue is shaken and stripped in the same manner as before and the resulting residue is shaken with ether, filtered and dried. The dried material is combined with water and that mixture is heated until the solid dissolves. The solution is allowed to cool to room temperature over a 16-hour period and the crystals which form are separated by filtration and dried, affording DL-2-amino-2,5-dimethylhexanoic acid melting at 250° to 260°C with decomposition and possessing antiulcer, hypocholesterolemic and antifungal activity.

Example 3: To approximately 99 parts of methanol, cooled to –10°C, is added dropwise 7.2 parts by volume of thionyl chloride. 8.0 parts of DL-2-amino-2,5-dimethylhexanoic acid is added and the reaction mixture is allowed to reflux for approximately 3 days. The solution is then stripped, shaken with ether, stripped again and dried in vacuo. The resulting oil solidifies slowly and is then triturated with n-hexane to give DL-2-amino-2,5-dimethylhexanoic acid methyl ester hydrochloride tritohydrate. That product, which exhibits antiulcer activity, melts at 98° to 101°C and has the formula:

$$
\begin{array}{c}
CH_2CH_2CH(CH_3)_2 \\
| \\
H_2N{-}C{-}COO{-}CH_3{\cdot}HCl{\cdot}\tfrac{1}{3}H_2O \\
| \\
CH_3
\end{array}
\qquad (2)
$$

Example 4: 17.8 parts of N-benzyloxycarbonyl-L-aspartic acid β-benzyl ester is dissolved in 200 parts by volume of methylene chloride. The resulting solution is cooled to –20°C and 5.3 parts of N-methylmorpholine is added, followed by the dropwise addition of 7.0 parts of isobutyl chloroformate. The mixture is stirred and a solution containing 11.0 parts of DL-2-amino-2,5-dimethylhexanoic acid methyl ester hydrochloride, 100 parts by volume of methylene chloride and 5 parts of N-methylmorpholine is added. The resulting mixture is refrigerated for 16 hours, then filtered and stripped of solvent.

The product is dissolved in ethyl acetate, washed three times with 1 N aqueous hydrochloric acid, twice with water, twice with 2 M aqueous potassium carbonate, and four more times with water, then dried over anhydrous sodium sulfate and filtered. The filtrate is stripped to afford, as an oil, β-benzyl-N-benzyloxycarbonyl-L-α-aspartyl-DL-2-amino-2,5-dimethylhexanoic acid methyl ester. That compound possesses antiulcer activity.

Example 5: 14.75 parts of β-benzyl-N-benzyloxycarbonyl-L-α-aspartyl-DL-2-amino-2,5-dimethylhexanoic acid methyl ester is warmed in 200 parts by volume of 90% aqueous acetic acid until dissolved. 1.4 parts of palladium black catalyst is added and the mixture is hydrogenolyzed at room temperature and a pressure of approximately 4 atmospheres for 2½ hours. The catalyst is removed by filtration and washed on the filter with 90% aqueous acetic acid. The combined filtrates are stripped to dryness, the residue is stripped once with water and the oil thus obtained is triturated with ether and then dried under reduced pressure to remove final traces of acetic acid. That glass-like material is again dissolved in ether and allowed to crystallize.

The crystals are removed by filtration and washed with anhydrous ether, which causes the crystals to soften. That gummy material is allowed to harden, thus affording L-α-aspartyl-DL-2-amino-2,5-dimethylhexanoic acid methyl ester hemihydrate, specific rotation +8.3° in acetic acid at 25°C and is represented by the structural formula:

```
                    CH2CH2CH(CH3)2
                    |
(3)   H2NCHCONHCCOOCH3·½H2O
          |          |
          CH2        CH3
          |
          COOH
```

L-Aspartyl-L-Aspartic Acid Esters

Esters of dipeptides of L-aspartic acid with another molecule of L-aspartic acid or aminomalonic acid have been developed by *N. Nakajima, H. Aoki, M. Fujino, O. Nishimura, M. Wakimasu and M. Mano; U.S. Patent 3,801,563; April 2, 1974; assigned to Takeda Chemical Industries, Ltd., Japan* as sweetening agents. These are compounds of Formula (1):

```
           COOH        COOR1
           |           |
           CH2         (CH2)n
           |           |
(1)   H2N—CH—CONH—C*H—COOR2
          (L)
```

where R^1 is a branched or cyclic alkyl group of 3 to 8 carbon atoms, R^2 is a lower alkyl group of 1 to 2 carbon atoms, n is an integer 0 or 1 and C* represents that the radicals attached to the carbon atoms show an L configuration when n is 1. All of these compounds have an excellent ability for sweetening foods.

These compounds can be prepared as shown by the following example, or by other known methods of peptide synthesis such as by the reaction of the appropriate, 2,5-oxazolidinedione with the amino group of the other amino acid.

Example: Production of L-Aspartyl-L-Aspartic Acid α-Methyl β-tert-Butyl Diester — 2.5 parts by weight of L-aspartic acid α-methyl β-tert-butyl diester hydrochloride is suspended in 100 parts by volume of dichloromethane, followed by the addition of 1.54 parts by volume of triethylamine, 3.6 parts by weight of N-benzyloxycarbonyl-L-aspartic acid benzyl ester and 2.3 parts by weight of dicyclohexylcarbodiimide under cooling with ice. The reaction mixture is stirred at room temperature overnight and filtered to remove dicyclohexylurea. The filtrate is washed with 100 parts by volume of a 1 N hydrochloric acid and with 100 parts by volume of a 4 weight percent aqueous sodium bicarbonate solution, and concentrated. The residue is crystallized by the addition of 50 parts by volume of petroleum ether. The crystals are recrystallized from ethyl acetate-petroleum benzin to obtain N-benzyloxy-carbonyl-(β-benzyl)-L-aspartyl-L-aspartic acid α-methyl β-tert-butyl diester melting at 76° to 77°C. The yield is 2.6 parts by weight (48%). $[\alpha]_D^{23}$ = –3.6° (c = 1, methanol).

2.0 parts by weight of N-benzyloxycarbonyl-(β-benzyl)-L-aspartyl-L-aspartic acid α-methyl β-tert-butyl diester is dissolved in 50 parts by volume of methanol, followed by the addition of 0.2 part by weight of palladium black. Hydrogen gas is passed vigorously through the mixture under stirring for 4 hours and after the hydrogenation the reaction mixture is filtered. The filtrate is concentrated under reduced pressure at room temperature to obtain L-aspartyl-L-aspartic acid α-methyl β-tert-butyl diester as needles melting at 123° to 125°C $[\alpha]_D^{23}$ = 5.5° (c = 1, methanol).

Compounds of Formula (1) where the groups are defined by the following table were tested for sweetness.

R^1	R^2	n	Configuration of the C* Atom in Above Formula	Abbreviation of the Compound
1-Ethylpropyl	CH_3	0	Not fixed	A
Cyclopentyl	CH_3	0	Not fixed	B
Cyclohexyl	CH_3	0	Not fixed	C

Each of the test compounds was dissolved and diluted with water and the sensory thresholds of the compounds were measured by the maximum dilution technique. Discriminable concentration in distilled water: 0.0004% A, 0.0003% B and 0.0002% C. Concentration at which the compound tastes sweet: 0.0010% A, 0.0008% B and 0.0005% C.

Derivatives of L-Aspartic Acid Amides

Derivatives of aspartic acid amides, many having N-trifluoroacetyl groups have been disclosed by *M. Lapidus and M.M. McGettigan; U.S. Patents 3,818,077; June 18, 1974; and 3,725,453; April 3, 1973; both assigned to American Home Products Corporation* for use as artificial sweeteners. Overall, the compounds of this process have the following formula:

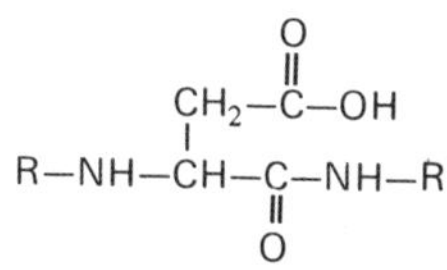

where R is hydrogen, trichloroacetyl or trifluoroacetyl; and R^1 is phenyl, p-chlorophenyl, p-fluorophenyl, p-bromophenyl, p-cyanophenyl, L-1-(lower)alkoxycarbonyl-2-phenethyl, L-1-(lower)alkoxycarbonyl-2-(p-hydroxyphenyl)ethyl or L-1-(lower)alkoxycarbonyl-3-(methylthio)propyl, with the proviso that when R is hydrogen, R^1 is phenyl, p-chlorophenyl, p-fluorophenyl, p-bromophenyl or p-cyanophenyl; and the nontoxic pharmaceutically acceptable salts of these compounds.

Of particular interest are compounds where R is Cl_3CCO— or F_3CCO— such as N-(L-α-carboxyphenethyl)-3-(2,2,2-trifluoroacetamido)-L-succinamic acid, N-methyl ester; and the dicyclohexylamine salt; L-3-(2,2,2-trifluoroacetamido)succininanilic acid; and L-4'-chloro-3-(2,2,2-trifluoroacetamido)succinanilic acid; and the dicyclohexylamine salt.

Also of interest are compounds where R' is phenyl or the p-substituted phenyls such as L-3-amino-4'-chlorosuccinanilic acid and L-3-amino-4'-cyanosuccinanilic acid.

These compounds are prepared by reacting an appropriate L-aspartic anhydride with an amine meeting the requirements of the formula as shown in the following examples.

Example 1: N-Trifluoroacetyl-L-Aspartic Acid Anhydride — L-aspartic acid, 212.8 grams (1.6 mols) is placed in a 2 liter round bottom flask fitted with a stirring motor, dropping funnel, and water condenser. Flask is cooled in a dry ice-acetone bath and trifluoroacetic anhydride, 840 grams (4 mols), is added with continuous stirring of the mixture over a period of 45 minutes. Bath is removed and the contents of flask are permitted to warm up. Reaction becomes vigorous at 14°C with the evolution of heat.

After reaction has subsided, the contents of the flask are refluxed for two hours and then cooled to room temperature. The contents of the flask are transferred to a beaker, washed with 1 liter of dry petroleum ether and the product collected. Washing is repeated again with petroleum ether and finally with ether. Finally, the product is pulverized and dried overnight in vacuo at 40°C. Yield: 333 grams (98 %); MP 134° to 135°C.

Example 2: L-4'-Chloro-3-(2,2,2-Trifluoroacetamido)Succinanilic Acid, Dicyclohexylamine Salt — Dissolve 10.5 grams (0.05 mol) of N-trifluoroacetyl-L-aspartic acid anhydride and 6.4 grams, (0.05 mol) of p-chloroanaline in 100 ml of dry tetrahydrofuran. After standing overnight at room temperature, the solvent is removed. The residue is dissolved in methanol, reacted with 9 grams (0.05 mol) of dicyclohexylamine, and the product crystallized from methanol-isopropyl ether. Yield: 9.7 grams (38%); MP 217°C with decomposition.

Esters of L-Aspartyl-D-Alaninol

Sweeteners which are stable in hot water and do not leave an unpleasant aftertaste have been synthesized by *I. Chibata, M. Miyoshi, H. Ito, T. Fujii and K. Kawashima; U.S. Patent 3,971,822; July 27, 1976; assigned to Tanabe Seiyaku Co., Ltd., Japan.* These compounds have the formula:

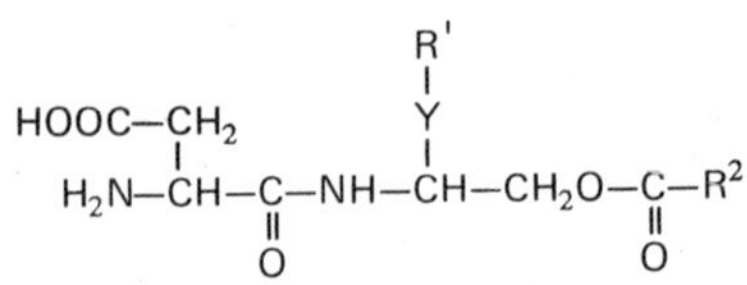

where R' is hydrogen or hydroxy, R^2 is alkyl of one to five carbon atoms, alkenyl of two to three carbon atoms, cycloalkyl of three to five carbon atoms or methylcycloalkyl of four to six carbon atoms, and Y is alkylene of one to four carbon atoms.

Among the above ester compounds those having L-D or L-DL stereochemical configuration have specifically potent sweetening properties. The ester compound can be prepared by condensing aspartic acid with

$$H_2N-\underset{}{\overset{\overset{\displaystyle R^1}{|}}{\overset{\displaystyle Y}{|}}}{C}H-CH_2OH$$

where R^1 and Y have the same meaning as defined above, followed by subjecting the resultant compound to esterification with a compound represented by the formula: R^2COOH where R^2 has the same meaning as defined above.

Each of the compounds shown in the following table was dissolved in water until the sweetness of the solution reached a comparable level to an aqueous solution containing 4% of sucrose. Sweetness of the solutions were judged by a taste panel consisting of 10 persons. Potency of sweetness of the compounds tested was compared by relative ratio of the concentration to 4% aqueous sucrose solution. The results are shown in the following table in terms of sweetness ratio which was calculated by:

$$\text{Sweetness ratio} = \frac{\text{Concentration of sucrose solution}}{\text{Concentration of test compound at which the sweetness is comparable to sucrose solution}}$$

Test compound No.	$-Y-R^1$	R^2	Stereochemical configuration	Sweetness ratio
1	$-CH_3$	$-CH_3$	L-D	75
2	$-CH_3$	$-CH_2CH_3$	L-D	120
3	$-CH_3$	$-CH_2CH_2CH_3$	L-D	60
4	$-CH_3$	$-CH(CH_3)_2$	L-D	145
5	$-CH_3$	$-C(CH_3)_3$	L-D	240
6	$-CH_3$	$-CH_2CH(CH_3)_2$	L-D	50
7	$-CH_3$	$-\underset{CH_3}{\underset{\mid}{C}}HCH_2CH_3$	L-D	150
8	$-CH_3$	$-\underset{CH_3}{\underset{\mid}{C}}HCH_2CH_2CH_3$	L-D	50
9	$-CH_3$	$-CH=CH_2$	L-D	100
10	$-CH_3$	$-\underset{CH_3}{\underset{\mid}{C}}=CH_2$	L-D	130
11	$-CH_3$		L-D	200
12	$-CH_3$	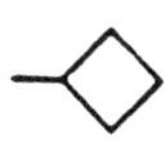	L-D	220

(continued)

Test compound No.	—Y—R¹	R²	Stereochemical configuration	Sweetness ratio
13	$-CH_3$		L-D	60
			L-D	60
14	$-CH_3$	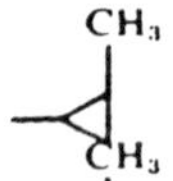	L-D	160
15	$-CH_3$	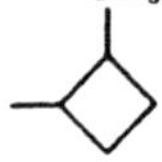		
16	$-CH(CH_3)_2$	$-CH_3$	L-D	60 - 75
17	$-CH(CH_3)_2$	$-CH_2CH_3$	L-D	60 - 75
18	$-CH_2OH$	$-CH_2CH_3$	L-D	160
19	$-CH_2OH$		L-D	120
20	$-CH_2OH$	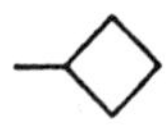	L-D	65

Example 1: 4.8 grams of N-benzyloxycarbonyl-β-benzyl-L-aspartic acid α-p-nitrophenyl ester were added to 30 ml of dimethylformamide containing 0.75 gram of D-alaninol and 1.5 grams of N-hydroxybenzotriazole at a temperature below 15°C. The mixture was stirred at 20°C for 30 minutes. The reaction solution was poured into ice-water and the mixture was extracted with ethyl acetate. The extract was washed with 2% hydrochloric acid, 5% sodium carbonate, 3% aqueous ammonia and an aqueous sodium chloride solution, successively. Then, the solvent was evaporated from the extract under reduced pressure. The crystalline residue thus obtained was recrystallized from a mixture of ethyl acetate and petroleum ether, whereby 3.0 grams of N-benzyloxycarbonyl-β-benzyl-L-aspartyl-D-alaninol were obtained. Yield: 72.5%; MP 110°-111°C.

Example 2: 3.0 grams of N-benzyloxycarbonyl-β-benzyl-L-aspartyl-D-alaninol were dissolved in 20 ml of dried pyridine. 1.75 grams of acetyl chloride were added dropwise to the solution at a temperature below 15°C. The mixture was stirred at 15° to 20°C for 2 hours. Pyridine was removed by evaporation from the reaction solution under reduced pressure and the residue was extracted with ethyl acetate. The extract was washed with 2% hydrochloric acid, 5% sodium carbonate and an aqueous sodium chloride solution, successively. Then, the extract was dried and concentrated under reduced pressure. The crystalline residue thus obtained was recrystallized from a mixture of ethyl acetate and petroleum ether, whereby 2.2 grams of the ester of N-benzyloxycarbonyl-β-benzyl-L-aspartyl-D-alaninol and acetic acid were obtained. Yield: 66.5%; MP 101°-103°C.

Example 3: 1.8 grams of the ester of N-benzyloxycarbonyl-β-benzyl-L-aspartyl-D-alaninol and acetic acid were dissolved in 50 ml of methanol and 0.1 gram of 5% palladium-charcoal was added to the solution. Hydrogen gas was introduced into the mixture until the absorption of hydrogen was completed. Then, the catalysts were removed by filtration and the filtrate was evaporated to remove solvent under reduced pressure. The residue thus obtained was recrystallized from a mixture of methanol and ether. 0.76 gram of the ester of L-aspartyl-D-alaninol and acetic acid was obtained. Yield: 82%; MP 146°-159°C.

L-Aspartyl-L-Phenylglycine Esters

Compounds possessing sweetening properties are provided by *C.L. Moriarty and G.L. Tritsch; U.S. Patent 3,972,860; August 3, 1976* which are the lower alkyl esters of L-aspartyl-L-phenylglycine. Of these, the methyl ester is preferred. Such compounds are useful for sweetening foods and beverages, and also other materials intended to be taken into the mouth and in which taste is important, e.g., chewing gum, dentifrices, mouth washes. Also disclosed are methods for the manufacture of such compounds and related materials of peptide structure.

An amino blocking group, such as a carbobenzoxy group is used to protect the free amine of an amino acid having another carboxylic acid amide group, after which the compound resulting is reacted with the amino group of an amino acid ester and the product resulting has the carbobenzoxy group removed and the other amide group converted to the corresponding acid. In a preferred embodiment the amino acid in which the amino group is protected is L-asparagine, and the other amino ester is an L-phenylglycine ester.

These sweeteners are generally of relatively low acidity, with a 1% solution usually usually exhibiting a pH of 3 to 7. Accordingly, they are suitable for sweetening various acidic beverages, such as soft drinks, ades, tea and coffee, as well as more neutral beverages, including milk and noncarbonated flavored water. Although highly basic or acidic aqueous media tend to promote hydrolysis of ester bonds, the present compounds are of sufficient stability to make them acceptable as sweeteners for the usual beverages.

Example: 1.31 parts L-asparagine are stirred with 100 parts water, and 1.7 parts carbobenzoxy chloride and 2.5 parts of 4 N NaOH are added. After 1 hour of stirring, the mixture is extracted with ether, the aqueous layer is acidified and the product is filtered off. It is recrystallized from hot water and the purified product, carbobenzoxy L-asparagine, melts at 163° to 165°C (uncorrected).

2.65 parts of L-phenylglycine, obtained by resolution of the D,L acid with hog kidney acylase, are suspended in 100 parts methanol, cooled to -10°C, and an excess, 1.6 parts of thionyl chloride, is added dropwise. The mixture is allowed to warm to room temperature and is boiled at reflux for 2 hours. The solvent is evaporated in vacuo and the product, L-phenylglycine methyl ester hydrochloride, is crystallized from ethanol. It has a melting point of 156° to 157°C.

2 parts of the L-phenylglycine methyl ester hydrochloride are added to 50 parts of dimethylformamide. 1.4 parts of triethylamine are added and after cooling to 0°C, 2.66 parts of carbobenzoxy L-asparagine, dissolved in 50 parts dimethylformamide, are mixed with the ester solution. While maintaining cooling with an ice bath, 2.06 parts of dicyclohexylcarbodiimide, dissolved in 10 parts of dimethylformamide, are then reacted with the mixture. After 10 minutes, the ice bath is removed and the solution is allowed to reach room temperature, and is maintained there overnight. The formed by-product, dicyclohexyl urea, is evaporated in vacuo. The product is filtered off and washed with cold water, followed by a cold 5% $NaHCO_3$ aqueous solution, cold 10% HCl aqueous solution and cold water.

The washed product, carbobenzoxy-L-asparaginyl-L-phenylglycine methyl ester, is then dried and recrystallized from hot ethyl acetate. It is a white powder with melting point of 191° to 194°C, obtained in almost stoichiometric amount.

The carbobenzoxy group is removed by hydrogenating 1 part of the above dipeptide in 50 parts methanol with 0.1 part of palladium black on charcoal catalyst at atmospheric pressure and room temperature for 2 hours. After filtration and evaporation, the product is suspended in 50 parts of 0.2 N HCl and heated at 100°C for 2 hours. The resulting L-aspartyl-L-phenylglycine methyl ester is obtained in almost stoichiometric amount. Its MP is 88°C (decomposes). The composition is verified by amino acid analysis. It is a free-flowing white powder, about 100 to 1,000 times as sweet as sucrose, with a slight citrus flavor. It is not bitter and gives no aftertaste.

The corresponding lower aliphatic esters of L-aspartyl-L-phenylglycine, e.g., the ethyl and isopropyl esters are prepared by similar methods. The ethyl and isopropyl esters may be used as flavoring agents for foods, beverages and oral preparations but the methyl ester is much sweeter, being estimated to be from 100 to 1,000 times as sweet as sucrose. The sweetness of the methyl ester is a pleasant one, having a very slight citric flavor, making it ideal for use as a sweetener for soft drinks.

Aspartyl Amides of Aliphatic, Cycloaliphatic and Phenylalkyl Amines

Three series of aspartic acid amides, termed 3-amino-N-succinamides, have been developed by workers at Searle. These compounds have use as sweetening agents and as anti-inflammatory agents.

The first group of compounds as disclosed by *R.H. Mazur, J.M. Schlatter and A.H. Goldkamp; U.S. Patent 3,803,223; April 9, 1974; assigned to G.D. Searle & Co.* can be represented by the formula:

$$\mathrm{HOOC{-}CH_2{-}\underset{\displaystyle NH_2}{\underset{|}{CH}}{-}\underset{\displaystyle NRR'}{\underset{|}{C}}{=}O}$$

where R is hydrogen or a methyl radical and R' is C_nH_{2n+1} and n a positive integer greater than 6 and less than 9.

The second group of compounds disclosed by *A.H. Goldkamp, R.H. Mazur and J.M. Schlatter; U.S. Patent 4,011,260; March 8, 1977; assigned to G.D. Searle & Co.* has the formula:

$$\mathrm{\underset{\displaystyle COOH}{\underset{|}{CH_2}}{-}\overset{\displaystyle NH_2}{\overset{|}{CH}}{-}\overset{\displaystyle O}{\overset{\|}{C}}{-}\overset{\displaystyle R}{\overset{|}{N}}{-}Alk{-}R'}$$

where R is hydrogen or a lower alkyl radical, Alk is a lower alkylene radical and R' is a carbocyclic radical such as cyclohexyl.

The third group of compounds disclosed in *A.H. Goldkamp, R.H. Mazur and J.M. Schlatter; U.S. Patent 4,025,551; May 24, 1977; assigned to G.D. Searle & Co.* has the formula shown on the following page. In the formula Alk is an alkylene radical containing 2 to 6 carbon atoms and X and Y are selected from hydrogen, halogen, hydroxy and lower alkoxy, with the proviso that when Alk contains 2 carbon atoms, then the phenyl ring must be substituted by at least one halogen, hydroxy or lower alkoxy radical.

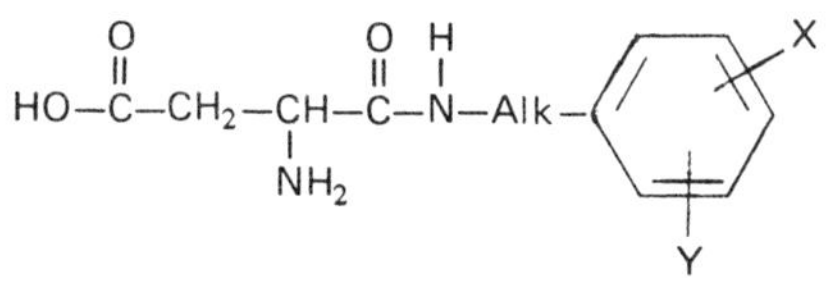

Particularly preferred compounds of this process are those where Alk is an alkylene radical containing 3 to 6 carbon atoms and X and Y are H.

These compounds are conveniently obtained by a two-step process involving coupling of a suitable L-aspartic acid derivative, where protecting groups are attached to the amino and β-carboxy groups, with the appropriate amine, followed by removal of the protecting groups, suitably by catalytic hydrogenolysis.

Examples of suitable starting materials are β-benzyl-N-benzyloxycarbonyl-L-aspartate and active esters thereof such as N-benzyloxycarbonyl-L-aspartic acid α-p-nitrophenyl, β-benzyl diester. When the starting material contains the free α-carboxy group, the reaction is carried out in the presence of an alkyl chloroformate and a suitable organic base.

Removal of the protecting groups of the aforementioned intermediates is most conveniently achieved by catalytic hydrogenolysis, preferably by use of a palladium black catalyst. These steps are illustrated in the following examples.

Example 1: To a solution of 4.53 parts of DL-1-methyl-4'-hydroxyphenethylamine in 22.5 parts of dimethylformamide is added 15.06 parts of N-benzyloxycarbonyl-L-aspartic acid α-p-nitrophenyl, β-benzyl diester and the resulting reaction mixture is stirred for a few minutes until homogeneous, then is allowed to stand at room temperature for about 16 hours.

At the end of that reaction period the mixture is diluted with ethyl acetate, then washed successively with dilute hydrochloric acid, water, dilute aqueous potassium bicarbonate and dilute aqueous sodium sulfate, and dried over anhydrous sodium sulfate, then stripped of solvent by distillation under reduced pressure. The residue thus obtained is triturated with cyclohexane to afford the crude product, benzyl-3-benzyloxycarbonylamino-N-DL-1'-methyl-4"-hydroxyphenethyl-L-succinamate which, after recrystallization from cyclohexane, melts at about 106° to 117°C and exhibits an optical rotation, in methanol, of -9.5°.

Example 2: A mixture containing 4.9 parts of benzyl-3-benzyloxycarbonylamino-N-DL-1'-methyl-4"-hydroxyphenethyl-L-succinamate, 0.5 part of palladium black and 120 parts of methanol is shaken with hydrogen at room temperature and approximately 4 atmospheres pressure until 2 molecular equivalents of hydrogen have been absorbed. The catalyst is then removed by filtration and the methanol distilled under reduced pressure to afford the residual crude product. Purification of that material by trituration with ether yields pure 3-amino-N-DL-1'-methyl-4"-hydroxyphenethyl-L-succinamic acid, which softens at about 160°C and exhibits an optical rotation, in water, of +5°.

An alternate process for production of these compounds uses the N-carboxy anhydride of L-aspartic acid as the starting material. That substance is prepared from L-aspartic acid by reaction with phosgene in dioxane according to the general procedure described by A.C. Farthing, *J. Chem. Soc.,* 1950, 3213. The N-carboxy anhydride is coupled with the appropriate amine in aqueous medium at a pH of about 10 according to the general procedure described by Hirschmann et al., *J. Org. Chem.*, 32, 3415 (1967).

L-ASPARTYL-L-PHENYLALANINE SWEETENERS—FORMULATIONS

The lower alkyl esters of aspartylphenylalanine would appear to have great commercial applications as sweetening agents since they range up to 200 times the sweetness level of sucrose, have no discernible unpleasant aftertaste, and can be produced from naturally occurring amino acids. Problems have arisen, however, with the use of these compounds in that their rate of solution into water is markedly slower than sucrose, and that the compounds tend to decompose, losing substantially all their sweetness upon heating at temperatures above 80°C. The following processes disclose formulations developed to improve the solubility, stability and sweetness of these dipeptides.

Drying Processes for Increased Solubility

In the process developed by *M. Glicksman and B.N. Wankier; U.S. Patent 3,761,288; September 25, 1973; assigned to General Foods Corporation* readily soluble materials are obtained by co-drying solutions of the dipeptide with an edible bulking agent.

Surprisingly, these aqueous solutions may be dried, such as by drum drying and spray drying, at conditions of elevated temperatures (i.e., above 100°C without significant adverse effects on the sweetening level of the dipeptide compounds.

The solid edible bulking agents suitable for use in this process may be any of the known nontoxic substances or combination of substances used for this purpose including the organic acids such as citric, adipic, fumaric acids, and the hydrolyzed starch materials such as dextrins and sugars. Especially suitable among these solid bulking agents are the dextrins of the class commonly known as corn syrup solids, since with certain of these materials it is possible to produce a low calorie, low bulk density, nonhygroscopic sweetening composition which has the appearance of table sugar and which will dissolve readily in cold water.

The dextrin material used for the production of these sugar substitutes must dissolve easily in water to produce a clear solution and must be nonhygroscopic. In this regard the DE (dextrose equivalent) of the dextrin material has been found to be a critical parameter. The dextrin material must have a sufficiently low molecular weight to be easily soluble in water and to produce a clear solution so that the final product will have the essential reflecting surfaces in order to give the appearance of a crystalline product. On the other hand the molecular weight of the dextrin material must be high enough so that hygroscopicity is avoided.

Accordingly, it has been determined that for the production of table sugar substitutes the dextrin material should have a DE preferably in the range of about

5 to 10. Additionally it has been found that the best results are obtained if the dextrin material contains little or no monosaccharide (i.e., glucose) and contains an irregular distribution of the other lower (one to eight saccharide units) saccharides with a preponderance of the hexamer and heptamer. Such corn syrup dextrins have been produced by means of enzymatic hydrolysis of starch and are typified by the products available as Mor-Rex.

Example 1: Three grams of citric acid and one gram of L-aspartyl-L-phenylalanine methyl ester are dissolved in 50 ml of water with stirring. The resulting solution is spread on a stainless steel tray (2.1 ft^2) and allowed to dry at ambient conditions for about two days. The dry material was then scraped from the tray and ground with a mortar and pestle. One-half gram samples of this ground material were added with stirring, to beakers containing 200 ml of water at 40°F. The material completely dissolved in an average time of about 55 seconds yielding solutions which were sweet with a slight acid taste.

Example 2: A solution was prepared containing 800 grams of water (80°F), 241 grams of 5 DE Mor-Rex and 5.95 grams of L-aspartyl-L-phenylalanine methyl ester. This solution was placed in a tray at a 1.5 inch thickness and freeze dried in a Stokes Freeze Drier for 48 hours. The material was then ground to a fine powder using a Waring blender at a high speed.

Tests have shown that varying the level of dipeptide in the sweetening compositions up to the level of about one part dipeptide per part of bulking agent does not have any appreciable or predictable effect on the rate of solution. All samples prepared in accordance with this process were found to dissolve in water as cold as 40°F in less than 2 minutes, whereas complete solution of equivalent amounts of the dipeptide material taken alone requires a time of about 30 min.

It has also been found that the elevated temperatures, in excess of 100°C, which are employed during the spray drying and drum drying operation do not affect the sweetness of the final product.

M. Glicksman and B.N. Wankier; U.S. Patent 4,007,288; February 8, 1977; assigned to General Foods Corporation have also used vacuum drum drying to produce readily soluble sweetening compositions.

According to this process, a table sugar substitute, having the appearance of crystalline sucrose and having a rate of solubility greater than if spray or freeze-dried, may be prepared by forming a uniform aqueous solution of a corn syrup dextrin material and a dipeptide sweetening agent and vacuum drum drying this solution in a manner so as to yield an expanded product which may then, if necessary, be ground to size. It is preferred that the vacuum drum drying process be conducted at controlled low temperatures not only because it serves to further reduce the bulk density of the composition, thus permitting use of higher molecular weight dextrins, but due to the fact that the dipeptide sweeteners tend to undergo partial degradation when exposed to increased temperatures.

From the production standpoint, the absence of air increases the efficiency of the process in that foaming commonly associated with dry artificial sweeteners, when placed in solutions, is eliminated. Thus, this process which functions to dry aqueous dipeptide sweetening solutions at temperatures below 100°C is accomplished by introducing water at below boiling point, or vacuum steam, into the rolls of the vacuum drum dryer. The roll temperature of the dryer may vary from about 24° to almost 100°C with a preferred range of 40° to 80°C. The pressure within the vacuum chamber of the drum dryer preferably should be maintained from 50 to 400 mm mercury to insure low temperature drying and to properly degas the material being dried.

Example: Four grams of dextrin having a DE of 10 and 3 grams of L-aspartyl-L-phenylalanine methyl ester are dissolved in 100 ml of water and maintained at 115°F. The solution is mixed at this temperature until it loses its milky appearance and becomes essentially clear.

Prior to introducing the feed solution into the vacuum drum dryer itself, ambient temperature water is circulated through the rolls of the drum dryer and the chamber evacuated to about 24 mm Hg pressure, the water then having a temperature of 24°C. When the solution is completely dry, the material is removed from the chamber and particulated by passing it through a 20 mesh screen.

Readily Soluble Dipeptide-Acid Formulations

It has been disclosed by *J.H. Berg and J. Trumbetas; U.S. Patent 3,868,472; February 25, 1975; assigned to General Foods Corporation* that dipeptide sweeteners, specifically L-aspartyl-L-phenylalanine methyl ester, exhibit improved solubility when placed in intimate contact with an acid. Explicitly, L-aspartyl-L-phenylalanine methyl ester demonstrates a significantly increased rate of solubility when coground with an acid in the presence of an organic solvent in which the dipeptide is insoluble, after which the solvent is evaporated to produce an intimate mixture, this method serving to effectively coat the surface of the low-solubility dipeptide with an enclosing, soluble, transport layer of acid.

Due to the fact that this method involves a type of commingling of the solid acid and dipeptide, there is no chemical interaction between the same, i.e., there is no protonation of the dipeptide due to the absence of protons in the dry system and ostensibly no salt formation. Consequently, the dispersibility of the dipeptide is increased while at the same time remaining structurally independent. Essentially, therefore, the rate of solubility of the dipeptide is increased by relying upon the innate solubility of the acid, the acid and dipeptide being in sufficient proportion to one another to permit the acid to coat the dipeptide thereby serving to partially or completely enclose the same in a soluble transport system.

The acids preferably used in this process are those dry acids which are suitable for incorporation into foodstuffs intended for human or animal consumption such as malic acid, citric acid or the like. However, the process is not circumscribed by the solid food acids as such and is limited only to those dry acids having sine qua non, sufficient solubility.

The components in the range of 1:2 to 1:4 weight ratio of sweetener to acid, 1:3 being preferred for better solubility, are coground in a volatile organic solvent in which the dipeptide is insoluble for a period of time sufficient to render a partial or complete coating of the sweetener with the acid.

Examples of such solvents are chloroform, ether, ethanol or the like, ethanol being preferred among these. The solvent is subsequently evaporated off to afford an intimate mixture. The organic solvent functions both to reduce the electrostatic forces between the container vessel and the ground material therein and in some cases, such as in the case of citric acid which has a slightly soluble character in organic solvents, to partially solubilize the acid thus effecting a more uniform coating of the dipeptide.

Example 1: Malic acid and L-aspartyl-L-phenylalanine methyl ester (1:4 weight ratio of sweetener to acid) were ground together in ethanol using a mortar and pestle for approximately 1 hour. The ethanol was subsequently evaporated off to afford a completely dry mixture.

Approximately 4.5 grams of the mixture was placed in a typical beverage formulation, whereby it comprised approximately 90% of the solid formulation of which 1.0 gram or 0.5% of the dry mixture was the dipeptide. In comparison to the two minutes solubility time of sucrose in the beverage system, the dipeptide commingled with the acid was found to dissolve completely in 25 seconds.

Example 2: Three samples of L-aspartyl-L-phenylalanine methyl ester were prepared and the following systems evaluated for their solubility, the overall concentration of L-aspartyl-L-phenylalanine methyl ester being in all three instances, 1%. The solubility tests were carried out in room temperature water with the same amount of stirring being applied to each sample.

Samples	H_2O Solubility
L-aspartyl-L-phenylalanine methyl ester, min	4.5
L-aspartyl-L-phenylalanine methyl ester + malic acid (1:3) added separately but simultaneously, min	4.25
L-aspartyl-L-phenylalanine methyl ester coground in ethanol with malic acid (1:3), sec	35

Effervescent Dipeptide Formulations

An effervescent dipeptide sweetener tablet for use primarily as a table-top sweetener in beverage systems has been developed by *T.P. Finucane; U.S. Patent 4,009,292; February 22, 1977; assigned to General Foods Corp.*

A readily soluble form of L-aspartyl-L-phenyl methyl ester (APM) is combined with an effervescent calcium carbonate system and citric acid. When compressed into table form, this combination produces an effervescent APM tablet having superior solubility and stability and is devoid of characteristic lingering sweet aftertaste.

A preferred embodiment of this process is an APM tablet formulated by combining a 4:3:1.5 weight ratio of anhydrous citric acid to dry carbonation system (50% $CaCO_3$/50% Mor-Rex) to soluble APM (60% APM/40% Mor-Rex). The Mor-Rex is a 10 DE hydrolyzed cereal solid available commercially as Mor-Rex 1918. Typically, this means that an APM tablet of a size conventionally associated with table-top tableted sweeteners (7/32 x 1/4 inch diameter) will contain about 0.08 gram anhydrous citric acid, 0.06 gram of the dry carbonation system (0.3 gram $CaCO_3$/0.3 gram Mor-Rex) and 0.03 gram of the APM system (0.018 as long as the above considerations are adhered to; namely, that (1) there is an less than 1,000 psig is required, but the tablet size is not critical. Nonappreciable variances from the above ratios and amounts of each component are not critical as long as the above considerations are adhered to; namely, that (1) there is an effective amount of citric acid present to react with the calcium carbonate and improve upon the solubility of the APM system, and (2) the APM and Mor-Rex slurry is homogenized and dried prior to combining with the dry carbonation system in order to shield the APM from reacting with the calcium carbonate during storage.

Example: An aqueous solution containing 60 parts APM and 40 parts 10 DE Mor-Rex was co-spray-dried in a Niro Spray Dryer at an air pressure of 5.2 kg/cm^2, an air inlet temperature of 160°C, and an air outlet temperature of 75°C with a rate of solution flow of 15 cc/min. 0.030 gram of the resultant dry material was set aside for incorporation into the tablet.

50.0 grams Purecal U calcium carbonate (average primary particle size being 33 to 40 millimicrons) is mixed with 50.0 grams Mor-Rex 1918 (10 DE hydrolyzed cereal solids) in 100.0 grams water. The colloidal mill gap is then set at No. 4 and the mixture recycled for 5 minutes to produce a milky dispersion of calcium carbonate. This dispersion is spray-dried in a Niro dryer at an inlet temperature of 175°C. The resulting dry powder is passed through a 140 U.S. mesh screen to remove large agglomerates and contaminants. The sample is then ground in a Mikro-Sauplmill hammer mill with 0.01 herringbone slot retaining screen insert and passed through a 400 U.S. mesh screen (0.038 mm) to obtain a dry carbonation source for use in this process.

0.060 gram of the above carbonation source was then dry blended with the 0.03 gram of the APM system and 0.080 gram of anhydrous citric acid. After blending to obtain a homogenous dry powdered mixture, the total mix was placed in a die having a ¼ inch diameter.

The powder prior to compression occupied a height of $2^{25}/_{32}$" and after compression at 50 lb/in^2 (1,000 spi), occupied a height of $2^{18}/_{32}$". The tablet had a dimension of $^{7}/_{32}$" x ¼" diameter. The tablet dissolved instantly in one 8-ounce cup of hot coffee and in less than 1 minute in the same amount of cold water (44°F) with a minimum of stirring.

Encapsulation with Dextrin

A dry, stable form of the dipeptide sweetener useable in foods such as dry beverage mixes has been developed by *L.D. Pischke and M.D. Shoaf; U.S. Patents 3,962,468; June 8, 1976 and 4,059,706; November 22, 1977; both assigned to General Foods Corporation.* A slurry-forming substantial level of discretely sized hydrated L-aspartic acid sweetening compound, preferably one milled to reduce the particle size of undissolved ester crystals, is encapsulated in a dextrinous water-soluble agent by drying a uniform dispersion as by spray drying to yield a high-density readily soluble powderous sweetening composition which is readily blendable and uniform in sweetening power and resists caking.

The dextrin material used in this process should dissolve easily in water and preferably produce a clear solution and should be relatively nonhygroscopic. In this regard the dextrin should have a sufficiently low molecular weight to be easily soluble in water and yet be high enough so that hygroscopicity is avoided in large measure. A dextrin material having a dextrose equivalency (DE) in solution of 5 to 15 is useful. It is believed that best results will be obtained if dextrin contains little or no monosaccharides, i.e., glucose, and contains a predominant (major) distribution of oligosaccharides, i.e., saccharides having 1 to 8 saccharide units, with a significant presence of hexamers and heptamers; a useful dextrin is an enzymatic corn syrup hydrolyzate (Mor-Rex).

Example: A 25% solids slurry is prepared for processing through a Fryma mill, the stones of the mill being adjusted to a spacing of about 75 microns. A solution is prepared consisting of 60% APM solids and 40% corn syrup solids (maltodextrin), 10 DE. The dispersion at 70°F enters the mill and issues at a temperature of approximately 85°F, the mill being jacketed with water maintained at 160°F. The relatively cool aerated slurry is collected in a stainless steel kettle at a density of 0.75 g/cc and is pumped by a positive displacement pump and fed under 250 psig spray pressure through a 40/27 nozzle and produces droplets which when dried have a size of 50 to 300 microns. The droplets are contacted by inlet drying air approximately 330°F entering a 16 ft diameter vertical drying tower, the air being fed at 14,000 cubic feet per minute and the outlet air temperature being at 210°F.

The droplets dry as spherical beads collected at a moisture content generally less than 5% and typically 2 to 3%. The dry particles have the following particle size distribution:

Particle Size (U.S. Standard Sieve)	Weight %
+50	2.57
-50+70	27.33
-70+120	45.25
-120+140	10.60
-140+200	11.38
-200+300	1.74
10 Pan	1.13

The particles have a uniform particle size distribution, are free-flowing, white, and dissolve in 60 seconds in 45°F water when in the presence of a beverage acid such as citric acid at pH 3.5. The droplets had a loose bulk density of 0.246 g/cc and packed density of 0.282 g/cc with percent packing after tapping of 12.8%.

The particles can be readily blended to a flowable condition with beverage mix ingredients such as dry citric acid or equivalent acidulent powder, flavors and colors and will be stable against caking when packaged in a water vapor transmission packaging barrier such as a polyethylene-coated foil having a water vapor transmission of about 0.04 gram per 100 in^2 in 24 hours at 100°F (95% RH).

Dipeptide Formulations with Improved Sweetening Properties

The dipeptide sweetener although about 150 times as sweet as sugar, does not have the natural sweetness taste of ordinary sugar. The dipeptide sweetener exhibits a flat sweetness along with a slight bitter aftertaste. Moreover, when used for the sweetening of beverages, for example, the sweet taste is slightly delayed.

M.E. Eisenstadt; U.S. Patent 3,875,312; April 1, 1975; assigned to Cumberland Packing Corporation has disclosed methods for mixing these dipeptides with other materials to produce a composition having the more natural sweetness of sucrose.

These compositions are prepared by mixing (a) the dipeptide sweetner with (b) cream of tartar and/or sodium bicarbonate or potassium bicarbonate, and also with (c) lactose and/or dextrose in a ratio of one part of (a) to 0.1 to 10 parts of (b), most preferably 0.5 to 5 parts of (b) and with 5 to 30 parts of (c), most preferably 10 to 20 parts of (c). All parts herein are parts by weight.

When these components are used in these proportions, the most desired effect of sweetness approaching that of natural sugar without any flatness of taste and without any cloying sweetness or any bitter aftertaste is obtained.

It is also possible to substitute up to 50% by weight of the dipeptide sweetener with a saccharin artificial sweetener. It is noted that too great an amount of lactose can result in the masking of the sweetness of the dipeptide sweetener as well as the masking of the taste of the food or beverage to which it is applied.

In percentages by weight, the composition of the process consists essentially of 80 to 98% of lactose and/or dextrose (preferably 90 to 97%), 1 to 4% of the dipeptide sweetener (up to 50% of which may be substituted by saccharin), preferably 2 to 3%, and about 0.1 to 6%, preferably 1 to 2% of any one or a combination of cream of tartar, sodium bicarbonate and potassium bicarbonate.

Example:1: 37.5 pounds of lactose (powdered), 3 pounds of dipeptide sweetener (aspartyl phenylalanine methyl ester) and 0.5 pound of cream of tartar are thoroughly mixed to provide a uniform mixture.

The resulting mixture is approximately ten times as sweet as sugar and can be used in the place of sugar to give a sweetening effect with low caloric intake. This composition can be used to sweeten beverages or in cooking, in all quantities, even to highly sweeten beverages, without causing any bitter aftertaste and without adversely affecting the taste of the food or beverage to which it is applied.

Example 2: A sweetening composition is prepared as in Example 1, however using 20 pounds of dextrose and 17.5 pounds of lactose. This composition may be used in the same manner as Example 1.

In further work by *M.E. Eisenstadt; U.S. Patent 3,875,311; April 1, 1975; assigned to Cumberland Packing Corporation,* a sweetening composition is prepared from (a) the dipeptide sweetener mixed with (b) cream of tartar and/or sodium bicarbonate and/or potassium bicarbonate, and also with (c) glucono-delta-lactone and/or sodium gluconate and/or potassium gluconate in a ratio of one part of (a) to 1/25 to 6 parts of (b), most preferably 1/4 to 2 parts of (b) and with 5 to 50 parts of (c), most preferably 10 to 20 parts of (c). All parts herein are parts by weight.

It is possible to substitute up to 50% by weight of the dipeptide sweetener with a saccharin artificial sweetener in these formulations. It is noted that too great an amount of (c) can result in a composition which has a tendency towards sourness and even more important will actually cause milk to curdle so that the composition cannot be used for the sweetening of coffee, for example. It is therefore essential that the amount of (c) be within the above limits. Furthermore, if (c) were used alone in these compositions then even the maintaining of (c) within the above limits would not be sufficient to obtain a complete masking of the undesired unnatural sweetness of the dipeptide sweetener without the disadvantages of possible sourness and curdling of milk. For this reason, it is necessary to use (b) along with (c) in the compositions. However, if (b) were used alone, that is without (c) there would be no masking of the undesired taste characteristics of the dipeptide sweetener.

In percentages by weight, these compositions consist essentially of 2 to 12% of (a), preferably 5 to 7%, 0.5 to 12% of (b), preferably 2 to 8%, and 65 to 97% of (c), preferably 82 to 92%.

Example 1: 53.9 pounds of glucono-delta-lactone, 36.3 pounds of sodium gluconate, 5.5 pounds of dipeptide sweetener, 1.07 pounds of cream of tartar and 3.23 pounds of sodium bicarbonate are thoroughly mixed to provide a uniform mixture.

The resulting mixture is approximately ten times as sweet as sugar and can be used in the place of sugar to give a sweetening effect with no carbohydrate or caloric intake whatsoever. This composition can be used to sweeten beverages or in cooking, and all quantities, even to highly sweetened beverages, without causing any bitter aftertaste and without adversely affecting the taste of the food or beverage to which it is applied. Furthermore, the composition does not curdle milk.

Example 2: 6.2 pounds of dipeptide sweetener, 85.8 pounds of glucono-delta-lactone and 8.0 pounds of potassium bicarbonate are thoroughly mixed to provide a uniform mixture. The resulting mixture can be used in the same manner as the composition of Example 1.

ASPARTIC ACID—OTHER DERIVATIVES

Preparation of Asparagine and Glutamine

An improved process for the production of asparagine and N-substituted glutamine derivatives has been developed by *P. Hirsbrunner and R. Bertholet; U.S. Patent 3,979,449; September 7, 1976; assigned to Société d'Assistance Technique pour Produits Nestle SA, Switzerland.* This process enables an asparagine or a glutamine, that is a compound corresponding to formula (1), with the exception of glutamine itself, to be obtained in high yields in only a few hours. The process is distinguished by the fact that a compound corresponding to formula (2)

(1) 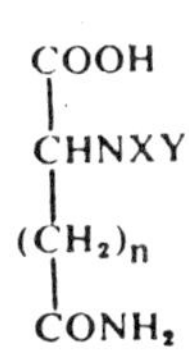

(2) 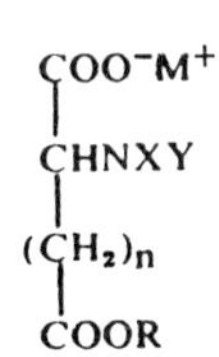

in which n is 1 or 2, M^+ is a cation, X and Y are any substitutents that will not interfere with the reaction, except in cases where n = 2 when they cannot be hydrogen atoms, and R is a C_{1-3} aliphatic group, is reacted with ammonia in aqueous medium and the compound corresponding to formula (1) subsequently liberated by acidification of the reaction medium.

Similar prior processes have acid alcoholic solutions of ammonia and required days to obtain acceptable yields. The conditions used in this process are shown in the following examples.

Example 1: 18.3 grams of β-methyl-L-aspartate hydrochloride are dissolved in 80 ml of a solution containing 20 grams of ammonia. After reaction at ambient temperature for 3 hours, during which the reaction mixture is slowly stirred, the excess ammonia is removed under a partial vacuum of 12 mm Hg. The pH-value is then reduced to pH 5.4 by the addition of 10 grams of 32% hydrochloric acid. The L-asparagine is precipitated in monohydrate form. 200 grams of methanol are then added to the mixture, followed by cooling to 0°C. The crystals are collected by filtration, washed with 20 ml of a saturated (approximately 2%) aqueous solution of L-asparagine and then dried at 70°C. L-asparagine monohydrate is thus obtained in a yield of 12 grams (80%). The product is a white crystalline powder; $[\alpha]_D^{20}$ = +34.1° (c = 5; 6 N HCl); MP = 235°C; which characteristics correspond to a purity of at least 99%.

Example 2: 10 grams of γ-methyl-L-glutamate are dissolved in 70 ml of methanol and 30 ml of water, followed by the addition of 10 ml of acetic anhydride. After reaction at 25°C for 2 hours, during which the reaction mixture is stirred, the pH-value of the medium is increased to pH 8 by the addition of 10 ml of concentrated ammonia. The methanol is then removed by evaporation under a partial vacuum of 50 mm Hg and at a temperature of 30°C, and 10 grams of ammonia gas are dissolved in this solution, freed from methanol, by controlled bubbling through under atmospheric pressure.

After stirring for 5 hours at 25°C, the excess ammonia is removed at 30°C under a partial vacuum of 50 mm Hg. The pH value is then reduced to 2.5 by the addition of 12 ml of 50% sulfuric acid, resulting in the precipitation of α-(N-acetyl)-L-glutamine. The α-(N-acetyl)-L-glutamine is then collected by filtration, washed three times with 10 ml of cold water and then oven dried. 10 grams of a white crystalline powder are obtained (yield 85%); $[\alpha]_D^{20}$ = -12.2° (c = 3; H_2O); MP = 194.5°C; which characteristics correspond to a purity of at least 99%.

Asparagine Derivatives Useful for Treating Hair

Compounds useful for softening and conditioning hair have been developed by *G. Kalopissis, A. Viout and G. Vanlerberghe; U.S. Patents 3,873,688; March 25, 1975; and 3,534,032; October 13, 1970 both assigned to L'Oreal, France.* These compounds also possess remarkable detergent properties, and show a very pronounced thickening and gelling effect.

These asparagine derivatives are compounds having the formula

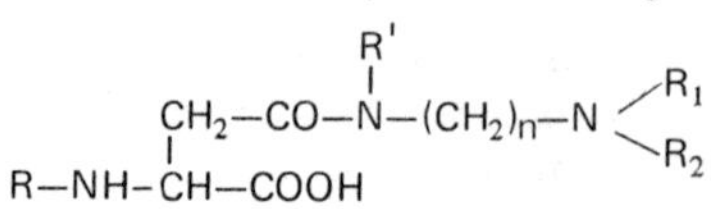

where R and R' are each alkyl or alkenyl of 10-18 carbons which can be the same or different; R_1 and R_2 are lower alkyl, which can be the same or different, or together with the nitrogen atom to which they are attached constitute a morpholino residue and n is 2-5, and Na, K, NH_4 and acid addition salts, such as lactate, phosphate, citrate and acetate salts, thereof. Preferred compounds are those in which R_1 and R_2 are C_{1-4} alkyl. The compounds can be prepared by condensing maleic anhydride with an alkylene diamine having the formula: $R'NH-(CH_2)_n-NR_1R_2$. where R', R_1, and R_2 and n have the meanings given above, modifying the reaction product to obtain the corresponding Na, K or NH_4 salt, and reacting this salt with a primary fatty amine of the formula: $R-NH_2$ where R has the meaning given above.

Excellent results have been obtained using as primary amine a mixture of fatty amines derived from copra acid or tallow acid, which yields a mixture of compounds of formula (1) differing only in the value of the substituent R, the various R substituents being the various alkyl radicals present in tallow or copra acids respectively. Preparation of the sodium salt of N^2-[alkyl(tallow)]-N-(dodecyl)-N-(N',N'-diethylaminoethyl)asparagine is described below.

Example 1: (1) Preparation of N'N'-Diethyl-N-Dodecyl-Ethylenediamine — There are mixed with stirring 2 mols of dodecyl bromide and 10 mols of N',N'-diethylethylenediamine (96.6%) the temperature being held below 50°C. After several hours, the reaction is complete. There are then added 4.3 mols of potassium hydroxide in pellet form, with vigorous stirring. The excess of potassium hydroxide and the potassium bromide formed are separated by filtration. The excess of N',N'-diethylethylenediamine is then evaporated under reduced pressure, and the crude product obtained is washed with water and dried in vacuo on the water bath. N',N'-diethyl-N-dodecyl-ethylenediamine is isolated in an almost quantitative yield and in a purity of 98%.

(2) Preparation of N-Dodecyl-N-(N',N'-Diethylaminoethyl)Maleamic Acid — One mol of N',N'-diethyl-N-dodecyl-ethylenediamine is dissolved in 300 ml of ethyl acetate. To this solution is added with stirring, in an hour and a half, 1 mol of maleic anhydride in solution in 300 ml of ethyl acetate, the temperature being maintained between 35° and 40°C. The solid which precipitates is separated, washed with ethyl acetate and dried in vacuo, and N-dodecyl-N-(N',N'-diethylaminoethyl)maleamic acid is isolated as a white, water-soluble powder in a yield of 80%.

(3) Addition of the Primary Amine Derived From Tallow Fatty Acids — 0.7 mol of N-dodecyl-N-(N',N'-diethylaminoethyl)maleamic acid is dissolved in 300 ml of ethyl alcohol, and the solution is neutralized with a calculated quantity of concentrated aqueous sodium hydroxide solution. There is then added 0.7 mol of a mixture of primary amines derived from tallow fatty acids, and the mixture is heated at 65°C for 11 hours. After evaporation of the alcohol, the product obtained takes the form of an oil dispersible in water, which contains 4% of unreacted primary amine.

Example 2: There is prepared by conventional procedure cream containing: 5 grams sodium salt of N^2-[alkyl(tallow)]-N-(ethyl)-N-(N',N'-diethylaminoethyl)-asparagine, 3 grams cetyl alcohol, 3 grams of the condensate of 1 mol of stearyl alcohol and 10 mols of ethylene oxide, and 70 cc water. The mixture is allowed to cool to 40°C and acidified by the addition of citric acid until a pH of 4 is reached, whereafter it is made up to 100 cc with water.

From 5 to 10 grams of this cream are applied to hair sensitized by a discoloration, which is massaged for several minutes and then rinsed. The hair thus acquires softness and a gloss and is easy to comb.

Aspartic Acid Salts for Treatment of Fingernails

Ungual lamellar dystrophy, or splitting fingernails, is a not uncommon condition which occurs mainly in women. It is characteristically manifested by a layered splitting, like birch bark, with flaking at the ends of the nails. Typically, the splitting extends proximally one to three millimeters. The cause of ungual lamellar dystrophy is unknown and, heretofore, no drug or vitamin has been found effective in treating the condition.

Soluble salts of aspartic acid have been disclosed by *D.W. Aiken; U.S. Patent 3,904,765; September 9, 1975* for effectively treating ungual lamellar dystrophy. The pharmaceutical value of alkali and alkaline earth metal aspartates has previously been recognized in U.S. Patent 3,009,859 as antifatigue drugs and also beneficial in treating ailments evidencing a fatigued or exhausted condition of cells, such as liver deficiency, and the sequelae or splanchic vasomotor disturbance, e.g., due to shock from injury or severe burns, states of respiratory deficiency, myocardial infarction, and other states. When used as an antifatigue composition the aspartate may be employed as an aqueous solution containing about 10% by weight aspartate salt and optionally containing from 1 to 5 parts by weight per 1,000 parts by weight aspartate salt of adenosine triphosphate.

Alternatively, the aspartates may be administered orally in solid form such as powder or tablets. One preferred tablet formulation comprises essentially 250 milligrams magnesium aspartate and 250 milligrams potassium aspartate and includes small quantities, e.g., 2 milligrams, of adenosine triphosphate.

The aspartate compositions taught in U.S. Patent 3,0009,859, and particularly the magnesium-potassium aspartate combination, have been found to be effective in the treatment of ungual lamellar dystrophy.

Example 1: Ten women with ungual lamellar dystrophy were treated with a mixture of potassium and magnesium aspartates. The dosage of each woman was two 500 milligram tablets (each containing 250 milligrams of potassium aspartate and 250 milligrams of magnesium aspartate) taken orally 3 or 4 times daily, preferably after meals and at bedtime. The nails of each of the women were periodically examined. In all but one patient a response was noted in 3 to 4 weeks (the precise time of response being impossible to determine in view of the slow growth rate of nails) with an observable cessation of splitting.

Inasmuch as normal regrowth of an entire fingernail usually takes 6 to 12 weeks it is believed that the aspartate composition functioned either by accelerating fingernail growth or by influencing the nail structure at a point distal to the germinal growth bed, or both. Following the favorable response, adminstration of the aspartate composition in some of the patients was terminated. Splitting of the nails returned in each case.

Example 2: Two teenage girls whose nails were neither cut nor bitten, yet were very short, suffered from fingernail hypoplasia, i.e., their nails did not grow at all. The girls were treated by orally administering to each two 500 mg tablets as in Example 1. In one patient all fingernails (and toenails) began growing in a short time. In the other patient some nail growth also occurred, but it was asymmetric.

The observed effectiveness of aspartate salts, and particularly magnesium and potassium aspartates, in treating ungual lamellar dystrophy, and the favorable growth response in the patients with hypoplastic nails has never before been reported or suggested in the medical literature. Moreover, the known use of these aspartate compositions, as an antifatigue drug, has no apparent relevance whatever to their use in preventing nail splitting and in promoting growth of fingernails. Further, there have been clinical indications that these same aspartate compositions may be effective in controlling splitting ends of hair and in accelerating hair growth rate.

GLUTAMIC ACID—PREPARATIVE PROCESSES

This five carbon amino acid can also be named 2-aminopentadioic acid, α-aminoglutaric acid or 1-aminopropane-1,3-dicarboxylic acid. It is one of the nonessential amino acids and is best known in the form of its sodium salt which imparts meat flavor to food.

Using Mutants Resistant to Sulfa Drugs

F. Yoshinaga, T. Tsuchida, K. Kikuchi and S. Okumura; U.S. Patent 3,886,039; May 27, 1975; assigned to Ajinomoto Co., Inc., Japan found that mutant strains having resistance to sulfa drugs are derived from parent strains having glutamine-producing ability, the amount of glutamine produced being considerably increased.

The microorganisms employed in this method may combine a nutrient requirement and/or resistance to other reagents with resistance to sulfa drugs. The sulfa drugs to which the mutants employed in this process are resistant, have the following general features and are generally known as antagonists to p-aminobenzoic acid.

(1) They contain in their molecule a group of the general formula:

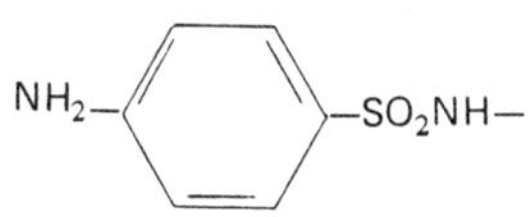

and have antimicrobial action as a general rule. (2) The antimicrobial action against usual wild strains is suppressed by addition of p-aminobenzoic acid. Such sulfa drugs having the foregoing features include

sulfapyridine
sulfathiazole
sulfadiazine
sulfaguanidine
sulfadimethoxine
sulfamethomidine
sulfamethoxypyridazine
sulfisoxazole
acetosulfamine
sulfanylamide
sulfisomezole
sulfamethizole
sulfaethidole
sulfapyrazine
irgafen
irgamide

The most effective glutamine-producing mutants found so far are *Brevibacterium flavum* FERM P-1684, *Brevibacterium flavum* FERM P-2371, *Corynebacterium glutamicum* FERM P-2372, *Corynebacterium glutamicum* FERM P-2373, and *Microbacterium flavum* FERM P-2374.

Example 1: A fermentation medium of pH 7 was prepared from the following:

Glucose, %	10
Ammonium sulfate, %	1
KH_2PO_4, %	0.25
$MgSO_4 \cdot 7H_2O$, %	0.04
$FeSO_4 \cdot 7H_2O$, mg %	1
Thiamine hydrochloride, γ/l	350
Biotin, γ/l	5
Soybean protein hydrolyzate, ml/dl	0.5

300 ml batches of the medium were introduced in glass jar fermentors, and were sterilized by steam. The media were inoculated with *Brevibacterium flavum* FERM P-1684 (mutant strain having resistance to sulfaguanidine), which had previously been cultured on bouillon slants for 24 hours at 31.5°C. The fermentation was carried out at 1,200 rpm at 31.5°C for 30 hours while aerating with ¼ volume of air per minute and volume of broth. The pH of each medium was held at about 6.5 by addition of gaseous ammonia. 3.9 g/dl of L-glutamine were found in the broth after 30 hours' cultivation.

The cultured broth was centrifuged to remove the cells, and one liter of the supernatant was stripped of L-glutamine by passage over a column of an ion exchange resin. 15.5 grams of crystalline L-glutamine was recovered from the broth by a conventional method.

On the other hand, only 0.85 g/dl of L-glutamine were found in the broth in which *Brevibacterium flavum* ATCC 14067 was cultured in the same way.

Example 2: Corynebacterium glutamicum FERM P-2372, *Corynebacterium glutamicum* FERM P-2373, *Microbacterium flavum* FERM P-2374, *Brevibacterium flavum* FERM P-2371 and their parent strains *Micrococcus glutamicus* ATCC 21010, *Brevibacterium flavum* ATCC 14067 and *Microbacterium flavum* ATCC 10340 were cultured as described in Example 1.

After 30 hours' cultivation, the amounts of L-glutamine produced in the respective broths were as shown in the following table.

Microorganism Employed	L-Glutamine
Corynebacterium glutamicum FERM P-2372	3.80
Corynebacterium glutamicum FERM P-2373	3.75
Micrococcus glutamicus ATCC 21010	0.90
Microbacterium flavum FERM P-2374	3.60
Microbacterium flavum ATCC 10340	0.92
Brevibacterium flavum FERM P-2371	4.20
Brevibacterium flavum ATCC 14067	1.10

Improved Fermentation Medium Containing Ammonium Acetate

It is the object of the process developed by *G.M. Miescher; U.S. Patent 3,929,575; December 30, 1975; assigned to Commercial Solvents Corporation* to provide an improved nutrient medium for the production of glutamic acid at shortened fermentation times. It has been discovered that the presence of ammonium acetate at concentrations significantly less than taught by Phillips in U.S. Patent 3,277,625, provides a fast start of growth of the organism which leads to an earlier start of glutamic acid synthesis at a higher production rate than the previous processes. During the fermentation the carbon dioxide content of the exhaust gas is monitored to prevent it rising above 6.5% by volume. Yields of as much as 100 to 110 g/l are obtained in 30 to 35 hours. This high yield is very surprising and unexpected, especially in such a short fermentation time.

Example: Brevibacterium divaricatum NRRL B-2311 was cultivated for 16 hours at 35°C on a rotary shaker at 385 rpm in a seed culture medium of the following composition:

Glucose, g	40
K_2HPO_4, g	1
$MgSO_4 \cdot 7H_2O$, g	0.5
BYF-100 (yeast extract), g	1
Urea, g	8
Tap water, ml	1,000

A fermentation medium was prepared with the following ingredients:

Glucose, g	1,210
KH_2PO_4, g	12
K_2SO_4, g	12
$MgSO_4$ (anhydrous), g	6
$FeSO_4 \cdot 7H_2O$, ppm	6
$MnSO_4 \cdot H_2O$, ppm	6
Tap water to make, ml	4,800
Antifoam agent, ml	1

The antifoam agent used was a polypropylene glycol type compound marketed under the designation Hodag K-67. This medium was sterilized by heating for 30 minutes at 15 psig of steam. To it was added the following: 50 grams ammonium acetate and 5,000 ml tap water.

This portion was sterilized as described above and to it was added the following growth starter which had been sterilized by heating for 20 minutes at 15 psig of steam: 60 grams corn sugar molasses, 60 ml water and the pH was adjusted with NH_3 to 8.3.

The mixture prepared as above was cooled to 32° to 33°C and transferred to a fermentor equipped with an agitator, a temperature controller, an automatic pH controller system and a CO_2 detector to monitor the CO_2 content of the exhaust gas. Aeration was provided by introducing compressed, sterile air at the bottom of the fermentor through a sparger.

Agitation was begun and 600 ml of the seed culture prepared as described above was added to the fermentation medium. The pH was adjusted to 8.5 with ammonia and 6.5 ml oleic acid was added. The automatic pH control system was then set to maintain a pH of 7.8. The CO_2 content of the exhaust gas was at all times held below 6.5% (by volume) and usually not above 4.5% by adjusting the air flow. After 14 hours the temperature was raised to 38°C and additional glucose feeding was commenced when the glucose content in the medium fell below 0.5 to 2% by incrementally adding a sterilized solution of 1,600 grams of glucose dissolved in sufficient tap water to make 2,250 ml. The glucose was introduced at the rate of 5 ml at every 2 min interval to the end of the fermentation. The glutamic acid production was monitored by periodical assay. After 28.5 hours the fermentation was complete at a yield of about 100 g/l.

The foregoing experiment was repeated in all essential details except that a solution of urea, 20 grams in 5,000 milliliters of water, was substituted for ammonium acetate and 5.5 ml of oleic acid was used. The CO_2 content was not monitored and the air flow remained constant. After 29 hours of fermentation glutamic acid production was 75 g/l and at 36 hours it was only 83 g/l. Accordingly the fermentation was judged to be complete.

Use of Culture Media Containing Methanol

K. Nakayama, M. Kobata, Y. Tanaka, T. Nomura and R. Katsumata; U.S. Patent 3,939,042; February 17, 1976; assigned to Kyowa Hakko Kogyo Co., Ltd., Japan have disclosed an improved process wherein L-glutamic acid is produced in high yields from methanol. Various mutants of known methanol-utilizing microorganisms of the genera Pseudomonas and Protaminobacter, particularly, of those having an ability to produce L-glutamic acid, were examined for productivity of L-glutamic acid. As a result, it has been found that a mutant having at least one property selected from: (a) a requirement of L-methionine; (b) a requirement for L-isoleucine; (c) a requirement for L-phenylalanine; and (d) a resistance to DL-lysine hydroxamate has a remarkably improved productivity of L-glutamic acid as compared with its parent strain.

As for the parent strain's ability to produce L-glutamic acid, there is no critical lower limit. However, it is desirable that the parent strain be able to produce at least 100 γ/ml of L-glutamic acid when cultured in a medium, such as disclosed in the example below, under usual culturing conditions for 24 to 48 hr. Examples of particularly suitable parent strains are *Pseudomonas insueta* ATCC 21276, *Pseudomonas methanolica* ATCC 21704 and *Protaminobacter thiaminophagus* ATCC 21371.

A parent strain selected in a manner described above, which is capable of utilizing methanol and of producing L-glutamic acid and belongs to the genus Pseudomonas or Protaminobacter is mutated to obtain mutants having at least one property selected from: a requirement for L-methionine; a requirement for L-isoleucine; a requirement for L-phenylalanine; and a resistance to DL-lysine hydroxamate. Such mutants are then suitable for this process.

Particularly preferred mutants are *Pseudomonas insueta* K-015 ATCC 21966 (requiring L-methionine), *Pseudomonas insueta* K-038 ATCC 21967 (requiring L-isoleucine–leaky type), *Pseudomonas methanolica* LHX-8 ATCC 21968 (resistant to DL-lysine hydroxamate) and *Protaminobacter thiaminophagus* K-224 ATCC 21969 (requiring L-phenylalanine), which are derived from the above-noted three parent strains.

Example: In this example, various strains shown in the following table are used. Each of the strains is inoculated into 20 ml of a medium having the following composition and a pH of 7 in 250 ml Erlenmeyer flasks.

Methanol, ml	20
$(NH_4)_2SO_4$, g	10
KH_2PO_4, g	2
K_2HPO_4, g	7
$MgSO_4 \cdot 7H_2O$, g	0.5
$FeSO_4 \cdot 7H_2O$, mg	25
$MnSO_4 \cdot 4H_2O$, mg	8
Thiamine hydrochloride, mg	1
Biotin, μg	10
Polypeptone, g	10
Phenol red, mg	10
Water to a volume of 1 liter	

Culturing is carried out with shaking at 30°C for 3 days, while feeding 1% (v/v) of methanol after 16 hours of culturing and 2% (v/v) of methanol after 27 and 40 hours of culturing. During culturing, the pH is adjusted to around neutral with 3 N aqueous ammonia three times a day. At the completion of culturing the yield of L-glutamic acid is measured as well as the degree of growth. The results are shown in the following table.

Strains	Growth OD (660 mμ) (X40)	Yield of L-Glutamic Acid (mg/ml)
Pseudomonas insueta		
ATCC 21276 (parent)	0.240	1.2
K-015 ATCC 21966 (mutant)	0.180	9.7
K-038 ATCC 21967 (mutant)	0.150	12.9
Pseudomonas methanolica		
ATCC 21704 (parent)	0.325	0.7
LHX-8 ATCC 21968 (mutant)	0.200	7.4
Protaminobacter thiaminophagus		
ATCC 21371 (parent)	0.316	2.1
K-224 ATCC 21969 (mutant)	0.170	11.2

From the above table, it is apparent that a yield of L-glutamic acid of 5 to 10 times that obtained by the parent strains is obtained by any of the mutants.

Use of Mutants Sensitive to N-Palmitoylglutamic Acid

The process developed by *K. Takinami, T. Tanaka, M. Chiba and Y. Hirose; U.S. Patent 3,971,701; July 17, 1976; assigned to Ajinomoto Co., Inc., Japan* relates to a method for producing L-glutamic acid in a medium which contains much biotin.

With fermentation in a culture medium containing much biotin, the addition of a surfactant was necessary for suppressing the excessive biotin activity. However, the surfactant involves a significant rise in the cost of the glutamic acid.

In this process, glutamic-acid-producing mutants, sensitive to N-palmitoylglutamic acid, when cultured in a medium containing an excessive amount of biotin, permit the amount of surfactant necessary for suppressing the excessive biotin to be greatly reduced without decreasing the yield of glutamic acid.

The mutants of this process include *Brevibacterium flavum* AJ 3612 (FERM-P 2308) and *Brevibacterium lactofermentum* AJ 3611 (FERM-P 2307).

These mutants were induced from *Brevibacterium flavum* ATCC 14067 and *Brevibacterium lactofermentum* ATCC 13869 respectively. The following glutamic-acid-producing microorganisms can be used also as the parent strains:

Brevibacterium divaricatum NRRL B-2311
Brevibacterium saccharoliticum ATCC 14066
Brevibacterium roseum ATCC 13825

The method of mutation is conventional such as exposure of the parent strain to gamma-rays, x-rays, or ultraviolet light, or to chemical mutagenic agents such as N-methyl-N'-nitro-N-nitrosoguanidine. The mutants mentioned above were induced by exposure to 250 γ/ml N-methyl-N'-nitro-N-nitrosoguanidine at 30°C for 30 minutes.

Usually, the mutant sensitive to N-palmitoyl glutamic acid is also sensitive to the following compounds:

Fatty acids having 6 to 19 straight carbon chains or their salts—lauric acid, myristic acid, stearic acid, palmitic acid and so on;

Esters of the fatty acids—esters of glycerol, sorbitan, sucrose, polyethyleneglycol, polyethylenepolypropyleneglycol, polyoxyethylene sorbitan ether or polyoxyethylenepolyoxypropylene sorbitan ether.

N-acyl derivatives of α-amino acids in which acyl is the acyl group of the fatty acids, the amino acids being, glycine, alanine, valine, leucine, threonine, serine, aspartic acid, glutamic acid, ornithine, lysine, arginine, phenylalanine, tyrosine, tryptophan, methionine or other α-amino acids.

Example: An aqueous culture medium was prepared to contain, per milliliter, 100 mg cane molasses (as glucose), 1 mg KH_2PO_4, 1 mg $MgSO_4 \cdot 7H_2O$, and 0.10 μg thiamine hydrochloride, and 30 ml batches of the culture medium were placed in 500 ml flasks, and heated at 115°C for 10 minutes.

Brevibacterium lactofermentum AJ 3611, and ATCC 13869 were inoculated in the aqueous culture medium and cultured at 31.5°C with shaking. The pH of the medium was maintained within the range from 6.5 to 8.0 by feeding 40% urea solution.

When the optical density at 562 mμ of each culture medium diluted 26 times reached 0.30, polyoxyethylene-sorbitan-monopalmitate was added in the amounts as shown in the following table. After 30 hours' cultivation, the amounts of glutamic acid shown in the table were found in the culture broths.

Polyoxyethylene-Sorbitan-Mono-Palmitate Added (mg/ml)	. . . Yield of Glutamic Acid (%) . . .	
	AJ 3611	ATCC 13869
0	1.1	2.3
0.5	27.5	6.1
1.0	44.5	22.5
2.0	55.0	40.1
4.0	53.1	52.0

Use of *Micrococcus glutamicus* Mutant

The known process for producing L-glutamic acid using *Micrococcus glutamicus* VNIIGenetika 490 has low productivity per cell and the fermentation process takes relatively much time. This strain also has low resistance to biotin.

N.I. Zhdanova, L.M. Evstjugov-Babaev, R.M. Balitskaya, A.F. Sholin, T.B. Kasatkina and N.K. Kuznetsova; U.S. Patent 4,054,489; October 18, 1977 have developed a mutant of this microorganism for production of glutamic acid.

The mutant strain is *Micrococcus glutamicus* VNIIGenetika 3144 obtained by multistep selection with mutagenic factors (diethyl sulfate vapor, ultraviolet radiation) from the starting strain *Micrococcus glutamicus* VNIIGenetika 490. The selection of strains was done on a medium containing 20% by weight of molasses. Strains possessing highest productivity were selected at each step.

The mutant strain *Micrococcus glutamicus* VNIIGenetika 3144 differs from the parent strain by some valuable properties that produce a significant effect on the fermentation and isolation of L-glutamic acid, namely:

(1) The resistance of the synthesis of L-glutamic acid to the inhibiting action of biotin increases three times;
(2) The concentration of cells in a unit volume of the culture fluid decreases three times;
(3) The precipitate of the biomass can be easily separated by filtration;
(4) The productivity of a cell increases six times;
(5) L-glutamic acid is intensively synthesized at a lower (three times less) concentration of dissolved oxygen in the nutrient medium; and
(6) The fermentation process is continued for a period of time 30 to 35 hours less than in the known process.

Example: One-day-old culture of *Micrococcus glutamicus* VNIIGenetika 3144, grown on Hottinger slant agar, is used to inoculate the seeding medium in 750 ml flasks. The quantity of the nutrient medium is 75 ml, and it has the following composition, in percent by weight:

	Percent by Weight
Molasses	8
NH_4Cl	0.5
Corn steep liquor	0.3
K_2HPO_4	0.05
$MgSO_4 \cdot 7H_2O$	0.03
$CaCO_3$	1.0
Water to make 100	

The pH of the medium is adjusted with a 60% solution of NaOH to 7.2. The medium is sterilized for 30 minutes at 0.8 atmosphere. The seeding material is grown for 15 to 17 hours on a reciprocating shaker (220-240 rpm).

The seeding material (75 ml) is transferred into a 5 liter fermentation vessel containing three liters of nutrient medium of the following composition, in percent by weight. The pH of the medium 6.8 to 7.0.

	Percent by Weight
Molasses	20
KH_2PO_4	0.5
$MgSO_4 \cdot 7H_2O$	0.3
Urea	0.8
$CaCl_3$	1.0
Water to make 100	

The fermentation medium (without urea) is sterilized in the fermentation vessel at 124° to 126°C for an hour, and then cooled to 30°C.

Urea, a 40% solution in water, is sterilized separately in an autoclave at 0.5 atm for 30 minutes and added to the medium immediately before inoculation.

The fermentation process is effected at 30°C with continuous stirring (700 rpm) and aeration (0.8 volume of air per volume of medium per minute). The sulfite value is 2.4 grams O_2/l/hr. As the pH of the medium falls below 7.0, a 40% solution of urea is automatically added. In 35 hour fermentation, 37 g/l of L-glutamic acid are accumulated in the culture fluid.

The specific gravity of the culture fluid is 1.05; the refractive index 1.3600; the pH 6.8. One liter of the fluid containing 37 grams of L-glutamic acid, is heated to 50°C and calcium oxide is added in the quantity of 5 grams per liter. The fluid is stirred for 10 minutes, and then concentrated (73%) orthophosphoric acid is added gradually with stirring to adjust the pH to 5.5.

The solution is heated to 80°C for 10 minutes, then cooled to 70°C, and filtered under pressure through calico and belting fabric. The precipitate is washed with 100 ml of water and the washings are added to the main filtrate.

The optical density, D, of the filtrate is 2; λ = 535 nm. The filtrate is passed through an ion-exchange column packed with 200 ml of macroporous condensation type resin on the basis of meta-phenylene diamine, formaldehyde, and phenol (resorcinol) at a rate of 200 ml per hour. The first 70 ml portion of the solution that emerges from the column is discarded. The next portions of the solution containing L-glutamic acid, are collected together with the residual portion of the starting solution, that is displaced from the column with 300 milliliters of water acidified to pH 5 with hydrochloric acid. The clarified solution is evaporated in vacuum at a temperature of 55°C to the residual volume of 200 ml.

Concentrated hydrochloric acid is added with stirring to the evaporated solution to adjust the pH to 3.2. The solution is now cooled in a crystallization vessel to 10°C and kept for 20 hours with slow stirring. The precipitated crystals are separated on a filter, washed with water acidified to pH 3.2 with hydrochloric acid, and dried. The yield of L-glutamic acid is 31.45 grams (85%). The assay is 98%.

Crystallization Process for Recovery of Glutamic Acid

It is known that 1-glutamic acid occurs in two crystal forms, designated α and β. When a saturated aqueous solution of 1-glutamic acid at 70°C is cooled rapidly, granular or prismatic crystals, the α-form, are obtained.

α-Crystals are hexagonal in cross-section with a diameter several times their thickness. They would be ideal for industrial processes because of their ease of handling. However their formation on a large scale is difficult and they rarely occur under processing conditions.

On the other hand when a saturated solution of glutamic acid at 90°C is cooled gradually so that crystallization occurs at 40° to 50°C, needle-like or flaky crystals, the β-form, are obtained. It is the one commonly encountered in industrial processing. The β-crystals are long and flat, i.e., the width and length are much greater than the depth.

These plate-like β-crystals pack easily during filtration and hinder additional filtration. Also they do not centrifuge well because they tend to break up, pack and trap impurities in the crystal cake and are difficult to wash. Consequently there is a need for a method of changing the growth habit of glutamic acid crystals to a form which can be more easily filtered or centrifuged.

The process developed by *J.S. Gallagher; U.S. Patent 3,957,864; May 18, 1976; assigned to Commercial Solvents Corporation* changes the growth habit of the crystals of 1-glutamic acid to a form designated as the γ-form.

The glutamic acid solution used is preferably but not necessarily, a fermentation broth, filtered or unfiltered. The concentration of the glutamic acid in solution should be above 60 grams per liter for good recovery. It if is less than that, the solution is concentrated, e.g., by evaporation. The temperature is brought to about 20°C, if necessary, and the pH is adjusted to 4.5 at such a rate of acid addition that the ambient temperature is maintained. The pH adjustment can be effected with any suitable acid, sulfuric acid is usually preferred. If the pH is below 4.5, it is adjusted upward with a suitable alkaline material, ammonia is preferred.

The solution is maintained at pH 4.5 at about 20°C, with constant agitation until it is determined that seed crystals have begun to form. Then it is heated to 50°C. Care is taken to avoid raising the temperature above 55°C, because it has been found that crystals formed after heating above that temperature are the β-form. The pH is then lowered to the isoelectric point, within pH 2.8 to 3.4, preferably 3.2. The temperature is then reduced to 20°C with agitation during which time the glutamic acid continues to crystallize. Additional adjustments of pH are made as necessary to maintain the range 2.8 to 3.4. Most, i.e., about 70 to 80%, of the crystals will be of the new γ-form, with about 20 to 30% in the old, platelet β-form.

After crystallization is complete, the crystals are recovered preferably by centrifugation or filtration. The glutamic acid so recovered is then converted to food-grade monosodium glutamate by any suitable method.

GLUTAMIC ACID DERIVATIVES

N-Succinyl Glutamic Acid

Compounds useful for treating microbic infections have been disclosed by *G. Hagemann and F. Clemence; U.S. Patent 3,825,664; July 23, 1974; assigned to Roussel-UCLAF, France.* These contain as the active ingredient racemic or optically active isomers of N-succinyl-glutamic acid and its nontoxic, pharmaceutically acceptable salts with mineral or organic bases.

The usual daily dose is 0.5 to 4 grams of the active compound with an individual dose of 0.1 to 2 grams for the adult depending upon the method of administration. The detoxicant effects of the compositions are profitable in association with certain antibiotics as it reduces fatigue effects on the organism attributable to prolonged administration of the antibiotic and the liberation of toxins resulting from its action on bacteria responsible for the infection.

N-succinyl-L-glutamic acid, N-succinyl-D-glutamic acid or N-succinyl-DL-glutamic acid can be used without any difference in activity. The salts of the acids can be a salt of one or more of the carboxylic functions.

Examples of suitable nontoxic, pharmaceutically acceptable salts are alkali metal salts such as sodium or potassium, ammonium salt; alkaline earth metal salts such as calcium, strontium or magnesium; aluminum; or salts of organic bases such as alkylamines such as ethylamine, diisopropylamine, ethanolamine, diethylaminoethanol; arylamines such as N,N-dimethyl-phenylethylamine; or amino acid salts such as those of leucine, serine, valine, histidine, arginine, phenylalanine or lysine.

Example 1: Preparation of N-Succinyl-L-Glutamic Acid — Step A: A solution of 25.4 grams of ethyl L-glutamate in 100 cc of benzene was added to a suspension of 13.10 grams of succinic acid anhydride in 100 cc of benzene and the mixture was stirred for 4 hours at room temperature. The solvent was evaporated off under reduced pressure and the residue consisting of ethyl N-(hydroxycarbonylpropionyl)-L-glutamate was taken up in ethanol. The mixture was heated at reflux for 4 hours in the presence of a few drops of sulfuric acid and the alcohol was evaporated off. The residue was crystallized from isopropanol to obtain an 87% yield of ethyl N-(β-carbethoxypropionyl)-L-glutamate in the form of colorless crystals melting at 45°C. IR spectrum: Bands at 3310, 1740, 1175 and 1025 cm^{-1}.

Step B: 15 grams of ethyl N-(β-carbethoxypropionyl)-L-glutamate in 1 N sodium hydroxide were heated at 30°C for 2 hours and the resulting solution was passed through a commercial cationic sulfonic acid exchange resin (Dowex) and the water was evaporated from the solution under reduced pressure. The residue was crystallized from a benzene-ether mixture to obtain a 90% yield of N-succinyl-L-glutamic acid melting at 76°C and having a specific rotation $[\alpha]_D^{25} = -14°$ (c = 4% in water).

The product occurred in the form of colorless crystals soluble in water and alcohol and insoluble in ether. IR spectrum (potassium bromide): Bands at 2020, 1700, 1540, 1000 and 800 cm^{-1}.

Example 2: Preparation of Sodium N-Succinyl-L-Glutamate — 1.32 grams of N-succinyl-L-glutamic acid were dissolved in 30 cc of water and 5 cc of a 3 N sodium hydroxide solution were progressively added with cooling. The reaction mixture was then evaporated to dryness under reduced pressure and the residue was taken up in a minimum of hot 90% ethanol. The solution was cooled to form a crystalline precipitate which was vacuum filtered and dried to obtain 1.25 grams of sodium N-succinyl-L-glutamate.

Pharmacological Study — Lots of 20 mice were given intraperitoneal injections of 0.5 mg of endotoxin extracted from *Escherichia coli* $O_{26}B_6$ by the method of Westphal (Difco 3121). Immediately after the toxin injection, the mice were given intraperitoneally 8 mg of N-succinyl-L-glutamic acid in solution in 0.5 cc of water and the percentage of survivors was determined after 24 hours and after 7 days. The results are reported in the table.

Groups	Survivors, % 24 Hours	7 Days
Controls	30	10
Treated	80	80

Glutamic Iron Chelate Additives in Mink Food

Feeding mink with cleaned, raw marine fish of various types as single source of protein results in a hypochromic iron-deficiency anemia which again results in high kit mortality, fur depigmentation (i.e., white underfur) and greatly reduced quality of fur at the normal time of skinning. It has further been proved that a diet including boiled marine fish, on the other hand, does not provoke anemia or, occasionally, only a slight degree of anemia.

The antianemic effect of boiled fish food is assumed to be the result of denatured protein compounds such as polypeptides, preferably the lower molecular peptides such as di- and tripeptides, possibly free amino acids, which are to be found both in cooked and in raw protein food. To protect the animals against anemia by daily boiling of fish for a long period diet, with the addition of usual inorganic iron compounds ($FeSO_4$), is elaborate and time consuming and is also expensive for the breeder of fur-bearing animals.

The object of the disclosure by *F. Ender; U.S. Patent 3,911,117; October 7, 1975* is to provide a diet for fur-bearing animals, particularly mink fed with raw marine fish, which ensures a supply of organic ferric iron. The improvement comprises an organic chelate additive containing ferric iron chelated with an organic acid selected from glutamic acid and ribonucleic acid.

The results of some feeding experiments are set forth in the following table, where, as basic feed, raw, cleaned coalfish was used with various additions to the basic diet, and average values of hemoglobin value, color, underfur and fur quality, total impression at the time of skinning, are set forth.

Normal hemoglobin values on skinning are about 17.5 to 19, these being much lower during the summer. Evaluation numbers for the color of the underfur and the total impression of the fur quality are given, the higher the number, the better the quality.

It is clear from the table that the hemoglobin values for mink fed with raw, cleaned coalfish without addition to basic diet of food additives according to the process, are as far down as 14.8 to 15.5. Addition to the basic diet of ferrosulfate or ferrihydroxide has no sure, advantageous effect on this form of anemia either, while the addition of amino acid in the form of an iron complex or special ferri glutamate, or addition of lactic acid or ferridesoxyribonucleate, prevents this special form of anemia, the color of the underfur and the total impression of the fur quality being, at the same time, much higher.

Experimental Years	Basic Diet	Addition to Basic Diet	. . .Average Values at Time of Skinning . . . Hemoglobin Value	Color in Underfur	Total Impression of the Fur Quality
2	Raw cleaned coalfish	Ferridigluta-mate readily soluble	18.1	5.05	5.35
3	Raw cleaned coalfish	Ferridigluta-mate powder not so readily soluble	17.0	4.86	5.23
2	Raw cleaned coalfish	Ferridigluta-mate readily soluble + HCl or lactic acid or sour milk	18.4	5.05	5.4
1	Raw cleaned coalfish	Lactic acid	17.4	4.6	5.1
1	Raw cleaned coalfish	Ferridesoxyri-bonucleate	17.1	5.1	5.1
3	Raw cleaned coalfish	Ferrosulfate	14.9	4.0	4.6
1	Raw cleaned coalfish	Ferrihydroxide	13.6	3.0	3.9
3	Raw cleaned coalfish	–	14.8	3.4	3.9
1	Raw cleaned coalfish	–	15.5	4.1	4.4

N-Acetyl Glutamic Acid as Skin Moisturizer

Compositions containing N-acetyl glutamic acid which provide a super-moisturizing effect on the skin have been developed by *H.H. Friedman, L.K. Halik and M.H. Schwarz; U.S. Patent 3,932,622; January 13, 1976; assigned to General Foods Corporation.*

The dialkali metal salts of N-acetylglutamic acid are particularly effective, when applied to the skin in water-in-oil emulsions, in increasing the moisture level of the skin not only initially, but further, over extended periods of time. It is presently believed that the unique ability of these compounds to bind water is due to their electronic configuration when ionized.

Any of the dialkali metal salts of N-acetylglutamic acid which are suitable for cosmetic use can be used. The sodium and potassium salts, especially the potassium salts, are preferred. These salts can be easily prepared by neutralizing the acid with any suitable base containing the appropriate alkali metal cations. For example, the disodium salt can be prepared by neutralizing N-acetyl-L-glutamic acid with a slight excess of sodium hydroxide, and the dipotassium salt with a slight excess of potassium hydroxide.

These salts are used in the cosmetic composition in any amount which is effective to super-moisturize the skin, preferably, from 2 to 5%, by weight of the emulsion.

In preparing the emulsion, the oil and the water phases are both prepared separately in known manners and then combined with vigorous agitation to effect emulsification.

Example: Two moisturizing cosmetic lotion bases were prepared having the compositions shown in the table.

Ingredients	Sample A	Sample B
	. . . .parts by weight. . . .	
Oil phase		
Amerchol L-101 (lanolin alcohols extract)	1.8	1.8
Oleinate 288 (glycerol trioleate)	0.9	0.9
Stearic acid	1.5	1.5
White perfecta petrolatum	5.0	5.0
Cosmetic fluid No. 687 (petroleum distillate)	15.3	15.3
Arlacel No. 83 (sorbitan sesquioleate)	2.7	2.7
Standamul G (2-octadodecanol)	1.5	1.5
Peach kernel oil	0.5	0.5
Standamul 1414E (myristyl ethoxy myristate)	3.4	3.4
Bees' wax	2.0	2.0
Ceraphyl 230 (diisopropyl adipate)	1.8	1.8
Ceraphyl 375 (isostearyl neopentonate)	7.0	7.0
Ceraphyl 368 (2-ethylhexyl palmitate)	9.0	9.0
Water phase		
N-acetyl-L-glutamic acid	3.0	0
Lime water (saturated)	45.5	45.5
$CaCl_2$	0.17	0.17
Distilled water	1.7	1.7
KOH	2.07	2.07
Sodium borate (borax)	0.07	0.07
Methylparaben (methyl p-hydroxybenzoate)	0.1	0.1
Propylparaben (propyl p-hydroxybenzoate)	0.01	0.01

Included in the water phase of separate 100 gram portions of base formulation A was the indicated amount of N-acetyl-L-glutamic acid and sufficient KOH to form the dipotassium salt thereof.

These moisturizing cosmetic bases were then prepared as lotions using tritiated water and tested for the moisturizing effectiveness of the lotions on the abdominal skins of rabbits. Tritium counts were made on skin sections which showed the effectiveness of the above lotions as humectants.

GLUTAMINE AND DERIVATIVES

The 5-amide of glutamic acid or 2-aminoglutaramic acid is commonly termed glutamine.

N-Acetyl Glutamine Aluminum Salt

The N-acetyl-L-glutamine aluminum salts are a kind of organic acid salt, but N-acetylglutamine exhibits such strikingly different properties from the ordinary organic acids that it is difficult to prepare the aluminum salts according to the usual methods for preparing salts of organic acids.

T. Kagawa, K. Fuji, M. Tanaka and H. Tanaka; U.S. Patent 3,787,466; Jan. 22, 1974; assigned to Kyowa Hakko Kogyo Co., Ltd., Japan have disclosed a synthetic method for preparing aluminum salts of N-acetyl-L-glutamine having a high degree of purity.

These salts can be synthesized by allowing N-acetyl-L-glutamine to react with an aluminum alkoxide in the presence of a large amount of water. The actual volume of water used will depend upon the amount of N-acetyl-L-glutamine used in the reaction. The aluminum salt is obtained by crystallization from the aqueous solution.

Included among the aluminum salts of N-acetyl-L-glutamine, are the salts which form at various stages during the reaction and have varying molar ratios of N-acetyl-L-glutamine to aluminum. When the present method is used, salts having a molar ratio of N-acetyl-L-glutamine to aluminum of from 1:1 to 3:1 can be prepared merely by changing the molar ratio of N-acetyl-L-glutamine to aluminum alkoxide.

The preferred solvent for the reaction is water or a mixture of water and a water-miscible organic solvent. Organic solvents such as acetone or lower alkanols such as methanol, ethanol, and propanol may be used. The salt is prepared by reacting an aluminum alkoxide, either as a solid or as an alcoholic solution, with an aqueous solution of N-acetyl-L-glutamine. The reaction is carried out with stirring.

Any suitable aluminum alkoxide may be used in the reaction such as, e.g., lower alkoxides such as aluminum methoxide, aluminum ethoxide, aluminum tertiary-butoxide, and aluminum isopropoxide and corresponding alkoxides containing up to six carbon atoms in the alkyl group.

When the solvent employed for the reaction is water or a solution comprised mostly of water, the reaction between the aluminum alkoxide and N-acetyl-L-glutamine takes place quickly resulting in a solution of the N-acetyl-L-glutamine aluminum salt. The reaction can be carried out at a temperature ranging from the room temperature to 80°C.

The crystalline-like N-acetyl-L-glutamine aluminum salt can be obtained by concentrating the reaction mixture and drying the resultant solution of the salt into a dry mass or by precipitating the salt out of solution by the addition of a solvent such as alcohol or acetone in which the salt is insoluble to the concentrated solution.

The aluminum salts of N-acetyl-L-glutamine can be used in the treatment of gastric ulcers or they may be used for a prevention of gastric ulcer. The salts can be administered orally, subcutaneously, intraperitioneally or intravenously. Good results are obtained when the salts are administered at dose levels ranging from 0.5 to 2 g/kg/day but, since the salts have low toxicity, larger doses may be administered in severe cases. Any pharmaceutically acceptable carrier such as, for example, mannitol, lactose and the like may be employed to administer the salts.

Example: A mixture of 47.0 grams of N-acetyl-L-glutamine and 600 ml of water is heated to 60°C, and 32.6 grams of aluminum isopropoxide is added to the warm mixture with stirring. The stirring is continued for 30 minutes. The reaction mixture is filtered and the filtrate is concentrated and dried to a mass under reduced pressure. After drying, the crystalline-like aluminum salt of N-acetyl-L-glutamine (53.8 grams) is obtained.

N^{ω}-4,4'-Dimethoxybenzyl Derivatives of Glutamine and Asparagine

In the synthesis of asparagine- and glutamine-containing peptides in which the amide function is not protected, numerous side reactions occur. In such peptides the amide function can be dehydrated by peptide-forming reagents to yield the nitrile group, cyclic imides obtained as intermediates may cause the formation of isoasparagine or isoglutamine derivatives and the amide function may be split by acid or basic hydrolysis.

To avoid these side reactions, protective groups for the amide functions have already been developed. For example, N-carbobenzoxy-L-glutamine can be reacted with xanthydrol in glacial acetic acid to form N-carbobenzoxy-N^{γ}-xanthyl-L-glutamine. The xanthyl group can be again split off by means of hydrobromic acid in glacial acetic acid. However, the yield is small after splitting.

W. Konig and R. Geiger; U.S. Patent 3,790,618; February 5, 1974; assigned to Farbwerke Hoechst AG, Germany have disclosed that 4,4'-dimethoxybenzhydrol reacts with asparagine or glutamine as well as with N-carbobenzoxy-asparagine or N-carbobenzoxy-glutamine in glacial acetic acid in the presence of protonic catalysts to yield the corresponding 4,4'-dimethoxybenzhydryl-amides. The 4,4'-dimethoxybenzhydryl group can easily be split off again in an acid medium.

The amido-substituted asparagine and glutamine derivatives may be prepared either (1) by the reaction of N-carbobenzoxy derivatives of glutamine and asparagine with 4,4'-dimethoxybenzhydrol in glacial acetic acid with the addition of catalytic amounts of acid catalysts such as concentrated sulfuric acid, BF_3-diethyl etherate, $ZnCl_2$, $AlCl_3$ or an organic sulfonic acid such as toluenesulfonic acid, or (2) by the reaction of glutamine or asparagine with 4,4'-dimethoxybenzhydrol in glacial acetic acid with the addition of molar amounts of concentrated sulfuric acid or BF_3-diethyl-etherate.

The 4,4'-dimethoxybenzhydryl groups can also be introduced subsequently under the reaction conditions mentioned into the asparagine- and glutamine-containing peptides, in which case free sulfhydryl groups present are blocked and free amino, carboxy and hydroxy groups may be blocked. As blocking groups there may be mentioned the following groups that are stable under the reaction conditions:

(1) Free amino groups are blocked by carbobenzoxy, 4-nitrobenzyloxycarbonyl, phthalyl;
(2) Free sulfhydryl groups are blocked by benzyl, 4,4'-dimethoxybenzhydryl, acetyl, benzoyl and other acyl groups as well as carbamoyl groups;
(3) Free carboxyl groups are blocked by methyl, ethyl, isopropyl, benzyl, 4-nitrobenzyl; and
(4) Free hydroxy groups are blocked by benzyl.

The compounds may then be used according to the methods generally known in peptide chemistry for preparing asparagine- and glutamine-containing peptides.

Example 1: N^{ω}-N,N′-Dimethoxybenzhydryl-L-Glutamine — (A) 1.45 grams (10 mmol) of glutamine and 2.4 grams of 4,4'-dimethoxybenzhydrol (10 mmol) were introduced into 50 ml of glacial acetic acid. At room temperature, 0.54 milliliters of concentrated sulfuric acid was added thereto and the whole was stirred for about 2 hours. Subsequently 2 grams of anhydrous sodium acetate were added and the solution was concentrated. The residue was distributed between sodium acetate solution and acetic acid ester. After standing overnight in the refrigerator, a precipitate separated which was suction-filtered and washed with water. Yield: 1.145 grams (= 31% of theoretical), MP 205° to 206°C, $[\alpha]_D{}^{22}$ = +7.5° (c = 2, glacial acetic acid).

(B) The compound may also be obtained in the following manner: 5.1 grams (10 mols) of N-carbobenzoxy-N^{ω}-4,4'-dimethoxybenzhydryl-L-glutamine (prepared according to Example 2) were dissolved in a mixture of glacial acetic acid and methanol (1:1). Palladium catalyst was added thereto and hydrogen was fed in until no more CO_2 escaped. The catalyst was suction-filtered and the filtrate was concentrated. The residue was triturated with ether and dried. Yield: 365 grams (= 98% of theoretical). After recrystallization from water the yield was 2.81 grams (75.5% of theoretical), MP 205° to 206°C.

Example 2: N-Carbobenzoxy-N^{ω}-4,4′-Dimethoxybenzhydryl-L-Glutamine — 28 grams of carbobenzoxy-L-glutamine (0.1 mol) and 24 grams of 4,4'-dimethoxybenzhydrol (0.1 mol) were dissolved in 250 ml of glacial acetic acid. At room temperature 0.5 ml of concentrated sulfuric acid was added thereto, the whole was allowed to stand overnight and then poured into 750 ml of water. An oil precipitated which crystallized soon afterwards. The crystal mass was suction-filtered and dissolved hot in acetic acid ester. The acetic acid ester solution was shaken with water, dried with sodium sulfate and concentrated. The residue was triturated with ether and suction-filtered. Yield: 45.8 grams (= 90% of theoretical), MP 117° to 120°C. $[\alpha]_D{}^{22}$ = -6.75° (c = 2, dimethylformamide).

Glutamine in Interferon Production

Interferon can be produced by the growth of animal cells in monolayer cultures, and this can be used to produce interferon preparations of high titre. However, monolayer culture techniques do not lend themselves readily to industrial-scale production.

Although animals cells can grow readily in stirred suspension cultures, attempts to produce interferon by such culture processes have not been successful. The process disclosed by *I. Gresser and M.G. Tovey; U.S. Patent 3,951,740; April 20, 1976; assigned to Agence National De Valorisation de la Recherche (ANVAR), France* provides for the production of interferon in stirred suspension culture with L-glutamine in the culture medium. The steps of this process comprise:

(a) adding a priming dose of interferon, the interferon being specific to the species from which the cells are derived, to a culture medium of cells grown in suspension up to the saturation density of the cells in the medium, and enough L-glutamine to keep the cells in an active metabolic state, and effecting priming of the cells, with suitable stirring of the medium;

(b) stopping the stirring, and separating the cells by sedimentation from the supernatant liquid;
(c) resuspending the cells in a small volume of a highly concentrated suspension of a virus inducer, leaving the cells in contact with the virus, and thereafter removing the virus inducer by centrifugation;
(d) subjecting the cells to metabolic inhibitors and removing the inhibitors by centrifugation;
(e) resuspending the cells in a nutrient medium without serum, in which the cells are incubated; and
(f) separating the cells by sedimentation, collecting the supernatant liquid containing the interferon, and concentrating and purifying the interferon.

Example: Preparation of Human Interferon from Namalva Line of Burkitt Cells – A continuous line of human leucocytes, called Namalva Burkitt cell line, was used because of its high capability for producing interferon.

Namalva cells were sown at a concentration of 2×10^5 cell per ml in a RPMI 1640 nutrient medium supplemented by 10% of calf fetus serum. The cells were cultivated in suspension in a 20 liter tank at 37°C, with mechanical stirring, until the cell saturation density was obtained. This was 2.5×10^6 cells/ml. This saturation density was reached after growth for 72 hours.

A priming amount of 10 units of human interferon per ml and 0.5 mg/ml of L-glutamine to keep the cells in an active metabolic state were then added. Priming was continued for 18 hours, at 37°C.

During this stage, the cells were allowed to deposit under gravity, and after 18 hours the supernatant liquid was sucked off. The residual medium containing the cells was centrifuged, and the resulting cell mass was resuspended in 0.01 of the original culture volume of a suspension of virus inducer (live virus, Herz strain, of Newcastle disease) to provide a multiplicity of infection of 1.

After a short adsorption stage, the cells were separated by centrifuging the virus inducer which was recovered. The Burkitt cells were resuspended at a concentration of 10^7 cells/ml in RPMI 1640 medium, which had been supplemented by 2% of calf fetus serum and contained 10 micrograms per ml of cycloheximide.

After several hours exposure to the cycloheximide, the latter was removed by centrifugation, and the cell mass was resuspended in RPMI 1640 nutrient medium, without serum, at a concentration of 10^7 cells/ml, and incubated for 18 hours with stirring.

The resulting cells were allowed to deposit under gravity, and the supernatant liquid containing the interferon was sucked off, concentrated, and purified by known methods.

In this way, using Namalva cells, human interferon preparations were obtained with a titre of 1/1280. After concentration 100 to 1,000 times, an interferon was obtained with a titre of 1 million international reference units per ml on human cells.

γ-Glutamyl Amide of Dopamine

Dopamine has been reported to be useful in treating congestive heart failure and shock. In addition, certain amino acids of dopamine have been found useful as renal vasadilators and antihypertensive agents.

However known compounds which are used as renal vasodilators have the drawback that they also produce undesirable systemic hemodynamic changes.

J. Kyncl, K. Riley, Y.C. Martin and C.W. Ours; U.S. Patent 3,903,147; Sept. 2, 1975; assigned to Abbott Laboratories have developed compounds which are useful as renal vasodilators, and yet do not produce unwanted hemodynamic changes. These compounds are γ-glutamyl amides of dopamine of the formula:

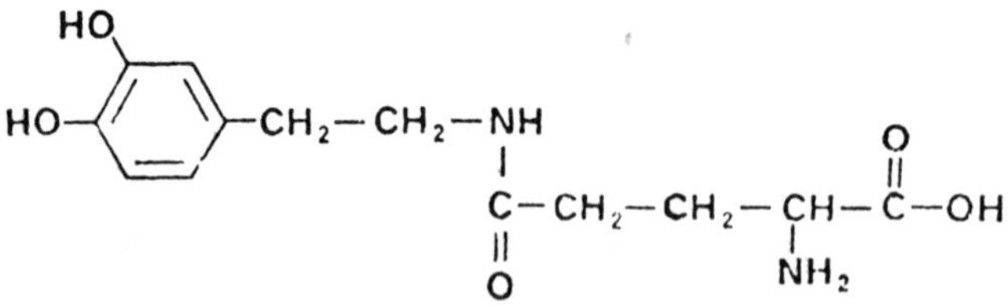

or pharmaceutically acceptable acid addition salts thereof.

These compounds may be prepared by a wide variety of methods. The following examples illustrate typical procedures:

Example: Preparation of [N⁵-β-(3,4-Dihydroxyphenyl)-Ethyl]-Glutamine Hydrochloride — (a) Preparation of N-CBZ-O⁵-Succinimido-L-Glutamic Acid O-Benzyl Ester: To a solution of N-CBZ-L-glutamic acid-α-benzyl ester (0.057 mol, 21.8 grams) and N-hydroxysuccinimide (0.057 mol, 6.6 grams) in 100 ml of dioxane was added with stirring at room temperature (0.057 mol, 12.0 grams) of dicyclohexylcarbodiimide. After 24 hours stirring at room temperature the mixture was filtered to remove dicyclohexylurea. The filtrate was concentrated in vacuo to give an oil. This oil was taken up in ethanol:ether (1:2) and cooled to give 22.4 grams (84%) of a crystalline solid, MP 104° to 106°C.

(b) Preparation of N-CBZ-[N⁵-β-(3,4-Benzyloxyphenyl)Ethyl]-Glutamine O-Benzyl Ester: A solution of the active ester (0.048 mol, 22.4 grams) and dibenzyldopamine·HCl (0.048 mol, 17.8 grams) in 100 ml of DMF was treated with triethylamine (0.1 mol, 14 ml) and stirred at room temperature for 4 hours. The solution was then diluted with water to give an oil which solidified on standing. The white solid was collected and washed well with water. The moist solid was dissolved in boiling absolute ethanol, filtered and allowed to crystallize. The solid was collected and washed with ethanol to give, after drying at 60°C overnight, 29.4 grams of the protected amide, MP 126° to 128°C.

(c) Preparation of the Glutamine Hydrochloride: A suspension of the protected amide (0.0365 mol, 25 grams) from step (b) was reduced in a Parr apparatus in ethanol containing 5.0 grams of 5% PdC and 1 equivalent of concentrated HCl. After uptake was complete, the catalyst was filtered and washed with ethanol. The filtrate was concentrated in vacuo to give, upon repeated azeotroping with methanol, a glass. This glass was dried under high vacuum at 60°C for 24 hours to give 11.2 grams (97%) of [N^5-β-(3,4-dihydroxyphenyl)ethyl]-glutamine·HCl.

In four anesthetized dogs, dopamine and the compound of Example 1, [N^5-β-(3,4-dihydroxyphenyl)ethyl]-glutamine hydrochloride were compared after intravenous injections. Whereas the renal vasodilator action of dopamine (150 μg/kg intravenous) was accompanied by pronounced increase in arterial blood pressure, heart rate, left ventricular systolic pressure and dp/dt, comparable increases in renal blood flow aftertreatment with the subject glutamine hydrochloride were obtained without such changes. In fact one could give a maximum single dosage of the glutamine hydrochloride up to 2,500 μg/kg to increase the renal blood flow for a period of 82.5±5.5 minutes without any systemic cardiovascular action.

HETEROCYCLIC AMINO ACIDS

The two nonessential amino acids that come under this category are proline (2-pyrrolidinecarboxylic acid) and hydroxyproline (4-hydroxy-2-pyrrolidinecarboxylic acid). These compounds are not truly α-amino acids in that the amino group and the carbon adjacent to the carboxylic group form a part of the pyrrolidine ring.

PROLINE

Production by Mutants Resistant to Sulfa Drugs

Certain auxotrophs belonging to the genus Brevibacterium produce proline in a medium, but at most 1.5 g/dl are produced under favorable conditions.

F. Yoshinaga, T. Tsuchida and S. Okumura; U.S. Patent 3,819,483; June 25, 1974; assigned to Ajinomoto Co., Inc., Japan have found that improved amounts of proline are accumulated in an aqueous culture medium by mutant strains resistant to sulfa drugs of Brevibacterium, or Corynebacterium. When the mutants are derived from proline-producing auxotroph parent strains, the production of proline is extremely improved. Moreover, mutant strains derived from parent strains incapable of producing proline can produce significant amounts of proline.

The method of mutation is conventional such as exposure of cells of the parent strains to ultraviolet light, x-rays, or gamma rays in mutagenic doses, or to sodium nitrate, nitrosoguanidine or diethyl sulfate solution in a manner conventional in itself. For example, the cells of a parent strain on an agar nutrient medium are exposed to ultraviolet light (2375 A) from a source which is placed 30 cm away, for 3 minutes.

The mutants resistant to sulfa drugs are selected from the exposed parent strain by conventional methods. The exposed parent strain is cultured on an agar medium or in an aqueous nutrient medium containing an amount of sulfa drugs which inhibits growth of the parent strain. Resistance of a mutant to sulfa drugs is determined by comparing the relative growth of the mutant strain on a sulfa drugs-containing medium with that of the parent strain. The most effective proline-producing mutants found (and the sulfa drugs employed in screening them) are *Brevibacterium flavum* FERM-P 1681 (sulfaguanidine), *Brevibacterium flavum* FERM-P 1682 (sulfaguanidine), *Corynebacterium acetoacidophilum* FERM-P 1683 (sulfamerazine) and *Corynebacterium glutamicum* FERM-P 1781 (sulfaguanidine).

The nutrient media fermented by means of the microorganisms of this process are conventional in themselves and contain assimilable sources of carbon and nitrogen, and inorganic salts. Minor amounts of organic nutrient such as vitamins and amino acids may be added.

For a good yield of proline, the fermentation is carried out aerobically with aeration and/or agitation. Best yield requires pH control within the range of 5 to 9. The desired pH may be maintained by adding gaseous or aqueous ammonia, calcium carbonate, alkali metal hydroxide, urea, organic or inorganic acids to the medium from time to time, some of which may also supply assimilable nitrogen. When the fermentation is carried out at 24° to 37°C, the concentration of proline in the broth reaches its maximum within 2 to 7 days.

Example: Relative growth in sulfaguanidine-containing medium of sulfa drug resistant mutant *Brevibacterium flavum* FERM-P 1681 which was screened on an agar medium containing 500 γ/ml sulfaguanidine, and that of its parent strain *Brevibacterium flavum* ATCC 15940 (isoleucine-requiring and proline-producing mutant) are shown in the following table.

Sulfaguanidine	Relative Growth of	
Added (γ/ml)	*Brevibacterium flavum* **FERM-P 1681**	*Brevibacterium flavum* **ATCC 15940**
0	100	100
100	98	95
200	95	88
300	90	42
400	68	0
500	40	0
600	11	0

The medium contained the following.

Glucose, g/dl	2
$(NH_4)_2SO_4$, g/dl	1
Urea, g/dl	0.1
Biotin, γ/l	50
Thiamine·HCl, γ/l	100
L-isoleucine, mg/dl	20
KH_2PO_4, g/dl	0.1
K_2HPO_4, g/dl	0.3
$MgSO_4 \cdot 7H_2O$, g/dl	0.1
Fe^{++}, ppm	2
Mn^{++}, ppm	2
Sulfaguanidine as shown in above table	

Cultivation was as follows: 3 ml of the medium are placed in 10 ml test tube, inoculated with 2 x 10^7 cells, and cultured at 31.5°C for 48 hours.

An aqueous medium was prepared with the following composition.

Glucose, g/dl	10
$(NH_4)_2SO_4$, g/dl	4
KH_2PO_4, g/dl	0.1
$MgSO_4 \cdot 7H_2O$, g/dl	0.04
Tartaric acid, g/dl	0.1
Ferric ion, mg/dl	0.2
Manganese ion, mg/dl	0.2
Biotin, γ/l	450
Thiamine·HCl, γ/l	1,000
L-isoleucine, mg/dl	10
Soybean protein hydrolyzate, ml/dl	0.1
$CaCO_3$, g/dl	5
pH adjusted to 7.0	

Three hundred milliliter batches of the medium were placed in 1,000 ml fermentation vessels, sterilized by steam, and inoculated with *Brevibacterium flavum* FERM-P 1681. Fermentation was carried out at 31°C for 72 hours. The pH of the medium was adjusted during the fermentation by introducing gaseous ammonia. After the cultivation, 2.89 g/dl proline was found in the fermentation broth. One liter of the combined broth was filtered to remove cells, and passed over an ion exchange resin. The resin was eluted with ammonium hydroxide solution, and the eluate was partly evaporated in a vacuum. Thereafter, 14.5 g of proline crystals were obtained by pouring ethyl alcohol into the eluate.

When *Brevibacterium flavum* ATCC 15940 was cultured by the same method as above, 1.50 g/dl proline was found in the fermentation broth.

L-Prolyl-L-Leucylglycine Amide for Treating Parkinson's Disease

L-prolyl-L-leucylglycine amide has recently been synthesized and the tripeptide structure is known as follows.

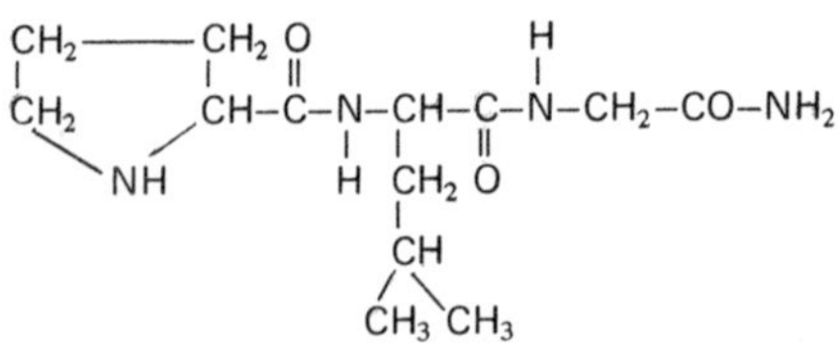

(See, for example, Celis, et al, *Biophysical Soc. Abstract* TPM-K 12, pp 98a, 1971.) For convenience this amide will be referred to as tripeptide.

N.P. Plotnikoff; U.S. Patent 3,795,738; March 5, 1974; assigned to Abbott Laboratories has found that this tripeptide provides excellent activity against Parkinson's disease at surprisingly low dosages. It has also been found that the dosage level for humans of 1 to 40 mg/kg of body weight daily is quite critical. If, for example, the dosage is less than this minimum, no effect is noted. However, on the other hand, if the dosage is in excess of the maximum stated amount the drug may exhibit an undesirable sedative effect. More often, the practical dosage range for best results is 1 to 20 mg/kg of body weight daily. The above tripeptide is particularly useful when used in combination with L-dopa.

When administered in a unit dosage form, the tripeptide will be present in an amount of 1 to 250 mg and the L-dopa will be present in an amount of 1 to 500 mg. The tripeptide potentiates the activity of L-dopa and small amounts of tripeptide in combination with L-dopa show activity where the same dosage of L-dopa is completely ineffective. Usually 0.1 to 40 mg/kg of body weight daily of tripeptide and 1 to 500 mg/kg of body weight daily of L-dopa are the dosage levels in the combination treatment.

The active ingredients, of course, will be accompanied by a pharmaceutically acceptable carrier. It is understood, of course, that the tripeptide alone may be administered to the patient as well as the combination of tripeptide and L-dopa. The tripeptide may also be administered in combination with other effective drugs used to treat Parkinson's disease.

For use in treating Parkinson's disease, the tripeptide alone or in combination with L-dopa or other drugs may be administered in either liquid or in solid form. Thus, the active compounds may be provided in granulation tablet, capsule, elixir and other dosage forms. Oral administration is preferred and is the most convenient. When administered orally in tablets, capsules and pills the tripeptide may be present in an amount of 1 to 500 mg.

Example 1: The most sensitive test for potential anti-Parkinson activity is the mouse antioxotremorine test (G.M. Everett et al, *Fed. Proc.,* 30, 2, 677, 1971). The effects of oxotremorine in mice closely resemble the symptoms of Parkinson's disease. The following method was employed. Normal male mice were pretreated with varying doses of the tripeptide 1 and 4 hours before administration of oxotremorine which was administered at 0.5 mg/kg IP. Four mice were tested at each of the dosages of the tripeptide. The responses were recorded by observation techniques and were compared with those observed in 16 intact mice which received only oxotremorine as a control.

As can be seen from the following table, the tripeptide was active in reducing the effects of oxotremorine in normal mice. The control animals receiving the oxotremorine exhibited marked signs of parasympathetic stimulations consisting of tremors, head twitch, decreased activity, ataxia, lachrymation, salivation and diarrhea.

Drug	Test Hours	Dose* and Route (mg/kg)	Tremors	Head Twitch	Ataxia	Lachrymation	Salivation	Diarrhea
Oxotremorine alone		0.5**	3	3	2	2	3	3
L-prolyl-L-leucyl-	1	0.5***	3	3	2	2	3	3
glycine amide		1***	3	2	2	1	2	2
		2***	2	2	2	0	1	2
		4***	2	1	2	0	1	1
		8***	2	1	1	0	1	0
		15***	2	1	1	0	1	0
		20***	2	1	1	0	1	0
		40***	2	1	1	0	0	0
L-prolyl-L-leucyl-	4	0.5***	3	3	2	2	3	3
glycine amide		1***	3	3	2	1	2	2
		2***	2	2	2	1	2	2
		4***	2	2	2	0	1	1

(continued)

Drug	Test Hours	Dose* and Route (mg/kg)	Tremors	Head Twitch	Ataxia	Lachry-mation	Saliva-tion	Diarrhea
L-propyl-L-leucyl-glycine amide	4	8***	2	1	2	0	1	1
		15***	2	1	1	0	1	0
		20***	2	1	1	0	1	0
		40***	2	1	1	0	1	0

Note: Degree of oxotremorine effects–3 is marked; 2 is moderate; 1 is slight; and 0 is none.

*Four mice per dose. **IP. ***Oral.

Example 2: The tripeptide was also observed to potentiate the effects of L-dopa in reducing the central and peripheral effects of oxotremorine. Thus, a minimally active dose of L-dopa (100 mg/kg IP) plus tripeptide (0.25 to 1 mg/kg) provided significant protection against the effects of oxotremorine as shown in the following table.

Oxotremorine Antagonism by L-Prolyl-L-Leucylglycine Amide and L-Dopa in Mice

	Tremors	Head Twitch	Ataxia	Lachrymation	Salivation	Diarrhea
Oxotremorine alone	3*	2	2	2	3	3
IP route, L-dopa alone, mg/kg						
100	2	2	2	1	2	2
200	1	1	1**	1	1	0
400	0	1	0**	0	0	0
L-prolyl-L-leucylglycine amide plus L-dopa, mg/kg						
0.1 plus 100	3	1	1	1	2	0
0.25 plus 100	1	1	0	0	1	0
0.5 plus 100	1	1	0	0	1	0
1.0 plus 100	1	1	0	0	1	0

*Degree of oxotremorine effects: 3 is marked; 2 is moderate; 1 is slight; and 0 is none.
**Increased motor activity.

AROMATIC AMINO ACIDS (OF PROTEINS)

PHENYLALANINE AND DERIVATIVES

As one of the essential amino acids not synthesized by the human body, phenylalanine or 2-amino-3-phenylpropionic acid, in its L-form, is needed for human growth. However, this amino acid can be supplied by vegetable protein.

In some children, the normal metabolism of phenylalanine is inhibited by the lack of the enzyme phenylalanine hydroxylase, resulting in phenylketonuria. In these cases a phenylalanine-free diet is used as disclosed in U.S. Patent 4,016,147, page 39.

Production Using *Microbacterium ammoniaphilum* Mutants

Among known methods for preparing L-phenylalanine by fermentation is a process using a microorganism which requires tyrosine for growth (Japanese Patent 6345/62). But this process can hardly be applicable to commercial production of L-phenylalanine because of its low yield.

K. Nakayama and H. Hagino; U.S. Patent 3,917,511; November 4, 1975; assigned to Kyowa Hakko Kogyo KK, Japan have found that strains of *Microbacterium ammoniaphilum* which exhibit resistance to at least one compound selected from tyrosine, phenylalanine and analogue thereof (such as e.g., 2-fluorophenylalanine, 3-methylphenylalanine, 2-hydroxyphenylalanine, 4-aminophenylalanine, 2-nitrophenylalanine, β-3-thienylalanine, 2-indole alanine, 1-naphthylalanine, 2-pyridylalanine, 3-thiazole alanine, 3-aminotyrosine, 5-hydroxy-2-pyridylalanine, phenylalanine hydroxamate, etc.) are capable of producing and accumulating a large amount of L-phenylalanine.

Cultures of these *Microbacterium ammoniaphilum* (ATCC 21645) were obtained by subjecting certain strains belonging to this bacteria to mutation such as e.g., irradiation by ultraviolet light, x-ray, cobalt 60, etc. and chemical treatment, etc. and then screening the strains for isolation from colonies capable of growing on agar plate media containing tyrosine, phenylalanine, or analogues thereof. For example, L-phenylalanine-producing strains which are resistant to 4-fluorophenylalanine are selected and isolated from colonies grown on a plate containing 100 to 10,000 μg/ml of 4-fluorophenylalanine. Among these strains are included some strains capable of growing in 4-fluorophenylalanine-containing media at high concentration.

Any synthetic or natural media may be used for this process, if they contain suitable amounts of nitrogen souces, carbon sources, inorganic materials as well as trace amounts of nutrients required for growth of the used strain, as shown in the example.

The strain is cultured under aerobic conditions, for example, by culturing with shaking, aeration-agitation, etc. The preferable culture temperature is 20° to 40°C. The pH of the medium is preferably maintained at 4 to 8 during the cultivation. The time for culture is usually from 2 to 5 days to accumulate a substantial amount of L-phenylalanine in the medium. After the completion of the cultivation, microbial cells are first removed from the fermented broth, and then L-phenylalanine is recovered in a suitable manner such as by using activated carbon treatment, ion exchange resin treatment, etc., as shown in the example.

Example: An L-phenylalanine-producing strain of *M. ammoniaphilum* (ATCC 21645) previously screened as being resistant to an analogue of phenylalanine (4-fluorophenylalanine) was cultured at 30°C for 24 hours in a seed medium containing glucose (2%), peptone (1%), yeast extract (1%) and NaCl (0.3%). The resultant seed (1 ml) was inoculated into a fermentation medium (10 ml) in a 250 ml Erlenmeyer flask and cultured with shaking at 30°C for 4 days to produce 4.7 mg/ml of L-phenylalanine from the culture broth.

The fermentation medium used comprised glucose (10%), K_2HPO_4 (0.05%), KH_2PO_4 (0.05%), $MgSO_4 \cdot 7H_2O$ (0.025%), ammonium sulfate (2%), NZ-amine (0.5%), biotin (30 mg/l) and $CaCO_3$ (2%). It had a pH of 7.2.

After the completion of the fermentation, 1 liter of the cultured broth was centrifuged to remove microbial cells and $CaCO_3$. The resultant liquor was mixed with active charcoal which had previously been treated with acetic acid and was thoroughly mixed to adsorb the phenylalanine. The charcoal was separated by filtration and was washed with water. The charcoal was eluted with 20% (v/v) acetic acid containing phenol (5% v/v) and the eluate was treated with ether. After the removal of phenol, the eluate was concentrated and was passed through a column of a strongly acidic cation exchange resin (Diaion SK-1A) (H^+ form) to adsorb phenylalanine which was then eluted with 0.3% aqueous ammonia. The phenylalanine fraction of the eluate was concentrated and precipitated with alcohol to recover 1.6 grams of L-phenylalanine.

Production Using *Corynebacterium glutamicum* Mutants

In a process related to U.S. Patent 3,917,511, *K. Nakayama and H. Hagino; U.S. Patent 3,759,790; September 18, 1973; assigned to Kyowa Hakko Kogyo KK, Japan* use mutants of *Corynebacterium glutamicum* for the production of phenylalanine. The mutants exhibit resistance to the same compounds as listed in U.S. Patent 3,917,511.

Example: L-phenylalanine-producing strains belonging to *Corynebacterium glutamicum* and having the properties shown in the table below were used.

Among these strains, those showing resistance to two or more analogues had been obtained by a series of mutation treatments. These strains were respectively cultured for 24 hours in each seed medium containing glucose (2%), peptone (1%), yeast extract (1%), and NaCl (0.3%). The resultant seeds (each 1 ml) were respectively inoculated each into 10 ml of fermentation medium in a large test tube and cultured at 30°C for 4 days with shaking. The amounts of L-phenylalanine obtained from their respective cultured broths are shown in the table below.

The composition of the fermentation medium was cane molasses (10% as glucose), K_2HPO_4 (0.05%), KH_2PO_4 (0.05%), $MgSO_4 \cdot 7H_2O$ (0.025%), $(NH_4)_2SO_4$ (2%), corn steep liquor (0.5%) and $CaCO_3$ (2%). It had a pH of 7.2.

After the completion of the fermentation, 1 liter of the cultured broth was centrifuged to remove microbial cells and $CaCO_3$. The resultant liquor was mixed with active charcoal which had previously been treated with acetic acid and was thoroughly mixed to adsorb phenylalanine thereon. The charcoal was separated by filtration and was washed with water. The charcoal was eluted with 20% (v/v) acetic acid containing phenol (5% v/v) and the eluate was treated with ether. After the removal of phenol, the eluate was concentrated and was passed through a column of a strongly acidic cation exchange resin (Diaion SK-1A) (H^+ form) to adsorb phenylalanine which was then eluted with 0.3% aqueous ammonia. The phenylalanine fraction of the eluate was concentrated and precipitated with alcohol to recover L-phenylalanine.

Strain	L-phenylalanine Produced, mg/ml
Tyr^-, PFP^R (ATCC 21668)	6.9
Tyr^-, TA^R (ATCC 21669)	6.4
Tyr^-, PFP^R, $3AT^R$ (ATCC 21670)	10.6
Tyr^-, TA^R, $3AT^R$ (ATCC 21671)	10.2
Tyr^-, PFP^R, TA^R, $3AT^R$ (ATCC 21672)	12.6
Tyr^-, PFP^R, PAP^R, $3AT^R$ (ATCC 21673)	13.2
Tyr^-, PFP^R, PAP^R, TA^R, $3AT^R$ (ATCC 21574)	13.8

Note: Tyr^- = tyrosine-requiring
PFP^R = resistant to 4-fluorophenylalanine
TA^R = resistant to β-2-thienylalanine
$3AT^R$ = resistant to 3-aminotyrosine
PAP^R = resistant to 4-aminophenylalanine

Production Using Brevibacterium Mutants

Mutant strains of *B. lactofermentum* and *B. flavum* have also been developed by *T. Tsuchida, H. Matsui, H. Enei and F. Yoshinaga; U.S. Patent 3,909,353; September 30, 1975; assigned to Ajinomoto Co., Inc., Japan* for production of L-phenylalanine. The growth of these mutants is not inhibited by tryptophan analogs nor by phenylalanine analogs in concentrations sufficient to inhibit the growth of the respective parent strains showing neither the typrosine requirement nor the resistance to tryptophan or phenylalanine analogs.

It has been found that mutants of Brevibacterium which resist one phenylalanine analog also resist other phenylalanine analogs, and that a similar relationship exists with respect to tryptophan-resistant mutant strains. The aqueous fermentation media in which these mutants produce L-phenylalanine are largely conventional. Aerobic conditions are maintained in the fermentation by aeration and/or agitation, and pH is held between 5 and 9 for good yields. When ammonia is used for pH control, it may also serve as a nitrogen source. The phenylalanine concentration in the broth reaches its maximum within 2 to 7 days if the fermentation is carried out at 24° to 37°C.

Example 1: B. lactofermentum No. 2256 (ATCC 13869) was exposed to 250 γ/ml N-methyl-N'-nitro-N-nitrosoguanidine and five tyrosine-requiring, phenylalanine-producing mutant strains were isolated. Strain AJ 3435 (FERM P-1912) did not show more resistance to phenylalanine or tryptophan analogs than the parent. Strain AJ 3436 (FERM P-1913) resisted p-fluorophenylalanine. Strain AJ 3432 (FERM P-1844) resisted 5-methyltryptophan. The strains AJ 3437 (FERM P-1914) and AJ 3699 (FERM P-2476) resisted both p-fluorophenylalanine and 5-methyltryptophan. Strain AJ 3699, in addition to tyrosine, required tryptophan and methionine for its growth.

Example 2: B. flavum No. 2247 (ATCC 14067) was exposed to 250 γ/ml N-methyl-N'-nitro-N-nitrosoguanidine, and four tyrosine-requiring, phenylalanine-producing strains were isolated. Strain AJ 3695 (FERM P-2472) resisted neither phenylalanine nor tryptophan analogs better than the parent. Strains AJ 3696 (FERM P-2473) and AJ 3697 (FERM P-2474) resisted p-fluorophenylalanine and 5-methyltryptophan respectively. Strain AJ 3698 (FERM P-2475) was resistant to both analogs.

Example 3: An aqueous culture medium was prepared to contain, per deciliter, 13 grams glucose, 1.0 gram $(NH_4)_2SO_4$, 0.15 gram KH_2PO_4, 5.0 μg biotin, 20 μg thiamine·HCl, 0.4 gram $MgSO_4 \cdot 7H_2O$, 1.0 mg $FeSO_4 \cdot 7H_2O$, 1.0 mg $MnSO_4 \cdot 7H_2O$, 40 mg L-tyrosine, 40 mg DL-methionine, 3.0 ml soy protein acid-hydrolyzate, 1.2 grams fumaric acid, 0.3 ml acetic acid, and 5.0 grams calcium carbonate, and adjusted to pH 7.0. 20 ml batches of the medium in 500 ml flasks were inoculated with seed cultures of the strains of *B. lactofermentum* listed in the table below and cultured at 30°C with shaking for 72 hours. The culture medium for strain AJ 3699 additionally contained 10 mg/dl L-tryptophan.

The broths were then analyzed for their phenylalanine contents, also listed in the table. Phenylalanine was recovered from the cultures of *B. lactofermentum* AJ 3437 and AJ 3699 in crystalline form by centrifuging each broth to remove cells and other solids, passing the cell-free liquid over an ion exchange resin, Diaion SK-1, eluting the adsorbed phenylalanine with ammonium hydroxide solution, and adjusting the pH of the eluate to the isoelectric point of phenylalanine (pH 5.5), as is conventional in itself. The yield was 65% of the phenylalanine originally found in the broth.

Strain	Phenylalanine, g/dl
No. 2256	0.00
AJ 3435	0.42
AJ 3436	0.71
AJ 3432	1.74
AJ 3437	2.20
AJ 3699	2.31

Example 4: The parent strain *B. flavum* No. 2247 and the four mutant strains described in Example 2 were cultured in the manner described in Example 3, and the amounts of phenylalanine accumulated in each broth after 72 hours were determined. They are listed in the table below.

Strain	Phenylalanine, g/dl
No. 2247	0.00
AJ 3695	0.23
AJ 3696	1.20
AJ 3697	1.40
AJ 3698	2.01

Production Using Enzymes Produced by *Flavobacterium aminogenes*

A method for producing L-amino acids, particularly L-phenylalanine, L-tryptophan and their derivatives from 5-benzylhydantoin, 5-indolylmethylhydantoin and their derivatives is disclosed by *K. Mitsugi, K. Sano, K. Yokozeki, K. Yamada, I. Noda, T. Kagawa, C. Eguchi, N. Yasuda, F. Tamura and K. Togo; U.S. Patent 4,016,037; April 5, 1977; assigned to Ajinomoto Co., Inc., Japan.*

The hydantoin compounds as mentioned above are produced by Strecker's reaction from aldehyde, and are intermediates for producing L-amino acids. For example, the hydantoin compounds are hydrolyzed to L-amino acids with the enzymes produced by *Flavobacterium proteus* ATCC 12848 (Japanese Patent 13850/1967).

It has been found that *Flavobacterium aminogenes* nov. sp. AJ-3912 (FERM P-3133) produces mainly in its cells an enzyme having remarkably high activity for converting hydantoin compounds of formula (1) where R is phenyl or substituted phenyl radicals or indolyl or substituted indolyl radicals

(1)

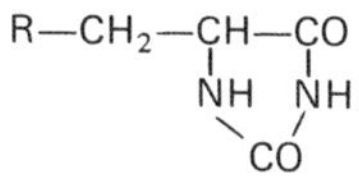

to L-amino acids of formula (2), where R is the same as in formula (1), in high yield.

(2)

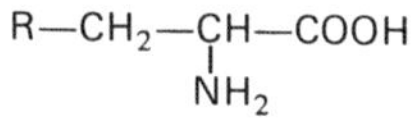

When *F. aminogenes* nov. sp. is cultured in conventional media, the enzyme for converting the hydantoin compounds to L-amino acid is produced mainly in the cells and slightly in the culture broth. The culture media used contain carbon source, nitrogen source, inorganic ions, and where required organic nutrients. When culture media contain the hydantoin compound of formula (1) such as 5-indolylmethyl-hydantoin, the enzyme activity is generally much higher. Cultivation is carried out at pH 6 to 9 at preferably 20° to 40°C under aerobic conditions. After 1 to 5 days cultivation, the enzyme is produced in the cells.

As the enzyme source, culture broth containing intact cells, homogenates of cells, sonicates of cells, freeze-dried cells, or cells dried with solvents and the like are preferably used. Protein fractions separated, for example, from the homogenate of cells, or the sonicate of cells by conventional methods such as gel-filtration or salting-out method can be also used as the enzyme source. It is theorized that a number of enzymes participate in the reaction of converting the hydantoin compounds to L-amino acids.

Example: An aqueous culture medium was prepared to contain 0.5 g/dl glucose, 1.0 g/dl yeast extract, 1.0 g/dl peptone, 0.5 g/dl NaCl, and 0.2 g/dl DL-5-indolylmethyl hydantoin. 50 ml batches of the aqueous culture medium were placed in 500 ml shaking flasks, heated with steam, and inoculated with *F. aminogenes* nov. sp. AJ 3912. Each flask was shaken at 30°C for 16 hours.

Cells were separated from 500 ml of culture broth by centrifuging, washed with physiological saline, and suspended in 500 ml of 0.1 M phosphate-buffer (pH 8.0). Each 5 ml of the suspension was added with 50 mg of the hydantoin compounds listed in the table below and 15 mg of sodium sulfite in the cases of DL-5-(3',4'-dihydroxybenzyl)-hydantoin, and D-5-(5'-hydroxyindolylmethyl)-hydantoin, and held at 37°C for 24 hours. The amount of L-amino acid shown in the table below was accumulated in the reaction mixture.

Hydantoin Compound	L-Amino Acid Accumulated	Amount (mg/ml)
DL-5-benzylhydantoin	L-phenylalanine	8.7
DL-5-(3',4'-dihydroxybenzyl)-hydantoin	L-3,4-dihydroxyphenylalanine	8.9
DL-5-(3',4'-methylenedioxybenzyl)-hydantoin	L-3,4-methylenedioxyphenylalanine	9.0
DL-5-(3',4'-dimethoxybenzyl)-hydantoin	L-3,4-dimethoxyphenylalanine	0.5
DL-5-(3'-methoxy-4'-hydroxybenzyl)-hydantoin	L-3-methoxy-4-hydroxyphenylalanine	8.7
DL-5-indolylmethylhydantoin	L-tryptophan	2.6
DL-5-(5'-hydroxyindolylmethyl)-hydantoin	L-5-hydroxytryptophan	0.72
D-5-(5'-hydroxyindolylmethyl)-hydantoin	L-5-hydroxytryptophan	0.73

Phenylalanine Derivatives by Hydroformylation Reactions

Derivatives of phenylalanine and other amino acids have been prepared by *T. Yukata, N. Yamakami, M. Honma, Y. Komachiya and H. Wakamatsu; U.S. Patent 3,996,288; December 7, 1976; assigned to Ajinomoto Co., Inc., Japan* from aldehydes produced by hydroformylation.

It has been found that alcohols and their hydrochloric and hydrobromic acid esters are converted to aldehydes having one additional carbon atom by hydroformylation in the presence of amides of carboxylic acids in good yields, and that the aldehydes are formed faster and at lower temperatures than in the known methods.

The exact function of the amide in the aldehyde formation is not known. The amide appears to participate in the reaction leading to the aldehyde, and limited tests indicate that an amide group present in the alcohol or ester employed as a starting material has an analogous effect. If the amide has an active hydrogen atom attached to its amide nitrogen, it reacts with the aldehyde formed and with available carbon monoxide according to the reaction disclosed in U.S. Patent 3,766,266 to form an N-acyl-α-amino acid.

The catalysts for this process are those known to be useful in carbonylation and hydroformylation reactions, and include, as principal active element, one of the transition elements of Group VIII of the Periodic Table, namely the iron group metals (iron, cobalt and nickel) and the platinum group metals (ruthenium, rhodium, palladium, osmium, iridium and platinum). For industrial purposes, catalysts including one of the elements of the iron group, particularly cobalt, are preferred. They may be supplied to the reaction zone as carbonyls, as metals, preferably finely dispersed, or as salts or complex compounds which form the carbonyls under the reaction conditions. Better results are often obtained when compounds containing phosphorus, nitrogen, oxygen, or halogen, which are known promoters for hydroformylation and carbonylation reactions, are also used as ligands.

When an N-disubstituted amide is used for producing aldehydes in accordance with this process an excess of amide is not required, and the amide is effective when present in the reaction mixture in a concentration of at least 0.01 mol per liter. When an amide having an active hydrogen atom on its amide nitrogen is used for producing an N-acyl-α-amino acid, the mol ratio of the amide to the alcohol or halide is not critical, but an excess of the amide tends to increase the yield.

The amount of catalyst employed varies with the metal used, but it is usually 1 to 0.01 mol per mol of the product formed. The reaction temperature is preferably between 50° and 200°C. The reaction pressure is usually 10 to 500 atmospheres, as in any carbonylation and hydroformylation reactions.

The reaction is preferably carried out in a system which includes a solvent as a diluent. The solvents normally employed in carbonylation and hydroformylation reactions are also effective for this process. When a halide ester is used as a starting material, especially together with a neutralizing agent (an acid acceptor), a mixture of water and an organic solvent sparingly soluble in cold water is preferred as a reaction medium for convenient operation and good yield. Suitable organic solvents include hydrocarbons such as toluene, benzene and hexane; ketones such as methyl isobutyl ketone and methyl ethyl ketone; ethers such as diethyl ether and dibutyl ether; nitriles such as butyronitrile and benzonitrile; and esters such as ethyl acetate and butyl acetate.

Details of the preparation of N-acetylphenylalanine by this process are shown in the following example. A 100 ml stainless steel pressure vessel equipped with an electromagnetic stirrer was used as a reactor for this process.

Example: 6.4 grams (50 mmols) benzyl chloride, 5.9 grams (100 mmols) acetamide, 4 grams dicobalt octacarbonyl, 47.5 ml acetone, and 2.5 ml water were placed in the reactor which was then charged to a gas pressure of 200 kg/cm^2 with a mixture of one volume carbon monoxide and 2 volumes hydrogen.

The contents of the reactor were stirred at 120°C for 20 minutes, whereby 67.6 mmols of the gas were absorbed. The reactor was cooled, and a blue liquid was poured from the reactor. It was found to contain 40 milliequivalents of a carboxylic acid. From a sample of the reaction mixture, the carboxylic acid was adsorbed on an anion exchange resin (Amberlite-4B), and the material recovered by elution was identified as N-acetylphenylalanine by elementary analysis, melting point, and infrared and mass spectra.

The amount of N-acetylphenylalanine in the reaction mixture was determined by gas chromatography of a sample after treatment with diazomethane to be 38.6 millimols, which corresponds to a 77.2% yield based on the benzyl chloride employed.

In a similar manner, N-acetyl-O-methyltyrosine was prepared from anise alcohol and acetamide, N-acetyltyrosine from p-hydroxybenzyl alcohol and acetamide, and N-acetylveratrylglycine from veratryl alcohol and acetamide.

Halobenzenesulfonyl-DL-Phenylalanines as Antiinflammatory Agents

For some time there has been continuing interest in the development of more effective, nonsteroidal, antiinflammatory agents which can be administered to mammals in effective dosages with minimal side effects. There is also an economic need for such agents which are relatively simple to produce from readily available, noncostly reagents. Compounds having the formula

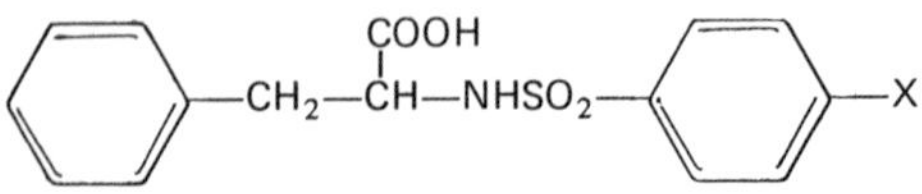

where X is F, Cl, or Br and the nontoxic, pharmaceutically acceptable inorganic and organic salts thereof have been developed by *R.F. Borne; U.S. Patents 3,852,469; December 3, 1974 and 3,772,377; November 13, 1973; both assigned to The University of Mississippi* for such use. Preferably the halogen substituent is bromine and is in the para position. These compounds have marked antiinflammatory activity when administered to mammals and can be given in effective, nontoxic doses. The compounds are preferably administered mixed with a nontoxic, pharmaceutically-acceptable diluent.

These compounds are prepared as, for example, by reacting the monohalogen substituted benzenesulfonyl chloride with phenylalanine in dilute aqueous sodium hydroxide, acidifying the product, and then separating the resulting solid. Optimum yields are generally obtained when the pH of the basic solution of phenylalanine is maintained between 9.5 and 10.5 during addition of the substituted benzenesulfonyl chloride.

The salts can be formed by conventional techniques by reaction of the acid form of the compounds with a therapeutically acceptable inorganic base, such as sodium, potassium, or ammonium hydroxide, or an organic base, such as an amine, e.g., methylamine, dimethylamine, triethylamine, ethanolamine, diethanolamine, triethanolamine, ethanol-diethylamine, or the like. Such salts are often preferred because of their greater solubility.

Example: 1.2 mols of a solution of p-bromobenzenesulfonyl chloride in dioxane was added to a solution of D,L-phenylalanine dissolved in 2 N NaOH at such a rate as to maintain the pH between 9.5 and 10.5. After addition, the resulting solution was stirred until no further pH decrease was observed. The solution was then acidified with concentrated HCl to a pH of between 2 and 3 and vigorously stirred. The resulting solid N-p-bromobenzenesulfonyl-D,L-phenylalanine was collected, dried, and recrystallized from the appropriate solvent system.

N-p-fluorobenzenesulfonyl-D,L-phenylalanine, N-p-chlorobenzenesulfonyl-D,L-phenylalanine, N-p-bromobenzenesulfonyl-D-phenylalanine and N-p-bromobenzenesulfonyl-L-phenylalanine were also prepared by the process of the above example. Additional details of the preparation and physical properties of these compounds are given in the table below.

X	Configuration	Yield (%)	MP (°C)	Recrystallization Solvent	$[\alpha]_D^{22}$
F	D,L	60	107-109	C_6H_6	–
Cl	D,L	77	138-140	$EtOH\text{-}H_2O$	–
Br	D,L	64	165-166	$EtOH\text{-}H_2O$	–
Br	D	50	136-138	$EtOH\text{-}H_2O$	+18.2°*
Br	L	60	137-139	$EtOH\text{-}H_2O$	-18.9°*

*EtOH.

The antiinflammation activity of these compounds were tested by two techniques, one an experimental in vitro technique described by J.H. Brown et al, *Proc. Soc. Exptl. Biol. Med.,* 125, 837 (1967) and by the standard in vivo rat paw edema technique described by C.A. Winter et al, *J. Pharmacol. Exp. Ther.,* 141, 369 (1963). The in vitro technique has been employed to screen antiinflammatory compounds by determining their ability to inhibit heat-induced hemolysis of red blood cells. The procedure, as described by J.H. Brown et al, above, was followed except that fresh human blood from fasted Type O+ donors was used rather than blood from anesthetized mongrel dogs. Results are given below.

The LD_{50} of the more active compounds was determined by the method of J.T. Litchfield et al, *J. Pharmacol. Exp. Ther.,* 96, 99 (1949) using four dose levels for each compound. Results are summarized in the table below.

	Percent Inhibition of Heat-Induced Hemolysis . . .Concentration (n)* . . .			Percent Inhibition of Edema . . . . Dose (n)*		
Compound	10^{-3} M	10^{-4} M	10^{-5} M	60 mg/kg	120 mg/kg	LD_{50} (mg/kg)
F-D,L	0 (9)	0 (9)	0 (9)	50 (16)	19 (16)	–
Cl-D,L	65 (9)	0 (9)	0 (9)	37 (32)	31 (32)	–
Br-D,L	80 (15)	32 (15)	5 (9)	43 (32)	74 (32)	385
Br-D	79 (18)	23 (18)	0 (9)	26 (16)	32 (16)	350
Br-L	67 (9)	11 (9)	11 (9)	31 (16)	60 (16)	–
Phenylbutazone	75 (15)	50 (15)	11 (15)	34 (32)	52 (32)	336**

*n is number of determinations per concentration or dose level.
**J. Ben-Bassat, E. Peretz and F.G. Sulman, *Arch. Int. Phamacodyn Ther.* 122, 434 (1959).

It will be noted that the compounds of this process were more effective than phenylbutazone in either or both the in vitro and in vivo assays. The N-p-bromobenzenesulfonyl-D,L-phenylalanine derivative showed the highest potency in both assays. This compound and the N-p-bromobenzenesulfonyl-D-phenylalanine derivative also exhibited lower toxicity than phenylbutazone in a significant degree.

TYROSINE AND DERIVATIVES

p-Hydroxyphenylalanine or 3-(4-hydroxyphenyl)-2-aminopropionic acid is a non-essential amino acid. Many of the compounds closely related in structure to L-tyrosine, such as L-dopa, are of current interest for use in the treatment of neurological disorders and other medical problems.

Tyrosine Produced Using *Corynebacterium glutamicum* Mutants

L-tyrosine has been produced for example by fermentation using a phenylalanine-requiring strain of *Corynebacterium glutamicum* such as disclosed in Japanese Patent 14395/63. However, it would be desirable to obtain higher yields of L-tyrosine than are produced in this prior process.

Mutant strains of *C. glutamicum* have been developed by *H. Hagino, K. Nakayama and H. Yoshida; U.S. Patent 3,787,287; January 22, 1974; assigned to Kyowa Hakko Kogyo Co., Ltd., Japan* which produce large amounts of L-tyrosine. These strains of *Corynebacterium glutamicum* show a resistance to at least one compound selected from analogs of tyrosine and phenylalanine such as 3-aminotyrosine, p-fluorophenylalanine, m-fluorophenylalanine, p-aminophenylalanine, β-2-thienylalanine or o-fluorophenylalanine.

Such property may be given to a strain by mutation induced by conventional methods such as irradiation by ultraviolet light, x-ray, ^{60}Co, etc., chemical treatment, etc. When a strain is resistant to one of the listed compounds, it is sometimes also resistant to other compounds. In fact, such cross resistance often occurs between p-fluorophenylalanine and m-fluorophenylalanine or o-fluorophenylalanine. However, a cross resistance hardly occurs between 3-aminotyrosine and p-fluorophenylalanine; 3-aminotyrosine and m-fluorophenylalanine; 3-aminotyrosine and p-aminophenylalanine; 3-aminotyrosine and β-2-thienylalanine; etc. Therefore, when a strain having a resistance to two or more of the listed compounds is desired, a series of mutation treatments is conducted on the parent strain to obtain the desired resistance step by step.

Production of L-tyrosine is preferably carried out by fermentation, under aerobic conditions, of an aqueous nutrient medium such as by shaking or aeration and stirring of the culture. It is preferred to maintain the temperature between 20° to 40°C and the pH approximately neutral during culturing in order to obtain high yields. However, the temperature and pH conditions may vary according to the specific strain of microorganism used.

Example 1: An L-tyrosine-producing strain, *C. glutamicum* (ATCC 21568) resistant to 3-aminotyrosine is used. 1 ml of seed culture obtained by culturing this strain in a seed medium containing 2% glucose, 1% peptone, 1% yeast extract and 0.3% NaCl at 30°C for 24 hours is inoculated into a 250 ml Erlenmeyer flask containing 10 ml of a fermentation medium. The composition of the fermentation medium used is: 10% sucrose, 0.05% K_2HPO_4, 0.05% KH_2PO_4, 0.025% $MgSO_4{\cdot}7H_2O$, 2% ammonium sulfate, 0.5% NZ-amine, 30 μg/l of biotin and 2% $CaCO_3$ (pH 7.2). Culturing is carried out at 30°C for 4 days with shaking and 5.7 mg/ml of L-tyrosine is formed.

After the completion of the fermentation, one liter of the culture liquor is adjusted to pH 10 with aqueous ammonia and then centrifuged. The microbial cells and $CaCO_3$ are removed from the liquor. The liquor is then concentrated under a reduced pressure to 50 ml, adjusted to pH 7.0 with hydrochloric acid and left in a cold room. The resulting crystals of L-tyrosine are recovered by filtration, yield 3.8 grams.

Example 2: L-tyrosine-producing strains of *C. glutamicum* having the properties shown in the following table are used. The strains which show a resistance to two or three analogs are obtained by adding a resistance to each of the analogs step by step by a series of mutations. The phenylalanine-requiring property is added prior to the addition of the resistance. These strains are respectively cultured in a seed medium described in Example 1 for 24 hours with shaking. One milliliter of each of the thus prepared seed cultures is inoculated into a large test tube containing 10 ml of a fermentation medium comprising: 10% cane blackstrap molasses (calculated as glucose), 0.05% K_2HPO_4, 0.05% KH_2PO_4, 0.025% $MgSO_4 \cdot 7H_2O$, 2% ammonium sulfate, 0.5% corn steep liquor and 2% $CaCO_3$ (pH 7.2). Culturing is carried out at 30°C for 4 days. The results are also shown in the table below.

Strain	L-tyrosine Formed, mg/ml
Phe^-, $3AT^R$ (ATCC 21567)	6.7
Phe^-, $3AT^R$, PFP^R (ATCC 21569)	10.3
Phe^-, $3AT^R$, MFP^R (ATCC 21570)	8.1
Phe^-, $3AT^R$, PAP^R (ATCC 21571)	7.4
Phe^-, $3AT^R$, TA^R (ATCC 21572)	7.7
Phe^-, $3AT^R$, PFP^R, PAP^R (ATCC 21573)	12.3
Phe^-, MFP^R (ATCC 21562)	5.8
Phe^-, OFP^R (ATCC 21564)	6.0
Phe^-, PFP^R (ATCC 21565)	5.8
PFP^R (ATCC 21566)	4.9
MFP^R (ATCC 21563)	4.3

Note: Phe^- = requiring phenylalanine
$3AT^R$ = resistant to 3-aminotyrosine
PFP^R = resistant to p-fluorophenylalanine
MFP^R = resistant to m-fluorophenylalanine
PAP^R = resistant to p-aminophenylalanine
TA^R = resistant to β-2-thienylalanine
OFP^R = resistant to o-fluorophenylalanine

Tyrosine Derivatives for Treating Gastric Ulcers

Tyrosinol and phenylalanyltyrosine derivatives have been developed by *T. Miki, Y. Hosokawa, T. Miwa, H. Fujita, M. Asano and S. Aibara; U.S. Patents 4,025,644; May 24, 1977 and 4,057,629; November 8, 1977; both assigned to Daiichi Seiyaku Co., Ltd., Japan* for use in the treatment of gastric ulcers. These derivatives have the formulas

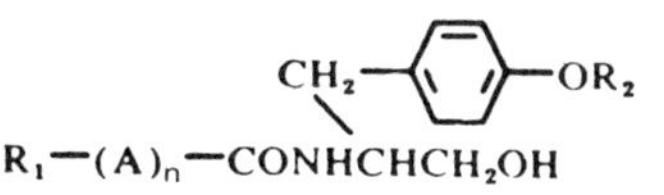

(1)

or

(2)

where R_1 is phenyl or cycloalkyl and the phenyl may have at least one substituent selected from lower alkyl, lower alkoxy, amino and nitro; R_2 is hydrogen or lower alkyl; A is $—CH_2O—$ or $—CH_2CH(O—$lower alkyl$)—$; n is 0 or 1; when R_1 together with $(A)_n$ is unsubstituted phenyl, R_2 is a lower alkyl; and R_3 is hydrogen or lower alkoxy.

Compound (1) can be synthesized by various reactions known per se, such as reduction of the corresponding carboxylic acid, ester or aldehyde, or acylation of tyrosinol or O-methyl-tyrosinol. Likewise, compound (2) can be synthesized by acylation of phenylalanyl-tyrosine, condensation or N-acylphenylalanine with tyrosine ester, and like processes.

When these tyrosine derivatives are used in humans for prevention or treatment of gastric ulcers, the daily oral dosage should normally be within the range of 200 to 1,200 mg/man, preferably 400 to 800 mg/man. As for acute toxicity, the LD_{50} values for these tyrosine derivatives were found to be over 4 g/kg body weight in mice and over 2 g/kg body weight in rats, when orally administered.

As can be seen from the table below, such tyrosine derivatives display remarkable effects against gastric ulcers and especially regarding experimental chronic ulcers are surprisingly superior to known drugs for treating these ulcers in humans.

Compound	Dose, mg/kg/day	Relative Therapeutic Effect
Compound (1) where		
$R_1 = CH_3O—$ 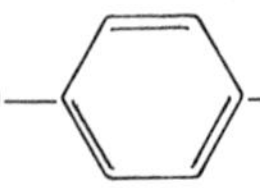 $—,$		
$R_2 = H$, n = 1, $A = —CH_2O—$	50	7.72
Compound (2) where		
$R_3 = CH_3O—$	50	7.85
Glutamine	1,000	1.00
Gefarnate	200	0.77

The therapeutic effect was calculated relative to the glutamine effect which was considered to be 1.00.

Example 1: In 150 ml of 50% aqueous methanol, 13.1 grams of sodium borohydride was added with stirring while cooling with ice. To the mixture was added dropwise 250 ml of a methanol solution containing 26.5 grams of methyl N-(4-methoxybenzyloxycarbonyl)-tyrosinate. The mixture was stirred while cooling with ice for 2 hours and at room temperature for four hours. After neutralization with 10% hydrochloric acid, the mixture was concentrated in vacuo and extracted with ethyl acetate. The extract was washed with water, dried and concentrated to dryness. The residue was crystallized from ethyl acetate-petroleum ether to give N-(4-methoxybenzyloxycarbonyl)-tyrosinol: yield 63%.

Example 2: In 500 ml of a mixture of chloroform-tetrahydrofuran (2:1), were dissolved 29.9 grams of N-(4-methoxybenzoyl)-phenylalanine, 23.2 grams of methyl tyrosinate hydrochloride, 10.1 grams of triethylamine and 11.6 grams of N-hydroxysuccinimide and the mixture was cooled with ice. To the cold solution was added 20.6 grams of N,N'-dicyclohexylcarbodiimide. The mixture was stirred for 2 hours and then allowed to sit at room temperature overnight. The precipitates were removed by filtration and the filtrate was concentrated. The residue was extracted with ethyl acetate. The extract was washed with 2% hydrochloric acid, 4% sodium bicarbonate and water, then dried and concentrated to dryness. The residue was crystallized from ethyl acetate to give 35 grams of methyl N-(4-methoxybenzoyl)-phenylalanyltyrosinate with a MP of 195° to 197°C in a yield of 74%.

Example 3: In 150 ml of methanol, was dissolved 30 grams of methyl N-(4-methoxybenzoyl)-phenylalanyltyrosinate. After addition of 135 ml of 1 N sodium hydroxide, the mixture was stirred at 50°C for 2 hours. The reaction mixture was neutralized with hydrochloric acid, concentrated and extracted with ethyl acetate. The extract was washed with water, dried and concentrated to dryness. The residue was crystallized from ethyl acetate to give 23.9 grams of N-(4-methoxybenzoyl)-phenylalanyltyrosine with a MP of 198° to 200°C in a yield of 82%.

TRYPTOPHAN PRODUCTION

1-α-aminoindole-3-propionic acid or tryptophan is one of the essential amino acids needed for human growth. There is commercial interest in this amino acid as an additive not only for the hydrolyzed protein formulations used for human nutrition but also for use in animal feeds.

Use of *Corynebacterium glutamicum* Mutants

The disclosure by *K. Nakayama and H. Hagino; U.S. Patent 3,849,251; Nov. 19, 1974; assigned to Kyowa Hakko Kogyo Co., Ltd., Japan* provides a process for the production of L-tryptophan by the fermentation of certain mutant strains of *Corynebacterium glutamicum.* These mutant strains of *C. glutamicum* which require phenylalanine and tyrosine for growth and which are resistant to at least one analogue of tryptophan and at least one phenylalanine or tyrosine analogue are cultured in a nutrient medium to produce increased amounts of extra-cellular L-tryptophan.

To obtain the mutant strains which are suitable for carrying out this process, bacteria belonging to the species *C. glutamicum* are artificially mutated into phenylalanine and tyrosine-requiring strains. These mutants are then again mutated by artificial means. The resulting mutants are then cultured on minimum agar medium mixed with at least one tryptophan analogue and at least one analogue of phenylalanine or tyrosine at a concentration not lower than a concentration which would normally inhibit growth of the parent strain. From colonies which have grown on this medium, the desired strains are isolated. For example, the derivation of mutants of the phenylalanine and tyrosine-requiring bacterium, *C. glutamicum* KY 9456, ATCC 21854, is shown below.

The strain (KY 9456, ATCC 21854) undergoes growth-inhibition by tryptophan analogues and phenylalanine analogues and tyrosine analogues when cultured in the minimum agar medium mixed with phenylalanine and tyrosine at concentrations as shown in the table below. In view of the minimum inhibitory concentration determination of the parent strain, the analogue-resistant strains are screened by scattering the cells of a further mutated strain of *C. glutamicum* KY 9456, ATCC 21854 on a minimum agar medium supplemented with a tryptophan analogue and a phenylalanine analogue or a tyrosine analogue at a concentration above the growth inhibitory concentration as listed in the table below.

Analogues	Minimum Concentration to Inhibit Growth (μg/ml)
DL-4-methyltryptophan	500
DL-5-methyltryptophan	500
DL-5-fluorotryptophan	500
DL-6-fluorotryptophan	3,000
L-tryptophan hydroxamate	500
DL-4-fluorophenylalanine	25
DL-3-fluorophenylalanine	50
DL-2-fluorophenylalanine	50
DL-β-2-thienylalanine	600
DL-phenylalanine hydroxamate	400
L-3-aminotyrosine	4,000
DL-tyrosine hydroxamate	600
DL-4-aminophenylalanine	3,000

After a suitable incubation period, cells are isolated from colonies which have been able to form on the medium. These cells are then suitably cultured to develop microbial populations for industrial purposes.

As a medium for screening the analogue-resistant mutant, usually a solid agar medium containing the analogues, a carbon source, a nitrogen source, inorganic salts, tyrosine, phenylalanine, biotin, vitamin B_1 and trace amounts of other nutrients necessary for growth of the parent bacterium, is employed. As for trace nutrients, natural materials which normally contain the nutrients may be employed.

Specific preferred mutant strains of *C. glutamicum* obtained by means of the above screening process have been accorded accession numbers ATCC 21842 to ATCC 21851.

Example: In this example, a L-tryptophan-producing mutant of *C. glutamicum* which has a tyrosine and phenylalanine requirement and a resistance to 5-methyltryptophan, 6-fluorotryptophan and 4-fluorophenylalanine, that is, *C. glutamicum* Trp-1, ATCC 21842 is used. The mutant is cultured in a seed medium containing 2% glucose, 1% peptone, 1% yeast extract and 0.3% NaCl at 30°C for 24 hours. 1 ml of the resulting seed culture is inoculated into 10 ml of a fermentation medium having the following composition in a 250 ml Erlenmeyer flask: 10% glucose, 0.05% KH_2PO_4, 0.05% K_2HPO_4, 0.025% $MgSO_4 \cdot 7H_2O$, 2% $CaCO_3$, 2% $(NH_4)_2SO_4$, 0.5% NZ-amine and 30 μg/l biotin with a pH of 7.2.

Culturing is carried out at 30°C for 4 days with shaking, whereby L-tryptophan is produced in a yield of 3.6 mg/ml.

2 liters of the culture broth resulting from the fermentation is centrifuged to remove the microbial cells and $CaCO_3$; and the resulting supernatant is passed through a bed of a strongly acidic cation exchange resin, Diaion SK-104 (H^+ form) to adsorb the L-tryptophan. After water washing, the resin is subjected to elution with 0.5 N aqueous ammonia, and then the resulting eluate is concentrated to obtain crude crystals of L-tryptophan. The crude crystals are dissolved in a small amount of hot 50% aqueous ethanol. The resulting solution is decolor-

ized with active carbon and cooled, whereby 3.5 grams of L-tryptophan is recrystallized. When the parent strain of the mutant, that is, *C. glutamicum* KY 9456, ATCC 21854 having a tryosine and phenylalanine requirement and *C. glutamicum* KY 9104, ATCC 21334 having a histidine requirement (disclosed in U.S. Patent 3,594,279) are cultured under the same conditions, the yields of L-tryptophan are 0.2 mg/ml, and 0.5 mg/ml, respectively.

Using Mutants of *Bacillus subtilis*

A prior process for the production of tryptophan has used a Bacillus mutant which requires anthranilic acid. However, the enzyme system is weakened resulting in reduced production of L-tryptophan when anthranilic acid is used as a starting material in a larger amount than that required for the growth of microorganism used regardless of the stages of addition of anthranilic acid, namely, even after the completion of the growth and formation of L-tryptophan-synthetic enzyme system.

In the process developed by *K. Arima, I. Nogami and M. Yoneda; U.S. Patent 3,801,457; April 2, 1974; assigned to Takeda Chemical Industries Ltd., Japan* a mutant which requires anthranilic acid for growth and at the same time is resistant to the inhibitive action of anthranilic acid is used.

The anthranilic acid-requiring microorganisms employed in this process are capable of growing in the presence of any substance that occurs in the biosynthetic path from the anthranilic acid to L-tryptophan, for example, indole or L-tryptophan itself and usually these microorganisms are of the genus Bacillus such as *B. subtilis, B. pumilus, B. circulans, B. licheniformis, B. polymyxa, B. firmus, B. brevis, B. sphericus,* etc. Among these microorganisms, two mutants of *B. subtilis* have been selected for deposit as IFO 13219/ATCC 21777 and IFO 13220/ATCC 21778 respectively.

This process may be carried out using indole in place of anthranilic acid as a starting material. However, the yield is considerably lower than the yield obtained by the addition of anthranilic acid. In the process, a microorganism of the genus Bacillus which requires anthranilic acid for its growth and is resistant to anthranilic acid is first cultivated in the presence of a growth-limiting amount of one or more members of the group consisting of anthranilic acid, indole and/or L-tryptophan and, after the microorganism has completed its growth under those conditions anthranilic acid is fed to the broth and the microorganism is further cultivated and allowed to elaborate L-tryptophan.

By growth-limiting amount is meant an amount which is not more than that required for the maximum growth of the microorganism used and, while such amount varies somewhat with the strains, cultural conditions and other factors, the amount may, for example, be 5 to 60 γ/ml anthranilic acid and 10 to 80 γ/ml of indole or tryptophan, respectively when they are used independently.

In addition to the above components, a known mono- and/or dicarboxylic acid(s) of low molecular weight is used with much better results in the process. Such carboxylic acids are those which have up to 7 carbons and do not inhibit the growth of the microorganisms used and are represented by the general formula shown on the following page.

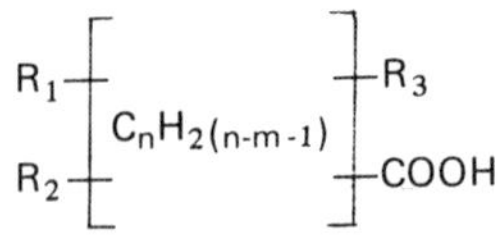

In the above formula R_1 is H or COOH, R_2 and R_3 are H or OH, n is whole number of 1 through 5 and m is 0 or 1 ($n>m$), respectively. Of the possible acids, succinic acid, tartaric acid, fumaric acid and maleic acid are particularly suitable and similarly, acetic acid is also quite suitable.

Desirably, the cultivation is carried out aerobically, for example, under agitation, or under aeration and agitation. The incubation is generally made at 20° to 40°C and at a pH preferably between 6.0 and 9.0.

The isolation and purification of L-tryptophan from the culture broth can be easily effected by conventional means as, for example, by a series of adsorption and elution steps with the use of activated carbon or an ion exchange resin.

Example: 20 parts by volume of a seed-culture medium composed of 2% soluble starch, 2% corn steep liquor, 0.3% dipotassium phosphate, 0.1% monopotassium phosphate, 30 γ/ml anthranilic acid and 1 γ/ml biotin (pH 7.2) is poured into a fermentor, (50 parts by volume in capacity) which is then sterilized and is inoculated with a culture of an anthranilic acid-requiring, anthranilic acid-resistant mutant (IFO 13219 ATCC 21777) derived from *B. subtilis* (IFO 3009). The inoculated medium is incubated at 37°C and aerated for 24 hours.

Separately, 200 parts by volume of a fermentation medium composed of 14% glucose, 3% corn steep liquor, 0.9% urea, 1% sodium succinate (0.0437% as free acid), 0.5% dicalcium phosphate, 0.5% tricalcium phosphate, 0.037% disodium ethylenediamine tetraacetate ($2H_2O$), 3% calcium carbonate, 1 γ/ml biotin and 20 γ/ml anthranilic acid (pH 7.2) is poured into a fermentor (500 parts by volume in capacity) and then sterilized. 20 parts by volume of the above-prepared seed culture is transferred to the fermentation medium and the inoculated medium is incubated at 37°C for 16 hours. Then, a neutralized solution of anthranilic acid is added in a concentration of 8 mg relative to 1 ml of the culture broth, followed by a further incubation at 37°C and aeration for 48 hours.

The resulting fermentation broth contains 10.4 mg/ml of L-tryptophan. The above broth (200 parts by volume) is adjusted to pH 6.5 and, after boiling at 100°C for 10 minutes, the cells are removed by centrifugation. The supernatant is treated with 50 parts by weight of activated carbon so that L-tryptophan is adsorbed thereon. After washing with water, the activated carbon is subjected to elution with 500 parts by volume of 50% ethanol to collect L-tryptophan. The eluate is concentrated under reduced pressure to 200 parts by volume and the concentrate is adjusted to a pH 4.5 and made up to 300 parts by volume. The solution is passed through 60 parts by volume of ion exchange resin (Dowex 50 x 8, Na-form) which has been buffered with 1,000 parts by volume of 0.1 M citrate buffer (pH 3.4).

Then, 1,500 parts by volume of 0.1 M citrate buffer (pH 5.0) is poured onto the column and, thereafter, the L-tryptophan is eluted with 400 parts by volume of 50% ethanol-0.28% ammonia. The eluate is concentrated under reduced pressure to 180 parts by volume and the concentrate is adjusted to pH 6.0 and made up to 250 parts by volume with water. The solution is treated again with 20 parts by weight of activated carbon, which is then eluted with 50% ethanol-0.28% ammonia. The eluate is concentrated under reduced pressure, whereupon 1.58 parts by weight of crystalline L-tryptophan is obtained.

Production Using *Candida humicola*

The process disclosed by *L.-E. Paulsson, L.I. Wiberger and A.W. Ohlsson; U.S. Patent 3,915,796; October 28, 1975; assigned to AB Bofors, Sweden* is based on the use of an isolate of *C. humicola* obtained after screening more than 600 organisms from different soil samples. The isolate is deposited in Centraal Bureau voor Schimmelcultures under number J IA CH XII (CBS 6434). The derivatives of tryptophan, which may be prepared by *C. humicola*, have the formula

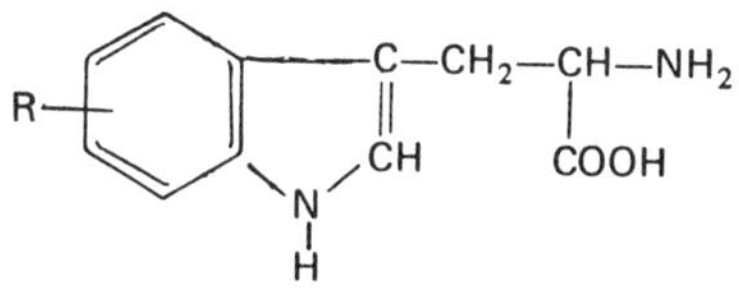

where R is hydrogen, hydroxy, halogen, lower alkyl or lower alkoxy. This process for the preparation of these tryptophan derivatives is characterized by culturing the *C. humicola* under aerobic conditions in a continuously supplied nutrient substrate with a separate continuous supply of the corresponding indole derivative of the formula

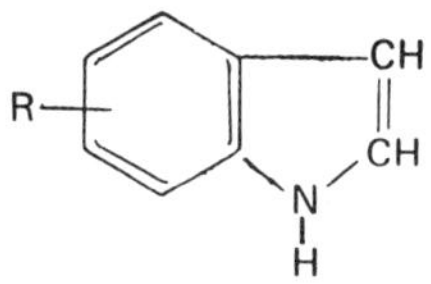

where R is hydrogen or hydroxy or halogen or alkyl or alkoxy. The microorganism used is characterized by a high productivity of L-tryptophan from indole. As the microorganism is sensitive to a high concentration of indole in the culture liquor, a continuous supply of the precursor must be made. The microorganism may be cultured on a completely defined substrate and besides carbohydrates and thiamine it does not require any other organic substances for its growth.

This isolate of *C. humicola* has a metabolism including tryptophan degradation started by tryptophan pyrolase activity. By changing the culture conditions and the substrate composition the degradation may however be controlled so that no decrease of the tryptophan yield may be seen. In order to stimulate the formation of mutants which give no tryptophan degradation, niacin may be added to the substrate in a concentration of 0.05 to 0.5 g/l of the substrate.

Comparatively low concentrations of free indole may decrease the growth velocity of the cells. The conversion of indole to tryptophan occurs however very rapidly and a supply of, e.g., 0.5% (w/v) of indole in relation to the substrate supply, will give slight amounts of free indole in the fermentor liquor. It is important to make the estimate of indole in the fermentor on substrate containing cells, as a very large amount of the free indole is intracellular. The feature that appears to have the greatest influence on the conversion velocity of indole to tryptophan and on the degradation of tryptophan via kynurenine, is the concentration of biologically available iron, e.g., in the form of ferrous citrate. The concentration of iron (Fe^{++}) in the culture liquor must be below the contents that will give degradation of tryptophan but be above the contents at which the growth of the culture will be severely disturbed and continuous culture thus be impossible.

The contents of biologically available oxygen in the fermentor liquor is obviously also important as regards the degradation velocity of tryptophan, since the tryptophan pyrolase is an enzyme containing a heme group and the degradation of tryptophan is thus an oxidative process.

To control the residual indole in the fermentor contents, a number of analyses are carried out. The titration frequency and the level of oxygen saturation will increase strongly at increased residual indole due to the fact that the aerobic metabolism is changed to an anaerobic one. These values are followed up and may be used to control the supply of ferrous citrate or the aeration.

The analyses on residual indole are carried out with unseparated fermentor contents on a gas chromatograph with o-nitrophenyl ethanol dissolved in methanol as an internal standard. Analyses on cell-free fermentor liquor do not give valuable information, since the indole is so rapidly absorbed by the cell. Too high contents of residual indole may also be used as a control signal to decrease the supply velocity of indole and/or to increase the iron supply and/or to increase the aeration.

To control the iron concentration, continuous analyses are carried out on a dextran gel column (Sephadex) with UV photometer and thereafter separate UV-analysis of the tryptophan fractions. These analyses give a good indication of metabolites of tryptophan, principally kynurenic acid. It has also been found that the capacity of the used strain to synthesize tryptophan is not limited by feedback inhibition. Thus the culture stands up to about 15 g/l of tryptophan without changing the synthesis velocity.

Example 1: Low Indole Supply and High Fe-Content – The fermentor is a Chemap GF 0014 with mechanical foam quencher. The conditions for the reaction, run at 26°C and a pH of 5.5, are shown in the table below.

Number of revolutions, rpm	2,000
Substrate velocity, ml/h	200
Volume, liters	4.8
Dilution velocity, %	4.1
Substrate	
Indole, g	1
p-Aminobenzoic acid (PABA), g	0.1
Yeast extract, g	25
Solution A (3% $CaCl_2 \cdot 2H_2O$), ml	200
Solution B (10% KH_2PO_4), ml	200
Solution C (5% $MgSO_4 \cdot 2H_2O$, 0.125% Fe-citrate, 0.015% $MnSO_4 \cdot 4H_2O$, 0.02% $ZnSO_4 \cdot 7H_2O$, 10% NaCl), ml	200
Glucose (separately autoclaved), kg	1
Distilled water, liters, qs	10
Aeration, liters per hour per liter of culture	80

Inoculation occurred with 50 ml of a 24 hour old culture of *C. humicola* which had been grown on a rotary shaker. The pH during the culturing was maintained at 6.5 by titration with NH_3. The culture was grown during 24 hours and then the pumping of substrate and indole was started. The velocity of the indole supply was 0.14% (w/v) of that of the substrate.

At this content of Fe-citrate, 24 mg/l, a good velocity of indole was obtained and no residual indole could be seen. The tryptophan production failed completely due to the fact that the tryptophan formed was at once degraded to kynurenic acid.

Example 2: Low Indole and Low Fe-Contents – The same conditions as in Example 1 were used but the substrate had the following composition.

NH_4Cl, g	50
Niacin, g	0.5
Fe-citrate, mg	10
Thiamine, mg	50
Solution K (0.068 g ammonium molybdate, 0.34 g $CuSO_4$), ml	10
Solution A (0.75% $CaCl_2{\cdot}2H_2O$), ml	100
Solution Bf (8% KH_2PO_4, 2.56% K_2HPO_4), ml	200
Solution Bk (10% KH_2PO_4), ml	200
Solution C (7.88% $MgSO_4{\cdot}7H_2O$, 0.11% $MnSO_4{\cdot}H_2O$, 0.02% $ZnSO_4{\cdot}7H_2O$, 10% NaCl), ml	200
Saccharose, kg	1
Distilled water, liters, qs	10

The inoculation, pH and culturing are as in Example 1. The velocity of the indole supply was 0.265% as calculated on the substrate supply. The production of tryptophan during the culture was 3.6 g/l, 24 hours (yield 78%). No degradation could be seen.

Fermentation Process Using Methanol in Culture Media

A large number of fermentative procedures involving different yeast and bacteria isolates are known for the production of L-tryptophan. Indole is used in most cases as precursor. The processes further involve the use of normal fermentative substrates, including for example yeast extract, meat extract, corn steep liquor, carbohydrate or some other complex nutritional preparation which as a rule constitutes difficulties in reproducibility together with limitations concerning the capacity for production.

The process developed by *S.V. Gatenbeck and P.O. Hedman; U.S. Patent 3,963,572; June 15, 1976; assigned to AB Bofors, Sweden* provides for the fermentative production of L-tryptophan and derivatives thereof with good reproducibility and production capacity by the use of methanol as an inexpensive source of carbon for the microorganisms.

However, a microorganism of a type that assimilates methanol in such a way that the methanol after oxidation reacts with intracellular glycine, whereby serine is formed, must be used. Yeasts and bacteria which metabolize methanol by the sugar phosphate pathway are poor tryptophan producers and should not be used. A preferable procedure for the fermentative production of tryptophan and its derivatives comprises growing under aerobic conditions such isolates of Pseudomonas and Methylomonas that are able to utilize methanol as the carbon source via serine. This process can be either batch culture or continuous culture. A mineral salt solution is used as the substrate as described in Example 1. No organic substrate component needs to be included with the exception of methanol and indole. The growth is accomplished preferably at a temperature of 25° to 30°C. The pH must be carefully regulated continuously during the cultivation, and ought to be held at 5.5 to 8.5, preferably between 6.5 to 7.5.

The supply of methanol should be added continuously regardless of whether it is batch or continuous culture. The methanol concentration should not exceed the marginal toxic value for the cultured isolate which lies around 5 volume percent methanol. The suitable concentration of methanol for optimal production of tryptophan lies between 0.1 and 1.0% and preferably around 0.5 volume percent methanol. It is thus necessary for a continuous automatic analysis of methanol. Several methods for such an analysis are described in the literature and the method which has been used here is based on gas chromatography.

In this process, indole or an indole derivative used as a tryptophan precursor is supplied continuously. Since the indole concentration is extremely critical for the process its concentration must also be regulated with the aid of continuous automatic analysis. The concentration of free indole in the culture medium must be kept below a value of 500 mg/l.

The production can be accomplished by an aerobic process, air being supplied by a conventional method and then distributed by a turbine agitator. The culturing should be carried out under oxygen limiting conditions, owing to the sensitivity of tryptophan for oxidative decomposition. The period of growth ought to lie between 30 and 90 hours with batch culture. After the cells have been separated an ion exchanger is used for the isolation and purification of produced tryptophan.

Example 1: Pseudomonas AM 1 is grown in 500 ml Erlenmeyer flasks on a mechanical shaker, at 25°C and containing 100 ml substrate of the following composition:

KH_2PO_4, g/l	1.4
Na_2HPO_4, g/l	5.4
$(NH_4)_2SO_4$, g/l	0.5
$MgSO_4 \cdot 7H_2O$, g/l	0.2
$CaCl_2$, mg/l	7.4
$FeSO_4 \cdot 7H_2O$, mg/l	5.0
$Na_2MoO_4 \cdot 2H_2O$, mg/l	2.5
$MnSO_4 \cdot H_2O$, mg/l	1.8
Methanol, g/l	5

After the microorganism has grown out, 11.2 mg of indole is added, and after 48 hours of growth, 3.3 mg of tryptophan is obtained which corresponds to an exchange of 17% calculated on added indole.

Example 2: The microorganism Pseudomonas AM 1 is grown aerobically in a fermentor on a substrate of the same composition as in Example 1. By titration with ammonia, the pH is kept at 6.7. The growing temperature is 28°C. To a

2 liter quantity of substrate, 510 mg of indole is added in two portions. After 72 hours of fermentation 490 mg tryptophan is obtained which corresponds to an exchange of 55% calculated on the added quantity of indole.

5-Hydroxytryptophan, 5-methoxytryptophan, 5-methyltryptophan and 7-methyltryptophan could also be produced from the corresponding indole precursors by the above processes. Also *Methylomonas methanolica* can be substituted for the Pseudomonas AM 1 in the production of tryptophan and its derivatives.

Condensation of Indoles with β-Haloalanine Using Tryptophanase

It is known that tryptophan and hydroxytryptophan are produced from indole and 5-hydroxy-indole, respectively by reaction with an α-amino acid such as serine or cysteine (Japanese Patent 46348/72) by the action of an enzyme produced by a microorganism of genus Proteus, Erwinia, Escherichia, Pseudomonas or Aerobacter.

S. Konosuke and K. Mitsugi; U.S. Patent 3,929,573; December 30, 1975; assigned to Ajinomoto Co., Inc., Japan have found that the same enzyme catalyzes the formation of tryptophan from indole and β-halogenoalanine, and hydroxytryptophan from 5-hydroxyindole and β-halogenoalanine. The yield of tryptophan or hydroxytryptophan is higher, and moreover, the reaction rate is very high as compared with the known methods. The above microorganisms are cultured in conventional media containing a carbon source, nitrogen source, inorganic ions, and preferably a minor organic nutrient. Addition of a small amount of tryptophan or 5-hydroxytryptophan promotes the production of enzyme activity.

Preferably 0.01 to 0.5 g/dl tryptophan or hydroxytryptophan are added to the medium to increase the enzyme activity. When the microorganisms are cultured in the above medium for 2 to 7 days at pH 4 to 9, and at 27° to 40°C, the enzyme capable of producing tryptophan and hydroxytryptophan is accumulated in the cells or in the broth. The cells or the broths themselves are used as an enzyme source. Freeze-dried cells, homogenate of cells, sonicate of cells, autolysate of cells, cells treated with acetone and enzyme preparation are also used. In this process the term cells includes those homogenate of cells and materials derived therefrom.

The reaction mixture contains the enzyme or a source of the enzyme, β-chloroalanine, and indole or 5-hydroxyindole. Pyridoxal-5-phosphate is necessary as the coenzyme when purified apo-enzyme is used. A reducing agent such as thiosulfate is preferably added to the reaction mixture to increase the yield of tryptophan or hydroxytryptophan. The reaction mixture is held at 5° to 60°C and at pH 5 to 10 until indole or 5-hydroxyindole is substantially converted to tryptophan or hydroxytryptophan. The tryptophan or hydroxytryptophan formed in the reaction mixture can be isolated and purified by conventional methods such as ion exchange chromatography, or precipitation at the isoelectric point.

Example 1: Proteus morganii IFO 3848 was cultured in 60 ml batches of the medium shown below held in 500 ml flasks with shaking at 30°C for 20 hours. Cells were collected by centrifuging from 1 liter combined culture broth thus obtained, and added to 1 liter of 0.1% NH_4Cl-NH_4OH buffer solution (pH 8.0) containing 20 grams indole and 20 grams β-chloro-L-alanine. The reaction mixture was held at 37°C for 20 hours with stirring. 24 grams of L-tryptophan was produced (by microbiological assay method).

Medium	Amount
L-tryptophan, g	2
KH_2PO_4, g	0.5
$MgSO_4 \cdot 7H_2O$, g	0.5
$FeSO_4 \cdot 7H_2O$, mg	10
$MnSO_4 \cdot 4H_2O$, mg	7
Corn steep liquor, ml	20
Yeast extract, g	2
Casein hydrolysate, g	5
Tap water, liter	1
pH (KOH)	7.0

After dissolving the precipitated tryptophan with NaOH, the reaction mixture was filtered to remove cells, adjusted to pH 4 with HCl and passed through a column packed with active carbon. L-tryptophan was eluted with 2 N NH_4OH, the eluate was concentrated, and 20.2 grams L-tryptophan crystals were obtained.

Cells of *Aerobacter aerogenes* ATCC 7256 or *Pseudomonas perlurida* ATCC 490 prepared as above were used in place of cells of *Proteus morganii* IFO 3848. In the case of ATCC 7256, 9.0 grams of L-tryptophan was produced, and in the case of ATCC 490, 19.1 grams of L-tryptophan was produced.

In comparison tests, β-chloro-L-alanine was replaced with 20 grams L-serine, and 22.3 grams, 8.0 grams and 18.4 grams of L-tryptophan were produced by the cells of *Proteus morganii* IFO 3848, *Aerobacter aerogenes* ATCC 7256 and *Pseudomonas perlurida* ATCC 490, respectively.

Example 2: Tryptophan in the medium of Example 1 was replaced with 2 grams of 5-hydroxy-L-tryptophan, and *Proteus morganii* IFO 3848 was cultured in the medium as in Example 1. The cells were collected by centrifuging and suspended in 1 liter of 0.1% NH_4Cl-NH_4OH buffer (pH 8.0) containing 20 grams 5-hydroxyindole and 20 grams β-chloro-L-alanine. The mixture was held at 37°C for 20 hours with stirring. The reaction mixture contained 14.1 grams 5-hydroxy-L-tryptophan. By the same method as in Example 1, 7.5 grams crystalline 5-hydroxy-L-tryptophan was recovered from the reaction mixture. When β-chloro-L-alanine was replaced with 20 grams L-serine, 13.6 grams 5-hydroxy-L-tryptophan were found in the reaction mixture.

Other specific microorganisms not mentioned in the above examples which can be used in the process include *Erwinia hervicola* ATCC 21433 and *Escherichia coli* ATCC 206.

TRYPTOHAN DERIVATIVES

Other tryptophan derivatives are discussed in U.S. Patent 3,963,572, page 303 and U.S. Patent 3,929,573, page 305.

5-Substituted Tryptophans Produced Using Tryptophanase

It is known that certain microorganisms produce L-tryptophan when cultured in a medium containing a precursor of tryptophan, such as indole or anthranilic acid. However, the concentrations of L-tryptophan produced are very low in the prior processes.

H. Enei, H. Nakazawa, H. Matsui, S. Okumura and H. Yamada; U.S. Patent 3,808,101; April 30, 1974; assigned to Ajinomoto Co., Inc., Japan have found that tryptophanase is also useful in forming 5-substituted tryptophans. It was found that a heterocyclic compound having the following formula

(1)

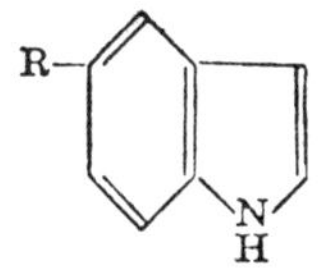

where R is hydrogen, hydroxyl, methyl or amino, reacts with ammonium ions and an acid selected from pyruvic acid, oxalacetic acid, malic acid, maleic oxime, fumaric acid, glyoxylic acid and lactic acid, in aqueous solution in the presence of an effective amount of tryptophanase or a source of tryptophanase to form high yields of an amino acid having the formula

(2)

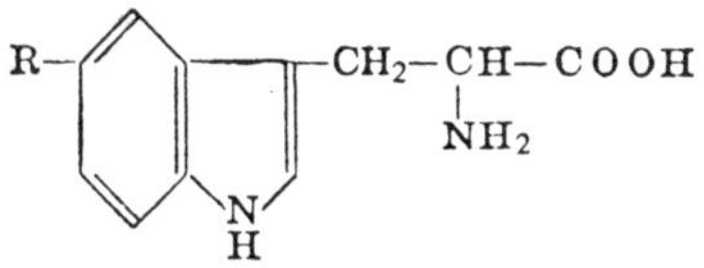

where R is as defined above. The enzyme tryptophanase is known to catalyze the decomposition of L-tryptophan to indole, ammonia and pyruvic acid. A pure enzyme preparation can be used in the process and sources of tryptophanase are preferably used. The tryptophanase source may be a broth in which a microorganism capable of producing the enzyme is cultivated, or a cell-free filtrate thereof, an aqueous suspension of the cells or of ground cells, or a filtrate of the suspension. Microorganisms capable of producing the enzyme are distributed widely and include those listed in Example 1.

The media employed for culturing the microorganisms may be conventional, containing sources of assimilable carbon and nitrogen and the usual minor nutrients and also include a proper amount of tryptophan, 5-hydroxytrytophan, 5-methyltryptophan or 5-aminotryptophan.

Tryptophan, 5-hydroxytryptophan, 5-methyltryptophan or 5-aminotryptophan should be added to the medium in an amount of more than 0.01 g/dl and preferably from 0.05 to 0.5 g/dl. The tryptophan may be in the form of organic nutrients, such as amino acid mixture, polypeptone or soybean extract.

When the microorganism capable of producing tryptophanase is cultured on the above media in the presence of a surface active agent, the enzyme activity of the tryptophanase obtained is very high. Useful surface active agents are nonionic, anionic, cationic and amphoteric surface active agents; nonionic surfactants are preferable.

The surface active agents may be generally present in the culture medium in an amount of 0.1 to 10 g/dl. The cultivation is carried out under aerobic conditions at 25° to 40°C. It is preferred to adjust the pH of the culture medium to 5.5 to 8.5 during cultivation. Culturing is generally carried out for from 10 to 72 hours.

The broth may be used as an enzyme source as it is, without removing the bacterial cells. A cell-free filtrate of the culture broth, a cell suspension and a suspension of crushed bacterial cells prepared by trituration, autolysis or ultrasonic oscillation can also be used as the enzyme source in this process. Crude enzyme and pure enzyme recovered by centrifugation, salting out and solvent precipitation can also be used. The amino acids of formula (2) may be prepared by adding the compound of formula (1), ammonium ions and the selected acid, salt or derivative to the aqueous medium during cultivation. The acid is preferably present in the reaction system in an amount of from 0.1 to 20% (weight/volume) as the free acid. The ammonium ions may be added in an amount of 0.1 to 20 g/dl.

The reaction is preferably carried out at pH 5 to 10 and at 10° to 50°C. The pH of the reaction system can be maintained by the addition of calcium carbonate, ammonia, caustic alkali or phosphate buffer solution.

Example 1: A culture medium containing 2 g/l L-tryptophan, 5 g/l KH_2PO_4, 0.5 g/l $MgSO_4 \cdot 7H_2O$, 2 ppm Fe and Mn ions, 2 g/l glycerol, 2 g/l yeast extract and 5 g/l casamino acid, of pH 7.0 was prepared. Each 60 ml batch of the medium was inoculated with a loop of bacterial cells as shown in the table below, and cultured at 31°C for 20 hours with shaking.

One liter of the cultured broth was centrifuged to collect the bacterial cells, the cells were added to 1 liter batches of respective solutions containing: (a) 20 grams indole, 24 grams sodium pyruvate and 40 grams ammonium acetate (pH 8.0), (b) 20 grams indole, 24 grams oxacetic acid and 40 grams ammonium acetate (pH 8.0), (c) 20 grams 5-hydroxyindole, 24 grams sodium acetate and 40 grams ammonium acetate (pH 8.0) and (d) 20 grams 5-hydroxyindole, 24 grams oxalacetic acid and 40 grams ammonium acetate (pH 8.0). Each reaction solution was stirred at 37°C for 20 hours. The amounts of L-tryptophan and 5-hydroxy-L-tryptophan found in the reaction mixtures are listed below.

	. . Tryptophan . .		5-OH-Tryptophan	
	. . . . Solution, g/l			
Bacteria Used	A	B	C	D
Proteus mirabilis ATCC 15290	8.0	6.2	4.8	4.2
Escherichia coli ATCC 3655	4.6	3.8	3.0	2.3
Escherichia coli ATCC 10798	5.2	4.1	3.2	2.5
Erwinia herbicola ATCC 21433	2.2	1.8	1.7	1.5
Erwinia herbicola ATCC 21434	2.4	1.9	1.7	1.6
Pseudomonas perlurida ATCC 490	3.4	2.7	2.3	2.1
Aerobacter aerogenes ATCC 8724	4.4	3.6	3.0	2.5
Aerobacter aeroge:. s ATCC 7256	4.0	3.3	2.7	2.4
Aerobacter liquefaciens ATCC 14460	6.0	5.1	3.0	2.4
Salmonella gallinarum ATCC 9148	2.1	1.8	1.5	1.2
Citrobacter freundii ATCC 8090	2.3	1.9	1.0	0.7
Citrobacter freundii ATCC 6750	2.4	1.9	1.1	0.7
Kluyvera citrophila ATCC 14236	3.4	2.5	1.3	0.9
Xanthomonas campestris ATCC 7381	1.8	1.5	1.3	0.5
Clostridium felsineum ATCC 13160	1.6	1.4	1.1	0.8

Aqueous sodium hydroxide solution was added to one liter reaction mixture from *E. coli* ATCC 3655 to dissolve precipitated tryptophan and the resulting solution was filtered in the presence of diatomaceous earth. The pH of the filtrate was adjusted to 4.0 with hydrochloric acid, the solution was introduced into a column packed with active charcoal, and L-tryptophan adsorbed on the active charcoal was eluted with 2 N ammonia water. The eluate was concentrated to dryness, the crystals obtained were washed with acetone and 5.2 grams pure crystalline L-tryptophan was obtained.

Example 2: Bacterial cells obtained from 10 ml culture broth of the bacteria indicated in the table below which had been cultured as in Example 1 were added to 10 ml of the following solutions (1) and (2), and were reacted at 37°C for 20 hours.

	Grams per Deciliter
Solution 1	
5-methylindole	2.0
Sodium pyruvate	2.0
Ammonium acetate pH 8.0 (with NH_4OH)	2.0
Solution 2	
5-aminoindole	2.0
Sodium pyruvate	2.0
Ammonium acetate pH 8.0 (with NH_4OH)	2.0

The amounts of 5-methyl-L-tryptophan produced in solution (1) and 5-amino-L-tryptophan produced in solution (2) are listed in the table below.

Bacteria Used	5-Methyltryptophan	5-Aminotryptophan
	(g/l)	
Erwinia herbicola ATCC 21433	1.2	1.5
Escherichia coli ATCC 7009	2.5	2.8
Proteus mirabilis ATCC 15290	3.0	3.5
Aerobacter aerogenes ATCC 8724	1.5	1.8
Kluyvera citrophila ATCC 14236	2.0	2.5

The yield can be greatly increased by adding vitamin B_6, such as pyridoxal phosphate, pyridoxal or pyridoxine and/or a reducing agent to the reaction system. Examples of suitable reducing agents are sulfurous acids, sodium sulfite, potassium sulfite, ammonium sulfite, thiosulfuric acid, potassium thiosulfate, sodium thiosulfate, ammonium thiosulfate, cysteine, β-mercaptoethanol and ascorbic acid. The vitamin B_6 and the reducing agents are generally used in an amount of from 0.01 to 1 g/dl. The yield of the end product can be greatly increased by adding inosine to the reaction system. L-tryptophan and inosine may react in the reaction system to form a precipitate of a tryptophan-inosine complex.

5-Hydroxytryptophan by Microbiologically Hydroxylating Tryptophans

5-Hydroxy-L-tryptophan is believed to be the transport form of 5-hydroxytryptamine (serotonin), produced from the amino acid by enzymatic decarboxylation. Serotonin is an important biogenic amine and plays a significant part as a transmitter substance in the central nervous system. As a serotonin deficiency can be overcome by supplying 5-hydroxy-L-tryptophan, a need has arisen for a commercially feasible process for the preparation of this substance.

A prior process involving the fermentation of L-tryptophan with *Chromobacterium violaceum* (ATCC 553) gave low yields of 5-hydroxyindole compounds, predominately 5-hydroxy-L-tryptophan. However, a pure product could not be isolated and the use of *Chromobacterium violaceum* is furthermore hazardous because of the pathogenic properties of this organism.

J. Daum and K. Kieslich; U.S. Patent 3,830,696; August 20, 1974; assigned to Schering AG, Germany have developed a microbiological process for the preparation of 5-hydroxy-L-tryptophan which can be isolated using microorganisms which are nonpathogenic to humans.

It has been found that gram-positive unicellular bacteria, in particular the microorganisms which are not human-pathogenic, of the families Micrococcus, Pseudomonas, Corynebacterium, Enterobacter and Bacillus are capable of converting L-tryptophan and ω-N-acetyl-L-tryptophan into 5-hydroxy-L-tryptophan in good yields. Of these families, a strain of bacillus, *Bacillus subtilis* ATCC 21733, has proven to be especially advantageous for the microbiological hydroxylation. Other specific microorganisms which might be used include *Sarcina lutea, Micrococcus lysodeikticus, Protaminobacter alboflavus, Pseudomonas facilis, Corynebacterium hoagii* and *Escherichia coli.*

When the substrate is ω-N-acyl-L-tryptophan, the acyl group is preferably aliphatic hydrocarbon carboxyl of one to five especially one to three carbons, for example, acetyl, or aromatic hydrocarbon carboxyl such as benzoyl and toluoyl. The microbiological hydroxylation is conducted in accordance with conventional methods. For example, there are usually first conducted preliminary experiments to ascertain the optimum fermentation conditions.

In this connection, it was found that the yield is increased by an addition of Fe^{++} ions to the fermentation solution preferably in a concentration of 5 to 50 mg/l, and that ascorbic acid preferably in a concentration of 40 to 80 mg/l accelerates the microbiological conversion. It was furthermore observed that it is advantageous to add the substrate toward the end of the main growth period of the microorganism, for example, 8 to 15 hours after inoculation. It is also to be noted that concentrations of 80 to 250 mg of substrate per liter of nutrient medium result in the highest yields.

As further preferred conditions, the culture is incubated at 25° to 40°C, the pH is adjusted to 6 to 8, and 1 to 5 liters of air per minute are added. The conversion of the substrate is suitably controlled by a thin-layer chromatographic analysis of sample extracts. After the fermentation has been terminated, the culture broth is freed of the bacterial cells by filtration or centrifugation.

Example 1: An Erlenmeyer flask containing 500 ml of a sterilized aqueous medium, consisting of 0.1% peptone, 0.2% corn steep liquor, 0.5% D(+)-glucose and 0.5% yeast extract (nutrient solution), adjusted to pH 8.0 is inoculated with 3 drops of supernatant broth of a tilted agar culture of *B. subtilis* (ATCC 21733) and shaken for 24 hours at 30°C.

These cultures are then centrifuged for fifteen minutes at 6,000 to 8,000 *g,* and respectively two cultures are suspended in 100 ml of sterilized nutrient solution. The suspension is transferred into an inoculation flask containing 300 ml of nutrient solution, as well as minor amounts of $FeSO_4·7H_2O$ and L-tryptophan. By means of this bacterial suspension, 10 liters of sterilized nutrient solution, containing 325 mg of tryptophan and 18.07 mg of $FeSO_4·7H_2O$ and adjusted to pH 7.5, is inoculated and incubated for 24 hours at 30°C at an aeration of 2 liters per minute and an agitating speed of 220 rpm. From this preliminary fermentation stage, 0.9 liter is transferred into a main fermentor charged with 15 liters of sterilized nutrient solution. During the main fermentation, the same technical conditions are used as in the preliminary fermentation. The pH is adjusted to 6.8. Ten hours after inoculation, 3.3 grams of L-tryptophan and 495 mg of $FeSO_4 \cdot 7H_2O$, dissolved in 635 ml of sterile water are added. The progression of the microbiological conversion is determined by analysis of samples taken at intervals. The substrate has been converted, except for trace amounts, 68.5 hours after the addition of the substrate.

The culture broth is freed of the bacillus cells by continuous centrifuging. The cell-free culture broth is mixed with 750 grams of freshly regenerated cation exchanger in the H^+ form and agitated for 45 minutes. Thereafter, the exchange resin is filtered off.

Respectively 20 grams of the exchange resin, stored in the cold state, is agitated several times with 25 ml aliquots of 1 M ammonium acetate solution for 10 minutes and then filtered. The combined eluates are lyophilized and then chromatographed in the form of an aqueous solution over a Sephadex column with 0.1 M ammonium bicarbonate solution at 10°C. After combining and lyophilizing the main fractions, 2.56 grams of 5-hydroxy-L-tryptophan is obtained, MP 268° to 270°C (under decomposition).

Example 2: This example differs from Example 1 merely in the concentration of the substrate. Ten hours after inoculation of the main fermentor, a solution is added consisting of 1.65 grams of L-tryptophan and 46 mg of iron(II) sulfate in 635 ml of sterile water. Measurements of the 5-hydroxy-L-tryptophan content after a total fermentation period of 50, 58 and 66 hours, respectively, indicate, in the same sequence, final yields of 100, 96 and 99.4% of theory.

5-Alkoxytryptophans from 5-Alkoxyindoles Using Microorganisms

Chemical processes for producing a L-5-lower-alkoxytryptophan, particularly L-5-methoxytryptophan referred to as 5-MTP, have been reported using 5-methoxytryptamine and acetamide as starting materials [J.W. Cook et al, *J. Chem. Soc.*, 1203 (1951)] and using formamide malonate ester [Asami, *Nagasaki Medical Report,* 30 490 (1955)]. These processes, however, are complex and give the DL form and not the L form.

A process for producing a L-5-lower-alkoxytryptophan from a 5-lower-alkoxyindole, with or without serine, using a microorganism has been developed by *N. Kitajima, S. Watanabe and I. Takeda; U.S. Patent 3,915,795; October 28, 1975; assigned to Asahi Kasei Kogyo KK, Japan.* A L-5-lower-alkoxytryptophan is an intermediate for use in the production of L-5-hydroxytryptophan referred to as L-5-HTP.

Microorganisms used in the process are selected based upon the degree of L-5-MTP-producing ability which was measured in the following manner. A liquid containing 2 mg/ml of 5-methoxyindole (5-MI), 10 mg/ml of L-serine, 0.1 millimol per milliliter of phosphate buffer (pH 8.0) and 80 mg/ml of freeze-dried test strain was allowed to react for a definite period of time. From a definite amount of the reaction liquid, unreacted 5-MI was removed by toluene extraction to obtain a fraction. This fraction was subjected to colorimetric determination according to the xanthydrol method [S.R. Dickman et al, *J. B. C.*, 220, 957 (1956)]. Subsequently, the concentration of L-5-MTP in the reaction liquid was calculated from the standard curve. According to the abovementioned determination method, the strains Corynebacterium sp. No. 14001 NRRL B-5395 and *Escherichia coli* ATCC 9637 gave L-5-MTP in amounts of 1.65 mg/ml and 2.50 mg/ml, respectively. Accordingly, microorganisms usable in the process may be those which have an ability of producing L-5-MTP in an amount of 0.1 mg/ml or more as measured by the above method.

The medium used in this process contains 5-methoxyindole (5-MI) together with or without serine, in addition to a carbon source, a nitrogen source, inorganic salts and natural organic nutrients which are suitable for the good growth of the microorganism used and for the smooth production of L-5-MTP. Details of the process and the microorganisms that may be used in the process are shown in the following examples.

Example 1: A microorganism Corynebacterium sp. No. 14001, NRRL B-5395 was inoculated in a shaking flask containing 50 ml of a medium (pH 7.0) containing 1.0% of peptone, 1.0% of meat extract, 0.4% of sodium chloride and 0.1% of yeast extract and was cultured while shaking the flask reciprocally at 30°C for 24 hours. The thus cultured microorganism was collected and suspended in 60 mmol of a phosphate buffer (pH 8.5). To the resulting suspension were added 20 mg of 5-MI, 20 mg of L-serine and 120 mg of glucose and the total volume of the suspension was made 20 ml. Subsequently, the suspension was allowed to react under stirring at 30°C for 16 hours, while maintaining the suspension at a pH of 8.5. As a result, 1.35 mg/ml of L-5-MTP was accumulated in the reaction liquid.

Subsequently, the same reaction as above was effected to obtain a reaction liquid. This reaction liquid was mixed with the aforesaid reaction liquid to obtain 950 ml of a mixed liquid. The mixed liquid was subjected to active carbon treatment to isolate 470 mg of crude crystals of L-5-MTP. 400 mg of the crude crystals were allowed to react, together with 4 ml of hydriodic acid and 4 ml of acetic anhydride, at 50°C for 90 minutes while introducing nitrogen, to produce 15.7 mg/ml of L-5-HTP.

Example 2: The microorganisms shown in the table below were individually inoculated in a shaking flask containing 50 ml of a liquid medium (pH 7.0) containing 3% of glucose, 0.6% of urea, 0.7% of peptone, 0.3% of yeast extract, 0.1% of KH_2PO_4, 0.1% of K_2HPO_4, 0.05% of $MgSO_4 \cdot 7H_2O$ and 0.001% of $MnSO_4 \cdot 4H_2O$ and cultured while shaking the flask reciprocally at 30°C. At the 32nd hour of the cultivation, 1.0 mg/ml of 5-MI and 2.0 mg/ml of DL-serine were added to the medium, and the cultivation was continued for an additional 24 hours. As a result, L-5-MTP was produced and accumulated in the medium in the amounts shown below.

Microorganism	L-5-MTP, mg/ml
Brevibacterium ammoniagenes ATCC 6872	0.21
Corynebacterium melassecola ATCC 17965	0.39
Arthrobacter globiformis ATCC 8010	0.20
Microbacterium ammoniaphilum ATCC 15354	0.33
Micrococcus luteus ATCC 21102	0.30

(continued)

Microorganism	L-5-MTP, mg/ml
Sarcina lutea ATCC 15176	0.35
Staphylococcus citreus ATCC 4012	0.22
Bacillus subtilis ATCC 14593	0.28
Serratia marcescens IFO 3046	0.46
Escherichia coli ATCC 9637	0.57
Aerobacter aerogenes IFO 3317	0.58
Proteus mirabilis IFO 3849	0.40
Erwinia carotovora IFO 3057	0.58
Pseudomonas fragi IFO 3458	0.45
Xanthomonas pruni IFO 3511	0.49
Protaminobacter alboflavus IFO 3707	0.41
Flavobacterium arborescens ATCC 4358	0.54
Achromobacter petrophilum FERM P No. 239	0.35
Agrobacterium tumefaciens IFO 3058	1.20
Azotobacter agilis IFO 3741	0.25
Nocardia asteroides IFO 3384	0.48
Streptomyces aureus ATCC 3309	0.25
Aspergillus oryzae IFO 4075	0.25
Penicillium chrysogenum ATCC 15241	0.24

L-DOPA

Recently, 3,4-dihydroxy-L-phenylalanine or L-dopa, known for a long time, has been employed successfully in the treatment of the Parkinson's disease. The L-dopa eliminates the disorders caused by this disease, when administered in relatively high doses, i.e., 5 to 10 grams per day. Its employment for such a purpose was authorized in the United States of America (*Drug Trade News* 1970, June 15, pages 1 and 18). Chemically, L-dopa is L-2-amino-3-(3,4-dihydroxyphenyl) propionic acid, also named 3-(3,4-dihydroxyphenyl)-L-alanine.

PRODUCTION OF L-DOPA

L-dopa can be synthesized from intermediates, such as piperonyl aldehyde, vanillin, tyrosine and catechol, or can be obtained from seeds such as velvet bean seeds by extraction. L-dopa is also biochemically produced from catechol and L-tyrosine by means of β-tyrosinase produced by a microorganism belonging to the genus Escherichia.

Isolation from Velvet Bean

L-dopa is extractible from materials such as vetch, broad beans, especially Georgia velvet bean (*Vicia faba)*, and the African Sommerset bean, as well as the seeds of *Mucuna puriens (L) DC Leguminosae*. In the extraction of ground velvet bean with acidified water, difficulty has been experienced in separating the aqueous L-dopa extract by filtration from the extracted ground velvet bean due to the solubilization of the water-soluble constituents. These water-soluble constituents cause difficulties when it is attempted to separate the aqueous extract phase from the extracted ground velvet bean, such as by filtration or centrifugation or settling, or to concentrate the aqueous extract phase.

G.J. Riccardi; U.S. Patent 3,941,835; March 2, 1976; assigned to Bio-Derivatives Corporation has developed an improved process for the extraction of L-dopa from materials, such as whole or ground velvet bean. In this process, L-dopa is extracted from L-dopa-containing materials, by contacting the material with water, desirably at temperatures of 15° to 25°C. Following the extraction, the resulting L-dopa-containing aqueous extract is separated from the original material by centrifugation and/or filtration. The resulting aqueous L-dopa-containing extract or filtrate is then concentrated for the removal of at least 50%, preferably up to 90%, of the water by vacuum concentration, at a pressure below 760 mm Hg absolute, such as 2 to 500 mm Hg absolute. The concentrated L-dopa extract can also be spray dried or freeze dried or lyophilized, if desired.

Following the concentration operation the concentrated aqueous extract which may be at a temperature of 40° to 50°C, depending upon the pressure at which the concentration operation was carried out, is decolorized by passing the concentrated aqueous extract through a mass of decolorizing agent, such as a bed of activated carbon or charcoal or by slurrying the concentrated extract with activated carbon or charcoal and filtering. The decolorized extract is then cooled to 0° to 5°C, to precipitate or crystallize the L-dopa. The L-dopa crystals are then filtered to separate the L-dopa-containing mother liquor which is then advantageously returned to the extraction operation or to the concentration operation to contact additional fresh L-dopa-containing material for the extraction of L-dopa.

L-dopa crystallization can also be effected by adding a salt, such as sodium chloride to the concentrated L-dopa solution to "salt out" the L-dopa. The separated L-dopa crystals are desirably washed with cold (0° to 5°C) water, such as ice water to remove any adhering mother liquor and the resulting wash liquor is also advantageously used to extract L-dopa from fresh L-dopa-containing feed material, or returned to the concentration operation. When it is desired to obtain pure, pharmaceutical grade L-dopa product, the L-dopa crystals after washing with ice water are again washed with alcohol or an alcohol-ether wash, again preferably at a reduced temperature. The resulting washed L-dopa crystals are then dried and recovered.

In the extraction operation, the water used may be tap water, deionized water or distilled water or other readily available water, such as substantially acid-free municipal water having a dissolved solids content below 500 ppm by weight. The extraction step may be a continuous or a batch type operation. For example, the L-dopa extraction may be carried out in a continuous countercurrent extraction operation wherein a mass of ground velvet bean is moved countercurrently with respect to a stream of water. If desired, the extraction may be carried out in a continuous concurrent operation wherein a stream of water and a stream of ground velvet bean are supplied to one end of an extraction zone and the extracted velvet bean and the resulting L-dopa-containing extract separately recovered from the other end of the extraction zone.

The extraction operation may also be carried out in a multistage batch extraction operation wherein in a single or plural stage contacting zone, water and a mass of ground velvet bean, such as fixed bed or a plurality of fixed beds of velvet bean are contacted for a sufficient length of time, e.g., 1 to 5 hours, depending upon the extraction temperature, the particle size of the L-dopa-containing material being extracted, and the proportions of the water to the velvet bean being extracted. The resulting aqueous extract is separated from the extracted velvet bean and the extraction operation continued or carried out until the amount of L-dopa in the extract or in the velvet bean residue is negligible or insufficient to warrant further extraction. In the extraction operation, it is preferred that the velvet bean be ground to a meal having a particle size from 10 to 20 mesh, i.e., a particle size which passes through a 10 mesh screen but is retained on a 20 mesh screen.

Example: Ground velvet bean and tap water in the proportion 1 part by weight ground velvet bean to 5 parts by weight water were mixed and the resulting mixture permitted to stand overnight at room temperature (about 20°C). The resulting extract was slightly brown in color. The mixture was filtered through a cake of filter aid (Celite) prepared from a mixture of 2 parts by weight Celite to 1 part by weight saturated SO_2 solution. The extract phase filtered rapidly through the Celite cake and the resulting L-dopa solution was clear and gave a strong postive test for L-dopa.

The filtered L-dopa-containing extract was concentrated and L-dopa was crystallized therefrom when the concentrated filtrate was reduced in temperature. The above extracted velvet bean was again extracted with water in the proportion 1 part by weight bean to 5 parts by weight water and the resulting aqueous extract phase was again filtered and concentrated. The resulting concentrated extract phase gave a positive test for L-dopa indicating that additional L-dopa was recovered from the velvet bean in the second extraction operation. Both filtration operations were carried out without difficulty and the admixtures of velvet bean and water at room temperature filtered rapidly.

Use of β-Tyrosinase

An improved process for the enzymatic production of L-dopa and related compounds has been disclosed by *K. Ogata, H. Yamada, H. Enei and S. Okumura; U.S. Patent 3,791,924; February 12, 1974; assigned to Ajinomoto Co., Inc.* The related compounds include L-tyrosine, one of the essential amino acids, and 2,4-dihydroxyphenyl-L-alanine, useful as a nutrient and food additive.

In this process a phenol derivative of the formula:

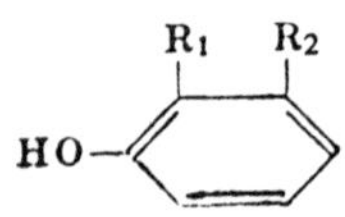

where R_1 and R_2 are hydrogen or hydroxyl, not more than one of the R_1, R_2 being hydroxyl, reacts with cysteine, cystine, serine, S-lower-alkyl cysteine or, alanine and salts or esters of these amino acids, or with ammonia and pyruvic acid, oxalacetic acid, malic acid, fumaric acid, maleic oxime, glyoxylic acid, or lactic acid and salts or esters thereof, in the presence of β-tyrosinase, to form high yields of phenylalanine derivatives of the formula:

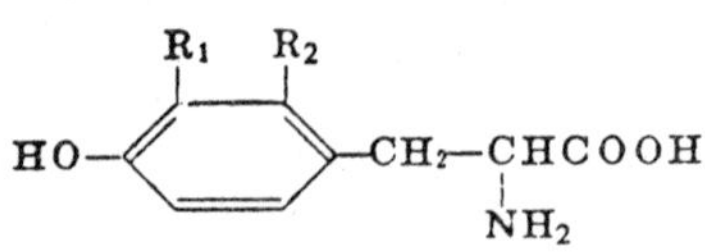

where R_1 and R_2 have the above meaning.

In the process, a suitable source of β-tyrosinase is mixed with an aqueous solution containing a phenol derivative and cysteine, cystine, serine, alanine and their salts or derivatives, or with an aqueous solution containing a phenol derivative, ammonia, or pyruvic acid, oxalacetic acid, malic acid, fumaric acid, maleic oxime, glyoxylic acid, or lactic acid and their salts or derivatives, and the mixture stands at pH 4 to 11 until the phenylalanine derivative is produced.

The microorganisms capable of producing β-tyrosinase belong to the genera Escherichia, Aerobacter, Erwinia, Pseudomonas, Bacillus, Xanthomonas, Proteus, Citrobacter, Paracholobactrum, Salmonella and Alcaligenes. Preferred microorganisms include *Citrobacter freundii* ATCC 6750, *Citrobacter freundii* ATCC 8090, *Proteus morganii* IFO 3848, *Proteus mirabilis* ATCC 15290, *Pseudomonas perlurida* ATCC 490, *Aerobacter aerogenes* ATCC 7256, *Salmonella gallinarum* ATCC 9148, *Erwinia herbicola* ATCC 21433, *Erwinia herbicola* ATCC 21434, *Paracolobactrum coliforme* ATCC 11605 and *Alcaligenes faecalis* ATCC 8315. The form (i.e., culture broth, cell suspension, isolated crystals, etc.) of the β-tyrosinase used is not critical.

The media for culturing the microorganisms may be conventional, containing sources of assimilable carbon and nitrogen and the usual minor nutrients and also include a small amount of tyrosine. Tyrosine, 0.05 to 0.5% (w/v), should be added to the medium as organic nutrients, i.e., amino acid mixture, polypeptone or soybean extract. Amino acids (0.1 to 5% by weight as free amino acids), e.g., methionine, glycine and alanine; and vitamins (>0.5 mg/dl), e.g., pyridoxine, pyridoxal and pyridoxal phosphate are also added to improve enzyme activity.

When catechol and resorcin are the phenol derivatives used to produce L-dopa and 2,4-dihydroxyphenyl-L-alanine, the yield can be increased by adding a reducing agent and/or chelating agent in an amount of 0.01 to 1% (w/v). Reducing agent examples are sulfurous acid, sodium sulfite, thiosulfuric acid, potassium thiosulfate, sodium thiosulfate, ammonium thiosulfate, cysteine, β-mercaptoethanol and ascorbic acid; chelating agent examples are EDTA, ethylenediamine, N-oxyethylenediamine, aspartic acid, N-dioxyethylglycine, nitrilotriacetate, and citric acid.

Example: An aqueous culture medium having the following composition was prepared and adjusted to pH 6.5.

L-tyrosine, g/dl	0.2
K_2HPO_4, g/dl	0.1
$MgSO_4$, g/dl	0.05
Citric acid, g/dl	0.2
Glycerol, g/dl	0.2
Yeast extract, g/dl	1.0
Pyridoxine, mg/dl	10.0
Glycine, g/dl	0.1
Alanine, g/dl	0.3
Methionine, g/dl	0.15

60 ml batches of the solution were placed in 500 ml shaking flasks. The aqueous media were inoculated with *Erwinia herbicola* ATCC 21434 cultured on bouillon agar slants at 28°C for 20 hours and maintained at 31°C for 20 hours with shaking. The microbial cells in one liter of the cultured broth were harvested by centrifuging, and washed with distilled water. The cells were suspended in a solution having the following composition, a pH of 8.8, and one-

half of the volume of the original cultured broth.

DL-serine, g/dl	1.5
Catechol, g/dl	0.75
Na_2SO_3, g/dl	0.2
NH_4NO_3, g/dl	0.2
EDTA, g/dl	0.1

The reaction was carried out at 30°C for 20 hours. 0.92 g/dl of L-dopa was found in the reaction solution. Two liters of the solution were adjusted to pH 3.5 with hydrochloric acid and centrifuged to remove the microbial cells. The supernatant solution was concentrated to 250 ml, adjusted to about pH 1.0, and applied to a column containing 130 g of dried active carbon. After the column was washed with 0.1 N hydrochloric acid to remove residual L-serine, L-dopa was eluted with diluted aqueous ammonia. Crude crystalline L-dopa was obtained by concentrating the eluate, and was recrystallized three times. The yield of L-dopa was 13.6 g. It was entirely in the L-form as determined by polarimeter.

Oxidation of N-Acyltyrosine

A process for the preparation of N-acyldopa compounds by reacting an N-acyltyrosine with an endiol compound in the presence of ferrous ion and oxygen has been developed by *H. Nakano and T. Suzuki; U.S. Patent 3,804,893; April 16, 1974; assigned to Kyowa Hakko Kogyo Co., Ltd., Japan.*

An L-tyrosine compound where the amino group has been acylated has a solubility at least ten times as great as the unacylated tyrosine. This N-acylated tyrosine can be treated according to the Udenfriend process resulting in the formation in good yield of an acylated L-dopa compound. It has further been found that the acylated L-dopa produced is stable even if the reaction time is prolonged up to 100 hours.

Upon completion of the reaction, the reaction solution contains only unreacted N-acyltyrosine, N-acyldopa and oxidized ascorbic acid compounds. Other substrate and product decomposition products are rarely observed. The desired N-acyldopa and the unreacted substrate can be easily separated and the latter reused, if desired. The N-acyldopa can be readily converted to L-dopa by known deacylation procedures.

In carrying out the process, for example, an N-acyltyrosine is dissolved in a 0.2 M phosphoric acid buffer solution to give a concentration of about 1 mg/ml to 30 mg/ml of N-acyltyrosine. Ferrous ion and disodium ethylenediaminetetraacetate (EDTA) are added to give a concentration of ferrous ion (shown as $FeSO_4 \cdot 7H_2O$) of 0.1 to 5 mg/ml and a concentration of EDTA of 10 to 100 μmol/ml, respectively. Further, L-ascorbic acid is added in the range of 1 mg/ml to 200 mg/ml just before the reaction or during the reaction, preferably intermittently.

The reaction is carried out at 5° to 90°C, typically about 30°C, in a pH range of 3 to 7 with vigorous aeration and/or stirring. The reaction time is 1 to 48 hours and the reaction is deemed complete when the maximum yield of N-acyldopa is obtained. After the completion of the reaction, a Sephadex (a hydrophilic, insoluble molecular-sieve chromatographic medium, made by crosslinking dextran) column is used to separate the unreacted substrate and reaction products from the reaction solution. The reaction solution is passed through a column having a volume 5 to 20 times as great as the volume of the reaction solution. The column is washed with sufficient amounts of water to wash off the oxidation products of the ascorbic acid or other endiol compounds. Then, a buffer solution such as phosphoric acid buffer is passed through the column as an eluting agent, whereby the unreacted substrate and the reaction product can be separated. The unreacted substrate fraction can be recycled if desired.

Example: 2 g of N-formyltyrosine, 5 g of L-ascorbic acid, 100 mg of ferrous sulfate and 200 mg of EDTA are dissolved in 100 ml of 0.2 M phosphoric acid buffer solution (pH = 6.0) and the reaction is carried out at 35°C for 24 hours in a reactor flask having a capacity of 500 ml. Aeration is carried out at a rate of 200 ml/min while stirring at 800 rpm. As a result of quantitative analysis with sodium tungstate, it is confirmed that 0.95 of N-formyldopa is formed.

The reaction solution is then passed through a column (diameter, 7 cm; height, 100 cm) packed with Sephadex G-10 and the oxidation products of ascorbic acid are washed away by passing sufficient amounts of water through the column. Then, elution is carried out by passing a 0.05 M phosphoric acid buffer solution through the column at a flow rate of 20 ml/min. The relative amounts of substrate and reaction product contained in the eluate are determined from the ultraviolet absorption spectrum. 400 ml of the substrate elution fraction are utilized as a portion of the next reaction solution, while 450 ml of the reaction product fraction are immediately concentrated at 45°C under a reduced pressure.

About 20 ml of the concentrated aqueous solution are acidified with acetic acid and held at 15°C for 8 hours to remove the formyl group. The pH is then adjusted to about 5 to 6. Free L-dopa is recovered as an insoluble precipitate and this precipitate is filtered, washed with cold water and dried to give 0.72 g of L-dopa. After recrystallization, the precipitate is confirmed to be L-dopa by means of IR spectrum, UV absorption spectrum, and paper chromatography.

Enzymatic Oxidation of L-Tyrosine

Attempts to prepare L-dopa from L-tyrosine by biocatalytic oxidation of an L-tyrosine substrate with tyrosinase insolubilized on a cellulosic support material have achieved only marginal success according to Wykes et al., (*Nature* 230, 187, 1970). Tyrosinase insolubilized in this manner proved to be unstable, losing about 75% of its original activity during a twenty-four hour period under the reported reaction conditions.

An improved process of using an insolubilized tyrosinase has now been disclosed by *F.E. Semersky; U.S. Patent 3,812,009; May 21, 1974; assigned to Owens-Illinois, Inc.* In brief, the process involves the enzymatic oxidation of L-tyrosine or an N-substituted derivative thereof with insolubilized tyrosinase followed by subsequent reduction and hydrolysis of the resultant oxidation products where necessary. Representative of the materials which can be oxidized in the synthesis of L-3,4-dihydroxyphenylalanine and N-derivatives thereof are L-tyrosine, N-acetyl-L-tyrosine, N-formyl-L-tyrosine, N-propionyl-L-tyrosine, N-monofluoroacetyl-L-tyrosine, N-monochloroacetyl-L-tyrosine, N-dichloroacetyl-L-tyrosine, N-chloropropionyl-L-tyrosine, N-benzoyl-L-tyrosine, and N-(p-chlorobenzoyl)-L-tyrosine. When N-blocked-L-tyrosine derivatives are used in this process, hydrolysis of the N-substituents on the oxidized products can be carried out by either synthetic or enzymatic means (in a few limited instances). For example, when N-acetyl-L-tyrosine is the substrate material, its corresponding o-dihydric compound, N-acetyl-L-3,4-dihydroxyphenylalanine, can be hydrolyzed enzymatically by adjusting the pH of the reaction medium to neutrality and then contacting this compound with either carboxypeptidase A or acylase.

Details of the preparation and insolubilization of the enzyme and use in the oxidation of N-acyl-L-tyrosine are given in the following example.

Example: One hundred milligrams of tyrosinase in the form of a crude lyophilized powder is dissolved in 20 milliliters distilled water and then combined with 9.44 grams solid ammonium sulfate. The pH of the resultant solution is adjusted to neutrality with 0.01 N NaOH and mildly agitated at 0°C for 30 minutes. The suspension formed is centrifigued at 30,000 times gravity for 30 minutes at 0°C; the supernatant discarded; and sufficient 0.01 M phosphate buffer, pH = 7, added to redissolve the precipitated material present.

The solution is then applied to a column (1.5 x 100 cm) of Sephadex G-50 gel (Aktiebolaget Pharmacia, Upsala, Sweden) and eluted with 0.01 M phosphate buffer. Those fractions obtained from the eluate having ultraviolet absorption ratios consistent with those of purified mushroom tyrosinase are retained for further use. The specific activities of the highly purified tyrosinase obtained by this technique are 23 mmol p-cresol converted to catechol per minute per milligram enzyme and 440 mmol catechol converted to 4-methyl-1,2-benzoquinone per minute per milligram enzyme, or 19:1 catecholase:cresolase ratio.

To 7.0 milliliters of phosphate buffered aqueous solution, pH = 7, containing 20 milligrams of highly purified tyrosinase (obtained as above) are added an equivolume solution containing 0.98 gram acrylamide monomer and 0.051 gram bisacrylamide. After the two solutions are thoroughly mixed, the mixture is further diluted by the addition of 0.06 milliliter N,N,N',N'-tetramethylethylenediamine. The resulting solution is transferred to a 100 mm petri dish to provide a greater surface area and shallow fluid level for ease of penetration of UV light.

Once transferred, 0.001 gram riboflavin (free radical initiator) is dispersed throughout the solution and the petri dish and its contents irradiated with a Canalco Preparation Light provided with a 14-watt General Electric Daylight Bulb placed 4 inches from the surface of the fluid. Ordinarily, 20 minutes of exposure to UV followed by an additional 60 minutes in natural sunlight provides the requisite energy to initiate and promote complete polymerization of the acrylamide monomer.

The hydrogel-enzyme complex is then cut into pieces 1 centimeter square, loaded into the barrel of a standard 5 milliliter Luer-lock syringe, the plunger reinserted, and the hydrogel-enzyme complex crushed by forcing it through the syringe nozzle into a beaker of distilled water. The crushed acrylamide-tyrosinase complex is allowed to settle for about 2 minutes and then the supernatant decanted off. This washing-decantation procedure is repeated with eight 50-milliliter volumes of distilled water. The above procedure (dispersement, washing, and decantation) is then repeated to further reduce the size of the gel particles.

One gram of the above hydrogel-enzyme complex is added to 10 milliliters of a saturated aqueous solution of N-acetyl-L-tyrosine, pH 5 (0.1 M acetate buffered) and the vessel containing these materials mildly agitated for 120 to 180 minutes while the temperature is maintained at a constant 20°C. The hydrogel-enzyme complex is then decanted from the supernatant into an empty beaker.

The supernatant is transferred to a stainless steel bomb containing a palladium catalyst on charcoal; the bomb pressurized to 3 atmospheres; and hydrogen gas bubbled through the supernatant for 3 to 5 hours. After reduction of the substrate oxidation product to the degree desired, the supernatant containing this product is removed from the bomb and placed in a beaker.

Crystallization of L-3,4-dihydroxyphenylalanine is accomplished by the addition of 30 milliliters of n-butyl acetate to the supernatant followed by extraction with 30 and then with 10 milliliters of water. The two aqueous extractions are collected in a single vessel, pH adjusted to 5.0 with 25% NH_4OH solution containing minor amounts of $NaHSO_3$ whereupon crystallization of the desired product occurs. After refrigeration of the supernatant overnight at ~4°C, the crystals are recovered by filtration, washed with distilled water and dried. The crystals thus recovered can be further purified by recrystallization in distilled water containing minor amounts of $NaHSO_3$. The process may also be accomplished using an agarose-enzyme conjugate as the insolubilized enzyme.

Electrochemical Reduction of Oxazolones

A process for preparing α-(N-acyl)-amino acids on a large scale basis and in high yields by reducing corresponding oxazolones (such as in formula 1) has been developed by *P.B. Berntsson, V. Frolunda, J.O. Gaarder and B.R. Lamm; U.S. Patent 3,838,027; September 24, 1974; assigned to Aktiebolaget Astra, Sweden.* The oxazolones and α-(N-acyl)-amino acids are compounds of formula 2.

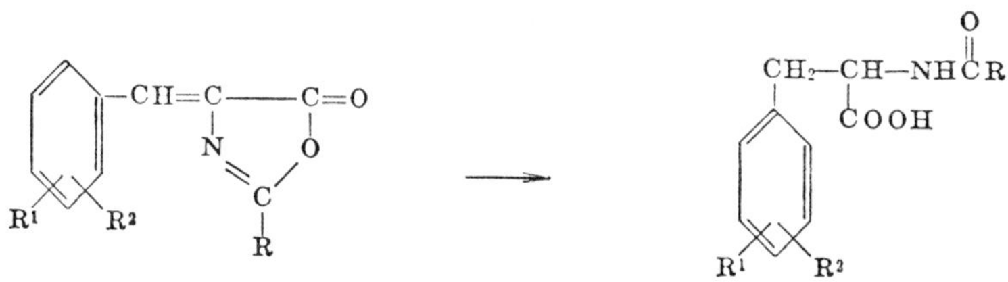

(1) (2)

In both (1) and (2), R is a straight or branched alkyl or cyclodialkyl having at most six carbon atoms or phenyl, and R^1 and R^2, which may be substituted in the 2, 3, 4, 5 or 6 position on the ring structure, are the same or different and selected from hydrogen, alkyl, cyclodialkyl, and alkoxy, each having at most six carbon atoms, halogen, hydroxy, and acyloxy. R^1 and R^2 may also together denote a divalent radical $—OR^3O—$ bound to the benzene nucleus in the 2 to 3 or 3 to 4 position, where R^3 is an alkylene group having 1 to 3 carbon atoms. These α-(N-acyl)-amino acids are suitable intermediates in the preparation of other amino acids, particularly L-dopa, p-chlorophenyl alanine and tyrosine.

According to a preferred method of the process, the electrochemical reduction of compounds of formula (1) above is carried out under alkaline conditions. The reduction may also be carried out in an electrolytic cell having a mercury cathode so that a greater potential difference can be used across the cell. An electrolytic cell is preferably used in which the electrodes are separated with a cation exchange membrane.

Example 1: Preparation of N-Benzoyl-4'-Chlorophenylalanine – 2.75 kg of 2-phenyl-4-(4'-chlorobenzylidene)oxazolone-5 was mixed with a hot (70°C) solution of 12.5 liters of ethanol in 12.5 liters of 1 M NaOH. This mixture was reduced in the cathode chamber of an electrolytic cell, wherein the cathode was a mercury cathode and the current density was 0.1 A/dm^2. The cathode and anode of the electrolytic cell were separated with a membrane which permits a selective passage of small cations, but not the product which is to be reduced or has been reduced.

A 2 M NaOH aqueous solution was used as anode liquid. The current was switched off when 125% of the estimated current amount had passed through the electrolytic bath, which happened after about 4 hours. The hot solution was filtered and made acid with concentrated HCl and then added to pH about 3. After 18 hours at room temperature the precipitate obtained was filtered, washed with water and then with cold 50% ethanol and dried. A yield of 2.38 kg of N-benzoyl-4'-chlorophenylalanine, melting point 168°C was obtained.

Example 2: Preparation of L-Dopa via N-Acetyl-3'-Methoxy-4'-Hydroxyphenylalanine – 100 g of 2-methyl-4-(3'-methoxy-4'-acetoxybenzylidene)oxazolone-5 was added to 2 liters of 0.5 M NaOH for 30 minutes. This mixture was electrolyzed in the same way as in Example 1. As anodic liquid 2 liters of 2 M NaOH, separated from the cathode by means of an ion exchange membrane, was used as in Example 1. The electrolysis was carried out with a current density of 4 A/dm^2, whereby a total of 30 Ah (50% excess) was used.

After finishing electrolysis the cathode liquid was made acid to pH 5.5 using acetic acid, filtered clear and evaporated in vacuo. The oil obtained was dissolved in 330 ml of water and made acid with 195 ml concentrated HCl and then extracted four times with 400 ml ethyl acetate. The collected extract was dried over anhydrous Na_2SO_4, filtered and evaporated in vacuo.

The residue, N-acetyl-3'-methoxy-4'-hydroxyphenylalanine, in a yield of 80% (73.5 g), smelling of acetic acid, was used without further purification in an acid hydrolysis to the free amino acid, 3'-methoxy-4'-hydroxyphenylalanine. 3'-methoxy-4'-hydroxyphenylalanine was then reacted in a known manner with phenyl acetylchloride to form N-phenylacetyl-3'-methoxy-4'-hydroxyphenylalanine. The latter was then subjected to enzymatic cleavage, whereby the L-form of 3'-methoxy-4'-hydroxyphenyl alanine was obtained, which then is reacted with HBr to form 3,4-dihydroxyphenylalanine.

Use of Optically Active Hydrogenation Catalysts

Optically active hydrogenation catalysts are used by *W.S. Knowles, M.J. Sabacky, and B.D. Vineyard; U.S. Patent 4,005,127; January 25, 1977; assigned to Monsanto Company* in the preparation of L-dopa. In the overall chemical process the starting material is obtained by reacting 3-alkoxy-4-hydroxybenzaldehyde and acetylglycine with acetic anhydride to form 2-methyl-4-(3'-alkoxy-4'-acetoxybenzal)-5-oxazolone (1). Alternately and preferably, the reaction can be carried out in the presence of sodium acetate. (1) is then subjected to mild hydrolysis to obtain α-acetamido-4-hydroxy-3-alkoxycinnamic acid acetate (2). It is critical that the hydrolysis be conducted so as to avoid the removal of the acetyl group in the 4-position on the benzal group. (2) is then subjected to the catalytic asymmetric hydrogenation of this process, resulting in N-acetyl-3-(4-hydroxy-3-alkoxyphenyl)-alanine acetate (3). This compound is present as the two enantiomorphs, L and D, the L enantiomorph being present in a major amount and the D enantiomorph being present in a minor amount.

L-dopa can be obtained from the L enantiomorph of (3) by known chemical methods. The homogeneous, optically active coordination metal complex hydrogenation catalysts comprise a metal selected from rhodium, iridium, ruthenium, osmium, palladium and platinum in combination with at least one optically active phosphine or arsine ligand. These catalysts are soluble in the reaction mass and are therefore referred to as "homogeneous" catalysts. The optically active phosphines and arsines used in the catalyst can be prepared according to the procedure of Mislow and Korpiun (*J. Am. Chem. Soc.* 89, 4784, 1967).

It has been found that catalysts prepared from soluble rhodium compounds and optically active phosphine ligands containing an o-anisyl group are particularly preferred in carrying out the catalytic asymmetric hydrogenation step of the process. In particular, the soluble rhodium compounds in combination with optically active (+) methylcyclohexyl-o-anisylphosphine provide excellent results.

Example: Asymmetric Hydrogenation of α-Acetamido-4-Hydroxy-3-Methoxycinnamic Acid Acetate – To an autoclave was added 596 parts of methanol, 0.9 part of triethylamine and 149 parts of α-acetamido-4-hydroxy-3-methoxycinnamic acid acetate. The slurry was heated to 50°C and thoroughly purged to remove all traces of oxygen and finally adjusted to 50°C and 40 psig of hydrogen.

A catalyst solution made by dissolving 0.036 part of rhodium 1,5-cyclooctadiene chloride in 1.9 parts of benzene followed by 0.075 part of (+) methylcyclohexyl-o-anisylphosphine in 1 part of methanol was added and the mixture treated with hydrogen for 5 minutes. The resulting clear solution was then forced into the hydrogenation apparatus under hydrogen pressure. Hydrogenation begins immediately and is complete in about 2 hours. Assay of the reaction mixture shows the optical purity to be 72% corresponding to an 86/14 L/D mixture.

One hundred parts of N-acetyl-3-(4-hydroxy-3-methoxyphenyl)-L-alanine acetate can be obtained by distilling off 422 parts of methanol and crystallizing followed by a washing of the resulting crystals with methanol. The product melts at 180°C and has a specific rotation $[\alpha]_D^{20}$ = +40.5° (methanol). Using a similar procedure with (–) methylcyclohexyl-o-anisylphosphine there was obtained the D analog in 65% optical purity. Thus, by proper choice of phosphine either enantiomorph can be obtained in major amounts.

PURIFICATION AND ISOLATION OF L-DOPA

To obtain the maximum beneficial effects of L-dopa in the treatment of Parkinson's disease, the L-dopa should be chemically as pure as possible. In view of the repeated use of this compound in relatively high doses to treat Parkinson's disease, even small amounts of chemical impurities could cause undesirable side effects. L-dopa usually contains water-insoluble contaminates such as α-amino acids which were used as starting materials. The most common water-insoluble starting materials are tyrosine and phenylalanine.

For instance, in the manufacture of L-dopa from L-tyrosine by fermentation [*J. Am. Chem. Soc.*, 91, (1969)] the L-dopa is produced in admixture with L-tyrosine. An approximately quantitative preparative isolation of L-dopa from these mixtures has only been possible with considerable effort (for example, by chromatography on ion-exchange columns or by forming derivatives thereof) and high costs. Therefore, there exists a need for providing a simple method for isolating L-dopa from mixtures containing, e.g., L-tyrosine and/or phenylalanine as impurity.

Complexing With Boric Acid or Borates

A process has been developed by *A. Kaiser, W. Koch, M. Scheer and U. Wolcke; U.S. Patent 3,799,975; March 26, 1974; assigned to Hoffmann-La Roche Inc.* for isolating L-dopa from aqueous solutions by forming water-soluble complexes of L-dopa. It has been found that L-dopa can be readily isolated in substantially pure form from a mixture with water-insoluble impurities by first treating the mixture in an aqueous medium with boric acid or a borate salt at a pH of at least 7.0 to form a complex of L-dopa and boric acid which is water soluble and then separating the water-soluble impurities from the aqueous medium containing the complex dissolved therein. The L-dopa can be isolated from the aqueous medium in substantially pure crystalline form by hydrolysis with an acid.

In the case where the impurities are tyrosine or phenylalanine, the pH of the aqueous medium during separation should be at 7.0 to 8.5. It is within this pH range that these impurities are insoluble in the aqueous medium. Where it is desired to separate other impurities from L-dopa, the separation step should take place at a pH above 7. The precise pH above 7 will depend upon the pH range at which the impurities are water insoluble.

In carrying out the formation of the complex, boric acid or any borate salt can be used. The preferred borate salt is borax (sodium borate). In forming the complex, 1 mol of boric acid or the borate salt can be used per mol of the L-dopa contained in the mixture. It is preferred to use the boric acid or the borate salt in excess of the stoichiometric amount. In forming the complex, temperature and pressure are not critical and it can be formed at ambient temperature and pressure (preferably 0° to 70°C). The separation step is carried out at a pH above 7, sufficient to maintain substantially all of the impurities in water-insoluble form. In the case of tyrosine and phenylalanine, which become water-soluble at higher pH's, the pH should be maintained at 7.0 to 8.5. When the pH of the aqueous medium during the step of complex formation is above 7 and above the value where impurities are water-insoluble, the pH can be lowered by the addition of an acid.

After adjustment of pH to maintain substantially all of the impurities in water-insoluble form, impurities which do not form a water-soluble boric acid complex precipitate in the solution or remain undissolved. Complete precipitation of the impurities can be achieved in a conventional manner, e.g., cooling to 0°C.

Furthermore the aqueous medium containing these impurities can be allowed to stand at 0°C for 48 hours or longer. The precipitate obtained consists of the water-insoluble impurities such as the α-amino acids. Tyrosine and phenylalanine can be separated from the aqueous solution containing the dopa-boric acid complex by any conventional means such as decantation or filtration. The filtrate obtained can subsequently be subjected to acidic hydrolysis to liberate L-dopa from the complex. This acid hydrolysis is preferably carried out at a pH of about 3 to 5. By this hydrolysis, the dopa-boric acid complex is cleaved and L-dopa crystallizes out.

Example: 50 g (0.258 mol) of L-dopa and 50 g (0.275 mol) of L-tyrosine as well as 450 ml of 2 N aqueous sodium hydroxide are added at room temperature under an argon atmosphere to a solution of 24.5 g (0.064 mol) of borax in 600 milliliters of water. Carbon dioxide is introduced into the clear brown solution until the pH falls to 7.8. The L-tyrosine already begins to crystallize out during the introduction.

The mixture is then cooled to 4°C for 12 hours. The precipitated crystals are filtered off and successively washed with 100 ml of water, three times with 100 ml of ethanol and twice with 100 ml of diethyl ether and then dried to constant weight at 70°C per 12 millimeters mercury. 48.5 g (97% by weight) of L-tyrosine are obtained as colorless crystals which contain less than 1% by weight L-dopa, melting point 301° to 304°C (decomposition), $[\alpha]_D$= 10.6° (c = 1%, 1 N aqueous hydrochloric acid).

The filtrate, combined with the wash water, is brought to pH 4.8 with concentrated aqueous hydrochloric acid and cooled to 4°C for 16 hours. The precipitated crystals are filtered off and successively washed with 100 ml of water, twice with 100 ml of ethanol and twice with 100 ml of diethyl ether and then dried to constant weight at 70°C per 12 millimeters mercury. 42.2 g of L-dopa are obtained as grey colored crystals of melting point 287 to 289°C (decomposition); $[\alpha]_D^{25}$ = –11° (c = 1%, 1 N aqueous hydrochloric acid). A further 2 g of L-dopa identical with the main fraction can be isolated from the mother liquor. Overall yield 88% by weight. The dopa thus isolated contains about 3% by weight L-tyrosine.

Purification By Formation of Lead Complexes

Natural L-dopa has been isolated from *Vicia faba* by means of a lead complex which is not soluble in water and which, according to R.R. Sealock, (*Biochem. Pre.* 1, 25-28, 1949), must be decomposed by means of hydrogen sulfide in a very dilute aqueous suspension (300 parts water to 1 part of L-dopa). Subsequent to the filtration of PbS and the concentration to one-tenth of the initial volume, the material is crystallized in the presence of sulfur dioxide. However, this process is not reliable.

M. Brenner; U.S. Patent 3,849,488; November 19, 1974 has developed a process for purifying L-dopa via the formation of lead complexes. It was found that lead complexes of dopa (dopa-Pb) can be obtained not only from the soluble salts of lead but also from insoluble salts of lead, such as chloride or sulfate, and its decomposition does not require the action of H_2S but it can be effected even by a weak acid that does not precipitate lead, e.g., acetic acid.

The process may be carried out with advantage under the following conditions. The crude dopa in suspension or in aqueous solution which is neutral, slightly acid (acetic acid) or strongly acid (mineral acids), is treated with an equimolecular quantity of a lead salt that is insoluble or slightly soluble in water, e.g., chloride or sulfate, and, in addition, with a base such as ammonia, bicarbonate or carbonate of ammonium, to bring and maintain the pH, preferably at 7 to 7.5.

The complex may be formed in the temperature range between the freezing and boiling of the solvent, its precipitation being accelerated under heat. The duration of the reaction also depends on the physical condition of the lead salt used and on the intensity of agitation. A duration of reaction between 30 minutes and 6 hours has been observed. The complex formed is isolated either by filtration or by centrifugation; it is washed repeatedly with water or with an organic solvent miscible with water, until the impurities disappear completely. The mother liquors and the washing waters are eliminated; they contain only traces of dopa and lead salt.

The isolated complex which may still be wet is mixed under stirring with a concentrated solution of hydrochloric acid or sulfuric acid, in at least stoichiometric amount in relation to lead and to dopa. The complex is decomposed immediately, even at ambient temperature. The lead salt precipitates while the dopa passes into solution. The lead salt is then separated. It is used again, as is, as a Pb^{++} source in the following operation. Subsequently, if necessary, the small amount of Pb^{++} which may remain in the solution (e.g., in the case where lead was precipitated in the form of chloride) is removed from the solution in a customary manner, e.g., by the action of hydrogen sulfide.

The process then comprises two variants for the isolation of dopa: (a) The solution of dopa concentrated to a low volume (2 to 40 parts per part dopa) is neutralized at 0 to 50°C, to a pH of preferably 4.5, e.g., by means of ammonia or ammonium bicarbonate; the dopa is crystallized and isolated. The totality of the mother liquors and washing waters is recycled to a following operation, prior to the precipitation of the complex. For completing the purification, dopa is recrystallized in water according to the method described by R.R. Sealock [*Biochem. Prep.* 1, (1949)], but without a concentration of the mother liquors. After the first lot of dopa is isolated, the mother liquors are recycled to a following operation, prior to the precipitation of dopa·Pb complex.

(b) The acid solution (in particular hydrochloric acid solution) is concentrated; the gaseous hydrochloric acid is bubbled in a temperature range comprised between 50° and the freezing temperature of the solvent; the chlorhydrate of dopa is crystallized and isolated. The totality of the mother liquors and washing waters is recycled to a following operation, prior to the precipitation of the complex. The hydrochlorate of dopa thus isolated is dissolved in a minimal amount of water and the solution is neutralized to about a pH of 4.5; the pure dopa is crystallized and isolated; the mother liquors and washing waters are recycled together with the preceding liquors and waters.

Example: Eight liters of water, 80 ml crystallizable acetic acid and 100 g L-dopa at 85% (0.445 mol) are introduced into a 10 liter vessel equipped with a stirrer, a reflux condenser and a thermometer. The material is heated to 60 to 65°C while stirring, to form a (pale yellow) solution; there remains a small amount of insoluble material. There is added 228 g (0.06 mol) crystallized lead acetate ($3H_2O$); the mixture becomes slightly cloudy and darker. Then 5 g of activated carbon is added, the material is cooled to 20 to 25°C and filtered on fritted glass.

The pale yellow filtrate is poured into a 10 liter vessel and about 238 ml ammonia at 20% is added slowly under nitrogen atmosphere to a pH of 7.2 to 7.3. At the start, the pH is 4; the dopa·Pb complex is precipitated progressively as the ammonia is introduced and the reactive mixture thickens at the end. The material is still stirred for 30 minutes and filtered on two large filters of fritted glass having a diameter of 260 mm; the filtering operation lasts for about 40 minutes. The material is dried as much as possible, each cake is placed into 5 liters water, stirred for 15 to 30 minutes, filtered again on fritted glass and each cake is washed with 500 ml water.

The dopa·Pb complex is introduced in portions into a 4-liter vessel containing 250 ml water and 190 ml 10.5 N hydrochloric acid (2 mols), while the stirrer is operated. The complex decomposes rapidly and (white) lead chloride is precipitated. After the introduction is completed, the material is stirred for 30 minutes. The total volume is about 2 liters; the material is then concentrated under vacuum to a volume of 1 liter. Lead chloride is filtered on fritted glass (ready filtration) and the material is washed twice, each time with 50 ml 0.1 N hydrochloric acid. The pH of the filtrate is brought to 0.8 to 1.2 by the addition of ammonium bicarbonate and the material is saturated with hydrogen sulfide to precipitate Pb^{++} remaining in the solution.

After one hour, a test sample is filtered and the conclusion of the precipitation is verified. Stirring is continued for two hours under a low pressure of hydrogen sulfide, then 2 g of activated carbon are added, which accelerates the subsequent filtration. It is performed on fritted glass through a cellulose filter and the material is washed twice, each time with 30 ml hot water. The filtrate is heated to 70 to 80°C while subjected to stirring and while nitrogen is bubbled in to remove hydrogen sulfide. 5 g activated carbon is added and the material filtered while hot and washed again with 3 x 30 ml hot water. The filter has a yellow-green tint. The material is poured into a 3-liter vessel and the pH is adjusted to 4.5 by adding about 75 g ammonium bicarbonate with stirring.

The start of precipitation of the L-dopa is observed at a pH of 2.5. There is added 1.5 g ethylenediaminetetraacetic acid (EDTA); the pH drops to 3.5 to 4 and the solution is slightly discolored. The material is stirred for 2 hours at 20°C under nitrogen, and then filtered on fritted glass. The L-dopa is washed on the filter with 4 x 50 ml water to eliminate the chlorides, and then with 50 ml ethanol and the material is dried in a drier at 60°C. Thus, 71.2 g (A_1) is obtained. According to the colorimetric quantitative analysis of Arnow, the mother liquors still contain 9.60 g L-dopa to be recycled. The rate of recovery of L-dopa is:

$$\frac{71.2}{85 - 9.60} \times 100 = 94.4\%$$

To recrystallize the product, 65 g L-dopa (A_1) and 2,600 ml (40 volumes) of water are introduced into a 4-liter vessel, equipped with a stirrer and a reflux condenser, and heated to boiling. 1 g EDTA and 5 g activated carbon are added, the material is stirred for 5 to 10 minutes and filtered on fritted glass directly into a 4-liter vessel; the filtrate is colorless. About 30 ml of an aqueous solution of sulfur dioxide is added and the material is cooled to 20°C. The material is stirred for 5 to 10 hours and filtered on fritted glass. The L-dopa is well crystallized and white. The material is washed with 2 x 30 ml water, then with 30 ml ethanol and dried in a drier at 60°C. Thus, 50 g (A_1*) are obtained. The yield of recrystallization is 77%.

	A_1	A_1*
Nitrogen, total (theory, 7.11%)	7.04%	7.09%
$[\alpha]_D$ (c = 2%, N HCl)	$-10.8° \pm 0.5°$	$-11.5° \pm 0.5°$
Iron (mg/100 g)	<1	<1
Pb (polarography, mg/100 g)	1	1
Chromatography on paper		
Tyrosine	<0.2%	<0.1%
Other impurities	One, hardly visible for spot of 100 γ, having lower R_f than that of L-dopa	One, hardly visible for spot of 500 γ

Purification by Low Temperature Crystallization

L-dopa forms stable cubic crystals. When crude L-dopa is recrystallized from its supersaturated aqueous solutions in the usual manner, the crystals are gray, if the supersaturation was brought about by partial evaporation of a more dilute solution or by neutralization of an acidic solution. In the latter case, the conventionally recrystallized material is additionally contaminated with anions of the acid employed in preparing the solution. Active carbon and oxygen acceptors cannot entirely prevent the crystals from being slightly colored.

Perfectly colorless and pure needle-shaped crystals have been obtained by the process developed by *M. Miyazawa, I. Sasaji, T. Hachiya, K. Toi, T. Uzuki and N. Sato; U.S. Patent 3,880,918; April 29, 1975; assigned to Ajinomoto Co., Inc., Japan.* It has been found that the crystals of dopa, that is DL-dopa, L-dopa, or D-dopa, formed from the supersaturated aqueous solutions below 30°C are needle-shaped and purer than the cubic crystals formed from the same solutions above 30°C. Holding the needle-shaped crystals in contact with water above 30°C causes their conversion to the cubic form. Residual impurities may be expelled during this conversion. The precipitation of the needle-shaped crystals may be brought about by making an aqueous dopa solution supersaturated at any temperature below 30°C at which the solution is liquid. For practical reasons, precipitation should normally take place between 0° and 25°C.

The numerical concentration of dopa in the supersaturated solution is not critical. It is convenient to operate at concentrations between 10 and 20 grams per deciliter, but lower and higher concentrations are equally operative, and the changes in concentration will produce only the predictable minor effects on the crystallization rate, product purity, and other operating variables. Seed crystals may be employed to initiate crystallization in a known manner, and needle-shaped seed crystals of one of the optically active forms induce crystallization of pure needle-shaped crystals of the same enantimorph from a contaminated solution of the DL-form, if the crystallization takes place at a temperature below 30°C.

When the needle-shaped crystals are heated in the presence of water to above 30°C, they are rapidly converted to the cubic form. The conversion is complete at 45° to 50°C, practically independently of the heating rate at which this temperature is reached. If the conversion is carried out in the presence of a relatively great amount of water, as by dispersing the needle-shaped crystals at least in an equal weight of water, the cubic crystals obtained may be significantly purer than the starting material.

The amount of water needed for the conversion, however, is very small, a film of moisture retained on the needle-shaped crystals after filtering and washing on the filter being entirely adequate. Three to five weights of water are preferably mixed with the needle-shaped crystals prior to heating to the conversion temperature for best removal of residual impurities.

Example: 10 g reddish-brown, crude, solid L-dopa were suspended in 100 ml water, and 60 ml concentrated hydrochloric acid were added to almost entirely dissolve the solids. The solution was further mixed with 10 mg crystalline sodium bisulfite as an oxidation inhibitor, and a small amount of insoluble material was removed by filtration.

The filtrate was cooled to 15°C, and a cold, 40% sodium acetate solution was added with stirring drop by drop at a rate so as not substantially to raise the temperature. The pH reached 3.0 to 3.5 after 10 minutes, and stirring at 15°C and without further additions was continued for 1 hour, whereby needle-shaped, colorless crystals were precipitated. They were filtered out and suspended in 40 ml water containing 10 mg sodium bisulfite.

The slurry so obtained was gradually heated to 60°C with stirring over a period of 9 minutes and then cooled to ambient temperature, whereby a second crop of pure L-dopa crystals was precipitated. When filtered off, washed with a little water, and dried, the crystals weighed 9.2 g. They were colorless and cubic and contained less than 0.01% chlorine.

For comparison purposes, a solution of crude L-dopa in aqueous hydrochloric acid was prepared as described above. Prior to filtering, it was stirred with 1.0 g active carbon at 25° to 30°C for one hour. The filtrate was still faintly yellow. It was heated to 60° to 70°C, and the pH was adjusted to 3.0 to 3.5 with stirring within 1 hour by adding the sodium acetate solution. The cubic crystals thereby precipitated were filtered off, washed with a little water, and dried. They weighed 8.9 g and were gray. Their chlorine content was 0.09 to 0.10%. When these crystals were again subjected to the same purification treatment, the yield was 8.1 g but the color and chlorine content were unchanged.

Isolation Using a Liquid Ion Exchanger

An improved process of isolating L-dopa from an aqueous solution, irrespective of the source of the solution has been disclosed by *T. Broderman, P. Hietala, and J. Sundman; U.S. Patent 3,928,431; December 23, 1975; assigned to Oy Medica Ab, Finland.* According to the process, L-dopa is extracted from the aqueous solution with the aid of a liquid ion exchanger comprising a sulfonic acid dissolved in an organic solvent. As a result of the ion exchange, the L-dopa is bound to the sulfonic acid as a salt, from which it is eventually liberated by neutralizing the solution with a base, which precipitates the L-dopa. The precipitate is separated from the solution, which is then regenerated by washing with a mineral acid solution. The sulfonic acid solution thus obtained is used again for a subsequent extraction of L-dopa.

The sulfonic acid used as the ion exchanger must be one which, when partitioned between the organic and aqueous phases, favors the organic solvent used. In addition, the sulfonic acid must be capable of forming, together with a suitable base, such as ammonia, hydrazine, alkylamines, such as methyl, ethyl, propyl or butyl amine or alkanolamines, such as methanolamine, ethanolamine, diethanolamine or triethanolamine, a salt that is soluble in the organic solvent used; and also capable of decomposing when washed with an acid, so that the sulfonic acid in the organic solvent is liberated and the liquid ion exchanger is regenerated.

Examples of suitable sulfonic acids include alkyl- and arylsulfonic acids. A preferred sulfonic acid is dodecylbenzenesulfonic acid, which is inexpensive and readily available. Estersulfonic acids may also be used, for example, laurylsulfonic acid. However, estersulfonic acids are less suitable because ester sulfonates are slowly decomposed by hydrolysis in acid solutions. Gardinol type detergents, "Teepol" or "Tergitol" may be used. Suitable organic solvents are alcohols, ketones, and esters which are insoluble or sparingly soluble in water. A preferred organic solvent is methyl isobutyl ketone (hexone).

The extraction apparatus used in this process is illustrated by Figure 10.1. The starting material is a vegetable extract or other L-dopa-containing solution from which most of the calcium content is removed by precipitation with oxalic acid. The precipitate, together with any turbidity present in the starting material is filtered or removed by centrifuging. The solution is then pumped into a container **1** and stirred. By recycle, there is introduced the aqueous solution obtained when the mother liquor remaining after crystallization of L-dopa is washed with an acid in liquid-liquid contactor **8**.

Uncrystallized L-dopa and the mineral acid salt of the base which is used for neutralization, remain in the mineral acid-containing wash solution. Sodium hydroxide solution is added to the solution in **1** to raise the pH of the outflowing solution to 1.5 to 2.5, preferably automatically with a pH meter. From container **1** the liquid flows into contactor **2**, where it is countercurrently extracted with an organic solvent. From contactor **2** the organic solvent flows into a washer **4**, where washing with an excess of an alkali solution is continuously effected. Most of the impurities in the organic solvent are removed by the alkali solution and withdrawn as waste.

From washer **4** the organic solvent flows into another washer **5**, where it is washed with water. This, too, is a continuous and automatic process. From washer **5** the organic solvent flows into a storage and pumping container **6** into which make-up organic solvent is introduced when necessary. The organic solvent is conveyed from container **6** to contactor **2** by pump **7** and the flow rate thereof is adjusted. The aqueous solution flowing out of washers **4** and **5** is saturated with organic solvent, as is the aqueous solution which flows from liquid-liquid contactor **2** to liquid-liquid contactor **3**. This aqueous solution still contains L-dopa. The organic solvent circulates continuously through contactor **2**, washers **4** and **5**, container **6**, and pump **7**.

FIGURE 10.1: FLOW DIAGRAM

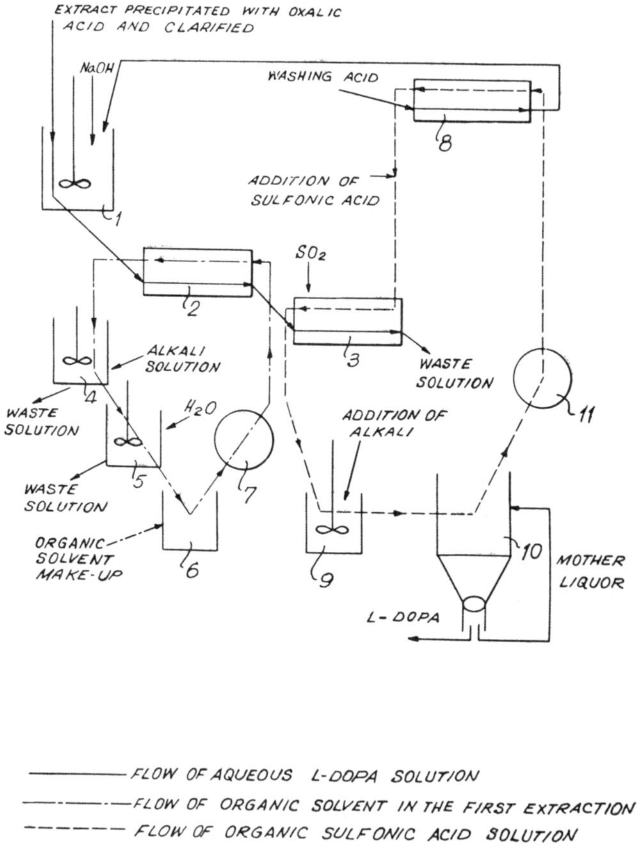

Source: U.S. Patent 3,928,431

In liquid-liquid contactor **2** the organic solvent extracts impurities from the aqueous solution flowing into contactor **2**, which impurities are removed from the organic solvent in washers **4** and **5**, whereafter pump **7** returns the solvent into circulation, with any required make-up being added to container **6**.

The extraction of L-dopa from the aqueous solution into the sulfonic acid-containing organic solvent takes place in liquid-liquid contactor **3**, into which the aqueous solution flows from contactor **2**. The organic solution thus obtained, which contains L-dopa as a sulfonic acid salt, flows into container **9**, where stirring is effected and where the acidity of the solution is adjusted to pH 4 to 5 by addition of a suitable base. The base added to the solution must be one which, together with the sulfonic acid, forms a salt that is soluble in the organic solvent, without causing the water in the solution to separate into a different liquid phase. After this, the liquid is transferred into container **10**, a large vessel where it remains for the time required for the crystallization of L-dopa, i.e., 2 to 4 hours. Container **10** is provided with an immersed filter and a stirrer. In the cone-shaped lower part of the container a tap is provided. Stirring is necessary to accelerate crystallization.

The L-dopa is separated from the mother liquor by periodically opening the tap at the bottom of the container, so that the crystalline suspension is let out. The suspension is filtered and the mother liquor returned to container **10**. A pump **11** pumps the mother liquor from container **10** through the immersed filter into liquid-liquid contactor **8**, where the liquid ion exchanger is regenerated and its content of uncrystallized L-dopa washed back with a mineral acid such as sulfuric or hydrochloric acid of suitable concentration.

Besides the L-dopa, the mineral acid extracts from the organic phase the base used for neutralization in container **9**. The wash solution is returned to container **1**. The regenerated sulfonic acid solution, to which sulfonic acid has been added, if necessary, returns to liquid-liquid contactor **3**. The addition of sulfonic acid makes up for the sulfonic acid removed from the system with the waste solution flowing out of contactor **3**. Liquid-liquid contactors **2, 3** and **8** are of the mixer-settler-box type. They each consist of 10 units with an extraction efficiency of 70 to 80%. Washers **4** and **5** are separate units of the same type.

Example: The starting material was a water extract of the whole vegetative unit of green horse bean plant (*Vicia faba*). The solution contained 75 g of dry substance and 2.4 g of L-dopa per liter. A saturated solution of oxalic acid at 40° to 50°C (3 g/l) was added to precipitate any calcium in the solution. The precipitate was separated from the solution in a filtering centrifuge.

At first, sulfuric acid was added to the filtrate to adjust the pH to 1.5. Subsequently, when the L-dopa was crystallized and the mother liquor washed with sulfuric acid, this acid solution was used again by adding it to the vegetable extract after centrifugation of the oxalic acid precipitation, after which the pH of the solution was adjusted to 1.5 by adding the necessary amount of a 70% NaOH solution. The solution thus obtained was extracted with methyl isobutyl ketone in liquid-liquid contactor **2**, which is a 10-unit mixer-settler-box-type contactor.

Each unit has a 5-liter mixing section and a 20-liter settling section. The flow rate of the organic solvent through the apparatus was 120 liters per hour and that of the aqueous solution 90 to 120 liters per hour. The methyl isobutyl ketone extract, which contained 3.0 to 3.5 g/l of dry substance was first washed in washer **4** with 2 N sodium hydroxide solution, the flow rate of which was 6 liters per hour, then in washer **5** with water at a flow rate of 6 liters per hour. Washers **4** and **5** were each separate units of the extraction apparatus, the mixing part of which had a volume of 5 liters and the settling part a volume of 20 liters.

The washed, circulating methyl isobutyl ketone contained 0.3 to 0.4 g of dry substance per liter. Pump **7** is a piston pump, the stroke length and frequency of which can be adjusted so as to give the desired flow rate. This type of pump was used for flow rate adjustment and dosage of all the solutions used in the extraction. The capacity of container **6** was 50 liters. The extraction of the prewashed aqueous solution with sulfonic acid-containing methyl isobutyl ketone was carried out in container **3**, which was similar to contactor **2**. The methyl isobutyl ketone contained 40 to 50 g/l of dodecylbenzenesulfonic acid. The flow rate of the aqueous solution was 90 to 120 liters per hour and that of the liquid ion exchanger was 120 liters per hour.

To the first unit of the extraction apparatus, sulfurous acid was added as an antioxidant in the form of an aqueous solution of sodium bisulfite in an amount of 1 gram per liter of solution to be extracted. The methyl isobutyl ketone-dodecyl benzenesulfonic acid extract coming from contactor **3** had an average L-dopa content of 3.7 g/l and a dry substance content of 65 to 75 g/l. Container **9**, where the neutralization took place, had a volume of 25 liters. The neutralization was carried out with gaseous ammonia, which was conducted into the solution from a gas cylinder.

The pH of the solution was kept at 4.5 and was continuously controlled with a pH meter. The liquid volume in the crystallization container **10** was 500 liters, i.e., the liquid remained in this container for about 4 hours on the average. The final product was separated from the mother liquor by periodically opening the tap at the bottom of container **10**, to release the crystalline suspension; the precipitate was filtered off, and the mother liquor was returned to container **10**. A pump **11** continuously sucked the solution through a filter immersed in the container, and pumped it into contactor **8**, where the mother liquor was washed with 2.5 N sulfuric acid solution.

The mother liquor contained about 2.0 g/l of L-dopa. The volume of the wash solution was ¼ that of the mother liquor. After washing, the sulfuric acid solution contained 5.2 g and the mother liquor 0.5 g of L-dopa per liter. The waste solution from contactor **3** contained per liter, about 0.24 g of L-dopa, 1 g of dodecylbenzenesulfonic acid and 15 g of methyl isobutyl ketone.

The extract used as the starting material contained 2.4 kg of L-dopa per 1,000 liters. From this extract there were separated 2.3 kg of a product consisting of 90% L-dopa. In addition to L-dopa, the product contained 3% tyrosine, and less than 1% each of leucine, phenylalanine and aspartic acid, as well as ammonium sulfate, ammonium sulfite, and dodecylbenzenesulfonic acid salt.

Using Polymers of p-Vinylbenzeneboronic Acid

A chromatographic separator process using polymers containing p-vinylbenzeneboronic acid has been disclosed by *W.L. Stanley, C.A. Ellinger and B.G. Chan; U.S. Patent 3,972,918; August 3, 1976; assigned to the U.S. Secretary of Agriculture.* It was found that when dopa and tyrosine are applied to a polymer of p-vinylbenzeneboronic acid, dopa is preferentially adsorbed by the polymer. Thus, this process involves applying a mixture of dopa and tyrosine to a solid polymer of p-vinylbenzeneboronic acid (hereinafter referred to as VBA). The treated polymer is washed with dilute aqueous alkali to remove the tyrosine while leaving the dopa adsorbed on the polymer. The dopa is then recovered by washing with dilute aqueous acid. The primary advantage of the process is that total separation of dopa from tyrosine is attained. Another advantage of the process is that it can be adapted to a continuous process. For example, where a mixture of dopa and tyrosine is prepared by contacting tyrosine hydroxylase, the mixture can be passed through a column containing the polymer. The effluent containing tyrosine, free of dopa, may be recycled to the enzyme treatment to produce more of the dopa-tyrosine mixture, which may then be passed through the column.

The recycling mentioned above also provides another advantage. In the enzymatic reaction of tyrosine to produce dopa, various by-products are produced in small amounts. These impurities deactivate the enzyme, thus "poisoning" it and halting the reaction. If the conversion to dopa can be maintained at a low level, for example, 2%, the impurity level is so low as to have little detrimental effect on the enzyme. Ordinarily, however, only 2% conversion to dopa would be impractical. Since the method of the process can be maintained on a continuous basis in conjunction with the above enzymatic conversion, 2% conversion to dopa is feasible.

The monomer (VBA) can be prepared by known methods from p-bromostyrene or p-chlorostyrene as shown by Lennarz et al, *J. Amer. Chem. Soc.,* 82, 2169-2171 (1960). Homopolymers of VBA or copolymers with styrene or acrylamide may be used in this separation process. These polymers are prepared by conventional polymerization techniques. For example, VBA per se or admixed with styrene or acrylamide is heated at 80° to 120°C in a suitable solvent such as diglyme (dimethyl ether of diethylene glycol) containing a small proportion of a polymerization initiator such as α,α'-azobisisobutyronitrile.

Example 1: Preparation of VBA Polymer on Polystyrene Beads – A solution of 1 g of p-vinylbenzeneboronic acid in 25 ml of diglyme containing 0.01 g of azobisisobutyronitrile was added to 5 g of porous polystyrene beads (80 to 100 mesh). The mixture was evacuated to remove entrained air from the polystyrene beads and draw the reaction solution into them. Excess solution was removed with a pipette, and the moist beads were placed in a rotary evaporator where they were heated to about 90°C for 20 minutes. The beads were then vacuum dried at 120°C for 1 hour.

To remove unreacted reagents, the dried beads were then treated as follows: The beads were suspended with stirring in 50 ml of dimethylformamide containing 5% water and warmed to about 80°C for 1 hour. The beads were collected by filtration, washed successively with methanol and ether, and dried in vacuum. Yield of the VBA-polymer-polystyrene beads was 5.36 g.

Example 2: A column (1 x 21 cm) was packed with the VBA-polymer-polystyrene beads prepared as in Example 1. The column was equilibrated by passing 0.1 M potassium phosphate buffer at pH 8 through the column for about 1 hour. A total of 60 ml of buffer solution was used. A solution of 20 micromols each of tyrosine and dopa was applied to the column which was then eluted with aqueous 0.1 M phosphate buffer at pH 8 at the rate of 60 ml per hour. The eluate was collected in 5 ml fractions and analyzed.

It was found that essentially all the tyrosine (20 micromols) was eluted in the first 20 to 30 ml of eluant. An additional 200 ml of eluate was collected without containing any dopa. The eluting liquid was then changed to aqueous 1 M acetic acid at pH 2.5 which was passed through the column at 60 ml/hr. Essentially all of the dopa (20 micromols) was eluted immediately and contained in the first 20 to 30 ml of the acid eluate.

RESOLUTION OF DOPA AND DERIVATIVES

Use of Cinchonine Salts of N-Benzoyl Dopa

It is known [(Yamada et al., *Chem. Pharm. Bull.* (Tokyo 10, 680-688 (1962)]. that racemic N-acetyl-3,4-methylenedioxyphenylalanine has been converted to the cinchonine salt and separated into the optical antipodes by crystallization from alcohol. In this process the L compound is obtained as a solution in alcohol, while the D compound crystallizes out.

Another prior process [*Helv. Chim. Acta* 35, 1776 (1952)] shows that the solubility behavior of the salts is completely altered by the use of a differently substituted starting material. In this process, the optical antipodes are separated by crystallizing the cinchonine salt of racemic N-benzoyl-3-methoxy-4-hydroxyphenylalanine out of pure acetone, the L compound being obtained as a difficultly soluble salt.

H. Schubel, P. Janssen, and H. Ratz; U.S. Patent 3,794,655; February 26, 1974; assigned to Dynamit Nobel AG, Germany have developed a process for resolving DL-N-benzoyl-3,4-dihydroxyphenylamine using cinchonine. The use of the 3,4-dihydroxy compound has the advantage that the later splitting off of ether that is necessary in prior processes to form L-dopa can be eliminated. Surprisingly it has been found that, in contrast to the cinchonine salt of the corresponding 3-methoxy compound, the cinchonine salt of the 3,4-dihydroxy compound cannot be separated by crystallization from pure acetone, since both antipodes are substantially soluble in pure acetone.

Whereas in the prior dissociation of the 3-methoxy-4-hydroxy derivative by crystallization from acetone the L salt crystallizes, when this process is applied to the corresponding 3,4-dihydroxy derivative, the optical antipodes remain in solution or form syrupy precipitates which cannot be isolated. Nevertheless, an easy cleavage of the racemates can be accomplished with an excellent yield if the cinchonine salt of N-benzoyl-3,4-dihydroxyphenylalanine is crystallized out of a homogeneous mixture of a lower aliphatic ketone, alcohol or ether, or several such compounds, and water, especially a mixture of acetone and water.

The amount of water present in the solvent mixture is to be sufficient to provide at least 1 mol of water per mol of the N-benzoyl-dopa. An excess of water, however, has no adverse effect on the separation of the racemate. Especially suitable for the process are mixtures of aliphatic ketones, alcohols and/or ethers having 2 to 6 carbon atoms, and water. A medium that is especially preferred for the process is a aqueous acetone. Good results are also achieved with aqueous n-butanol, aqueous methyl ethyl ketone to which a small amount of isopropanol has been added, an aqueous mixture of isopropyl ether and isopropanol, aqueous isopropanol and aqueous ethanol.

Further details of the resolution and conversion to L-dopa are shown in the following examples.

Example 1: In an Erlenmeyer flask 159.5 g racemic N-benzoyl-dopa monohydrate (0.5 mol) and 147 g of cinchonine (0.5 mol) are dissolved in 750 ml of methanol and the solution is filtered into a 4-liter round flask, and the solvent is removed by heating in a water-jet vacuum. The syrupy salt residue is dissolved with vigorous shaking or stirring in a boiling mixture of 2.5 liters of acetone and 250 ml of water.

Immediately the cinchonine salt of D-N-benzoyl-dopa begins to crystallize. To complete the crystallization the mixture is allowed to stand overnight at room temperature or in a refrigerator. The crystals are suction filtered and washed with thoroughly chilled aqueous acetone (approximately 10% water). Yield: 150.25 g (= 98% of the theory). The substance is sufficiently pure for further processing to D-N-benzoyl-dopa and therefore does not need to be further purified. For analysis, a portion of the substance was recrystallized from the acetone-water mixture. It then showed the following characteristics: MP = 168°C; $[\alpha]_D^{20}$ = +85°±2° (c = 1; methanol).

Example 2: If the mother liquor obtained from Example 1 above, after the suction filtering and washing of the D salt crystals, is evaporated in the water jet vacuum to the dry state and the residue is washed with ethyl acetate, the crude cinchonine salt of L-N-benzoyl-dopa is obtained in a yield of 95 to 99% of the theory. The compound is pure enough for further processing to L-N-benzoyl-dopa and L-dopa. For analysis a portion of the substance was recrystallized from acetic ester. The following characteristics were determined: MP = 154° to 155°C; $[\alpha]_D^{20}$ = 135°±2° (c = 1; methanol).

Example 3: 100 ml of semiconcentrated hydrochloric acid and 400 ml of acetic ester are added to 150 g of the cinchonine salt of L-N-benzoyl-dopa obtained in Example 2. The acetate layer is separated, and the aqueous layer is shaken 5 times with 200 ml of ethyl acetate each time. The combined acetate ester extracts are washed with water, dried with Na_2SO_4 and concentrated in a water-jet vacuum. As the residue, one obtains L-N-benzoyl-dopa which is sufficiently pure for further processing to L-dopa. Yield 72 g (= 96% of the theory). A portion of the substance was recrystallized from water to determine the analytical data. MP = 172°C; $[\alpha]_D^{20}$ = -30°±2° (c = 1; methanol).

Example 4: 50 g of L-N-benzoyl-dopa is boiled for 2 hours with about 225 ml of 20% hydrochloric acid under shielding gas (nitrogen). During the cooling that follows, a portion of the liberated benzoic acid crystallizes and is removed by filtration. The remainder is extracted from the solution with diethyl ether. The aqueous solution is concentrated under shielding gas (nitrogen) to the dry state.

The L-3,4-dihydroxyphenylalanine hydrochloride is dissolved in water and the solution is adjusted to the isoelectric point with a base such as triethylamine. After relatively long standing in the refrigerator (overnight is best), 31 g of L-dopa crystallizes (= 92% of the theory). After recrystallization from water containing SO_2, 30.0 g are obtained of L-dopa (= 89% of the theory); MP 290°C (decomposition). $[\alpha]_D^{20}$ = -11.3°±0.5° (c = 1; 1 N HCl). The cinchonine salt of D-N-benzoyl-dopa can be converted to D-dopa by the processes of Examples 3 and 4.

H. Schubel, P. Janssen, and H. Ratz; U.S. Patent 3,796,750; March 12, 1974; assigned to Dynamit Nobel AK, Germany have modified the process to obtain additional L-dopa by racemizing the D-dopa derivative to the racemate.

The D-form of N-benzoyl-3,4-dihydroxyphenylalanine which occurs as a by-product of U.S. Patent 3,794,655 is treated with acetic anhydride, with heating, until a solution is formed. For this purpose, the temperature can be raised to the boiling temperature of the acetic anhydride, but lower temperatures preferably 60 to 90°C also serve the purpose. The amount of acetic acid anhydride used is at least 3 mols per mol of D-N-benzoyl-3,4-dihydroxyphenylalanine. Usually, however, the use of a greater excess of acetic acid anhydride is preferred.

The solution obtained is then treated with enough water to form a precipitate. For this purpose the still warm solution can be poured into water or water can be added to the solution. It has been found that treatment with water is an important feature of this process. The water should be used in a mol ratio of 5,000 to 20,000 per mol acetyl derivative. The temperature of the precipitation step should be preferably 30° to 40°C. For if an attempt is made to perform removal of the acetic anhydride by evaporation in vacuo, a product is obtained which cannot be used for the further stages of the process.

The compound produced in this stage is believed to be N-benzoyl-3,4-diacetoxyphenylalanine. The acetyl radicals of this product (intermediate) are saponified by known methods, as for example by treatment with aqueous alkalies, especially aqueous NaOH. In this manner, racemic N-benzoyl-3,4-dihydroxyphenylalanine is produced, which is reused in the first stage of the process of U.S. Patent 3,794,655.

Example: In a 250 ml round flask, 30.1 g D-N-benzoyl-3,4-dihydroxyphenylalanine (0.1 mol) is heated with stirring at about 80°C with 100 ml of acetic anhydride. After about 15 minutes this D-isomer had dissolved. Stirring is continued at this temperature, and then the still hot contents of the flask are poured into about 1 liter of water with vigorous stirring. A colorless precipitate forms, which is suction filtered, washed with a large amount of water, and finally boiled for about 1 hour in about 1 liter of water.

Then the precipitate is heated with an aqueous solution of about 10% NaOH, whereupon it goes into solution. After about 15 minutes the solution is cooled and acidified with HCl or H_2SO_4. The hydrolyzed product that crystallizes out is filtered and recrystallized from water. According to elemental analysis, infrared spectrum and TLC, this substance is racemic N-benzoyl-3,4-dihydroxyphenylalanine monohydrate. MP = 197°C; $[\alpha]_D^{20}$ = 0° (c = 1; methanol). Yield: 29.7 grams (= 93% of the theory). This product is then resolved as by the process described in the previous U.S. Patent 3,794,655.

Use of Dehydroabietylamine as Resolving Agent

Salts of dehydroabietylamine with derivatives of racemic DL-dopa are used for separating the L- and D-antipodes in the process developed by *A. Kaiser, M. Scheer, W. Häusermann, and L. Marti; U.S. Patent 3,969,397; July 13, 1976; assigned to Hoffmann-La Roche Inc.* The derivatives used are of a racemate of the formula below where R_1 is methoxy; R_2 is hydroxy, and R_3 is benzoyl.

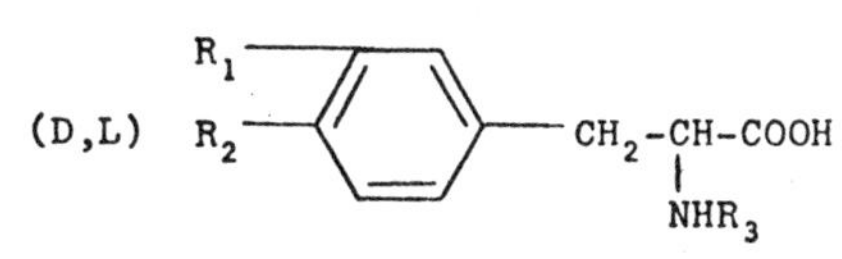

A stable salt is formed by reacting the above racemate with the dehydroabietylamine in an inert organic solvent preferably a lower alkanol solvent or in a mixture of water and a lower alkanol solvent. The preferred lower alkanol solvents are methanol and ethanol. This reaction can be carried out at 20°C to the reflux temperature of the reaction medium.

Upon cooling to room temperature (35°C) or below (0°C), the L-salt crystallizes out of solution with the D-salt remaining in solution. After cooling, the L-salt can be recovered in high yield from the mother liquor by filtration. It is preferred that for every mol of the racemate of the above formula, from 0.4 to 1 mol of dehydroabietylamine be used. The L-salt can be converted to L-dopa by hydrolysis, whereas the D-salt can be racemized back to the D,L-acid of the above formula.

The starting material of the above formula can be prepared by standard methods from vanillin and hippuric acid (N-benzoylglycine) with subsequent hydrolysis and reduction of the 2-phenyl-4-(o-acetylvanillylidene)-2-oxazolin-5-one intermediate formed. Details of the resolution and formation of L-dopa from L-N-benzoyl-3-(4-hydroxy-3-methoxyphenylalanine are shown in the following examples.

Example 1: 31.5 g of D,L-N-benzoyl-3-(4-hydroxy-3-methoxyphenyl)-alanine and 28.5 g of dehydroabietylamine are placed in 80 ml of absolute methanol and heated. On boiling, all of the reactants dissolve. The resulting solution is then allowed to cool so that a precipitate is formed. After 22 hours at room temperature the precipitate is filtered off and rinsed portionwise with a total of 50 ml of ice-cold methanol, then with diethyl ether, and dried at 60°C under a vacuum.

For recrystallization, this salt is dissolved in 800 to 1,000 ml of boiling methanol, filtered off hot, concentrated to 100 ml (whereby the substance begins to crystallize), left to stand at room temperature for an hour, thereupon treated with 700 ml of diethyl ether and left to stand overnight at 0°C. On the next morning, it is filtered, washed with ether and dried. 28.6 g of dehydroabietylamine salt of the L-antipode are obtained in the form of white crystals with a melting point of 232° to 233°C; $[\alpha]_D^{24}$ = +55.0° (c = 1 in methanol). A second recrystallization does not raise the rotation.

Example 2: 27 g of the dehydroabietylamine salt of the L-antipode prepared in Example 1 are treated in a shaking funnel with 800 ml of cold methanol and 200 ml of 2 N caustic soda. The resulting clear solution is extracted with two portions each of 600 ml of petroleum ether. The methanol is evaporated off from the aqueous-methanolic layer. The residue is made congo-acidic with concentrated hydrochloric acid. To this acidified residue, ice is added and then the residue is extracted twice with 600 ml of ethyl acetate each time.

The ethyl acetate layers are washed twice with saturated aqueous sodium chloride solution, dried over anhydrous sodium sulfate and evaporated to an oil. This residue is sprinkled with isopropyl ether and allowed to stand at room temperature to form crystals. The residue is then treated with a little petroleum ether and allowed to stand overnight at 0°C. After filtering off and rinsing with petroleum ether, for recrystallization this substance is dissolved in hot ethyl acetate, then left to cool to room temperature (whereupon recrystallization begins) and treated gradually with a little petroleum ether. After standing overnight at 0°C, the crystalline precipitate is filtered off and rinsed with petroleum ether. There are obtained 11.7 g of pure L-N-benzoyl-3-(4-hydroxy-3-methoxyphenyl)-alanine with a melting point of 154°C and a rotation of $[\alpha]_D^{24}$ = –32.7° (c = 1 in methanol). Further recrystallization does not alter the rotation value.

Example 3: 5 g of L-N-benzoyl-3-(4-hydroxy-3-methoxyphenyl)-alanine are treated under a nitrogen atmosphere with 40 ml of colorless aqueous solution containing 48% by weight hydrobromic acid and heated for 2½ hours in a 140°C bath. The dark-colored residue is cautiously evaporated on the rotary evaporator. After evaporation there resulted a mass permeated with crystals. This mass is added to 80 ml of distilled water and extracted with 2 portions of diethyl ether. In this manner, the benzoic acid formed is extracted. The aqueous layer is evaporated to an oil, so that as much hydrobromic acid as possible is removed. A little distilled water is then added and the mixture is once more vigorously evaporated. This operation is repeated.

The resulting dark-colored oil is dissolved in 20 ml of distilled water and 10 ml of acetonitrile to form a dark clear solution. This solution is cooled to room temperature and then treated with 5 ml of propylene oxide. After 15 minutes, the pH is tested, and if necessary adjusted to 5.5 to 6 with further propylene oxide. At this pH almost colorless crystals precipitate. 100 ml of acetonitrile are gradually added and the mixture is left to stand overnight at 0°C. The crystals are filtered and washed first with distilled water/acetonitrile 1:5 by volume and then with acetonitrile and finally then with diethyl ether. The crystals are recrystallized by dissolving them in distilled water in a sulfonation flask under constant stirring while saturated with SO_2 gas.

The contents of the flask are warmed during this procedure. After all of the crystals have been dissolved, the heat source is removed and an inert nitrogen gas stream is led through the solution whereupon the crystallization begins. The mixture is left at 0°C and filtered off on the next morning. The crystals are then rinsed with distilled water/acetonitrile (1:1 by volume mixture), then with acetonitrile alone and finally with diethyl ether and dried. There are obtained 2.2 g of L-3-(3,4-dihydroxyphenyl)-alanine (L-dopa) with a melting point of 281° to 282°C. $[\alpha]_D^{24}$ = –11.8° (c = 1, 1 N HCl).

Resolution of Dopa Esters Using Chymotrypsin

Optical antipodes of high purity are obtained by the resolution process developed by *N.L. Benoiton, J.H.Y. Tong, A. D'Iorio and J.C. Petitclerc; U.S. Patent 3,813,317; May 28, 1974; assigned to Canadian Patents and Development Limited, Canada.* Broadly this process provides for the resolution of racemic mono- and diring-substituted phenylalanines, where the substituents are Cl, F, OH and OCH_3. This process comprises: forming an ester of an alcohol containing 1 to 4 carbon atoms with the racemate; subjecting the racemic ester to the action of a chymotrypsin, e.g., α-chymotrypsin at an acid pH either as such or attached to a solid support; and recovering the L-isomer acid in optical purity of at least 99.8% as a precipitate from the reaction mixture which is substantially free of the D-isomer ester.

Examples of the racemic mono- and di-ring-substituted phenylalanines resolvable according to this process include the following: 3,4-dihydroxyphenylalanine,

(L-dopa), 4-chlorophenylalanine, 4-fluorophenylalanine, 4-hydroxyphenylalanine (tyrosine), 3-hydroxyphenylalanine (m-tyrosine), 2-hydroxyphenylalanine (o-tyrosine), 2-methoxyphenylalanine. The ester formed is of an aliphatic alcohol of relatively small carbon chain length. As the chain length increases, the ester becomes less water soluble and tends to be slower to be hydrolyzed by the enzyme. Examples of suitable esters include methyl, ethyl, n-propyl, isopropyl, n-butyl, sec-butyl and isobutyl esters.

The enzyme preferably used is a chymotrypsin including α-chymotrypsin and active precursors, namely δ-chymotrypsin and γ-chymotrypsin. However, the enzyme preferred is α-chymotrypsin. The amount of the enzyme used is variable; however, a smaller amount of enzyme reduces cost. Furthermore, the hydrolytic reaction proceeds with as small an amount as possible of the enzyme. Amounts successfully used have ranged from 35 to 120 mg enzyme per gram of substrate. The enzyme may also be attached to any suitable solid support.

It is preferred to operate in an aqueous solution. The more dilute the solution, the more active is the enzyme. On the other hand, the more concentrated the solution, the easier is the recovery. The enzyme is most active at a temperature of 37°C. However, normal room temperature (e.g., 25°C) is preferably used. The use of higher temperatures, while favoring the activity of the enzyme, also increases the danger of spontaneous (i.e., nonenzymatic) hydrolysis, and this is undesirable. It is essential to carry out the enzymatic hydrolysis at an acidic pH of 5 to 6. As the pH approaches 7, the danger of nonenzymatic saponification increases.

Example: DL-dopa [3,4-dihydroxyphenylalanine (3.95 g; 0.02 mol)] was suspended in absolute ethanol (240 ml) and hydrogen chloride was bubbled through the mixture for 20 minutes. The next day, the clear solution was evaporated to dryness, and the evaporation was repeated three times after the addition of 50 milliliters of ethanol each time. The residue was dissolved in water (100 ml) and the pH adjusted to 5.5 with sodium hydroxide. α-Chymotrypsin (140 mg) was added and the mixture was kept at 37°C for 30 minutes while the pH was kept constant by the addition of 1 N NaOH from an automatic titrator.

The mixture was concentrated to 50 ml and the L-dopa which crystallized was filtered after cooling the mixture for 1 hour. Yield: 1.54 g (78%). The product was recrystallized by dissolving in a slight excess of 1 N HCl, filtering and neutralizing to pH 5.5 with sodium hydroxide. Yield: 1.38 g (70%). The original mother liquor was brought to pH 9.0 with sodium hydroxide, saturated with sodium chloride, and extracted three times with 100 ml of ethyl acetate. The extract was dried ($MgSO_4$) and filtered into 50 ml ethyl acetate containing hydrogen chloride. The solvent was removed under reduced pressure, and the residue was refluxed for 1 hour in 120 ml of 1 N hydrochloric acid.

The mixture was evaporated to dryness three times after the addition of 30 ml of water each time. The residue was dissolved in water (20 ml) adjusted to pH 5.5 with sodium hydroxide, and the crystallized D-dopa was filtered with the aid of ethanol. Yield: 1.18 g (60%). The identity and purity of the products was established as described above. The L-dopa contained less than 0.2% of the D-isomer. The D-dopa contained 0.5% of the L-isomer.

Resolution of Derivatives of Dopa Esters with Chymotrypsin

M.S. Matta, J.A. Kelley and M.F. Rohde; U.S. Patent 3,878,043; April 15, 1975; assigned to Southern Illinois University Foundation have disclosed a process similar to that of U.S. Patent 3,813,317, but one which gives high rates and high yields. It has been found that an N-acyl-L-alanine, which is readily convertible to L-dopa, may be rapidly obtained in high yield from the stereospecific enzymatic hydrolysis of an alkyl N-acyl-DL-alaninate ester precursor corresponding to the formula:

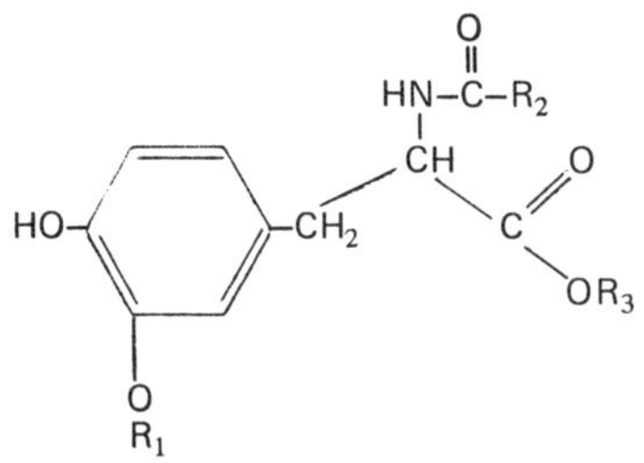

where R_1 is hydrogen, methyl or ethyl, R_2 is hydrogen, alkyl, aryl, furyl, thienyl, styryl, t-butyloxy or benzyloxy, and R_3 is methyl, ethyl, n-propyl or isopropyl.

A preferred precursor is methyl N-acetyl 3-(4-hydroxy-3-methoxyphenyl)-DL-alaninate, i.e., R_1, R_2 and R_3 are each methyl. In accordance with the process, a racemic mixture of the N-acyl-DL-alaninate ester precursor is placed in an aqueous medium and a proteolytic enzyme, preferably α-chymotrypsin, is added. The aqueous medium preferably includes a solvent compatible with water to increase the solubility of the precursor in the system.

Up to 30% by weight of solvent can be tolerated with 15% by weight being preferred. The solvent is preferably methanol but may be another polar solvent such as, for example, isopropanol, ethanol, n-propanol, p-dioxane, dimethylformamide, or dimethylsulfoxide. In the presence of α-chymotrypsin, the L-isomer of the precursor compound is selectively and rapidly hydrolyzed to form an N-acyl-L-alanine while the D-isomer of the ester remains substantially unchanged. The reaction proceeds most rapidly and completely at a pH of preferably 7.8.

Although the use of α-chymotrypsin is preferred, essentially any proteolytic enzyme may be used to catalyze the hydrolysis of the N-acyl-L-alaninate ester. Among the alternative enzymes which can be used are pronase and the various subtilisins. Proteolytic enzymes which have been immobilized by covalent bonding to an insoluble support may also be effective hydrolysis catalysts.

An improved three step method for the production of the preferred alkyl N-acyl alanine precursor, methyl N-acetyl-3-(4-hydroxy-3-methoxyphenyl)-DL-alaninate from benzyloxy-3-methoxy benzylaldehyde and acetyl glycine, is also described. In the first step, the azlactone 4-[4-(benzyloxy)-3-methoxybenzylidene]-2-methyl-2-oxazolin-5-one was prepared by reaction of 4-benzyloxy-3-methoxybenzaldehyde and acetylglycine in the presence of acetic anhydride.

The azlactone was then refluxed with methanol containing sodium carbonate to produce an 85% yield of colorless methyl α-acetamido-4-benzyloxy-3-methoxy cinnamate which was subsequently reduced with 5% palladium on carbon catalyst to form a 97% yield of colorless methyl N-acetyl-3-(4-hydroxy-3-methoxyphenyl)-DL-alaninate.

Example: A solution was prepared containing 4.0 g (15 mmol) of methyl N-acetyl-3-(4-hydroxy-3-methoxyphenyl)-DL-alaninate in 15 ml methanol. Deionized water (85 ml) was then added and the clear solution was adjusted to a pH of 7.8 by addition of 1 N sodium hydroxide solution. α-Chymotrypsin (Worthington lot CDI OBK, 20 mg) dissolved in water (2 ml) was introduced into the stirred aqueous methanol solution of the DL-alaninate. Hydrolysis proceeded in the presence of the α-chymotrypsin, and the pH was maintained at approximately 7.8 during the course of the reaction by the further addition of 1 N sodium hydroxide.

After 45 minutes, 7.45 ml of the sodium hydroxide solution had been consumed (99% of theoretical based on exclusive hydrolysis of one isomer) and the reaction was stopped. The reaction mixture was then continuously extracted overnight with ethyl acetate, the organic layer removed, and the organic solvent evaporated at reduced pressure to provide 1.95 g (98% recovery) of an oily liquid. This liquid was taken up in a mixture of chloroform and hexane and crystallized to produce 1.8 g (90% recovery) of colorless unchanged methyl N-acetyl-3-(4-hydroxy-3-methoxyphenyl)-D-alaninate having a MP of 127° to 128°C. A solution containing 5 g of this material in 100 ml of methanol exhibited a specific rotation at 25°C of –23.8°.

The aqueous layer obtained from the extraction was adjusted to a pH of 3 by addition of hydrochloric acid, and then evaporated to dryness at reduced pressure. The residue obtained from evaporation was dissolved in warm methanol (30 ml) and the resulting solution filtered. Solvent was removed in vacuo from the filtrate to provide 1.8 g (90% yield) of N-acetyl-3-(4-hydroxy-3-methoxyphenyl)-L-alanine, a brittle colorless glass exhibiting a specific rotation (5 grams per 100 milliliters methanol) of +35.8° at 25°C.

To confirm its optical purity, 1.0 g (3.96 mmol) of the N-acyl-L-alanine product was reesterfied and the ester obtained crystallized from a mixture of chloroform and hexane. 800 mg (76% yield) of the ester having a MP of 127° to 128°C was obtained. A 5 g/100 ml methanol solution of this ester had a specific rotation at 25°C of +23.6°, indicating that the products of the stereospecific hydrolysis were of high optical purity.

Use of *E. coli* Acylase on N-Acyl Dopa Compounds

P. Bamberg and B.O.H. Sjoberg; U.S. Patent 3,838,008; September 24, 1974; assigned to Astra Lakemedel AB, Sweden have disclosed a process whereby it is possible to resolve stereoselectively certain N-acyl dopa derivatives using an *E. coli* acylase in such a way that only the L-amino acid derivatives were deacylated. The starting materials for this stereoselective resolution are compounds of the formula shown on the following page.

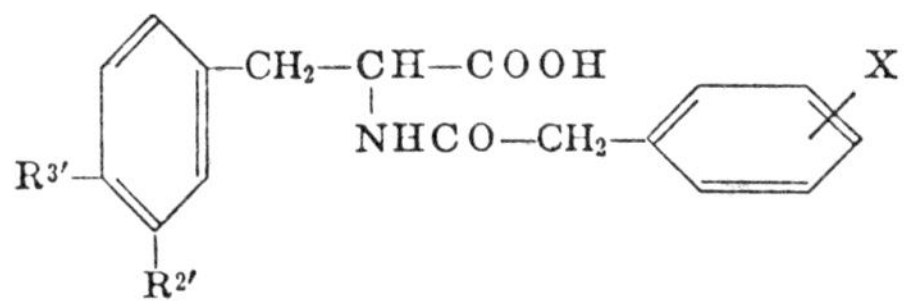

In the above formula X is hydrogen, fluorine, chlorine, bromine, or $-NO_2$; $R^{2'}$ is a hydroxy or alkoxy group with not more than 5 carbon atoms; and $R^{3'}$ is hydrogen, hydroxy, or alkoxy group with not more than 5 carbon atoms, or $R^{2'}$ with $R^{3'}$ form an alkylenedioxy group with one or two carbon atoms.

The racemic starting material is obtained by the known process such as the condensation of piperonal with N-phenylacetylglycine to form 2-benzyl-4-(3,4-methylenedioxybenzylidene)-oxazolone-5 which was then hydrolyzed to form α-phenylacetamido-3,4-methylenedioxy-cinnamic acid and hydrogenated to form N-phenylacetyl-3-(3,4-methylenedioxyphenyl)-alanine.

The resolution into the stereoisomeric antipodes of such starting material is carried out by incubating the racemate obtained at a pH of 6 to 8 with *E. coli* acylase during the necessary time interval, usually from 1 to 72 hours. Hereafter the pH value is lowered to between 1 and 3 whereupon nonreacted acylamino acid is extracted with ethyl acetate or other suitable solvents.

The acylase used in the process is obtained by cultivating *E. coli* under conditions suitable for producing stereospecifically acting acylase. Representative strains which may be used for this purpose include the *E. coli* ATCC 9637, ATCC 11105, BRL 1040, NCIB 9465 and Astra 1339. The *E. coli* is cultured in a conventional manner. For example, very suitable cultures are prepared under aerobic submerged conditions in a conventional nutrient medium of assimilable sources of carbon, nitrogen and inorganic salts. Recovery of the enzyme from the fermentation broth is carried out in the usual manner. As is industrially known, suspensions of *E. coli* cells containing the acylase are frequently used.

Example: L (–)-3-(3,4-methylenedioxyphenyl)-alanine and D (–)-N-phenylacetyl-3-(3,4-methylenedioxyphenyl)-alanine were obtained by incubating DL-N-phenylacetyl-3-(3,4-methylenedioxyphenyl)-alanine (5 g) with an *E. coli* acylase (6,000 units) in 50 ml water at pH 7.8 whereby for regulation of pH ammonia, sodium hydroxide or lithium hydroxide was used as a base. Various incubation experiments were performed and the time used varied between 1 and 72 hours and the incubation temperature varied between 25° and 37°C.

When the incubation had been finished, the pH was adjusted to 1.5 with 2 N hydrochloric acid or sulfuric acid, whereafter remaining nonresolved phenylacetylated amino acid was extracted with ethyl acetate. The remaining water phase was concentrated and gave, after adjusting the pH to 5.5, a L (–)-amino acid which was no longer protected at the amino nitrogen, in the form of white crystals with the melting point of 202° to 210.5°C (water) $[\alpha]_D^{20} = -28.12°$ (water, c = 1).

The ethyl acetate phase contained the N-phenylacetylated D (–)-amino acid which had a melting point of 143° to 157°C (ethanol), $[\alpha]_D^{20}$ = 59.46° (ethyl acetate, c = 1).

Resolution Via Protease Preparation of Anilide Derivatives

DL-N-benzoyltyrosine compounds are resolved in improved yields by the process disclosed by *E.-O. Renth, K. Schromm and A. Mentrup; U.S. Patent 3,855,062; December 17, 1974; assigned to Boehringer Ingelheim GmbH, Germany.* In this process a derivative of DL-N-benzoyltyrosine of the formula:

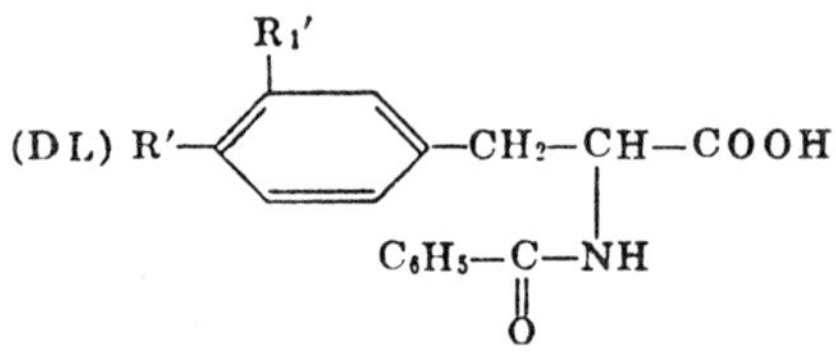

where R' is hydrogen, hydroxyl or lower alkoxy, and R_1 is hydroxyl or lower alkoxy, is treated with aniline or a ring-substituted aniline in a buffered aqueous solution having a pH of 5 to 7 and in the presence of a protease which is effective within this pH range. This reaction forms almost quantitative yields of the correspondingly substituted L-N-benzoyl-m- or p-tyrosyl-anilide, which can subsequently be hydrolyzed with a mineral acid of suitable strength.

Suitable proteases are especially vegetable proteases, such as bromelain, ficin, papain or the like, or also microbial (bacterial or fungal) proteases.

The synthesis of the L-N-benzoyl-tyrosyl-anilide intermediate is carried out with a pH of preferably 5.6 to 6.2, the desired pH of the reaction mixture being adjusted with the aid of conventional buffers, such as citrate or phosphate buffer. In place of aniline itself, ring-substituted derivatives thereof may also be used; the substituents may be lower alkyl, lower alkoxy, lower alkyl-mercapto, halo or sulfhydryl groups. The D-form of the starting material may be recovered virtually quantitatively from the mother liquor and reconverted into the racemate by conventional methods.

Example 1: 31.5 g (0.1 mol) of DL-N-benzoyl-3-methoxy-tyrosine were dissolved in 200 ml of warm 0.05 N sodium hydroxide, the resulting solution was mixed with 150 ml of aqueous 2 M sodium acetate, and the mixture filtered. The clear filtrate was mixed first with 300 ml of 0.1 N cisteinium chloride and finally with 18 ml of freshly distilled aniline.

The resulting mixture was combined with a solution of about 3 g of papain in 300 ml of 0.05 N citrate buffer solution, and the pH of the mixed solution was adjusted to 5.80 by addition of either glacial acetic acid or 1 N sodium hydroxide, as needed. After a short time a precipitate began to form, and after about 48 hours of standing the reaction mixture was filtered, and the filter cake was washed with 300 ml of water, dried and recrystallized from methanol/water. A quantitative yield of L-N-benzoyl-3-methoxy-tyrosyl-anilide, MP 280.0° to 208.5°C, $[\alpha]_D^{20}$ = +40.0°±0.15° (c = 1.000 in methanol), was obtained.

The mother liquor remaining after separation of the anilide was heated to 80°C, mixed with 100 ml of methanol, and the resulting solution was adjusted to pH 2.0 with concentrated hydrochloric acid. The acidic solution was then cooled, and the precipitate formed was collected by vacuum filtration and dried, yielding D-N-benzoyl-3-methoxy-tyrosine which was racemized by conventional methods and used as the starting material for a subsequent run.

Example 2: 9.75 g (25 mmol) of L-N-benzoyl-3-methoxy-tyrosyl-anilide were refluxed for seven hours with 50 ml of 48% hydrobromic acid. Thereafter, the reaction mixture was cooled, and the precipitated benzoic acid was separated by vacuum filtration and rinsed with a small amount of cold 48% hydrobromic acid. The slightly yellowish-brown filtrate was concentrated in vacuo and then evaporated to dryness; the residue was mixed with 50 ml of distilled water, and the mixture was again evaporated to dryness. The residue was dissolved in 50 milliliters of methanol, and the resulting solution was adjusted to pH 6.0 with concentrated ammonia, whereby a precipitate formed.

The mixture was allowed to stand in a refrigerator overnight, then vacuum-filtered, and the filter cake was dried at no higher than 50°C. The raw product thus obtained was dissolved in 180 ml of boiling water, the solution was filtered through charcoal, and the filtrate was first cooled to 20°C in an atmosphere of nitrogen, then allowed to stand at this temperature until crystallization set in, and finally allowed to stand overnight in a refrigerator.

The following day the mixture was vacuum-filtered, and the filter cake was dried at no higher than 50°C, yielding 3.4 g (69.2% of theory) of L-3-hydroxyl-tyrosine (L-dopa). The filtrate contained an additional 0.9 g of the same product, making a total yield of 4.3 g (87.4% of theory) of the compound having a melting point of 290°C (decomposition) and a specific rotation $[\alpha]_D^{20} = -12.15° \pm 0.13°$ (c = 4.000 in 1 N hydrochloric acid). The same result was obtained when bromelin or ficin was used as the protease in Example 1.

DERIVATIVES OF L-DOPA

N-Substituted L-Dopa Compounds

A. Kaiser, W. Koch, M. Scheer and U. Wolcke; U.S. Patent 3,852,338; Dec. 3, 1974; assigned to Hoffmann-La Roche Inc. have found that compounds of the formula:

(1)

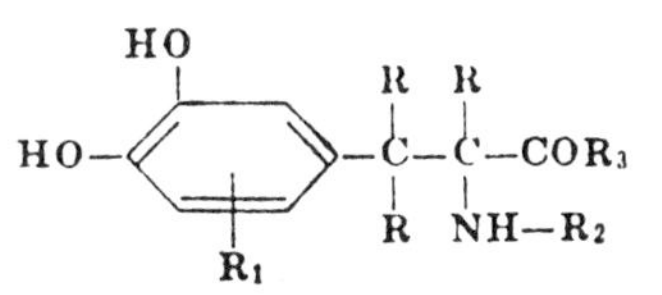

where R is hydrogen or alkyl; R_1 is hydrogen, halogen, alkyl, alkoxy, nitro or dialkylamino; R_3 is hydroxy, alkoxy, phenylalkoxy, amino, or alkylamino; R_2 is:

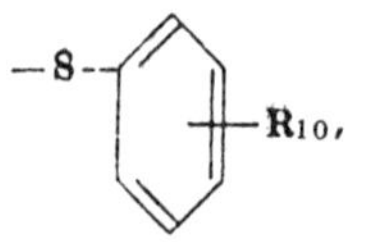

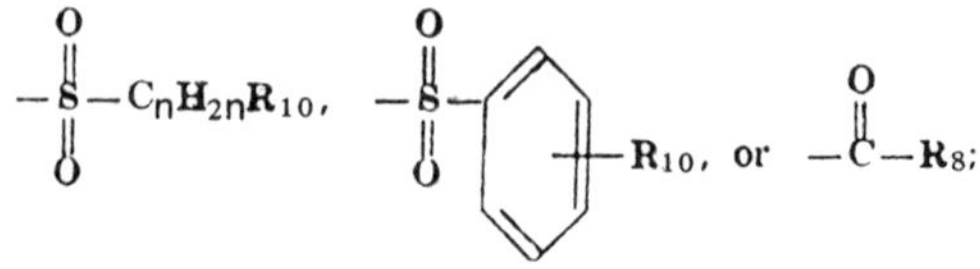

R_{10} is hydrogen, hydroxy, alkoxy, alkysulfonyl, halogen, nitro, cyano, alkylaminocarbonyl, alkanoyloxy or alkyl; R_8 is alkoxy, alkylamino or $-C_nH_{2n}R_{12}$; R_{12} is hydrogen, hydroxy, alkoxy, alkylthio, phenylthio, alkylphenylthio, halogen, cyano, carboxy, alkoxycarbonyl, alkylaminocarbonyl, or alkanoyloxy; and n is an integer from 1 to 17; or pharmaceutically acceptable salts thereof are useful as hypotensive, antipyretic and anti-Parkinson agents.

A preferred group of phenylalanine derivatives comprises those compounds of formula (1) in which R is a hydrogen or methyl group, especially hydrogen, R_1 is hydrogen and R_3 is a hydroxy group, and salts of these compounds. Interesting phenylalanine derivatives provided by this process are N-(t-butoxycarbonyl)-L-dopa and its methyl and benzyl ester, N-ethoxyacetyl-L-dopa; N-[(o-nitrophenyl)-thio]-L-dopa and its methyl ester and N-(p-toluenesulfonyl)-L-dopa; as well as pharmaceutically acceptable salts of these compounds.

These compounds are prepared using the boric acid complex as the protective agent for the phenyl hydroxyl groups while modifying the amino group as in the following examples.

Example 1: A mixture of 100 g (0.507 mol) of L-dopa, 200 g (0.525 mol) of borax and 1,000 ml of water is treated in an argon atmosphere with 2 N aqueous sodium hydroxide solution until it has a pH value of 9.5. The solution obtained, which contains the boric acid complex of L-dopa, is then treated at one time with half of the crude t-butoxycarbonyl azide prepared from 120 g of t-butoxycarbonyl hydrazide. The resulting mixture is stirred at room temperature for 6 hours, the pH being held between 9.3 and 9.8 by occasional addition of 2 N aqueous sodium hydroxide solution.

The remaining t-butoxycarbonyl azide is subsequently added and the mixture is further stirred for 14 hours, after which time the pH value has fallen to 8.5. The pH value is brought to pH 9.5 with 2 N aqueous sodium hydroxide solution and the mixture is stirred for a further 2 hours, then extracted with two 300 ml portions diethyl ether. The aqueous phase, which contains the boric acid complex of N-(t-butoxycarbonyl)-L-dopa is used in the next step without isolation of the product.

The aqueous, alkaline solution containing the boric acid complex of N-(t-butoxycarbonyl-L-dopa is then brought to pH 2.5 with a 10% by weight aqueous citric acid solution in an argon atmosphere, saturated with sodium chloride and extracted with two 300 ml portions of ethyl acetate. The combined organic extracts are washed ten times with 500 ml of water, dried over sodium sulfate and concentrated under reduced pressure. The glassy residue obtained is crystallized by digestion with cyclohexane and recrystallized twice from ethyl acetate/cyclohexane to give N-(t-butoxycarbonyl)-L-dopa of MP 148°C; $[\alpha]_D^{25}$ = +16.4°, (c = 1% in methanol).

Example 2: A mixture of 19.7 g of L-dopa, 40 g of borax, 50 ml of 2 N aqueous sodium hydroxide solution, 600 ml of water and 200 ml of dioxan is stirred in an argon atmosphere until all has gone into solution (about 15 minutes), then the solution obtained is simultaneously treated, dropwise at 20°C with a solution of 21 g of o-nitrophenylsulfenyl chloride in 50 ml of dioxan and with 2 N aqueous sodium hydroxide solution in such a way that the pH value is held between 9.5 and 10. After the addition the mixture is stirred at 20°C for a further 2 hours to give a solution containing the boric acid complex of N-[(o-nitrophenyl)thio]-L-dopa which is used in the next step without isolation of the product.

This solution is acidified to pH 1 in an argon atmosphere with ice-cooling with 6 N aqueous sulfuric acid and extracted with two 300 ml portions of ethyl acetate. The organic solvents are washed five times with water, then combined, dried over sodium sulfate and evaporated at 30°C/11 mm Hg with the addition of 20 g of dicyclohexylamine. The residue crystallizes on digestion with diethyl ether to give N-[(o-nitrophenyl)-thio]-L-dopa dicyclohexylamine salt of MP 126°C (with decomposition); $[\alpha]_D^{25}$ = +36°, (c = 0.1% in methanol).

4-Methoxy L-Dopa

It has been found by *A. Kaiser, W. Koch, M. Scheer and U. Wolcke; U.S. Patent 3,853,954; December 10, 1974; assigned to Hoffmann-La Roche Inc.* that the L-antipode of a compound:

(1)

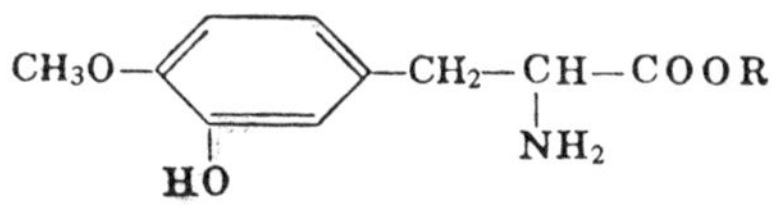

where R is hydrogen or lower alkyl; and pharmaceutically acceptable salts thereof are useful in reducing the symptoms of Parkinson's disease such as rigor, akinesia and tremor, without causing peripheral adrenergic effects and other undesirable secondary effects. The L-antipode of formula 1 can be prepared by resolving a racemic mixture of compound 1 above into its optical antipodes and isolating the L-antipode. On the other hand, compound 1 can be prepared from the L-antipode of a compound:

(2)

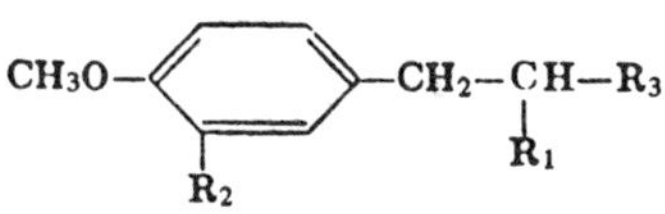

where R_1 is an amino group protected by a conventional amino protecting group convertible to an amino group; R_2 is hydroxy or hydroxy protected by a conventional hydroxy protecting group convertible to a hydroxy group; R_3 is carboxy or a conventional carboxy protecting group convertible to a carboxy group; or a pharmaceutically acceptable salt thereof, by simultaneously, or in any sequence, converting the protected amino group into free amino groups, converting the protected hydroxy group into free hydroxy groups, and/or converting the protected carboxy group into the free acid. The free acid can, if desired be converted into its salt form.

The following examples illustrate the preparation of N-acetyl-4-methoxy-L-m-tyrosine (the 4-methyl ether of L-dopa).

Example 1: 167 ml of titanium tetrachloride are added with stirring in the course of 5 minutes to a solution of 95 g of N-acetyl-3-(p-methoxyphenyl)-L-alanine methyl ester in 1,500 ml of nitrobenzene, the temperature rising to 38°C. The mixture is cooled to +23°C, 69.4 ml of dichloromethyl methyl ether are then added in the course of 3 minutes and the mixture is stirred at room temperature for 2 hours. The solution is cooled to +5°C and poured with stirring into 380 ml of ice-cold 3 N aqueous hydrochloric acid. 4,000 ml of ethyl acetate and 2,000 ml of tetrahydrofuran are added, the mixture is neutralized by the introduction with stirring of 1,200 g of anhydrous potassium carbonate and dried over sodium sulfate.

After filtration, the filtrate is evaporated (finally in a high vacuum at 70°C). The residue is dissolved in 500 ml of methylene chloride and applied to a chromatographic column containing 200 g of silica gel. Elution with ethyl acetate/methylene chloride (1:1, v/v) yields, after evaporation, 72 g of N-acetyl-3-(3-formyl-4-methoxyphenyl)-L-alanine methyl ester which can be further purified by recrystallization from ethyl acetate/hexane.

Example 2: A solution of 27.9 g (0.1 mol) of N-acetyl-3-(3-formyl-4-methoxyphenyl)-L-alanine methyl ester, 20 g of 40% peracetic acid in glacial acetic acid and 150 ml of glacial acetic acid is left to stand at 25°C for 23 hours and then evaporated under reduced pressure. 200 ml of saturated aqueous sodium hydrogen carbonate solution are added to the residue and it is extracted twice with 250 ml of ethyl acetate each time. The extracts are combined, dried over sodium sulfate and evaporated.

The residue consists of a mixture of N-acetyl-4-methoxy-L-m-tyrosine methyl ester and its formate, i.e., N-acetyl-3-(3-formyloxy-4-methoxyphenyl)-L-alanine methyl ester. The hydrolysis of the formate to N-acetyl-4-methoxy-L-m-tyrosine methyl ester is completed by chromatography of this mixture on a column containing 3,000 g of silica gel [elution with ethyl acetate/toluene (1:1 parts by volume)] and 16.8 g (63%) of the desired compounds are obtained after one recrystallization from ethyl acetate/ether. MP 97° to 98°C, $[\alpha]_D^{25}$ = +31.5°, (c = 2% in 95% ethanol).

Example 3: 3.3 g (13.3 mmol) of 4-methoxy-L-m-tyrosine hydrochloride are dissolved in 150 ml of ethanol and 12.5 ml of propylene oxide are added to the solution in several portions in the course of 3 hours. The mixture is then left to stand at 25°C for 20 hours and the precipitated crystals are filtered off. The pure 4-methoxy-L-m-tyrosine is obtained by one recrystallization from 40 milliliters of absolute ethanol to which a few drops of water are added.

Yield: 2.15 g; melting point 251° to 252°C; $[\alpha]_D^{25}$ = -34.2°, (c = 1% in water).

Example 4: A mixture of 5.3 g (19.9 mmol) of N-acetyl-4-methoxy-L-m-tyrosine methyl ester and of 100 ml of 3 N aqueous hydrochloric acid is heated at reflux in an argon atmosphere for 2 hours. The 4-methoxy-L-m-tyrosine hydrochloride crystallizes out on concentration of the solution under reduced pressure. Yield: 3.3 g; melting point 247° to 247.5°C; $[\alpha]_D^{25}$ = -12.75°, (c = 4% in water).

The 3-hydroxy group may also be introduced into N-acetyl-3-(p-methoxyphenyl)-L-alanine by reaction with acetyl chloride and subsequent hydrolysis of the 3-acetyl group to a 3-hydroxy group.

The compounds of formula 1 and their salts can be used as medicaments, for example, in pharmaceutical preparations which contain them in association with a compatible pharmaceutical carrier. The pharmaceutical preparations can be made up in solid form or in liquid form. The preparations may be sterilized and/or may contain adjuvants such as preserving, stabilizing, wetting or emulsifying agents, salts for varying the osmotic pressure or buffers. They can also contain other therapeutically valuable substances.

In comparison the L-dopa, a known anti-Parkinson agent, the compounds of formula 1 or salts thereof and the pharmaceutical preparations provided by the process possess a depot action (half-life 4 to 5 hours). The LD_{50} in the mouse amounts to more than 5,000 mg/kg per os in the 24-hour test. On administration of the pharmaceutical preparations provided by the process, the undesirable secondary effects which usually appear on administration of L-dopa, namely gastrointestinal complaints, involuntary movements and circulatory complaints, do not occur or only occur to a very slight extent.

Transient Pro-Drug Forms of L-Dopa

It is the object of the process developed by *N.S. Bodor, K.B. Sloan and A.A. Hussain, U.S. Patent 3,891,696; assigned to Interx Research Corporation* to provide transient, pro-drug forms of L-dopa, useful in the treatment of Parkinson's disease in warm-blooded animals. These pro-drug forms of L-dopa, cleave in such a manner as to enable the original drug (L-dopa) to be released in the blood-stream of a warm-blooded animal and to further permit the cleaved non-dopa moiety(ies) to be nontoxic and metabolized in a nontoxic fashion. Owing to their high solubility and/or better absorption properties, high blood levels of L-dopa can be attained, following cleavage of the pro-drug form. These transient, pro-drug forms of L-dopa having the following formula:

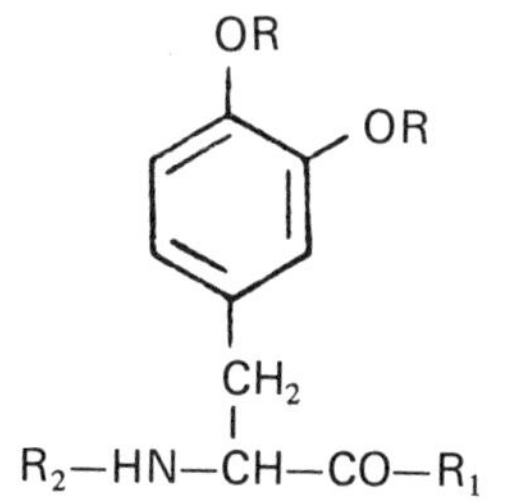

where R is hydrogen, $-COCH_3$, $-COC_2H_5$, $-COOC_2H_5$, or $-COC(CH_3)_3$; where R_1 is hydroxyl, –OM (where M is an alkali metal or an ammonium ion), $-OCH_2-C_6H_5$, $-OCH_3$, $-OC_2H_5$, or $-NH-CH(R_3)COOH$, where R_3 is the residue of any naturally occurring amino acid; and where R_2 is hydrogen, a formyl group, $-CO-C_6H_5$, –CO– pyridyl group, $-(R_5)CO-CH{=}CH(R_4)$ or $NH_2-CH(R_6)-CO-$ group, where R_4 is methyl and where R_5 is methyl, $-OC_2H_5$, or C_6H_5-, and where R_6 is the residue of any naturally occurring amino acid, with the proviso that R cannot be hydrogen when R_1 is $-OCH_3$ or –OH and when R_2 is hydrogen or $-COCH_3$, respectively; R cannot be hydrogen when R_1 is hydroxyl and R_2 is formyl or benzoyl; and R cannot be hydrogen when R_1 is hydroxyl and R_6 is the residue of L-leucine, respectively.

Some 102 specific compounds are cited coming under the following classifications:

Class I: The compounds of this class are those of the 3,4-diacylated-L-dopa type and their HX salts, where X represents here and in other classes a pharmaceutically acceptable acid addition salt anion, such as a chloride, bromide, perchlorate ion, and the like, R is an acyl group; R_1 is hydroxy; and R_2 is hydrogen. To prepare compounds of Class I L-dopa is reacted with an acylating agent (e.g., acyl chloride or acyl anhydride) in the presence of a solvent (e.g., acetic acid, benzene, or a halohydrocarbon) in the presence of a strong mineral acid, such as hydrochloric acid, perchloric acid, etc., over a period of 2 to 6 hours and at room temperature or that equivalent to the boiling point of the solvent.

Class II: These are carboxy esters of L-dopa and their respective HX salts, where R is hydrogen or an acyl group; R_1 is $-OCH_2-C_6H_5$, $-OC_2H_5$, or $-OCH_3$; and R_2 is hydrogen. To prepare these compounds, L-dopa is reacted with the corresponding alcohol in the presence of an acid catalyst (e.g., p-toluenesulfonic acid, benzenesulfonic acid, etc.), the water formed during the reaction being eliminated.

Class III: These compounds are the enamines derived from the compounds of Class I. With respect to these compounds, R is hydrogen or an acyl group; R_1 is hydroxyl or –OM where M is an alkali metal, or an ammonium ion, and R_2 is $-(R_5)CO-CH{=}CH(R_4)-$ group, where R_4 is methyl and R_5 is methyl, C_6H_5, or $-OC_2H_5$. To prepare the compounds, the beta, gamma-diketo derivative is reacted with L-dopa in methanol or ethanol in which an equivalent amount of MOH is introduced in a ratio 1:1.1 for several hours at a temperature ranging from room temperature to 60°C in the presence of a nitrogen atmosphere.

Class IV: These compounds constitute the amides derived from the compounds of Class I, where R is acyl; R_1 is hydroxyl or –OM, and R_2 is –COH, $-CO-C_6H_5$, or –CO– pyridyl group. These compounds are prepared by reacting L-dopa with a conventional acylation agent, such as acyl chloride or acyl anhydride, or in the alternative, a conventional "mixed anhydride" system can be used.

Class V: These compounds are the N-terminal dipeptides of L-dopa where R is hydrogen or an acyl group; R_2 is hydrogen; and R_1 is $-NHCH(R_3)COOH$ where R_3 is the residue of a naturally occurring amino acid and the HX salt thereof. These compounds and those of Classes VI through VIII are prepared by standard dipeptide synthetic methods.

Class VI: These compounds constitute the C-terminal dipeptides of L-dopa, where R is hydrogen or acyl; R_1 is hydroxyl, $-OCH_3$, $-OC_2H_5$, or $-OCH_2-C_6H_5$; and R_2 is $NH_2CH(R_6)$, where R_6 is the residue of a naturally occurring amino acid and the HX salt thereof.

Class VII: These compounds are compounds of Class VI which are basically dipeptides formed from 2 molecules of 3,4-dihydroxy-L-phenylalanine, where R is hydrogen or acyl, R_1 is hydroxyl, $-OCH_3$, $-OC_2H_5$, or $-OCH_2-C_6H_5$; R_2 is $NH_2-CH(R_6)-CO-$, and R_6 is the residue of 3,4-dihydroxy-L-phenylalanine or 3,4-L-diacetyloxy-phenylalanine.

Class VIII: This class basically constitutes the polypeptides of L-dopa per se, or its 3,4-diacetylated derivatives.

In the table below, values are shown relative to the blood concentration of L-dopa per se and selected L-dopa derivatives, following oral administration of 0.1 g of L-dopa and an equivalent amount of an L-dopa derivative to beagle dogs.

Compound	Time (hr)* 0.5	1	1.5	2	3	4	6	Area Under Curve (μg/hr/ml)	L-Dopa at Peak (μg/ml)	Time for Reaching Peak (hr)
	Concentration in Blood (μg/ml)									
L-dopa	0.14	0.64	–	0.87	–	0.22	0.11	2.6	0.9	1.8
I	1.57	1.75	1.38	1.10	0.35	0.13	0.06	7.20	1.8	0.8
II	2.16	2.29	1.61	0.70	0.40	0.28	0.20	9.07	2.4	0.9
III	1.33	0.18	0.60	0.39	0.26	0.12	0.05	4.60	1.6	0.8
IV	0.06	1.13	1.11	0.39	0.12	0.09	0.02	3.56	1.6	1.2
V	2.30	1.03	0.55	0.31	0.14	0.14	0.06	5.94	2.2	0.8
VI	1.71	1.30	1.08	0.55	0.20	0.14	0.05	5.96	2.3	0.5
VII	1.34	1.95	1.21	0.43	0.27	0.13	0.02	7.25	2.4	0.75
VIII	0.79	1.17	0.51	0.33	0.08	0.02	0	3.38	1.7	0.80
IX	0.26	1.30	1.62	0.80	0.23	0.03	0	5.30	1.8	1.2

*This refers to the values obtained from three beagle dogs for each compound tested, the beagle dogs being of both sexes and weighing from 12 to 14 kg.

Compound I is 3,4-diacetyloxy-L-phenylalanine hydrochloride.
Compound II is the potassium salt of 3,4-diacetyloxy-N-[1-methyl-2-acetyl-vinyl]-L-phenylalanine.
Compound III is 3,4-dihydroxy-L-phenylalanine-methyl ester hydrochloride.
Compound IV is 3,4-diacetyloxy-L-phenylalanine-methyl ester hydrochloride.
Compound V is 3,4-dihydroxy-L-phenylalanine-benzyl ester hydrochloride.
Compound VI is 3,4-diacetyloxy-L-phenylalanine-benzyl ester hydrochloride.
Compound VII is glycyl-3,4-diacetyloxy-L-phenylalanine hydrochloride.
Compound VIII is glycyl-3,4-diacetyloxy-L-phenylalanine-methyl ester hydrochloride.
Compound IX is 3,4-diacetyloxy-L-phenylalanyl-glycine hydrochloride.

Di- and Tripeptides of Dopa

Numerous di- and tripeptides have been prepared by *A.M. Felix; U.S. Patent 3,803,120; April 9, 1974; assigned to Hoffmann-La Roche Inc.* for use as anti-Parkinson agents. Those included compounds of the following formulas, of which formula I is an amide of L-dopa and dopamine, not a peptide.

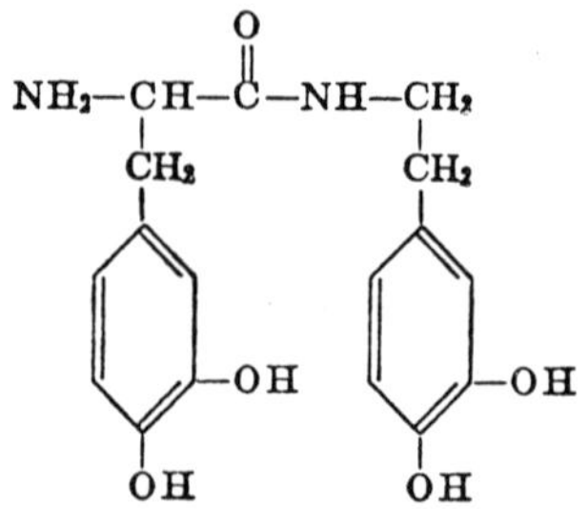

(I)

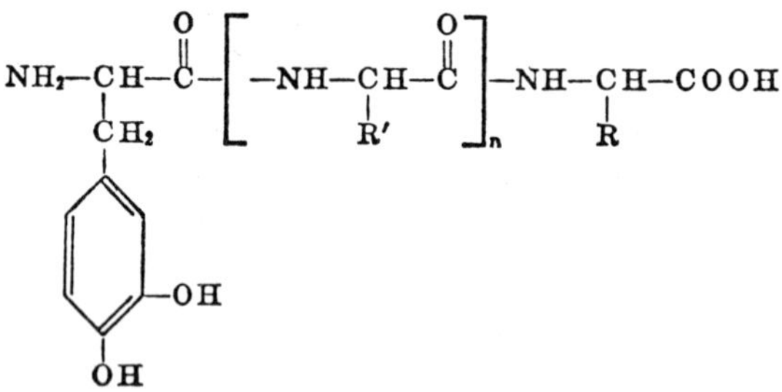

(II)

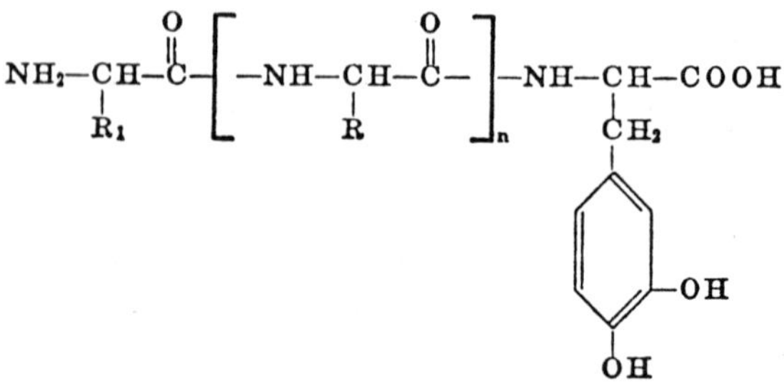

(III)

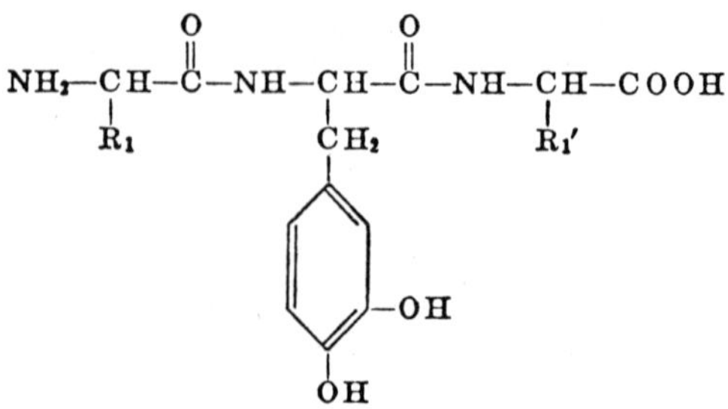

(IV)

where R and R' are

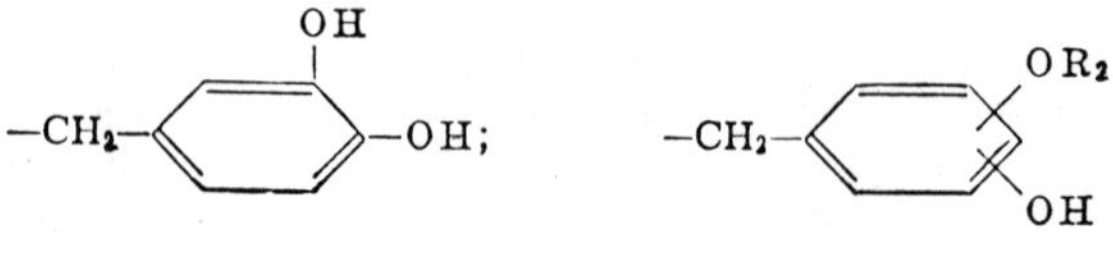

or the structural moiety which forms a natural protein occurring α-amino acid with the glycyl moiety; R_1 and R_1' are

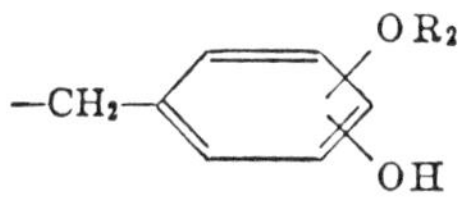

or also the structural moiety which forms a natural protein occurring α-amino acid with the glycyl moiety; n is 0 or 1; and R_2 is lower alkyl and salts thereof.

That the compounds of Formulas I, II, III and IV are effective anti-Parkinson agents can be seen by the fact that compounds such as L-3-(3,4-dihydroxyphenyl)alanyl-L-3-(3,4-dihydroxyphenyl)alanine, glycyl-L-dopa; and L-tyrosyl-L-dopa have ED_{50}'s less than 300 mg/kg IP when administered to mice having catatonia induced by reserpine in accordance with the procedure described by Chesare et al. in *Archives Internationals, Pharmacodynamie et de Therapic* 169, 26 (1967). On the other hand, the ED_{50} of L-dopa when administered by the same method was 430 mg/g IP.

These compounds are formed by the coupling of dopa with dopa, dopamine, a natural protein amino acid or any amino acid that will satisfy the Formulas I through IV. Although the above materials may be used in their D, L, or DL forms, it is preferred to use the L forms of the starting materials.

The dipeptides are formed by condensing a dopa derivative (V) containing a conventional amino protecting group and optionally aromatic hydroxyl protecting groups with another amino acid (VI) satisfying the formulas of Formulas I through IV which has a protected carboxyl group. Any conventional means of peptide coupling can be utilized to react the compound V with a compound VI to produce a dipeptide. Among the conventional means for forming peptides which can be used is reacting compound V with compound VI with the carboxylic acid group of compound V converted to a reactive functional derivative.

Any conventional reactive derivative of an acid which will form a peptide bond with an amino group can be utilized in carrying out these reactions. Among the preferred reactive functional acid derivatives are halides, azides, anhydrides and active esters such as N-hydroxy-succinimide ester, or para-nitrophenyl ester, or activation by a C–N or a C–C multiple bond such as N,N'-dialkylated carbodiimide.

A preferred method of carrying out this coupling is by reacting the compounds V and VI in the presence of a conventional peptide coupling agent. Among the preferred peptide coupling agents are included dicyclohexylcarbodiimide, carbonyldiimidazole and alkylchloroformate such as ethyl-chloroformate, isobutylchloroformate and other lower alkyl chloroformates as well as aryl chloroformates such as phenyl chloroformate. Generally, this reaction is carried out in an inert organic solvent. Preferred inert organic solvents include tetrahydrofuran, ethyl acetate, methylene chloride, chloroform, dioxane, and dimethylformamide.

In carrying out this reaction, temperatures of from –10° to 25°C are preferred. Generally, it is preferred to start the reaction at –10°C and slowly allow the temperature of the reaction to rise to 25°C.

Compound V can contain any conventional amino protecting group which can be easily removed by hydrogenolysis and hydrolysis. Among the preferred amino protecting groups are aralkoxycarbonyl groups such as benzyloxycarbonyl as well as tert-butyloxycarbonyl. The following example illustrates the preparation of L-dopyl-L-dopa according to this process.

Example 1: A solution containing 106 g (0.178 mol) of O,O'-dicarbobenzoxy-L-dopa benzyl ester hydrochloride in 400 ml CH_2Cl_2 was chilled in an ice bath and neutralized with 24.4 ml of triethylamine. Some solid triethylammonium hydrochloride formed at this point. Without too much delay, 112 g (0.178 mol) of N,O,O'-tricarbobenzoxy-L-dopa dissolved in 400 ml of chilled CH_2Cl_2 was added followed by 39.2 g (0.19 mol) of dicyclohexylcarbodiimide. The reaction mixture was stirred at 0° for 1 hour and 2 hours at room temperature. The insoluble dicyclohexylurea and triethylammonium hydrochloride were removed by suction filtration and washed several times with small portions of CH_2Cl_2.

The combined filtrate and washings were washed with water, dried over Na_2SO_4 and evaporated to dryness. The crude oil obtained was dissolved in a small volume of CH_2Cl_2 and filtered to remove some more dicyclohexylurea. Evaporation of the filtrate under reduced pressure gave clear oil. This oil was crystallized from a minimum volume of CH_2Cl_2 with petroleum ether (BP 30° to 60°C) to give N,O,O'-tricarbobenzoxy-L-dopa-O,O'-dicarbobenzoxy-L-dopyl benzyl ester melting at 102° to 108°C.

Example 2: N,O,O'-tricarbobenzoxy-L-dopa-O,O'-dicarbobenzoxy-L-dopa benzyl ester (144 g, 0.127 mol) was dissolved in a solvent mixture (800 ml) containing peroxide-free THF methanol and acetic acid in the ratio of 6:9:1 parts by volume. The catalyst (40 g, 5% Pd on $BaSO_4$) was added and the mixture hydrogenated in a Parr apparatus at 45 psi for 5 hours. The catalyst was then removed by filtration and the filtrate evaporated to give a hard oil which on treatment with 400 ml absolute ethanol started to crystallize slowly (2 to 3 days) giving crude di-dopa, MP 167° to 170°C. Recrystallization from hot ethanol gave L-dopyl-L-dopa melting at 174° to 176°C.

THERAPEUTIC FORMULATION OF L-DOPA

Preparation of Concentrated Dopa Solutions

Dopa (and methyldopa) are inconvenient to use as an injection owing to their low water-solubility. For example, the water-solubility of dopa is about 2.5 mg/ml and 20 ml of water is required for producing a preparation containing 50 mg of dopa as a dose. Therefore, a large ampoule is used but it is inconvenient to use. Under the circumstances, a high concentration preparation of these medicines has been demanded clinically. However, no preparation satisfying such a requirement has appeared.

T. Miki and S. Tanaka; U.S. Patent 3,911,137; October 7, 1975; assigned to Sankyo Company Limited, Japan have discovered that a dopa or tyrosine can be dissolved in a high concentration by the addition of certain water-soluble high molecular weight substances.

This water-soluble high molecular weight substance may be methylcellulose, hydroxypropylcellulose, hydroxypropylmethylcellulose or gelatin. By this procedure, dopas or tyrosine can be dissolved in water in a concentration of 4 to 5 times their water-solubility.

To carry out this process it is preferable that a dopa or tyrosine is previously dissolved in a high concentration by heating or pH adjustment and a solution of at least one of the above-mentioned high molecular weight substances is then added to the resulting solution. It is preferable that the high molecular weight substances are added so that the concentration of the substances in the final solution may be 0.05 to 1.0 w/v %, since the addition of a large amount of the substances causes troubles such as an increase in viscosity in pharmaceutical manufacturing procedure. The use of a mixture of 2 or more of these high molecular weight substances also gives a good result.

Example 1: 400 mg of sodium bisulfite and 100 mg of disodium ethylenediaminetetraacetate were dissolved in 20 ml of water for injection. One gram of dopa was added to the resulting solution and an aqueous sodium hydroxide solution was added dropwise to dissolve dopa. To the resulting aqueous dopa solution was added a solution of 100 mg of methyl cellulose (SM-10) in 30 ml of water for injection with stirring. The pH of the mixture was adjusted to 9 by the use of an aqueous sodium hydroxide solution. The mixture was made up to 100 ml with water for injection and heated at 80°C for 10 minutes to obtain a highly concentrated aqueous solution containing 10 mg/ml of dopa. (The water-solubility of dopa is 2.5 mg/ml.)

Example 2: The same procedure as in Example 1 was repeated except that 100 mg of methyl cellulose was replaced by 500 mg of hydroxypropyl cellulose. Thus, a highly concentrated aqueous solution containing 10 mg/ml of dopa was obtained.

Example 3: The same procedure as in Example 1 was repeated except that 100 mg of methyl cellulose was replaced by 500 mg of hydroxypropylmethyl cellulose. Thus, a highly concentrated aqueous solution containing 10 mg/ml of dopa was obtained.

Example 4: The same procedure as in Example 1 was repeated except that 100 mg of methyl cellulose was replaced by 20 mg of gelatin (isoelectric point 9). Thus, a highly concentrated aqueous solution containing 10 mg/ml of dopa was obtained.

Soluble Dopa Complexed With Polyvalent Metal Salts

Solubilization of L-dopa and methyldopa is also accomplished by the process disclosed by *R. Okada, T. Nitanai, and K. Sasahara; U.S. Patent 3,916,004; October 28, 1975; assigned to Sankyo Company Limited, Japan.* According to this process, a dopa preparation can be produced by dissolving dopa or methyldopa in water and adding to the resulting aqueous solution at least one nontoxic polyvalent metal salt such as a salt of Ca, Co, Fe, Mn, Ni, Zn, Al, Cu or Mg, and preferably a chloride, iodide, sulfate, phosphate, nitrate, etc., thereof, or an aqueous solution thereof.

The thus obtained dopa preparation is readily soluble in water and can be easily absorbed on administration to human body. According to this process, dopas can be dissolved in a remarkably high concentration. For example, 50 mg of dopas as a dose can be dissolved in only 2 ml of water although 20 ml of water has previously been required. Alternatively, a complex compound of a dopa with a polyvalent metal, which may be separated as amorphous fine particles, can be formulated into any suitable form such as tablets, granules, powder, etc. Also, a suspension containing the complex compound separated is particularly suitable for the preparation of an injection. This process is carried out by dissolving a dopa in water and adding to the resulting aqueous solution the polyvalent metal salt or an aqueous solution thereof to form a chelate compound.

Example 1: 3.54 g of dopa was suspended in 10 ml of water for injection and hydrochloric acid was added to dissolve dopa. Then, 2.16 g of aluminum chloride (hexahydrate) was added and 200 mg of sodium bisulfite was added with neutralizing with an aqueous sodium hydroxide solution. The pH of the mixture was adjusted to 7.5 by the addition of an aqueous sodium hydroxide solution and the mixture was made up to 100 ml with water for injection. Thus, a highly concentrated aqueous solution containing 35.4 mg/ml was obtained. (The solubility of dopa in water is 2.5 mg/ml.)

Example 2: 2.63 g of dopa and 400 mg of sodium bisulfite were suspended in 10 ml of water for injection and an aqueous sodium hydroxide solution was added to dissolve dopa. Then, 1.38 g of zinc chloride was added. The pH of the mixture was adjusted to 9 by the addition of hydrochloric acid and the mixture was made up to 100 ml with water for injection. Thus, a highly concentrated aqueous solution containing 26.3 mg/ml of dopa was obtained.

Example 3: 3.7 g of methyldopa was suspended in 10 ml of water for injection and hydrochloric acid was added to dissolve methyldopa. Then, 3.6 g of aluminum chloride (hexahydrate) was added. After dissolution, 200 mg of sodium bisulfite was added, with neutralizing with an aqueous sodium hydroxide solution, to adjust the pH of the mixture to 7.3 and the mixture was made up to 100 ml with water for injection. Thus, a highly concentrated aqueous solution containing 37.0 mg/ml of methyldopa was obtained. (The solubility of methyldopa in water is 10 to 15 mg/ml.)

Mixtures of L-Dopa and L-Tryptophan

It has been found by *W. Birkmayer and E. Neumayer; U.S. Patent 3,795,739; March 5, 1974; assigned to Hoffmann-La Roche Inc.* that the use of L-dopa with either L-tryptophan or L-5-hydroxytryptophan reduces or does away with the psychotic side effects often associated with the administration of L-dopa alone.

It has been found that on oral or parenteral administration to patients having Parkinson's disease, particularly those with idiopathic Parkinson's disease, of a pharmaceutical preparation containing L-dopa or a pharmaceutically acceptable salt thereof in combination with L-tryptophan or L-5-hydroxytryptophan or a pharmaceutically acceptable salt of either, the typical symptoms of the disease such as rigor, akinesia, and tremor can be significantly improved or abolished without the simultaneous occurrence of undesirable side effects.

Furthermore, it has been found that the treatment of the psychotic side effects of L-dopa with L-tryptophan or L-5-hydroxytryptophan or a pharmaceutically acceptable salt of either has considerable practical significance since the L-tryptophan or L-5-hydroxytryptophan or a pharmaceutically acceptable salt of either not only causes the psychotic transit phase to disappear, but can be continued in combination with the administration of L-dopa or a pharmaceutically acceptable salt thereof without reappearance of psychotic side effects.

The benefits can also be obtained by using a peripheral decarboxylase inhibitor in combination with the L-dopa and the tryptophan compound to treat Parkinsonism. This is advantageous since it permits the utilization of less L-dopa per dosage to achieve the same effect.

The L-tryptophan or L-5-hydroxytryptophan and the L-dopa as well as the optionally present decarboxylase inhibitor need not be administered simultaneously but can be administered separately. The undesirable side effects which occur when L-dopa is used alone or in combination with a decarboxylase inhibitor can be treated by separate administration of the tryptophan compound, i.e., L-tryptophan, L-5-hydroxytryptophan or a pharmaceutically acceptable salt of either, resulting in a reduction or removal of the side effects.

Example 1: 250 parts of L-tryptophan, 100 parts of L-3-(3,4-dihydroxyphenyl)-alanine and 100 parts of mannitol are mixed until homogeneous then sieved. The resulting powder mixture is well moistened with 300 parts of a 10% maize starch glue and subsequently granulated, dried and sieved once again. The resulting granulate is mixed with 3 parts of magnesium stearate and 17 parts of talcum. The mixture is then filled into number 0 capsules. Each capsule has the following composition.

	Mg.
L-tryptophan	250
L-3-(3,4-dihydroxyphenyl)alanine	100
Mannitol	100
Maize starch	30
Magnesium stearate	3
Talcum	17
	500

In the above procedure, an equal amount of L-5-hydroxytryptophan can be employed in place of L-tryptophan.

Example 2: 250 parts of L-tryptophan, 100 parts of L-3-(3,4-dihydroxyphenyl)-alanine, 25 parts of N^1-D,L-seryl-N^2-(2,3,4-trihydroxybenzyl)hydrazide hydrochloride (a decarboxylase inhibitor), 5 parts of citric acid and 105 parts of microcrystalline cellulose are mixed until homogeneous then seived.

The resulting powder mixture is well moistened with 600 parts of a 10% maize starch glue and subsequently granulated, dried and sieved again. The resulting granulate is mixed with 5 parts of magnesium stearate, homogenized and pressed into tablets having a weight of 500 mg. Each tablet has the composition shown on the next page. In this procedure, an equal amount of L-5-hydroxytryptophan can be used in place of L-tryptophan.

	Mg
L-tryptophan	250
L-3-(3,4-dihydroxyphenyl)alanine	100
N^1-D,L-seryl-N^2-(2,3,4-trihydroxybenzyl)-hydrazide-hydrochloride	25
Citric acid	5
Microcrystalline cellulose	105
Maize starch	60
Magnesium stearate	5
	500

Clinical examples on 2 males and 1 female with Parkinson's disease were given in some detail to show how the drug combinations of this process eliminated the psychotic disturbances of the patients produced by treatment with L-dopa alone or with a decarboxylase inhibitor.

OTHER AROMATIC AMINO ACIDS

PHENYLGLYCINE

2-Phenylacetic acid or phenylglycine is produced as a DL-racemate. D-phenylglycine is useful as a starting material for the preparation of α-amino benzyl penicillin while L-phenylglycine may be used to form a peptide sweetening agent with L-asparagine.

Enzymatic Resolution of DL-Phenylglycine Amide

It is known that amino acid amides may be hydrolyzed selectively by means of an enzyme. Aromatic amino acid amides, such as phenylalanine amide, may in general be hydrolyzed under the influence of enzymes, such as papain, bromelain or ficin. However, these enzymes do not hydrolyze phenylglycine amide.

A process for the resolution of DL-phenylglycine amide using an aminopeptidase, especially leucine aminopeptidase, has now been developed by *W.H.J. Boesten; U.S. Patent 3,971,700; July 27, 1976; assigned to Stamicarbon BV, Netherlands*. The preferred enzyme is leucine aminopeptidase (Enzyme Commission Number 3.4.1.1), which effects highly stereospecific hydrolysis of the racemic amide in a short time.

The reaction may be carried out under conventional conditions, such as described in Dixon & Webb, *Enzymes,* p. 247 (1965) for the hydrolysis of other racemic amino acid amides by means of aminopeptidase enzymes. Preferably the temperature is maintained at 20° to 40°C and the reaction is effected in a buffered aqueous solution at a pH of between 7 and 9.5. Activating compounds such as, for example, manganese salts and magnesium salts may be added.

The processing method for the isolation of the reaction products D-phenylglycine amide and L-phenylglycine is independent of whether or not use is made of free aminopeptidase or water-insoluble aminopeptidase. Since L-phenylglycine is poorly soluble in water, the process will have to be effected in dilute solutions when insoluble enzyme is used, in order to prevent crystallization of L-phenylglycine on the insoluble enzyme. In both variants of the process the L-phenylglycine can be separated from the D-phenylglycine amide by extraction, crystallization, or by means of one or more ion exchange resins. The L-phenylglycine may be removed as such or after racemization. DL-phenylglycine may be converted by means of an acid and an alcohol, into the salt of phenylglycine alkyl ester, which can be treated with ammonia to obtain DL-phenylglycine amide, which may be returned to the process.

Example 1: 3.0 mmol (560 mg) of DL-phenylglycine amide hydrochloride are dissolved at room temperature in 52.0 ml of a buffer solution (boric acid-potassium chloride-sodium hydroxide) with a pH of 9.0 in a reaction vessel provided with a stirrer.

After addition of 2.2 ml of 0.125 M $MgCl_2$ and 0.1 ml of 0.025 M $MnCl_2$, the pH is adjusted to a value of 8.5 by means of 2.0 ml of approximately 1 N sodium hydroxide. Next, 0.4 mg of leucine amino-peptidase (80 mg of Merck 25010 enzyme suspension, from pigs' kidneys) is added to this solution and stirred at room temperature and a pH of 8.5 for 1.5 hours.

A thin layer chromatographic analysis shows that the L-phenylglycine amide has been fully hydrolyzed to L-phenylglycine. After a reaction time of 20 or more hours the reaction mixture contains according to an amino acid analysis: 0.40% by weight of L-phenylglycine (efficiency = 100%) and 0.39% by weight of D-phenylglycine amide (yield = 99.5%).

Example 2: A solution of 5.0 mmol (930 mg) of DL-phenylglycine amide hydrochloride, 50 mg of $MgCl_2$ and about 0.2 mg $MnCl_2$ in 100 ml of water is adjusted to a pH of 8.0 by means of 1 N sodium hydroxide. After addition to this solution of 250 mg of leucine amino-peptidase (Merck 25010) covalently bonded to 3-amino-propyl-triethoxysilyl Bio-Glass with a load of 0.3% by weight, obtained as described in *Biotechnology and Bioengineering,* Vol. XVI, pp. 275 to 277, 1974, the solution was stirred at room temperature for 18 hours.

After filtration of the leucine amino-peptidase bonded to glass powder, the filtrate is subjected to a thin layer chromatographic examination. This showed that the DL-phenylglycine amide was converted into equimolar amounts of L-phenylglycine and D-phenylglycine amide by leucine amino-peptidase covalently bonded to glass.

The filtrate is adjusted to a pH of 10.0 with 1 N NaOH and then passed over 100 g of Amberlite IRC–50 ion exchanger in the H^+ form. Next 100 ml of water is passed over the ion exchanger, the eluate is evaporated to a volume of 15 ml in vacuo, and the L-phenylglycine is isolated by filtration. The yield of L-phenylglycine which is pure according to thin layer chromatography is 0.34 g (efficiency 90%).

The D-phenylglycine amide bonded to the Amberlite IRC–50 is extracted with 150 ml of 0.5 N sulfuric acid. The eluate is concentrated to 10 ml by evaporation in vacuo and then boiled for 1 hour. After cooling and neutralization with concentrated aqueous ammonia, the crystallized D-phenylglycine is filtered over a glass filter and washed on the filter with 5 ml of water. The yield of D-phenylglycine is 0.33 g (efficiency 90%).

p-HYDROXYPHENYLGLYCINE AND DERIVATIVES

2-(p-hydroxyphenyl)glycine, which can also be named 2-(4-hydroxyphenyl)glycine, α-amino-p-hydroxyphenylacetic acid or C-(p-hydroxyphenyl)glycine, has been prepared in racemic form by numerous workers. Its D-(–)- or levo-rotatory isomer has been of particular interest for coupling (after appropriate protection of the free amino group) with 6-aminopenicillanic acid (6-APA) to produce a penicillin and for coupling with 7-aminocephalosporanic acid (7-ACA) and 7-aminodesacetoxy-cephalosporanic (7-ADCA) to produce cephalosporin derivatives.

Resolution of N-Chloroacetyl p-Hydroxyphenylglycine with Dehydroabietylamine

A method has been disclosed by *C.T. Holdrege; U.S. Patent 3,796,748; March 12, 1974; assigned to Bristol-Myers Company* for obtaining the desired D-(–)-2-(p-hydroxyphenyl)glycine from the racemic mixture. In this process, racemic 2-(p-hydroxyphenyl)glycine is efficiently resolved by N-chloroacetylation followed by the addition of about 0.5 equivalent of dehydroabietylamine (DHA) in isopropanol:water (2:1 v/v) to precipitate the substantially pure, crystalline DHA salt of D-(–)-N-chloroacetyl-2-(p-hydroxylphenyl)glycine from which the desired D-(–)-2-(p-hydroxyphenyl)glycine may be obtained. Details of the process are given in the following examples. All temperatures are in degrees centigrade.

Example 1: DL-N-Chloroacetyl-2-(p-Hydroxyphenyl)Glycine – A solution of 12.0 g (0.3 mol) of sodium hydroxide in 30 ml of water was added to a suspension of 50.2 g (0.3 mol) of DL-2-(p-hydroxyphenyl)glycine in 510 ml of water giving about pH 10 and causing much of the amino acid to dissolve. The mixture was cooled to 5°. The cooling bath was removed and about ½ of the total amount of 102.5 g (0.6 mol) of chloroacetic anhydride was added all at once followed by aqueous sodium hydroxide (36 g of sodium hydroxide in 150 ml of water) as needed to maintain the system at pH 10. The remainder of the chloroacetic anhydride was added again followed by sodium hydroxide as needed.

After about 15 minutes the solution was adjusted to pH 8 (HCl), cooled and polish filtered. The filtrate was cooled in an ice bath and adjusted to pH 2 with concentrated HCl. Scratching induced the slow crystallization of the product. After stirring for 2.5 hours in an ice bath the crop of crystalline product was collected by filtration. The two crops of product were combined, recrystallized from 500 ml of water, air dried and then dried in vacuo over phosphorus pentoxide; 38.6 g (53.0%), MP 183° to 185° dec.

Example 2: Resolution of N-Chloroacetyl-2-(p-Hydroxyphenyl)Glycine with Dehydroabietylamine – The salt of D-N-chloroacetyl-2-(p-hydroxyphenyl)glycine with dehydroabietylamine is prepared as follows. A mixture of 5.0 g (0.0205 mol) of racemic N-chloroacetyl-2-(p-hydroxyphenyl)glycine, 3.52 g (0.01025 mol) of dehydroabietylamine acetate, 240 ml of 2-propanol and 170 ml of water was heated on the steam bath giving a solution. The hot solution was filtered and the filtrate was stored in an insulated container to cool slowly. Crystallization was allowed to proceed for 16 hours giving long needle-like crystals. The crystals were collected by filtration, washed sparingly with 2:1 2-propanol:water and dried at 40°; 2.6 g (28%), MP 206° to 212°.

The filtrate containing L-N-chloroacetyl-2-(p-hydroxyphenyl)glycine is preserved; heating it at alkaline pH, e.g., with added NaOH followed by cautious acidification, racemizes the L-isomer to the original DL racemate used as starting material.

D-N-chloroacetyl-2-(p-hydroxyphenyl)glycine is prepared as follows. A vigorously stirred suspension of 2.6 g of the D-dehydroabietylamine salt in 20 ml of water and 50 ml of methylene chloride was adjusted to pH 10.7 by the slow addition of 20% sodium hydroxide causing solution of the solid. The aqueous phase was separated, washed 3 times with methylene chloride, filtered and acidified to pH 2 with 6 N HCl. Scratching induced crystallization of the product in large clumps of fluffy crystals. After 1 hour in the refrigerator the product was collected by filtration, washed sparingly with water and dried at 40°; 0.73 g (61.2%) MP 193° to 196° dec. The dehydroabietylamine may be recovered from the solvent phase above by evaporation of the solvent or salt formation.

D-2-(p-hydroxyphenyl)glycine is prepared as follows. A mixture of 1.19 g of D-N-chloroacetyl-2-(p-hydroxyphenyl)glycine and 12 ml of 2 N HCl was heated at reflux for 1.5 hours. The mixture was concentrated to dryness. Water (5 ml) was added to the residue and the solution again concentrated to dryness leaving as residue, the crystalline hydrochloride salt of D-2-(p-hydroxyphenyl)glycine, MP 215° dec.

A solution of the residue in 10 ml of water was adjusted to pH 4.5 with 20% sodium hydroxide. A gelatinous mass formed. The mixture was warmed to about 50° on the steam bath whereupon another crystalline (granular) form started to separate. The mixture was concentrated to approximately one-half of its initial volume. After 1 hour at room temperature the flask was stored in the refrigerator overnight. The product was collected by filtration, washed sparingly dropwise with ice water and dried in vacuo at 65° over phosphorus pentoxide; 0.61 g of D-2-(p-hydroxyphenyl)glycine (81.3%), MP 223° to 226° dec. was recovered.

Resolution of Acyl Derivatives with Dehydroabietylamine

D.R. Palmer; U.S. Patent 3,869,505; March 4, 1975; assigned to Beecham Group Limited, England has disclosed that DL-2-(4-hydroxyphenyl)glycine is readily resolved by using the mono- or diacyl derivatives, preferably the acetyl derivatives. Salts of the derivative are formed with dehydroabietylamine which may be separated, then decomposed and hydrolyzed to obtain the optically active forms.

The acyl derivative for use in the process is normally made by reacting DL-2-(4-hydroxyphenyl)glycine with an excess of an acylating agent, such as an acid anhydride or acyl chloride to prepare the desired N-acyl or N,O-diacyl derivative. Such reaction may be carried out as the final stage of a process for preparing 2-(4-hydroxyphenyl)glycine without isolation thereof. Alternatively a 2-(4-acyloxyphenyl)glycine may be reacted with an acylating agent when, as is preferred, the N,O-diacyl derivative is to be used.

The preferred acyl derivatives are the acetyl derivatives, but other acyl derivatives, such as the formyl, propionyl, benzoyl and phenylacetyl derivatives, may also be used. When the N-mono-acyl derivatives are to be used, these are often conveniently prepared by partial hydrolysis with mild alkali of an initially prepared N,O-diacyl compound.

While a crude product of tree rosin containing dehydroabietylamine and other amines, for example, Hercules Amine D, can be used for present purposes, it is preferred that this should first be purified, for example, by the preparation of various salts, particularly the acetate, which are then recrystallized from suitable solvents, before decomposition with alkali followed by solvent extraction. It has, however, now been found that purification is very conveniently carried out by preparing a carbon dioxide adduct of dehydroabietylamine as shown in the examples.

Example 1: A crude extract of tree rosin containing dehydroabietylamine and other amines, (Hercules Rosin Amine D), (2 kg) was dissolved in methylated spirits (10 liters) and gaseous carbon dioxide passed into the solution until precipitation was complete. The precipitated solid was filtered, triturated with fresh methylated spirits (5 liters) and refiltered. After drying at room temperature a pure dehydroabietylamine-carbon dioxide adduct was obtained (650 g) of $[\alpha]_{546}{}^{20}$ = +42° (c = 1, methanol).

DL-2-(4-hydroxyphenyl)glycine (167 g, 1.0 mol) was dissolved in water (1 liter) containing sodium hydroxide (160 g, 4 mols), the solution cooled to 0° to 5°C and acetic anhydride (250 g, 2.5 mols) rapidly added while maintaining the temperature below 5°C. After a further 30 minutes stirring, the mixture was acidified and DL-N-acetyl-2-(4-acetoxyphenyl)glycine filtered off (220 g), 88% yield, MP 216°C.

The dehydroabietylamine-carbon dioxide adduct (330 g) was dissolved in methylated spirits (4 liters) by boiling and to this solution was added the DL-N-acetyl-2-(4-acetoxyphenyl)glycine (420 g) with stirring until solution was complete. This solution was seeded with an optically pure sample of the required salt, preferably the salt of (–)-N-acetyl-2-(4-acetoxyphenyl)glycine and dehydroabietylamine (42 g) and the mixture left overnight to crystallize. The resulting crystals were filtered and washed with methylated spirits to yield the required salt (270 g) of specific rotation $[\alpha]_{546}{}^{20}$ = –50° (c = 1, methanol). Recrystallization of the foregoing salt from a minimum of methylated spirits yielded the salt of (–)-N-acetyl-2-(4-acetoxyphenyl)glycine and dehydroabietylamine (200 g) of specific rotation $[\alpha]_{546}{}^{20}$ = –66° (c = 1, methanol).

This diastereomeric salt was stirred with 10% sodium carbonate in the presence of ether and filtered to remove the liberated dehydroabietylamine and any resulting carbon dioxide adduct and the resulting aqueous solution acidified to pH 2 to precipitate a material which after recrystallization from aqueous alcohol afforded pure (–)-N-acetyl-2-(4-acetoxyphenyl)glycine $[\alpha]_{546}{}^{20}$ = –216° (c = 1, methanol).

The foregoing pure (–)-N-acetyl-2-(4-acetoxyphenyl)glycine (25 g) was heated under reflux in 6 N hydrochloric acid (250 ml) for 2 hours and the solution evaporated to dryness. The resulting solid was dissolved in water (50 ml) and brought to pH 6.5 by the addition of 4 N sodium hydroxide. The resulting precipitate was washed with water and dried to yield pure (–)-2-(4-hydroxyphenyl)-glycine (13 g), $[\alpha]_{546}{}^{20}$ = –130° (c = 1, water).

Example 2: DL-N-acetyl-2-(4-hydroxyphenyl)glycine was prepared by boiling a solution of DL-N-acetyl-2-(4-acetoxyphenyl)glycine (250 g, 1.0 mol) as prepared in Example 1 in water (2 liters) containing sodium carbonate (150 g, 1.5 mols) for 1 hour. The solution was cooled and acidified with concentrated hydrochloric acid to pH 3.5 and the product filtered and washed with water to yield DL-N-acetyl-2-(4-hydroxyphenyl)glycine (150 g), 75% yield.

The adduct of dehydroabietylamine and carbon dioxide (16.4 g, 0.05 mol) was dissolved in boiling methanol and to this solution was added a hot solution of DL-N-acetyl-2-(4-hydroxyphenyl)glycine (10.4 g, 0.05 mol) and the mixture allowed to cool. The precipitated salt of (+)-N-acetyl-2-(4-hydroxyphenyl)glycine and dehydroabietylamine was filtered off and used for the preparation of (+)-2-(4-hydroxyphenyl)glycine as described below. The residual methanolic filtrate was evaporated to dryness to furnish a crude solid which was boiled with aqueous sodium carbonate to liberate the amine as a brown oil. The resulting aqueous liquors were adjusted to pH 2 with concentrated hydrochloric acid and the resulting precipitate filtered, washed with water and dried to yield optically pure (–)-N-acetyl-2-(4-hydroxyphenyl)glycine (5.2 g), 50% yield, $[\alpha]_{546}^{20} = -271°$ (c = 1, methanol).

The pure diastereomeric (+)-N-acetyl-2-(4-hydroxyphenyl)glycine dehydroabietylamine salt was stirred with aqueous alkali in the presence of ether to remove the liberated dehydroabietylamine and the resulting aqueous solution acidified to pH 2 to yield, after washing with water, pure (+)-N-acetyl-2-(4-hydroxyphenyl) glycine $[\alpha]_{546}^{20} = +271°$ (c = 1, methanol).

Resolution Using Ephenamine

In the process developed by *A. Weber and D. Bouzard; U.S. Patent 3,803,213; April 9, 1974; assigned to Bristol-Myers Company*, the deacetyl derivatives of racemic p-hydroxyphenyl)glycine is resolved with ephenamine. 1,2-diphenyl-2-methylaminoethanol, commonly called ephenamine, has the structure

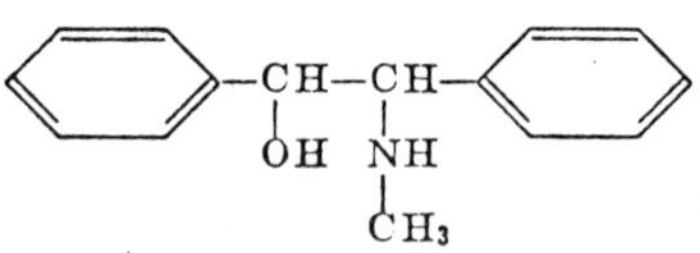

The compound is also named N-methyl-1,2-diphenyl-2-hydroxy-ethylamine or alpha, beta-diphenyl-beta-hydroxy-N-methyl-ethylamine or 1,2-diphenyl-2-methylamino-1-ethanol. This process utilizes only the levo-erythro-isomer. Methods for its preparation are described in U.S. Patents 2,645,638 and 2,768,081.

The process for the production of levo-erythro-ephenamine D-(–)-2-acetylamino-2-(4-acetoxyphenyl)acetate from racemic 2-acetylamino-2-(p-acetoxyphenyl)acetic acid [also called racemic N-acetyl-2-(p-acetoxyphenyl)glycine] comprises mixing the latter with levo-erythro-ephenamine in an amount of about one mol of the amine per mol of the amino acid in acetone or a lower alcohol and preferably in isopropyl alcohol or ethanol or acetone, preferably in about the minimum volume needed to effect solution upon warming the mixture, with sufficient application of heat to form a solution which is then cooled slowly to precipitate substantially pure, crystalline levo-erythro-ephenamine D-(–)-2-acetylamino-2-(4-acetoxy)acetate.

The process for obtaining D-2-amino-(p-hydroxyphenyl)acetic acid from levo-erythro-ephenamine D-(–)-2-acetylamino-2-(4-acetoxyphenyl)acetate comprises treating the latter with a strong base, e.g., sodium hydroxide, in aqueous media and then removing the free base levo-erythro-ephenamine by solvent extraction to leave an aqueous solution of D-N-acetyl-2-(p-acetoxyphenyl)glycine which is then heated, e.g., to reflux for several hours, at low pH in a strong acid, e.g., 2 N HCl, to obtain the corresponding acid addition salt of D-2-amino-2-(p-hydroxyphenyl)acetic acid, e.g., its hydrochloride, which is then converted to the zwitterion of D-(–)-2-amino-2-(p-hydroxyphenyl)acetic acid by treatment with alkali, e.g., aqueous sodium hydroxide or ammonia.

The D-(–)- or levo-rotatory isomer has been of particular interest for use as the side-chain in semi-synthetic penicillins and cephalosporins. This process is illustrated in more detail by the following example.

Example: DL-2-Acetamino-2-(4-Acetoxyphenyl)Acetic Acid – 561 g of crude DL-2-(4-hydroxyphenyl)glycine were dissolved in a solution of 538 g of sodium hydroxide in 2,200 ml of water and the solution was cooled to +5°C. 794 ml of acetic anhydride was added with stirring at 5° to 10°C during 1.5 hours. The mixture was stirred 0.5 hour after the end of the addition of the acetic anhydride and acidified below +10°C with 33% hydrochloric acid to pH 1; 1,100 ml of HCl were used. The precipitated DL-N-acetyl-(4-acetoxyphenyl)glycine was collected by filtration, washed with water and crystallized from ethanol (3.5 l); to give 428 g, MP 221°C and a second crop of 34 g, MP 216°C, yield = 55%.

Levo-Erythro-Ephenamine D-(–)-2-Acetylamino-2-(4-Acetoxyphenyl)Acetate – 113 g of DL-2-acetamino-(4-acetoxyphenyl)acetic acid and 102 g of levo-erythro-ephenamine was dissolved with heating in 450 ml of ethanol and let stand overnight at room temperature. The crystallized levo-erythro-ephenamine D-(–)-2-acetylamino-2-(4-acetoxyphenyl)acetate was collected and washed twice with ethanol to give 59.6 g (55.5%) which melted around 130°C and resolidified and melted at 180°C, $[\alpha]_D$ = –152° (c = 1, methanol).

D-(–)-2-(4-Hydroxyphenyl)Glycine – The above levo-erythro-ephenamine D-(–)-2-acetylamino-2-(4-acetoxyphenyl)acetate was suspended in 300 ml of methylene chloride, and stirred with a solution of 6 g of sodium hydroxide in 70 ml of water until all solid material disappeared. The layers were separated and the aqueous layer was washed twice with 50 ml of methylene chloride. 25 ml of concentrated hydrochloric acid were added to the aqueous layer and the mixture was refluxed for 1.5 hours, cooled, adjusted to pH 5 with ammonia (preferably at 40° to 50°C) and let stand overnight. 11.5 g of D-(–)-2-(4-hydroxyphenyl)glycine crystallized out; $[\alpha]_D$ = –104.4° (c = 1, H_2O). Water recrystallized material had $[\alpha]_D$ = –108°.

Resolution Using Dibenzoyltartaric Acid

In the process developed by *R.R. Lorenz; U.S. Patent 3,832,388; August 27, 1974* esters of racemic 2-(p-hydroxyphenyl)glycine are resolved using L-(–)- or D-(+)-dibenzoyltartaric acid as resolving agent. When the D-(+)- or L-(–)-dibenzoyltartaric acid and the glycinate were reacted in a mol to mol ratio (molar ratio of 1) or one mol to two mol ratio (molar ratio of ½), resolution of the diastereoisomeric salts could not be achieved.

However, it was found that when a molar ratio of D-(+)- or L-(–)-dibenzoyltartaric acid to the D,L-(±)-2-(p-hydroxyphenyl)glycinate of greater than ½ to less than 1 is used, preferential crystallization of the corresponding D-(–)-glycinate D-(+)-dibenzoylbitartrate or L-(+)-glycinate L-(–)-dibenzoylbitartrate respectively can be achieved in good yield. Treatment of the appropriate filtrate, from which the less soluble diastereoisomeric salt was separated, yields the L-(+)-glycinate D-(+)-dibenzoylbitartrate and D-(–)-glycinate L-(–)-dibenzoylbitartrate respectively.

The diastereoisomeric salt is then converted, on treatment with an equivalent of a strong acid, to the corresponding optically active 2-(p-hydroxyphenyl)glycine ester acid-addition salt which is then hydrolyzed in the presence of a strong acid to give the corresponding optically active 2-(p-hydroxyphenyl)glycine acid-addition salt which is treated with an acid-acceptor to give optically active 2-(p-hydroxyphenyl)glycine. The use of L-(–)-dibenzoyltartaric acid as resolving agent is preferred because of its ready availability from naturally occurring L-(+)-tartaric acid and hence relatively low cost.

Example: Resolution of D,L-(±)-Ethyl 2-(p-Hydroxyphenyl)Glycinate – A mixture of 100 g (0.5125 mol) of D,L-(±)-ethyl 2-(p-hydroxyphenyl)glycinate and 141.6 g (0.3755 mol) of L-(–)-dibenzoyltartaric acid monohydrate in 650 ml of absolute ethanol was stirred at room temperature until solution was complete (½ hour). The solution was filtered and the flask and filter were rinsed with 80 ml of absolute ethanol. To the clear filtrate was added 960 ml of water and the solution seeded with a trace amount of L-(+)-ethyl 2-(p-hydroxyphenyl)glycinate L-(–)-dibenzoylbitartrate and stirred until crystallization had well progressed.

After standing 1 hour at room temperature, the mixture was cooled overnight at 4°C. The precipitated L-(+)-ethyl 2-(p-hydroxyphenyl)glycinate L-(–)-dibenzoylbitartrate was filtered. The filter cake was pressed well and then washed twice with 50 ml of cold 25% ethanol (1:3, ethanol:water). The L-(+)-ethyl 2-(p-hydroxyphenyl)glycinate L-(–)-dibenzoylbitartrate, after drying at 60°C in vacuo, weighed 122 g (86%), MP 171° to 174°C (dec.), $[\alpha]_D^{25}$ –36° (c = 1%, MeOH). This salt can be obtained pure by dissolving it in 14 volumes of refluxing 25% ethanol and allowing the mixture to stand at room temperature for 24 hours. The recovery is 85%, MP 180° to 181°C (dec.), $[\alpha]_D^{25}$ –32.2° (c = 1%, MeOH).

The filtrate from above (including the 2 x 50 ml washes) was treated immediately with a solution of 51.8 g (0.1375 mol) of L-(–)-dibenzoyltartaric acid monohydrate in 125 ml of absolute ethanol with good stirring and seeded with a trace amount of D-(–)-ethyl 2-(p-hydroxyphenyl)glycinate L-(–)-dibenzoylbitartrate. Crystallization was heavy after 15 minutes.

The mixture was allowed to stand 1 hour at room temperature and then it was cooled at 0°C for 2 hours. The resulting precipitate was filtered, the filter cake was pressed well, then washed 3 times with 50 ml of cold 25% ethanol, and dried at 60°C in vacuo for 48 hours to give 124 g (87.5%) of D-(–)-ethyl 2-(p-hydroxyphenyl)glycinate L-(–)-dibenzoylbitartrate, MP 160° to 161°C, $[\alpha]_D^{25}$ –125.3° (c = 1%, MeOH). This product (40 g) was dissolved in 480 ml of 25% ethanol and allowed to crystallize by standing at room temperature for 24 hours. The crystals were filtered, pressed well and washed 2 times with 20 ml of 25% ethanol.

After drying at 60°C in vacuo for 20 hours, the optically pure D-(–)-ethyl 2-(p-hydroxyphenyl)glycinate L-(–)-dibenzoylbitartrate weighed 34.5 g (86.4% recovery), MP 161° to 163°C (dec.), $[\alpha]_D^{25}$ –131.6° (c = 1%, MeOH).

Resolution of Sulfonate Salts

A process of obtaining optically active p-hydroxyphenylglycine by selective crystallization has been developed by *T. Shirai, Y. Tashiro and S. Aoki; U.S. Patent 3,994,962; November 30, 1976; assigned to Nippon Kayaku Kabushiki Kaisha, Japan.* It was found that the racemate of the sulfonic acid derivative salt of p-hydroxyphenylglycine of the formula:

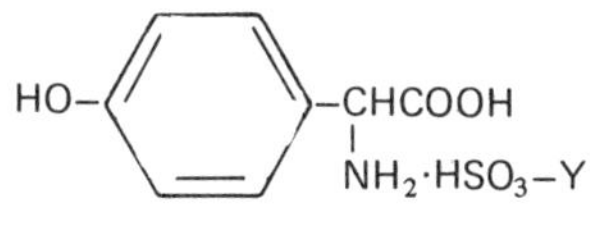

(where Y is a p-tolyl group or a 2,4-xylyl group) has a solubility in water greater than that of the optically active salt, as shown in the table below.

Solubility of Sulfonic Acid Derivative Salt of p-Hydroxyphenylglycine in Water (g/100 g of Water)

	p-Hydroxyphenylglycine p-toluene sulfonic acid salt		p-Hydroxyphenylglycine m-xylene sulfonic acid salt	
	Racemate of sulfonic acid	Optically active salt	Racemate of sulfonic acid	Optically active salt
10°C	13.15	9.62	59.26	17.33
30°C	24.89	13.99	93.27	34.94
50°C	41.30	18.78		

As a result, the sulfonic acid derivative salt of DL-p-hydroxyphenylglycine can be resolved in an aqueous solvent by the selective crystallization method. Thus, the present process is characterized by inoculating a supersaturated solution of DL-p-hydroxyphenylglycine sulfonic acid derivative salt represented in the formula above with an optically active p-hydroxyphenylglycine-sulfonic acid derivative salt to effect selective crystallization of the same type of optically active salt; or inoculating a supersaturated solution of a p-hydroxyphenylglycine sulfonic acid derivative salt having either of the two optically active salts in an excess amount over the other, with the optically active salt identical to the optically active salt existing in the excess amount in the solution to cause selective crystallization of the same type of optically active salt, collecting the selectively crystallized salt and thereafter subjecting the salt to treatment for the removal of the sulfonic acid derivative. Details of the process are given in the following examples.

Example 1: With 88.4 g of DL-p-hydroxyphenylglycine p-toluene sulfonic acid salt (DL-form salt) were mixed 17.9 g of L-p-hydroxyphenylglycine p-toluene sulfonic acid salt (L-form salt). The resultant mixture was dissolved in 200 ml of water by heating.

At 30°C the solution was inoculated with 0.05 g of L-form salt and the solution was gently agitated at the same temperature for 1.5 hours. The resultant crystals therein, were separated from the solution by filtration, washed with water and thereafter dried. There was obtained 27.16 g of crude L-form salt. $[\alpha]_D^{25}$ = 59.97° (c = 2, H_2O). Since pure L-form salt shows $[\alpha]_D^{20}$ = 69.7 (c = 2, H_2O), the crude L-form salt is calculated to have optical purity of 86.0%. When 25 g of this crude L-form salt were recrystallized from 75 ml of water, there were obtained 20.5 g of optically pure L-form salt, MP 230°C (dec.).

The L-form salt thus obtained was dissolved in 60 ml of water and then neutralized with a 10% sodium carbonate aqueous solution. Consequently, there was obtained 8.76 g of an optically pure L-p-hydroxyphenylglycine. $[\alpha]_D^{25}$ = +160.2° (c = 1, N HCl). The starting material DL-p-hydroxyphenylglycine p-toluene sulfonic acid salt was obtained as follows: With 2 ml of water, 1.67 g of DL-p-hydroxyphenylglycine and 1.72 g of p-toluene sulfonic acid were mixed and dissolved by heating. When the solution was left to cool, there ensued crystallization of DL-p-hydroxyphenylglycine p-toluene sulfonic acid salt. The yield was 2.9 g, MP 222°C (dec.).

Example 2: The mother liquor, remaining after the separation of crystals by filtration in Example 1, was mixed with 22.0 g of DL-form salt and dissolved in 8.5 ml of water by heating. The solution was then kept at 30°C and inoculated with 0.05 g of D-p-hydroxyphenylglycine p-toluene sulfonic acid (D-form salt), was agitated at the same temperature for 1.5 hours. By following the procedure of Example 1 from this point on, there was obtained 28.5 g of crude D-form salt, $[\alpha]_D^{21}$ = +59.9°C (c = 2, H_2O). When this salt was recrystallized from water, there were obtained 26.5 g of pure D-form salt, MP 229°C (dec.), $[\alpha]_D^{25}$ = -69.5° (c = 2, H_2O). When 25 g of the D-form salt were neutralized by following the procedure of Example 1, there were obtained 10.7 g of D-p-hydroxyphenylglycine, $[\alpha]_D^{25}$ = -160.5° (c = 1, N-HCl).

Resolution Using α-Phenylethylamine

T. Shirai, Y. Tashiro and S. Aoki; U.S. Patent 4,002,666; January 11, 1977; assigned to Nippon Kayaku Kabushiki Kaisha, Japan have found that optically active α-phenylethylamine effectively resolves a racemic mixture of p-hydroxyphenylglycine derivatives. These derivatives may be represented by the formula

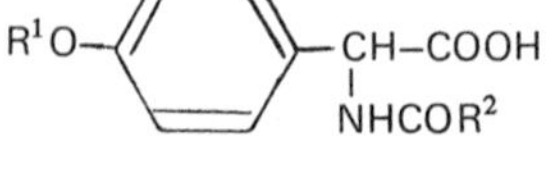

where R^1 is a hydrogen atom, a lower alkyl group or an acyl group, and R^2 is a hydrogen atom, a lower alkyl group, a phenyl group or a benzyloxy group with optically active α-phenylethylamine.

To carry out the present process, a racemic compound of a p-hydroxyphenylglycine derivative is reacted with optically active α-phenylethylamine in equivalent amount to yield a phenylethylamine salt of the p-hydroxyphenylglycine derivative. This reaction is preferably carried out in solvent at 50° to 100°C for 0.5 to 3 hours. The solvent is preferably a polar solvent, particularly water, a lower alcohol such as methanol, ethanol, propanol or the like, or a mixture thereof.

If the p-hydroxyphenylglycine derivative used is one where R^1 is hydrogen and R^2 is a phenyl or benzyloxy, the lower alcohol is preferred as solvent. On the other hand, if R^1 is a lower alkyl or an acyl and R^2 is hydrogen, a lower alkyl, phenyl or benzyloxy group, either water or the lower alcohol may be used.

In the production of a phenylethylamine salt of the optically active p-hydroxyphenylglycine derivative, a D-antipode can be obtained in the form of crystals of an optically active diastereomer salt with high purity when the p-hydroxyphenylglycine derivative is such that R^1 is hydrogen or acyl and R^2 is hydrogen, a lower alkyl, phenyl or a benzyloxy group, and is reacted with (+)-α-phenylethylamine. Furthermore, where the p-hydroxyphenylglycine derivative which is used is such that R^1 is a lower alkyl and R^2 is hydrogen, the lower alkyl group, phenyl or a benzyloxy group and is reacted with (–)-α-phenylethylamine, a diastereomer of D-antipode can also be formed.

In the mother liquor which allows the diastereomer salt to crystallize, there exists the other diastereomer salt in large amount, so that the other diastereomer can be obtained by subjecting the mother liquor to concentration under reduced pressure, separating the resultant crystals by filtration and recrystallizing the thus separated crystals with a suitable solvent several times.

Example 1: 10.0 g of DL-N,O-diacetyl-p-hydroxyphenylglycine, 5.28 g of (+)-α-phenylethylamine, and 100 ml of water were mixed and heated to about 80°C until the contents were completely dissolved. Then, the resultant solution was allowed to stand at 40°C for 1 hour to crystallize (+)-α-phenylethylamine-D-N,O-diacetyl-p-hydroxyphenylglycinate, MP 186° to 188°C, $[\alpha]_D^{25} = -98.0°$ (c = 1, MeOH). The resultant crystals were separated by filtration and then dissolved in 60 ml of water while heating, followed by allowing the resultant solution to stand at 40°C for 1 hour to obtain 4.2 g of purified crystals of D-antipode, MP 190.5° to 191.5°C, $[\alpha]_D^{24} = -102.4°$ (c = 0.3, MeOH).

10% aqueous sodium carbonate solution was added to 4.2 g of the crystals to adjust to about pH 9. The free (+)-α-phenylethylamine was extracted with ether, while the pH of the water layer was adjusted to 1 to 2 by means of hydrochloric acid to obtain 1.5 g of D-N,O-diacetyl-p-hydroxyphenylglycine, MP 213°C, $[\alpha]_D^{20} = -213.3°$ (c = 1, MeOH), which was then hydrolyzed with 2 N hydrochloric acid for 2 hours under reflux, followed by the treatment with a 10% aqueous sodium carbonate solution for making the pH of the resultant solution at 6 to 7 thereby to obtain 0.9 g of D-p-hydroxyphenylglycine, MP 240° to 242°C (dec.), $[\alpha]_D^{24} = -160.0°$ (c = 1, 1 N HCl).

Example 2: 1.0 g of DL-N-acetyl-p-methoxyphenylglycine and 0.54 g of (–)-α-phenylethylamine were mixed with 20 ml of ethanol and were dissolved under reflux. The resultant solution was allowed to stand in a refrigerator overnight to give 0.48 g of (–)-α-phenylethylamine-D-N-acetyl-p-methoxyphenylglycinate in the form of crystals, MP 195° to 197°C, $[\alpha]_D^{20} = -108.2°$ (c = 1, EtOH).

The crystals were dissolved in 15 ml of water, the pH adjusted to about 2 with concentrated hydrochloric acid, followed by cooling to crystallize D-N-acetyl-p-methoxyphenylglycine. The crystals were separated from the solution by filtration and were mixed with 5 mols of 48% hydrobromic acid under reflux for 3 hours thereby to obtain 0.17 g of D-p-hydroxyphenylglycine, MP 240° to 242°C (dec.).

Resolution Using d-3-Bromo-2-Oxo-10-Bornanesulfonic Acid

A process for resolving DL-2-(4-hydroxyphenyl)glycine without the need of protective groups has been developed by *K. Kariyone, H. Yagi and K. Yoshida; U.S. Patent 4,016,205; April 5, 1977; assigned to Fujisawa Pharmaceutical Co., Ltd., Japan.*

This process comprises treating DL-2-(4-hydroxyphenyl)glycine with d-3-bromo-2-oxo-10-bornanesulfonic acid to form a mixture of d-3-bromo-2-oxo-10-bornanesulfonic acid salts of D- and L-2-(4-hydroxyphenyl)glycine, treating the mixture with isobutyl alcohol or a mixture of lower alkanol and chloroform or toluene to selectively precipitate a solvate of the d-3-bromo-2-oxo-10-bornanesulfonic acid salt of L-2-(4-hydroxyphenyl)glycine with isobutyl alcohol, chloroform or toluene, collecting the precipitates, obtaining d-3-bromo-2-oxo-10-bornanesulfonic acid salt of D-2-(4-hydroxyphenyl)glycine from the mother liquor and then recovering pure, crystalline D-2-(4-hydroxyphenyl)glycine from the salt and L-2-(4-hydroxyphenyl)glycine from the solvate. Details of the process are shown in the following example.

Example: A mixture of DL-2-(4-hydroxyphenyl)glycine (20 g), d-3-bromo-2-oxo-10-bornanesulfonic acid (39.2 g) and isopropyl alcohol (150 ml) was stirred at 60°C for 30 minutes. After removing insoluble substances from the reaction mixture, the filtrate was concentrated under reduced pressure to make the amount of isopropyl alcohol equal to 120 g. 370 g of toluene was added to the mixture which was then seeded with a small amount (10 mg) of d-3-bromo-2-oxo-10-bornanesulfonic acid salt of L-2-(4-hydroxyphenyl)glycine. This acid salt was previously prepared in a manner similar to the above and crystallized from a similar mixture except for the presence of a greater amount of toluene and crystallizing from the similar mixture but containing further increased amount of toluene than that of the above.

The mixture was stirred at 15°C for 3.5 hours. The precipitating crystals were collected by filtration, washed with a mixture (50 ml) of toluene and isopropyl alcohol (4:1), and dried at room temperature for 12 hours under reduced pressure (5 mm Hg) to give the solvate (30 g) of d-3-bromo-2-oxo-10-bornanesulfonic acid salt of L-2-(4-hydroxyphenyl)glycine with toluene, MP 196° to 197°C (dec.), $[\alpha]_D^{20}$ = +109° (c = 1, H_2O).

A solution of the above solvate (30 g) and hot methanol (60 ml) was evaporated to dryness under reduced pressure. After dissolving the oily residue in isopropyl alcohol (50 g), toluene (200 g) was added and the solution was treated similarly to the above to give the pure solvate (29 g), MP 199°C (dec.). Differential thermal analysis of the solvate gave the endothermic peak at 150°C. $[\alpha]_D^{20}$ = +120° (c = 1, H_2O).

Toluene was released from the solvate by drying at 70°C for 24 hours under reduced pressure (1 mm Hg) to give d-3-bromo-2-oxo-10-bornanesulfonic acid salt of L-2-(4-hydroxyphenyl)glycine. $[\alpha]_D^{20}$ = +131° (c = 1, H_2O).

The solvate (29 g) of the d-3-bromo-2-oxo-10-bornanesulfonic acid salt of L-2-(4-hydroxyphenyl)glycine with toluene was dissolved in 90% methanol (75 ml) and the solution was adjusted to pH 7 with 28% ammonium hydroxide and then to pH 5 with acetic acid.

The solution was stirred under warming until precipitates were formed, and then methanol (2 times the volume of the solution) was added, and allowed to stand overnight under cooling. The precipitates were collected by filtration, washed with methanol and dried to give crystals of L-2-(4-hydroxyphenyl)glycine (8.4 g), MP 214° to 215°C (dec.). $[\alpha]_D^{20}$ = +108° (c = 1, H_2O).

The filtrate and the washings, which were obtained in the above by removing the crude solvate of d-3-bromo-2-oxo-10-bornanesulfonic acid salt of L-2-(4-hydroxyphenyl)glycine with toluene, were evaporated under reduced pressure and the oily residue was dissolved in isopropyl alcohol (50 ml). The solution was concentrated again under reduced pressure. Isopropyl alcohol was added to the oily residue such that the amount of isopropyl alcohol was equal to 37 g.

After adding toluene (130 g), the solution was seeded with the authentic specimen of d-3-bromo-2-oxo-10-bornanesulfonic acid salt of D-2-(4-hydroxyphenyl)-glycine which was prepared preliminarily by a conventional manner, and then stirred for 4 hours at 30°C. The precipitating crystals were collected by filtration, washed with a mixture of toluene and isopropyl alcohol (4:1), and dried to give d-3-bromo-2-oxo-10-bornanesulfonic acid salt of D-2-(4-hydroxyphenyl)-glycine (27.0 g), MP 208° to 209°C (dec.). $[\alpha]_D^{20}$ = +11.8° (c = 1, H_2O).

A solution of the salt (27 g) obtained above in hot methanol (50 ml) was evaporated under reduced pressure. The oily residue was dissolved in isopropyl alcohol (50 ml) and evaporated under reduced pressure. The oily reduced was dissolved again in isopropyl alcohol (30 g). 130 g of toluene were added to the solution. Then the solution was seeded with the authentic specimen of d-3-bromo-2-oxo-10-bornanesulfonic acid salt of D-2-(4-hydroxyphenyl)glycine and stirred for 3 hours at 30°C.

The precipitating crystals were collected by filtration, washed with a mixture of toluene and isopropyl alcohol (4:1) and then dried to give the pure d-3-bromo-2-oxo-10-bornanesulfonic acid salt of D-2-(4-hydroxyphenyl)glycine (26 g), MP 210° to 211°C (dec.). $[\alpha]_D^{20}$ = +5.6° (c = 1, H_2O). The salt (26 g) obtained above was treated in a similar manner to that for the L-form to give D-2-(4-hydroxyphenyl)glycine (8.4 g), MP 213° to 215°C. $[\alpha]_D^{20}$ = –108° (c = 1, H_2O).

Cephalosporin Derivatives

An extensive series of 7-substituted cephalosporin compounds containing the 7-(p-hydroxyphenyl)glycyl or 7-phenylglycyl moiety have been prepared by *G.L. Dunn and J.R.E. Hoover; U.S. Patent 3,867,380; February 18, 1975; assigned to SmithKline Corporation; and U.S. Patent 3,855,213; December 17, 1974; assigned to Smith Kline & French Laboratories.* These compounds are represented by the formula

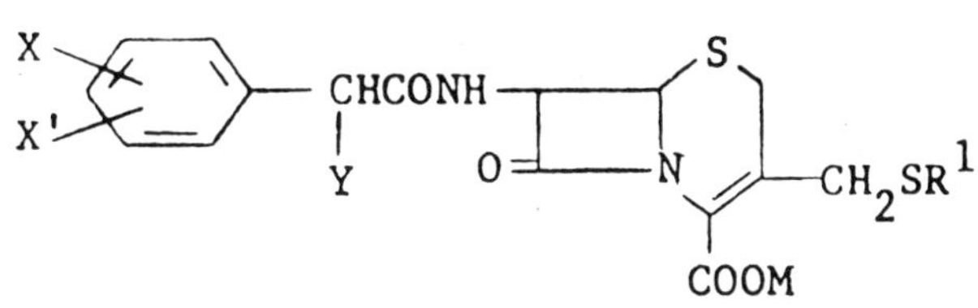

in which X and X' are each hydrogen, lower alkyl of 1 to 4 carbon atoms, lower alkoxy of 1 to 4 carbon atoms, hydroxy, hydroxymethyl, halo, nitro, amino, aminomethyl or trifluoromethyl; Y is NH_2 or NHR; R is hydrogen, aminoalkanoyl of up to 6 carbon atoms, where the alkanoyl group may be branched or straight chain and may be unsubstituted or substituted with hydroxy, mercapto, methylthio, carboxy, amino or phenyl; substituted or unsubstituted alkanoyl of up to 8 carbon atoms, where the substituent may be lower alkoxy of up to 4 carbon atoms, hydroxy, allyloxy, lower alkylthio of up to 4 carbon atoms, azido, halo, cyano, carboxy, carbalkoxy where the alkoxy has up to 4 carbon atoms, or phenoxy.

R may also be carbocyclic acyl unsubstituted or substituted with lower alkyl or alkoxy of up to 4 carbon atoms, halo, nitro, trifluoromethyl, hydroxy, amino, cyano, or dilower alkylamino of 2 to 8 carbon atoms; or heterocyclic acyl which may be unsubstituted or substituted with methyl. R^1 is a 5 or 6 membered heterocyclic ring containing one or more atoms of N, O or S, with or without ring substituents such as lower alkyl, cycloalkyl or alkenyl of up to 6 carbons, lower alkoxy of up to 4 carbons, lower alkoxyalkyl, CF_3, NH_2, alkylamino, dialkylamino, phenyl, methylthio, or halogen.

A preferred group of compounds within the scope of the formula is that in which X is p-HO, X' is H and Y is NH_2. These compounds have highly advantageous properties as antibacterial agents.

The compounds are prepared by reaction of a compound differing from one of the previous formula in that the 3-substituent is acetoxymethyl rather than R^1-thiomethyl, with a mercaptoheterocycle. The reaction is preferably conducted at a pH near neutrality. The solvent is preferably water.

The reaction may be carried out at temperatures from about room temperature to the boiling point of the solvent, the time of reaction varying with the particular temperature, solvent and reactants. The reaction product is isolated by careful acidification of the reaction mixture and extraction with an appropriate organic solvent.

The amino starting material should be protected with a readily removable group such as tert-butoxycarbonyl, carbobenzyloxy or trichloroethoxycarbonyl. The displacement at the 3-position is then conducted, followed by removal of the protective group in the conventional manner.

It is recognized that due to the asymetric α-carbon in the 7-acetamido group optical isomers will exist. The D isomer is the preferred isomer; however, the L isomer and the racemic mixture are also within the scope of this process.

The starting materials for preparing these compounds are known and readily preparable by known methods. The products are antibacterial agents active against Gram-negative and Gram-positive organisms such as *Staph. aureus, Strep. pyogenes, Strep. faecalis, Diplococcus pneumoniae, E. coli, Klebsiella pneumoniae,* Salmonella, Serratia sp., Shigella, and *Enterobacter aerogenes.*

Example: 7-(D-α-Aminophenylacetamido)-3-(5-Methyl-1,3,4-Thiadiazol-2-ylthiomethyl)-3-Cephem-4-Carboxylic Acid – A solution of 7-(D-α-tert-butoxycarboxamidophenylacetamido)cephalosporanic acid (1.82 g, 3.59 mmol) [J.L. Spenser et al., *J. Med. Chem.* 9:746 (1966)] and 2-mercapto 5-methyl-1,3,4-thiadiazole (0.50 g, 3.78 mmol) [J. Sandstrom and I. Wennerbeck, *Acta. Chem. Scand.* 20:57 (1966)] in 33 ml of phosphate buffer (pH 6.2) containing bicarbonate (0.315 g, 3.78 mmol) was heated at 57° for 23 hours. The cloudy solution was cooled to room temperature and extracted with ethyl acetate. The aqueous phase was cooled in ice, layered with ethyl acetate and acidified to pH 2.0 with 3 N hydrochloric acid.

The aqueous phase was extracted again with ethyl acetate and the combined, dried ($MgSO_4$) organic layers were evaporated in vacuo to give 0.79 g (38%) of desired product as the free acid of the tert-butoxycarboxamidophenylacetamido-cephalosporin. A portion of the crude product (0.5 g) was reprecipitated by dissolving it in 5% aqueous sodium bicarbonate solution followed by acidification to give the free acid (0.22 g) as an off-white solid: MP 160° (dec.).

The sodium salt was prepared by treating a solution of the acid in isopropyl alcohol with a solution containing a slight excess of sodium 2-ethylhexanoate in isopropyl alcohol. The tert-butoxycarbonyl derivative (0.22 g), prepared as described above, was dissolved in 3 ml of cold trifluoroacetic acid and the solution was stirred for 30 minutes at room temperature. The excess trifluoroacetic acid was evaporated in vacuo at room temperature and the glassy residue obtained was triturated with ether to give 77 mg of the trifluoroacetate of the title compound: IR (Nujol) 5.65 μ (β-lactam).

The trifluoroacetate salt (77 mg) was converted to the zwitterionic form by dissolving the salt in 3 ml of water and stirring the solution at room temperature with 3 ml of a 25% solution of Amberlite LA-1 (acetate form) in methyl isobutyl ketone for 2 hours. The layers were separated and the aqueous phase was evaporated at 30° in vacuo to give 47 mg of desired product: MP 160° (dec.); IR (Nujol) 5.65 μ (β-lactam). A related series of compounds is disclosed in U.S. Patents 3,891,634 and 3,972,919 later in this chapter.

PHENYLSERINE DERIVATIVES

1-amino-2-hydroxy-2-phenylpropionic acid could also be termed 3-hydroxyphenylalanine or more simply phenylserine. As there are two asymmetric carbon atoms in these compounds, there are four possible optical forms of phenylserine or its derivatives.

Resolution of 3-(3,4-Dihydroxyphenyl)Serine

The separation of rac-threo- and rac-erythro-3-(3,4-dihydroxyphenyl)serine through fractionation of 3-(3,4-dibenzyloxyphenyl)serine from tert-butanol and hydrogenolysis is described in *JACS*, 76, 1322 (1954) and *Ber*, 87, 892 (1954).

B. Hegedüs and A. Krasso; U.S. Patent 3,920,728; November 18, 1975; assigned to Hoffmann-La Roche Inc. have developed an improved process of separating and obtaining the four optical forms using 3-(3,4-dibenzyloxyphenyl)-N-carbobenzoxyserine (I) which is easily converted to the desired forms of 3-(3,4-dihydroxyphenyl)serine (II). These dihydroxyphenylserine compounds are precursors for the biogenic catecholamines or antagonists against the biogenic catecholamines. Also, they exhibit antihypertensive and psychotropic activity.

In the following discussion and examples, 3-(3,4-dibenzyloxyphenyl)-N-carbobenzoxyserine and 3-(3,4-dihydroxyphenyl)serine will be referred to as (I) and (II) respectively.

It has been found that the separation of the threo and erythro forms can be carried out in a simple manner by salification of rac-threo/erythro (I) with dicyclohexylamine in acetonitrile. The separated threo and erythro forms are subsequently converted into rac-threo- or rac-erythro (II) by hydrogenolysis. The separation can be effected because of the discovery that, on the one hand, the dicyclohexylamine salt of rac-threo (I) is relatively insoluble in acetonitrile and, on the other hand, the rac-erythro (I) is relatively insoluble in chloroform.

In addition, rac-threo- or rac-erythro (I) can be resolved into their four optical antipodes. Rac-threo (I), e.g., can be resolved into the corresponding (–)-threo or (+)-threo form by treatment in ethanolic solution with optically active ephedrine or optically active threo-1-(p-nitrophenyl)-2-amino-1,3-propanediol. Rac-erythro (I) can be separated into the corresponding (–)-erythro or (+)-erythro form by treatment in ethanolic solution with optically active threo-1-(p-nitrophenyl-2-amino-1,3-propanediol. The four stereomeric forms of (I) obtained are then hydrogenolytically converted into the corresponding (–)-threo-, (+)-threo-, (–)-erythro- or (+)-erythro (II). Compound (I) can be prepared by condensation of 3,4-dibenzyloxybenzaldehyde with glycine followed by reaction with carbobenzoxychloride.

Example 1: 215 g of rac-threo/erythro (I) are dissolved with warming in 2,500 ml of acetonitrile and treated with 78 g of freshly distilled dicyclohexylamine. After standing overnight at room temperature, the crystals obtained are filtered off and washed with ether. There is obtained rac-threo (I)-dicyclohexylamine salt, 144° to 145°C. The mother liquor (A) can be further processed as described in Example 2.

According to the NMR spectrum, the substance is uniform and contains no contamination by the rac-erythro form. The rac-threo (I)-dicyclohexylamine salt is converted into rac-threo (I) by reaction with hydrogen chloride gas in ethyl acetate as in Example 2. This compound is subjected to hydrogenolysis in the manner given in Example 7. There is obtained rac-threo (II).

Example 2: The mother liquor A of Example 1 is evaporated to dryness under reduced pressure, the residue stirred with 2,000 ml of ether and left to stand overnight. After filtration, there is obtained a mixture of rac-threo- and rac-erythro (I)-dicyclohexylamine salt in a ratio of about 1:1 (according to NMR spectrum). The separated salt is dissolved in 2,000 ml of ethyl acetate and treated with an approximately 5 N solution of hydrogen chloride gas in ethyl acetate up to a strongly acidic reaction.

After stirring for 2 hours, the precipitated dichlohexylamine hydrochloride is filtered off and rinsed with ethyl acetate. The ethyl acetate solution is washed twice with water. The ethyl acetate phase is dried over sodium sulfate and evaporated under reduced pressure. The residue is dissolved in 600 ml of boiling chloroform and left to stand overnight at room temperature. The crystals obtained are filtered off and washed with chloroform. There is obtained rac-erythro (I), MP 143° to 146°C, free of the corresponding rac-threo form. The rac-erythro (I) obtained is subjected to hydrogenolysis to form rac-erythro (II).

Example 3: 101 g of rac-threo (I) are dissolved in 2,000 ml of absolute ethanol and treated with 15.6 g of (+)-ephedrine. After allowing to stand overnight at room temperature, the crystals obtained are filtered off and washed with ethanol and ether. There is obtained the (+)-ephedrine salt of (–)-threo-(I), MP 189° to 190°C after recrystallization from 1,000 ml of methanol. $[\alpha]_D$ = +38.5° (c = 1, dimethylformamide).

Example 4: The mother liquor and ethanol washings of Example 3 are evaporated under reduced pressure, the residue dissolved in ethyl acetate and shaken with 3 N aqueous hydrochloric acid. After drying over sodium sulfate, the organic phase is evaporated under reduced pressure. The residue consists of optically impure (+)-threo-(I) which is purified further as follows:

The compound is dissolved in 2,000 ml of absolute ethanol and treated with adequate (–)-ephedrine to make the mixture react weakly alkaline. The mixture is left to stand 3 to 4 hours, filtered off and washed with ethanol and ether. The crude product obtained is recrystallized from 1,000 ml of methanol and yields optically pure (–)-ephedrine salt of (+)-threo-(I), MP 189° to 190°C. $[\alpha]_D$ = –38° (c = 1, dimethylformamide).

Example 5: 37 g of rac-erythro-(I) are dissolved in 20 ml of absolute ethanol and treated with a solution of 7.5 g of (–)-threo-1-(p-nitrophenyl)-2-amino-1,3-propanediol in 180 ml of boiling ethanol. The mixture is left at room temperature for 2 to 3 hours to crystallize, filtered off and washed with ethanol and ether. The product is crystallized from 200 ml of methanol. There is obtained the (–)-threo-1-(p-nitrophenyl)-2-amino-1,3-propanediol salt of (–)-erythro-(I), MP 159° to 160°C. $[\alpha]_D$ = –24.3° (c = 1, acetone), +9.8° (c = 1, acetonitrile).

Example 6: The mother liquor of Example 5 is concentrated to 20 ml under reduced pressure and treated with a boiling solution of 7.5 g of (+)-threo-1-(p-nitrophenyl)-2-amino-1,3-propanediol in 180 ml of ethanol. The mixture is left at room temperature for 2 to 3 hours to crystallize, filtered off and washed with ethanol and ether. After recrystallization from 300 ml of methanol, there is obtained the (+)-threo-1-(p-nitrophenyl)-2-amino-1,3-propanediol salt of (+)-erythro-(I) MP 163° to 165°C. $[\alpha]_D$ = +24.3° (c = 1, acetone), –9.4° (c = 1, acetonitrile).

Example 7: 50 g of (+)-ephedrine salt of (–)-threo-(I) are shaken with a mixture of 500 ml of ethyl acetate and 300 ml of 3 N aqueous hydrochloric acid in a separating funnel until formation of two clear layers. The ethyl acetate phase is separated, washed once with water and the aqueous phase again extracted with ethyl acetate. The combined ethyl acetate phases are dried over sodium sulfate, evaporated under reduced pressure and crystallized from a mixture of ethyl acetate, isopropyl ether and petroleum ether. There is obtained (–)-threo-(I), MP 134° to 135°C. $[\alpha]_D$ = –17.2° (c = 1, chloroform).

10.4 g of (–)-threo-(I) are dissolved in 400 ml of ethanol and 21 ml of 1 N aqueous hydrochloric acid and hydrogenated with palladium-carbon (5%). After termination of the hydrogen uptake, the catalyst is filtered off and the filtrate evaporated, the last traces of moisture being removed by repeated evaporation with ethanol. The residue is dissolved in 100 ml of absolute ethanol and, with cooling, adjusted to a pH of 5.5 to 6 with a dilute ethanolic diethylamine solution. A precipitate forms. The whole is left to stand overnight at –15°C, filtered and washed with ethanol and ether.

For further purification, the crystals are introduced into 200 ml of boiling water and subsequently filtered. The crystallization is completed by standing overnight at about 0°C. The crystals are filtered off, washed with ethanol and ether and dried under reduced pressure at about 80° to 90°C. There is obtained (–)-threo-(II), MP 232° to 235°C with decomposition. $[\alpha]_D$ = –39° (c = 1, 1 N aqueous hydrochloric acid).

In a manner similar to Example 7, the free acids of the other three optical forms of compounds (I) and (II) were obtained.

	(I) MP, °C	(II) MP, °C
(+)-threo	134-135 (dec)	232-235 (dec)
(–)-erythro	154-157	191-194 (dec)
(+)-erythro	154-157	191-194 (dec)

Production of DL-threo-β-p-Methylsulfonylphenylserine Derivatives

DL-threo-β-p-methylsulfonylphenylserine copper complex and DL-threo-β-p-methylsulfonylphenylserine esters are useful as intermediates for production of thiamphenicol, an antibacterial agent. DL-threo-β-p-methylsulfonylphenylserine copper complex is esterified with an alcohol to form a DL-threo-β-p-methylsulfonylphenylserine ester. The DL-threo-β-p-methylsulfonylphenylserine ester is subjected to optical resolution with d-tartaric acid to form the D-form ester, which is then reduced and dichloroacetylated. Thus, thiamphenicol can be obtained in a high yield.

H. Tobiki, T. Okamoto and H. Akiyama; U.S. Patent 3,927,054; December 16, 1975; assigned to Sumitomo Chemical Company, Ltd. have developed an improved process for producing DL-threo-β-p-methylsulfonylphenylserine copper complex and DL-threo-β-p-methylsulfonylphenylserine esters.

This process comprises reacting p-methylsulfonylbenzaldehyde with glycine copper complex in a suitable solvent under basic conditions to form DL-threo-β-p-methylsulfonylphenylserine copper complex and esterifying with an alcohol to form a DL-threo-β-p-methylsulfonylphenylserine ester. In the reaction for producing DL-threo-β-p-methylsulfonylphenylserine copper complex, water, a lower alcohol such as methanol, ethanol and propanol or a mixture of water and the lower alcohol may be used as a reaction solvent. The reaction temperature is preferably between room temperature and 100°C, and may proceed at a higher temperature. However, temperatures exceeding 100°C are not preferable since part of the resulting DL-threo-β-p-methylsulfonylphenylserine copper complex may be decomposed at such temperatures.

It is necessary to make the reaction system alkaline with an alkali carbonate, an alkali hydroxide or ammonia and preferable pH values are 8 to 11. The reaction proceeds advantageously when an amount of glycine copper complex is used is 1 mol or more per mol of p-methylsulfonylbenzaldehyde.

In the present process, it is not necessary to isolate the glycine copper complex. The copper complex may be reacted with p-methylsulfonylbenzaldehyde as such. Therefore, glycine may be first reacted with a copper salt and then p-methylsulfonylbenzaldehyde may be added to the reaction mixture. Alternatively, a mixture of glycine, a copper salt and p-methylsulfonylbenzaldehyde may be treated.

When a serine type compound is produced by the condensation of an aldehyde with glycine, an erythro isomer is generally formed as a by-product in addition to a threo form compound. According to this process the desired threo form compound can be selectively produced.

The esterification is carried out by heating the DL-threo-β-p-methylsulfonylphenylserine copper complex in a lower alcohol in the presence of an acidic compound such as hydrogen chloride, sulfuric acid or p-toluenesulfonic acid or a chlorinating agent such as thionyl chloride, phosphorus trichloride or phosphoryl chloride. The reaction temperature and reaction time depend upon the acidic compound used. When hydrogen chloride or sulfuric acid is used as an acidic compound, the reaction temperature is usually higher than room temperature, but it is most desirable to carry out the reaction at the boiling point of the alcoholic solvent used. The reaction is usually completed in 1 to 10 hours.

After the reaction, the resulting copper salt is filtered off or the solvent is removed by distillation under reduced pressure without filtering the copper salt. Thereafter, the copper salt thus separated is dissolved in water and copper is removed with, for example, hydrogen sulfide as copper sulfide. The residue is then neutralized with, for example, ammonia water to obtain the ester compound in the form of crystals.

Example: Production of DL-threo-β-p-Methylsulfonylphenylserine Ethyl Ester – A mixture of 15 g of glycine, 25 g of copper sulfate, 18.4 g of p-methylsulfonylbenzaldehyde, 600 ml of water and 17 g of 28% ammonia water was stirred at 45°C for 30 hours and at 70°C for 1 hour. The separated crystals were collected by filtration, washed with 300 ml of warm water and dried to yield 24.3 g of DL-threo-β-p-methylsulfonylphenylserine copper complex, melting point 232° to 234°C (decomposition).

To 20 g of this copper complex was added 120 ml of ethanol. Dry hydrogen chloride gas was passed through to saturate the mixture. The resulting mixture was refluxed for 4 hours and then concentrated under reduced pressure. To the residue was added 200 ml of water to dissolve the residue. Hydrogen sulfide gas was passed through the solution and the separated copper sulfide was filtered off and washed with water. The filtrate was combined with the washings and neutralized with ammonia water while cooling with ice. The separated crystals were collected by filtration, washed with water and dried under reduced pressure to give the objective ester. Yield, 17.6 g, MP 125°C.

MISCELLANEOUS AROMATIC AMINO ACIDS

α-Hydroxymethyl Aromatic Amino Acids

A process of preparing α-hydroxymethyl derivatives of amino acids has been developed by *R.A. Schnettler and J.T. Suh; U.S. Patent 3,856,854; December 24, 1974*.

This process may be illustrated by the reaction sequence shown on the following page in which R is an amino acid moiety which is unchanged by the process such as an alkyl group, phenyl or a substituted phenyl, e.g., methoxyphenyl, hydroxyphenyl or halophenyl.

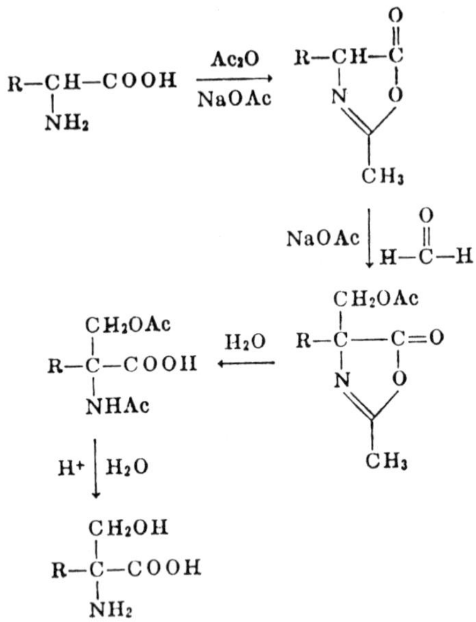

Representatives of the amino acids which may be employed as starting materials and the resulting products are given in the following table.

tyrosine	α-hydroxymethyl tyrosine
3,4-dihydroxyphenylalanine	α-hydroxybenzylserine
phenylalanine	α-hydroxymethyl-3,4-dihydroxyphenylalanine
4-chlorophenylalanine	α-hydroxymethyl-3,4-dichlorophenylalanine
4-fluorophenylalanine	α-hydroxymethyl-4-fluorophenylalanine
4-bromophenylalanine	

The process is further illustrated by the following examples.

Example 1: O,O,N-Triacetyl-p-Hydroxybenzylserine – To 510 g (5 mols) acetic anhydride is added 46 g (0.56 mol) sodium acetate and 100 g (0.55 mol) tyrosine. The mixture is stirred on the steam bath for 2 hours, at which time 30 g (1.0 mol) paraformaldehyde is added and the reaction mixture stirred for an additional 6 hours on the steam bath, cooled, and diluted with 700 ml water. The organic material is extracted into chloroform, washed with 2 liters water and without drying, the chloroform is evaporated.

The orange oily residue is heated and stirred with 200 ml water for 30 minutes and separated from the water with the aid of a small amount of chloroform. The chloroform is evaporated and the residue is dissolved in 200 ml water and 200 ml acetone and refluxed for 30 minutes and the acetone distilled. The residue is extracted into chloroform, washed with water and dried. On standing in chloroform, the product crystallized to give O,O,N-triacetyl-p-hydroxybenzylserine as a white solid which is recrystallized from methanol to give a purified product which has a MP of 224° to 226°C.

Example 2: p-Hydroxybenzylserine – To 20 ml 6 N hydrochloric acid solution is added 9.0 g (0.0267 mol) O,O,N-triacetyl-p-hydroxybenzylserine. The solution is heated on a steam bath under nitrogen for 1 hour and the water is removed by azeotropic distillation with toluene. A glassy residue is obtained which is dissolved in 50 ml of water and is stirred with 20 g Dowex 1 x 8. A precipitate forms which is decanted from the resin and collected on the filter. The white solid is recrystallized from water to give p-hydroxybenzylserine in the form of white crystals, MP 313° to 314°C.

meta-Tyrosine as Antidepressant

Compositions containing m-tyrosine are developed by *M. Sandler; U.S. Patent 3,787,581; January 22, 1974* for use in treating depression. According to the first aspect of the present process, there is provided a method of treating a patient having depression by administering an amount effective to ameliorate depression of at least one compound of the formula:

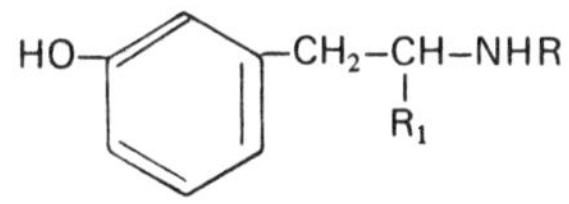

(I)

or a pharmaceutically acceptable salt thereof, in which R is hydrogen or alkoxycarbonyl and R_1 is hydrogen or $COOR_2$ where R_2 is hydrogen or alkyl, at least one of R and R_1 not being hydrogen. When R_1 is not hydrogen, the compound is preferably in the form of its L-isomer. In a second aspect, a method of treating depression is provided which comprises administering a compound of Formula (I) or a pharmaceutically acceptable salt thereof and concomitantly administering a compound of the formula:

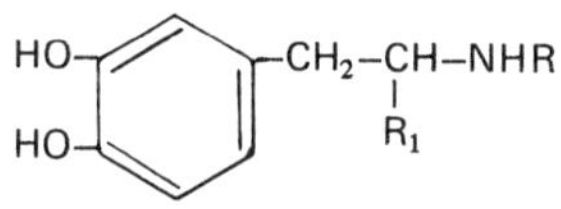

(II)

or a pharmaceutically acceptable salt thereof, in which R and R_1 have the same meanings as Formula I. A particularly preferred compound according to Formula II is L-dopa, L-3-(3,4-dihydroxyphenyl)alanine. The amounts of the compounds of Formula I and Formula II are so chosen as to produce an antidepression effect. The concomitant use of the Formula II compounds appears to lead to a synergistic effect.

Preferred compounds of Formulas I and II are those where R is hydrogen and R_1 is COOH. The compounds of Formulas I and II in which R is alkoxycarbonyl are generally more easily absorbed and/or produce fewer side effects than the free amine. They may readily be prepared by reaction of the free amine with the corresponding alkyl chloroformate. The compounds of Formula I in which R_1 is $COOR_2$ and R_2 is alkyl are readily prepared by esterification of the L-isomer of the free acid using, for example, the corresponding alkanol or alkyl halide.

When R_1 in Formulas I and II represents $COOR_2$ and R_2 is hydrogen, the pharmaceutically acceptable salts include basic salts and ion-exchange resin salts. When R in Formulas I and II is hydrogen, the pharmaceutically acceptable salts include acid addition salts and ion exchange resin salts.

Acid addition salts comprise especially pharmaceutically acceptable nontoxic addition salts with suitable acids, such as those with inorganic acids, for example, hydrochloric, hydrobromic, nitric, sulfuric or phosphoric acids, or with organic acids, such as organic carboxylic acids, for example, acetic, glycollic, maleic, tartaric, citric, o-acetyloxy-benzoic, nicotinic or isonicotinic acid, or organic sulfonic acids, for example, methane sulfonic, ethane sulfonic, 2-hydroxyethane sulfonic, p-toluene sulfonic or naphthalene 2-sulfonic acid.

Example 1: Tablets each having the following composition were made as follows: L-3-(3-hydroxyphenyl)alanine, 500 mg; dried maize starch, 100 mg; talc, 25 mg; sodium lauryl sulfate, 2 mg; calcium sodium alginate, 14 mg; and aluminum stearate, 1 mg. A starch paste (1 in 10) was prepared from a portion of the starch (20 mg per tablet) and water.

The L-3-(3-hydroxyphenyl)alanine and the balance of the starch were passed through a BSS No. 20 sieve and mixed. The mixed powders were massed with the starch paste and the mass granulated. The granules were dried, sieved and the mixture was compressed into tablets. Three of such tablets administered 3 times per day provide relief from or a reduction of depression in at least a proportion of patients suffering from depression.

Example 2: Capsules were prepared by mixing 300 mg of N-ethoxycarbonyl-m-hydroxyphenethylamine and 25 mg lactose and filling the mixture into gelatin capsules. A regime of two capsules 4 times per day controlled depression.

Example 3: Tablets having similar action to those of Example 1 were prepared by using the ethyl ester of L-3-(3-hydroxyphenyl)alanine instead of the free acid.

Example 4: Capsules were prepared as in Example 2 except that the active ingredients were 250 mg of L-dopa and 250 mg of L-3-(3-hydroxyphenyl)alanine. Three capsules administered 3 times per day produced in some patients a rapid antidepressive action without deleterious side effects.

2,5-Dopa as a Cardiac Stimulant

3-(2,5-dihydroxyphenyl)alanine (2,5-dopa) is a known compound which has the following structural formula:

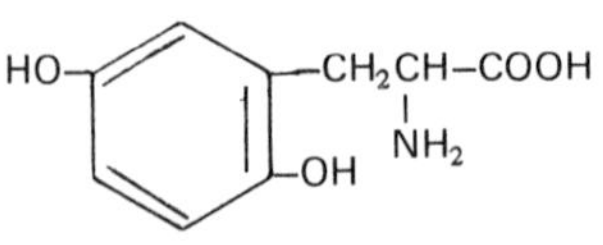

The preparation of 3-(2,5-dihydroxyphenyl)alanine is fully described in the *JACS,* 71, 3758 (1949) starting with 2,5-dihydroxybenzaldehyde and proceeding through the intermediates 3-acetamido-6-acetoxy coumarin or 2-phenyl-4-(2,5-diacetoxybenzal)-5-oxazolone.

W.S. Saari; U.S. Patent 3,976,782; August 24, 1976; assigned to Merck & Co., Inc. has found that 3-(2,5-dihydroxyphenyl)alanine has a surprising and high degree of cardiac stimulant activity. This activity has been found to be best when the active compound is administered orally.

Also included in the present process are the use of the pharmaceutically acceptable salts of 3-(2,5-dihydroxyphenyl)alanine. The salts may be either basic salts with the carboxylic acid function of the molecule or may be acid salts with the amine function of the molecule. Also included for use in this process are the individual enantiomers of the racemic (+)-3-(2,5-dihydroxyphenyl)alanine. The optical isomers result from the assymetric carbon atom bearing the amine and carboxy functions. The individual enantiomers may be used in compositions for the treatment of heart failure since both the (+) and (–) isomers possess substantial, although not identical, cardiac stimulant activity.

The racemic mixture may be resolved into its optical isomers by reaction with an optically active molecule, fractional crystallization of the product to separate the individual diastereomers and regeneration of the optically active reactant. It has been found preferable to initially protect the amine function of 3-(2,5-dihydroxyphenyl)alanine with an appropriate amide group. Any amide group will suitably protect the amine function and the formamide, acetamide, benzamide and others may be employed.

The benzamido group, however, is preferred. In addition, it is desirable to protect the phenolic hydroxyl groups during the resolution procedure. The preferred procedure uses the methyl ether as the phenolic protective group. The preparation of DL-2-benzamido-3-(2',5'-dimethoxyphenyl)propionic acid is described by Neuberger in *Biochemical Journal*, 43, 599 (1948). With the molecule suitably protected the resolution is effected by reacting an optically active amine with the carboxy group of the protected amino acid. Any optically active amine will suffice for this procedure, however, (–)-α-methylbenzylamine is preferred. The following examples show the resolution of the 2,5-dopa and the formation of tablets containing 2,5-dopa.

Example 1: Resolution of Racemic-2-Benzamido-3-(2,5-Dimethoxyphenyl)Propionic Acid – A solution of 59.7 g (0.18 mol) of (±)-2-benzamido-3-(2,5-dimethoxyphenyl)propionic acid and 21.8 g (0.18 mol) of (–)-α-methylbenzylamine in 360 ml of warm methanol is stirred at room temperature for 5 minutes and concentrated to dryness under reduced pressure. The residue is recrystallized 5 times from hot water to give the (–)-α-methylbenzylamine salt of (–)-2-benzamido-3-(2,5-dimethoxyphenyl)propionic acid with essentially constant rotation, $[\alpha]_D^{24}$ = –18.31° (c = 0.95, CH_3OH).

The resolved salt is dissolved in 600 ml hot water, made basic with excess 5 N sodium hydroxide solution and extracted with three 50 ml portions of chloroform. The aqueous extract is acidified with 6 N hydrochloric acid. The precipitate is removed by filtration and dried to give 20 g of (–)-2-benzamido-3-(2,5-dimethoxyphenyl)propionic acid, MP 170° to 172°C. $[\alpha]_D^{23}$ = –34.54° (c = 0.9, C_2H_5OH).

The mother liquors from the recrystallization of the (–)-α-methylbenzylamine salt are combined and concentrated under reduced pressure. The residue is dissolved in hot water and converted to the free acid by basification with 5 N sodium hydroxide solution, chloroform extraction and acidification with 6 N hydrochloric acid. A solution of this acid, 34.2 g, and 12.6 g of (+)-α-methylbenzylamine in 200 ml of warm methanol is concentrated to dryness under reduced pressure. After recrystallizing the residue 3 times from hot water, the rotation of the salt becomes essentially constant, $[\alpha]_D^{24}$ = +18.14° (c = 1.0, CH_3OH). Neutralization of this salt as before with 5 N sodium hydroxide solution, chloroform extraction and acidification with 6 N hydrochloric acid gives 16.5 g of (+)-2-benzamido-3-(2,5-dimethoxyphenyl)propionic acid, $[\alpha]_D^{23}$ = +34.14° (c = 1.75, C_2H_5OH).

Example 2: (+)-3-(2,5-Dihydroxyphenyl)Alanine Hydrate – A solution of 12 g (+)-2-benzamido-3-(2,5-dimethoxyphenyl)propionic acid in 120 ml of 48% hydrobromic acid and 120 ml of glacial acetic acid is heated at reflux under nitrogen for 5 hours. The reaction solution is concentrated to dryness under reduced pressure, 100 ml of water is added and the solution is reconcentrated. The residue is dissolved in water and extracted with 2 portions of ethyl ether. After concentrating the aqueous extract under reduced pressure, water is added to bring the volume to approximately 100 ml. The pH of the solution is adjusted to 5.0 to 5.5 by the addition of concentrated ammonium hydroxide solution. Cooling and filtering gives (+)-3-(2,5-dihydroxyphenyl)alanine hydrate, MP 237° to 245°C dec., $[\alpha]_D^{23}$ = +8.57° (c = 1.3, 1 N HCl).

Example 3: Tablets suitable for oral administration which contain the following ingredients may be prepared by conventional tabletting techniques. Such tablets are useful in treating heart failure at a dose of one or two tablets 2 to 4 times a day.

Ingredient	Weight (mg)
(+) 3-(2,5-dihydroxyphenyl)alanine	50
Tragacanth	10
Lactose	237.5
Corn starch	25
Talcum	15
Magnesium stearate	2.5

p-Hydroxymethylphenylglycine and Cephalosporin Derivatives

An extensive series of 7-(p-hydroxymethylphenylglycylamido)cephalosporins were prepared by *G.L. Dunn and J.R.E. Hoover; U.S. Patents 3,972,919; August 3, 1976 and 3,891,634; June 24, 1975; both assigned to SmithKline Corporation* for use as antibiotics.

The amino acid moiety, p-hydroxymethylphenylglycine, is prepared by first reducing one of the aldehyde groups of 1,4-benzenedicarboxaldehyde with lithium tri(tert-butoxy)aluminum hydride, condensing the remaining aldehyde group of the resulting hydroxymethyl aldehyde with ammonium carbonate and sodium cyanide to give a hydantoin, and hydrolyzing the hydantoin with barium hydroxide. Details of this step are given in the following procedure.

To a solution of 1,4-benzenedicarboxaldehyde (50.0 g, 0.373 mol) in 200 ml of dry tetrahydrofuran under nitrogen in an ice bath was added dropwise lithium tri(tert-butoxy)aluminum hydride (104.0 g, 0.410 mol) dissolved in 500 ml of dry tetrahydrofuran. After stirring for ½ hour in an ice bath, the reaction mixture was poured into 2 liters of ice cold 2 N hydrochloric acid. The aqueous solution was extracted with four 800 ml portions of ether. The combined ether layers were washed with 500 ml of ice cold 5% sodium bicarbonate solution and then with 500 ml of saturated sodium chloride solution.

After drying, the ether was removed under reduced pressure to give 46 g of crude p-hydroxymethylbenzaldehyde. The crude product was chromatographed over 1 kg of neutral alumina and the fractions eluted with ether and concentrated. Upon cooling there crystallized out p-hydroxymethylbenzaldehyde (17.6 g, 35% yield), MP 44.5° to 46°C.

To a stirred mixture of p-hydroxymethylbenzaldehyde (10.0 g, 0.0735 mol) and ammonium carbonate (17.1 g, 0.15 mol) in 110 ml of 60% ethanol heated to 50°C there was added dropwise sodium cyanide (4.0 g, 0.081 mol) dissolved in 10 ml water. The mixture was stirred and heated to 55° to 60°C for 3 hours and then the temperature raised to 85°C for 1 hour. After cooling in an ice bath, the pH of the solution was brought to 6 by the addition of concentrated hydrochloric acid. Upon overnight cooling, the solid which had precipitated was filtered, washed with cold water, and dried. The 5-(p-hydroxymethylphenyl)hydantoin (11.0 g, 72% yield), MP 189° to 196°C (dec) was used to prepare the amino acid without further purification.

A mixture of 5-(p-hydroxymethylphenyl)hydantoin (10.9 g, 0.053 mol) and barium hydroxide·$8H_2O$ (25.5 g, 0.081 mol) in 125 ml water was stirred and refluxed for 18 hours. After cooling in an ice bath the reaction mixture was diluted with 125 ml water. The solution was acidified with concentrated sulfuric acid to pH 1, the barium sulfate filtered, and the pH of the filtrate brought to 6 with lead carbonate. After filtration of the lead sulfate, the filtrate was saturated with hydrogen sulfide and the lead sulfide filtered. The aqueous solution was then concentrated to 100 ml by azeotroping with ethanol under reduced pressure. After cooling, there was precipitated p-hydroxymethylphenylglycine (5.2 g, 54% yield) MP 228° to 229°C (dec). After recrystallization from ethanol-water the compound had MP 230° to 231°C (dec).

The cephalosporin compounds of this process have the formula

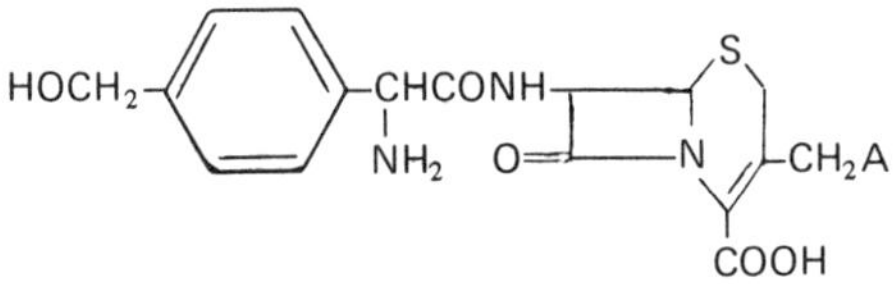

(I)

where A is acetoxy, hydrogen, methoxy, methylthio, or monocyclic heterocyclic thio. These compounds are prepared by acylating a compound of the formula

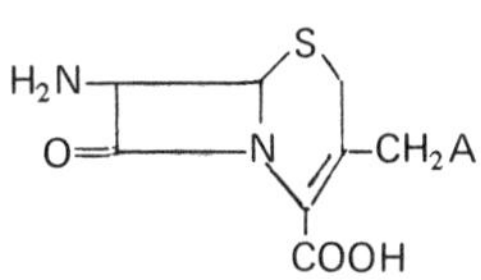

(II)

where A is defined above, with p-hydroxymethylphenylglycine, or an acylating or activated derivative thereof, the amino group being protected with an easily removable protecting group. Among the preferred compounds of formula I are those in which A is heterocyclic thio, e.g., 1,2,3-triazol-5-yl, 2-methyl-1,3,4-thiadiazol-5-yl, and 1-methyltetrazol-5-yl.

The aforementioned compounds are prepared by acylating the 7-amino group of a compound of formula II with p-hydroxymethylphenylglycine. Prior to acylation, it is desirable to protect the amino group on the glycine moiety with an easily removable protective group such as tert-butoxycarbonyl, benzyloxycarbonyl, trichloroethoxycarbonyl, or similar protective group commonly used in the synthesis of peptides. The carboxyl group can be activated for acylation by conversion to the acid chloride or to a mixed anhydride with, for example, a lower alkyl chloroformate. It can also be activated by conversion to the 2,4-dinitrophenyl or N-hydroxysuccinimidyl esters.

If an ester of the carboxyl group on the cephalosporin nucleus is used as an acylation substrate, e.g., a benzhydryl, tert-butyl, trichloroethyl, or benzyl ester, the amine-protected phenylglycine can be coupled directly to the 7-amino group by using a carbodiimide such as dicyclohexylcarbodiimide. Alternatively, the protected phenylglycine can be activated for condensation by reacting it first with carbonyl diimidazole or its equivalent. Following the acylation, the protective groups can be removed with an acid such as trifluoroacetic acid. The resulting salt is converted to the zwitterionic product by means of a basic ion exchange resin (Amberlite IR-45) or else by basification of an aqueous solution of the salt. Closely related compounds are also disclosed in U.S. Patents 3,855,213 and 3,867,380 earlier in this chapter.

Derivatives of α-Methyl-L-Dopa

Derivatives of L-dopa useful in the treatment of hypertension have been developed by *W.S. Saari; U.S. Patents 4,051,169; September 27, 1977; and 3,983,138; September 28, 1976; both assigned to Merck & Co., Inc.* These derivatives are compounds of the formula

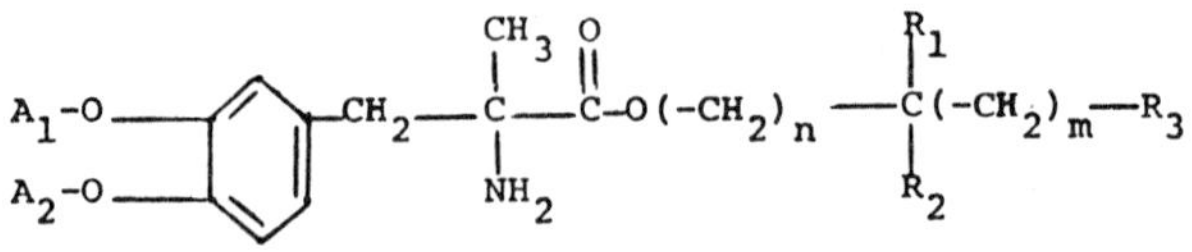

where n is 0, 1, 2 or 3; m is 0, 1, 2 or 3; A_1 and A_2 are individually H or a lower alkanoyl group; R_1 and R_2 are individually H or alkyl of 1 to 3 carbon atoms and R_3 is selected from (1) a monocyclic or bicyclic heterocyclic radical containing from 3 to 12 nuclear carbon atoms and 1 or 2 nuclear N or S atoms with at least 1 being N, and with each ring in the heterocyclic radical containing 5 to 6 members and (2) the radical $X–R_4$ where X is –O–, –S– or –NH– and R_4 contains up to 21 carbon atoms and is a hydrocarbon radical or an acyl radical of an organic acyclic or monocyclic carboxylic acid containing not more than 1 hetero atom in the ring or a pharmaceutically acceptable acid addition salt thereof.

In a preferred embodiment of the process, n and m are 0 or 1, A_1 and A_2 are H, R_1 and R_2 are hydrogen or methyl and R_3 is a heterocyclic ring which may be substituted by lower alkyl groups of 1 to 3 carbon atoms, or R_3 is $–X–R_4$ wherein X is –O– or –NH– and R_4 is an acyl radical from an alkanoic acid containing 2 to 5 carbon atoms. The preferred R_3 heterocyclic ring is shown on the following page.

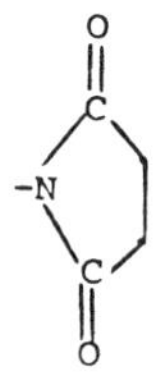

The L-form of these compounds is preferred for treatment of hypertension. When the compounds of the present process are tested orally, distinct anti-hypertensive activity is noted. The compounds also show anti-hypertensive activity when tested intraperitoneally. In some instances the compounds show substantially more activity than α-methyl-L-dopa. The following example illustrates the preparation of these compounds.

Example: (a) A mixture of 19.3 g (0.0777 mol) of L-3-(3,4-dihydroxyphenyl)-2-methylalanine hydrochloride [α-methyl-L-dopa hydrochloride] and 37 g (0.156 mol) of dichlorodiphenylmethane is immersed with slow stirring in a preheated oil bath at 190°C. After reaction has started, as evidenced by vigorous gas evolution, the reaction mixture is stirred rapidly for 6 minutes at 190°C, removed from the hot oil bath and allowed to cool to 25° to 30°C. The crude product from 12 runs is combined, slurried with 3 liters of diethyl ether, filtered, washed with an additional 2 liters of diethyl ether and dried at 30°C under 50 mm pressure. Recrystallization is accomplished by dissolving the product in ethanol and adding ethyl acetate to precipitate the product. The procedure gives 255 g (66.4%) of L-3-(3,4-diphenylmethylenedioxyphenyl)-2-methylalanine hydrochloride, (A) MP 267° to 268°C dec.

(b) A mixture of 175 g (0.425 mol) of (A), 1,750 ml of acetone and 1,750 ml of water is stirred under nitrogen at a temperature below 10°C while the pH is adjusted to 12.0 by the slow addition of a 10% sodium hydroxide solution. Carbobenzyloxy chloride, 93 g (0.545 mol) is added dropwise over 5 to 7 minutes to the reaction mixture at 20° to 30°C accompanied by the simultaneous addition of a 10% sodium hydroxide solution to maintain a pH of 12.0 to 12.2. After addition of the carbobenzyloxy chloride is complete, the reaction mixture is stirred at 25° to 30°C for 3 hours. Most of the acetone is then removed under reduced pressure at 25° to 35°C to precipitate the sodium salt of the desired N-carbobenzyloxy derivative.

The sodium salt is extracted into 1.5 liters of ethyl acetate, washed with 200 ml of 5% sodium hydroxide solution and 200 ml of a saturated sodium chloride solution and then dried over magnesium sulfate. After adding 17.5 g of decolorizing carbon and filtering through a magnesium sulfate pad, solvents are removed under reduced pressure at 25° to 35°C. The residue is slurried 2 times with 1 liter of a 20% ethyl ether-80% hexane (by volume) solution and filtered to give the sodium salt of the desired N-carbobenzyloxy derivative. This sodium salt is dissolved in 1.5 liters of ethyl acetate, cooled to 10°C and acidified to pH 2 with 6 N hydrochloric acid. The ethyl acetate extract is washed with 200 ml of a saturated sodium chloride solution, dried over magnesium sulfate, filtered and concentrated under reduced pressure at 25° to 35°C. The N-carbobenzyloxy derivative is dried further at 25° to 30°C and 0.2 to 0.3 mm Hg to give 169 g (78.0%) of L-N-carbobenzyloxy-3-(3,4-diphenylmethylenedioxyphenyl)-2-methylalanine (B).

(c) A solution of 13.5 g (0.0265 mol) of (B), 2.7 g (0.027 mol) of triethylamine and 5.19 g (0.029 mol) of N-bromomethylsuccinimide in 35 ml of dry dimethylformamide is stirred at 25° to 30°C for 16 hours. The reaction mixture is poured into 400 ml of ice water and the product extracted into 200 ml of a 50% chloroform-50% diethyl ether (by volume) mixture. The organic extract is washed with 50 ml of a dilute (5%) sodium carbonate solution and 50 ml of a saturated sodium chloride solution and then dried over anhydrous magnesium sulfate to remove water. After filtering and concentrating under reduced pressure, the residue is recrystallized. Recrystallization is accomplished by dissolving the product in ethanol and adding hexane to precipitate the product. The process gives 12.1 g (73.6%) of succinimidomethyl L-N-carbobenzyloxy-3-(3,4-diphenylmethylenedioxy)phenyl-2-methylalaninate (C), MP 143.0° to 145.0°C.

(d) A suspension of 6.6 g (0.0106 mol) of (C) in 180 ml of absolute ethanol and 9 ml of a 9.6 N ethanolic anhydrous hydrogen chloride solution is hydrogenated with 3.3 g of a 10% palladium-on-carbon catalyst at an initial pressure of 30 psi until hydrogen uptake is complete. After removal of catalyst by filtration, the filtrate is concentrated under reduced pressure. The residue is extracted with 50 ml of benzene and then 50 ml of ethyl acetate. The insoluble solid is then shaken with 50 ml of a 10% ethanol-90% ethyl acetate (by volume) mixture and 10 ml of a saturated sodium carbonate solution. After filtration, the filtrate is dried over anhydrous magnesium sulfate, filtered and concentrated under reduced pressure. The residue is redissolved in 25 ml of absolute ethanol, treated with 5 ml of a 9.6 N ethanolic anhydrous hydrogen chloride solution and concentrated under reduced pressure to give 2.5 g (62.7%) of succinimidomethyl L-3-(3,4-dihydrophenyl)-2-methylalaninate hydrochloride hydrate, homogeneous upon thin layer chromatography [fluorescent silica gel plate, 30% methanol-70% benzene (by volume) solvent] with an observed R_f = 0.5.

COMPANY INDEX

INVENTOR INDEX

U.S. PATENT NUMBER INDEX

NOTICE

INDUSTRIAL ENZYMES 1977
Recent Advances

by J. C. Johnson

Chemical Technology Review No. 85

Much of the current development work on enzymes is in the production of supported, immobilized, flocculated, bonded or water-insoluble enzymes to permit their recovery and reuse. This reduces the cost of processes using otherwise short-lived enzymes as biocatalysts and permits the recovery of purer products.

Industrial enzymes are now mostly obtained from microorganisms rather than being extracted from plant and animal tissues. Bacteria and fungi growing on non-pathogenic and non-toxic substrates can be manipulated genetically to excrete the required "tailored" enzyme. Proteolytic, starch-converting and glucose-isomerizing enzymes dominate the industrial, non-medicinal market at present.

For the future, enzymatic processes are being investigated for the purification of air and wastewater, the conversion of waste cellulose into glucose and ultimately fructose. Also promising is the use of nitrogen-fixation enzymes in nonleguminous crop production.

This review covers some 200 process descriptions from U.S. patents for the production and use of enzymes on an industrial scale. A partial and condensed table of contents follows here:

ISBN 0-8155-0655-4 **$39** **348 pages**

TEXTURED PROTEIN PRODUCTS 1977

by Marcia Gutcho

Food Technology Review No. 44

There is a major effort in the world today to produce nutritious, high protein, meat-like products of other than animal origin, suitable for human consumption.

Although long known in the Orient, American and European research efforts have focused on the production of meat analogs and extenders from soy protein.

The prime objective is to make soybean-derived protein fit international cuisines, appeal to personal tastes and balance the family diet without breaking the budget. Other sources are peanuts, cottonseed, high protein wheat, sesame and sunflower seeds, and in the near future perhaps, single-cell proteins such as yeasts, grown on cellulose or hydrocarbons.

Currently meat analogs are made from two methods: wet fiber spinning and thermo-vegetable proteins processed primarily by plastic extrusion. Improvements in methods using these techniques are discussed in the present volume. In addition, a variety of new, innovative protein-texturizing and fiber-producing methods are disclosed. A total of 176 processes are described. Included too, are formulations for a wide variety of meat analogs and extenders. Of interest also are procedures for making and using meat flavors. A partial and condensed table of contents follows. Numbers in parentheses indicate the numbers of processes per topic.

ISBN 0-8155-0681-3

355 pages

PHARMACEUTICAL MANUFACTURERS OF THE UNITED STATES 1977

This popular directory and business information source was first published in the late summer of 1970 and was an immediate success. It was followed in June 1972 by a much enlarged second edition which had to undergo several printings before being allowed to reach the out-of-print status in 1975.

By most insistent demands from very divergent areas of interest the book has now been re-edited and for all practical purposes is an entirely new book.

Painstaking care was taken to verify all data presented and each company listed in this compendium was contacted either by letter or telephone during late spring and early summer 1977 to make sure that the information given was up-to-date and correct.

It must be emphasized that all data appearing in this book were selected by the editor from each manufacturer's communications and descriptive literature (catalogs, annual reports, and form 10-K reports to the Securities and Exchange Commission) at no cost to, nor influence from, the manufacturers of the drugs and other materials.
Because of the nation's expanding $100 billion health care business, cosmetic firms have been omitted from this edition, but much more attention has been paid to firms in the ancillary fields of surgical and medical equipment and manufacture of diagnostic reagents for clinical laboratories. Supportive diagnostic aids are being used more and more as government and insurance companies take increasing interest in the dispensing and pricing of drugs.

The layout has been kept simple and easy to follow. Companies can be traced quickly and information can be extracted without having to take recourse to explanatory symbols and abbreviations.

The pharmaceutical industry of the United States is a large, well-established and affluent industry; this publication lists close to 400 leading pharmaceutical and ancillary firms. Their business falls into the following categories:

Prescription Drugs (Ethical Drugs)

Proprietaries [Over-the-Counter (OTC) Drugs] and Various Health Care Products

Biologicals (Substances which are prepared from organic origin, or those which effect immunity such as Vaccines, Pollen Extracts, etc.)

Veterinary Pharmaceuticals and Related Animal Health Care Products

Medical, Hospital and Laboratory Supplies (includes Diagnostic Reagents, Kits, etc.)

Dental Supplies

Custom Manufacture (Private Formulas, Generic Name Drugs) and customized packaging.

In order to be listed, each firm must do some manufacturing in the USA. The following useful information has been listed for each firm, whenever available:

Company Name
Address (with Zip code)
Telephone (with Area code)
Parent Company or Headquarters
Category (see above)
Annual Sales (1976 where available)
Number of Employees
Executives (Names and Titles)
Types of Products and well-known Brand Names
Plant Locations
Domestic Divisions, Subsidiaries and/or Affiliates
Foreign Subsidiaries

The category of products manufactured has been placed directly under the name and address of the firm for quick reference.
Pharmaceutical firms of Canada and Mexico are not listed, but wholly-owned subsidiaries of American firms in these and other foreign countries are enumerated.

The index lists the names of all divisions, subsidiaries, and affiliates which are involved in pharmaceutical and related businesses not listed separately in the book, with the name of the parent company.

This compendium allows you to concentrate your efforts in the most profitable direction:

Pinpoint your sales efforts toward the Big Buyers
Know whom to contact
Find out who owns whom
Prepare market reports
Increase sales effectiveness
Search for potential acquisitions
Use the book...
as a guide to employment and as a personnel guide.

ISBN 0-8155-0677-5

266 pages

ENERGY FROM BIOCONVERSION OF WASTE MATERIALS 1977

by Dorothy J. De Renzo

Energy Technology Review No. 11
Pollution Technology Review No. 33

One of the chief gaseous products of the anaerobic decomposition of organic matter is methane, CH_4. This is how natural gas was formed in prehistoric times along with other fossil fuels.

By applying this principle today in environmentally acceptable fashion it is possible to bioconvert municipal solid sewage, animal manure, agricultural and other organic wastes into substitute natural gas (95% CH_4). In its simplest essentials the process consists of loading the material into a digester (a closed tank with a gas outlet). Given favorable thermal and chemical conditions, the appropriate biological processes will then take their course.

The bioconversion of waste materials to methane provides at least partial solutions not only to the energy problem, but also to the solid waste disposal problem. The harvesting of heretofore undesirable vegetations, such as algae, water hyacinths, and kelp as "energy crops" offers unconventional opportunities for supplementary utilization of natural resources.

This book describes practical methods for the bioconversion of waste matter. It is based on reports of academic and industrial research teams working under government contracts. A partial and condensed table of contents follows here. Chapter headings and important subtitles are given.

Note: Each chapter is followed by bibliographic reference lists in order to provide the reader with easy access to further information on these timely topics.

ISBN 0-8155-0656-2 **223 pages**

YEASTS FOR FOOD AND OTHER PURPOSES 1977

by Jeanne C. Johnson

Food Technology Review No. 45

Yeasts are primitive, single cell plants that can proliferate without the benefit of sunlight, yet they contain many valuable nutrients including minerals, proteins, fats and B-complex vitamins.

Yeasts are also highly versatile microbial agents that can be manipulated for specific purposes. Brewer's yeast can ferment carbohydrates, producing ethyl alcohol. Other yeasts cannot induce fermentation, but can dissimilate carbohydrates into a variety of organic products such as carbon dioxide and acetic, lactic, malic and succinic acids.

Suitability cultivated yeast can contain as much as 67% protein on a dry basis. The bulk of the single cell yeast protein is in the form of enzymes and is a valuable source of them or of nutritional single cell protein.

This review covers over 160 processes. About a hundred of these deal with the growth of yeast cells and the use of yeast protein as food for humans and animals. Another 40 or more are concerned with the uses of yeast in enzymatically modifying other proteins or in fermentation processes for bakery products or in the manufacture of chemicals. Miscellaneous uses of yeast, as in condiments, encapsulating agents, etc., are included in some 20 process descriptions. A partial and condensed table of contents follows. Numbers in parentheses indicate the number of processes per topic.

ISBN 0-8155-0682-1

345 pages

DRUG AND COSMETIC PACKAGING 1975

by Roger C. Griffin Jr. and Stanley Sacharow
with chapters by D. A. Dean and P. L. Corby

This compact volume contains an enormous amount of practical, basic information, indispensable to pharmaceutical and cosmetic firms.

Every manufacturer of finished products in these fields must necessarily get involved in the development of successful packaging: selection of materials and package types, expiration dates and product compatibilities, the distribution cycle, and marketing requirements.

In the pharmaceutical area the packaging problems are stressed by the fact that the failure of a drug because of a package may literally mean the difference between life and death.

This book explains in great detail the tremendous variety of packaging now available to the drug and cosmetics manufacturer. The common non-actinic glass bottle of the early twentieth century has largely been replaced by sophisticated laminations, plastic tubes, and a multitude of other packaging innovations.

It is evident at once that the authors' efforts, based on years of experience, have been eminently successful in providing a basic source of information. Even solid waste disposal methods are discussed so as to help in establishing criteria for proper package design. A partial and condensed table of contents follows here:

ISBN 0-8155-0570-1 **267 pages**